I0473302

# Advances in High-Order Sensitivity Analysis

The high-order sensitivities of model responses with respect to model parameters are notoriously difficult to compute for large-scale models involving many parameters. The neglect of higher-order response sensitivities leads to substantial errors in predicting the moments (expectation, variance, skewness, kurtosis, and higher-order) of the model response's distribution in the phase space of model parameters. The author expands on his theory of addressing high-order sensitivity analysis in this book, *Advances in High-Order Sensitivity Analysis.*

The mathematical/computational models of physical systems comprise parameters, independent variables, and dependent variables. Since the physical processes themselves are seldom known precisely and most of the model's parameters stem from experimental procedures that are also subject to imprecision and/or uncertainties, the results predicted by these models are imprecise, affected by the uncertainties underlying the respective model.

In the particular case of sensitivity analysis using conventional methods, the number of large-scale computations increases exponentially. For large-scale models involving many parameters, even the first-order sensitivities are computationally very expensive to determine accurately by conventional methods. Furthermore, the "curse of dimensionality" prohibits the accurate computation of higher-order sensitivities by conventional methods.

Other books by the author, all published by CRC Press, include *Sensitivity and Uncertainty Analysis, Volume I: Theory* (2003); *Sensitivity and Uncertainty Analysis, Volume II: Applications to Large-Scale Systems* (Cacuci et al. 2005); *Computational Methods for Data Evaluation and Assimilation* (Cacuci et al. 2014); *The Second-Order Adjoint Sensitivity Analysis Methodology* (2018); and *Advances in High-Order Predictive Modeling Methodologies and Illustrative Problems* (2025).

**Advances in Applied Mathematics**
*Series Editor:* Daniel Zwillinger

*Delay Ordinary and Partial Differential Equations*
Andrei D. Polyanin, Vsevolod G. Sorkin, and Alexi I. Zhurov

*Clean Numerical Simulation*
Shijun Liao

*Multiplicative Partial Differential Equations*
Svetlin Georgiev and Khaled Zennir

*Engineering Statistics: A Matrix-Vector Approach with MATLAB®*
Lester W. Schmerr Jr.

*General Quantum Numerical Analysis*
Svetlin Georgiev and Khaled Zennir

*An Introduction to Partial Differential Equations with MATLAB®*
Matthew P. Coleman and Vladislav Bukshtynov

*Handbook of Exact Solutions to Mathematical Equations*
Andrei D. Polyanin

*Introducing Game Theory and Its Applications, Second Edition*
Elliott Mendelson and Dan Zwillinger

*Modeling Operations Research and Business Analytics*
William P. Fox and Robert E. Burks

*Decision Analysis through Modeling and Game Theory*
William P. Fox

*Advances in High-Order Predictive Modelling*
Dan Gabriel Cacuci

*Introduction to Financial Mathematics, Second Edition*
Kevin J. Hastings

*General Quantum Variational Calculus*
Svetlin G. Georgiev and Khaled Zennir

*Advanced Engineering Mathematics with MATLAB®, Sixth Edition*
Dean G. Duffy

*Exact Methods for Nonlinear PDEs*
Andrei D. Polyanin

*Markov Processes, Second Edition*
James R. Kirkwood

*Advances in High-Order Sensitivity Analysis*
Dan Gabriel Cacuci

https://www.routledge.com/Advances-in-Applied-Mathematics/book-series/CRCADVAPPMTH?pd=published,forthcoming&pg=1&pp=12&so=pub&view=list

# Advances in High-Order Sensitivity Analysis

Dan Gabriel Cacuci

CRC Press
Taylor & Francis Group
Boca Raton London New York

CRC Press is an imprint of the
Taylor & Francis Group, an **informa** business
A CHAPMAN & HALL BOOK

Designed cover image: Shutterstock



First edition published 2026
by CRC Press
2385 NW Executive Center Drive, Suite 320, Boca Raton, FL 33431

and by CRC Press
4 Park Square, Milton Park, Abingdon, Oxon, OX14 4RN

*CRC Press is an imprint of Taylor & Francis Group, LLC*



ISBN: 978-1-032-75211-2 (hbk)
ISBN: 978-1-032-76359-0 (pbk)
ISBN: 978-1-003-47817-1 (ebk)

DOI: 10.1201/9781003478171

Typeset in Times
by codeMantra

# Contents

# About the Author

**Dan Gabriel Cacuci** is a Distinguished Professor Emeritus in the Department of Mechanical Engineering at the University of South Carolina and the Karlsruhe Institute of Technology, Germany. He received his PhD in applied physics, mechanical, and nuclear engineering from Columbia University, New York City. He is also the recipient of many awards, including four honorary doctorates, Germany's Humboldt Preis, the Ernest Orlando Lawrence Memorial Award from the U.S. Department of Energy, and the Arthur Holly Compton, Eugene P. Wigner, and Glenn Seaborg Awards from the American Nuclear Society.

# Preface

The high-order sensitivities of model responses with respect to model parameters are notoriously difficult to compute for large-scale models involving many parameters. The neglect of higher-order response sensitivities leads to substantial errors in predicting the moments (expectation, variance, skewness, kurtosis and higher-order) of the model response's distribution in the phase space of model parameters.

The mathematical/computational models of physical systems comprise parameters, independent variables, and dependent variables. Since the physical processes themselves are seldom known precisely and most of the model's parameters stem from experimental procedures that are also subject to imprecision and/or uncertainties, the results predicted by these models are imprecise, being affected by the uncertainties underlying the respective model. The functional derivatives (also called *sensitivities*) of results (also called *responses*) produced by mathematical/computational models are needed for many purposes, including: (i) understanding the model by ranking the importance of the various parameters; (ii) performing *reduced-order modeling* by eliminating unimportant parameters and/or processes; (iii) quantifying the uncertainties induced in a model response due to model parameter uncertainties; (iv) performing *model validation* by comparing computations to experiments to address the question, "Does the model represent reality?"; (v) prioritizing improvements in the model; (vi) performing data assimilation and model calibration as part of forward *predictive modeling* to obtain best-estimate predicted results with reduced predicted uncertainties; (vii) performing inverse *predictive modeling* to *calibrate* the model's parameters while reducing their initial uncertainties; and (viii) optimizing the system.

As has been mentioned in the foregoing, neglecting the higher-order response sensitivities leads to substantial errors in predicting the moments of the model response's distribution in the phase space of model parameters. For example, a *first-order sensitivity and uncertainty quantification* will always produce a zero-valued third moment of the predicted response distribution, which is erroneous unless the unknown response distribution happens to be symmetrical in the phase space of model parameters. At least second-order sensitivities must be determined and employed to estimate the third-order moment (and, hence, the skewness) of the response distribution. With pronounced skewness, standard statistical inference procedures such as constructing a confidence interval for the expectation value of a computed/predicted model response will not only be incorrect, in the sense that the proper coverage level will differ from the nominal (e.g., 95%) level, but the error probabilities will be unequal on each side of the predicted mean.

For large-scale models involving many parameters, even the first-order sensitivities are computationally very expensive to determine accurately by conventional finite-difference and/or statistical methods. Furthermore, the computation of higher-order sensitivities by conventional finite-difference and/or statistical methods is subject to the *curse of dimensionality*, a term coined by Bellman (1957) to describe phenomena in which the number of computations increases exponentially in the respective phase space. In the particular case of sensitivity analysis using conventional finite-difference and/or statistical methods, the number of large-scale computations increases exponentially in the phase space of model parameters as the order of sensitivities increases.

It is known that the *adjoint method* of sensitivity analysis conceived by Cacuci (1981a, 1981b) is the most efficient method for computing exact first-order sensitivities, since it requires a single large-scale (adjoint) computation, independent of the number of model parameters. The mathematical underpinnings of the first-order adjoint sensitivity analysis, along with representative applications to large-scale systems, are detailed in the books by Cacuci (2003) and Cacuci, Ionescu-Bujor, and Navon (2005). The most important uses of these sensitivities are for data assimilation and model calibration, as illustrated by Cacuci, Ionescu-Bujor, and Navon (2014).

Cacuci (2015, 2016) has extended his first-order adjoint sensitivity analysis methodology by conceiving the second-order adjoint sensitivity analysis methodology, which enables the most efficient computation of all second-order sensitivities, requiring only as many large-scale (adjoint) computations as there are non-zero first-order sensitivities. The mathematical underpinnings of the second-order adjoint sensitivity analysis, along with representative applications to large-scale systems, are detailed in the books by Cacuci (2018, 2019).

Subsequently, Cacuci (2022, 2023a) has generalized his second-order adjoint sensitivity analysis methodology to enable the exact and efficient computation of response sensitivities of arbitrarily high order with respect to imprecisely known (i.e., uncertain) model parameters, model boundaries, and/or internal interfaces, for both nonlinear systems and for response-coupled forward/adjoint linear systems. These general methodologies are called the *n*th-Order Comprehensive Adjoint Sensitivity Analysis Methodology for Nonlinear Systems (*n*th-CASAM-N) and the *n*th-Order Comprehensive Adjoint Sensitivity Analysis Methodology for Response-Coupled Forward/Adjoint Linear Systems (*n*th-CASAM-L). Both the *n*th-CASAM-N and the *n*th-CASAM-L are formulated in linearly increasing higher-dimensional Hilbert spaces in the dependent-variables phase space (as opposed to exponentially increasing parameter-dimensional spaces), thus overcoming the curse of dimensionality in sensitivity analysis. The most important uses and impacts of the 2nd-, 3rd-, and 4th-order sensitivities on predictive modeling (which includes the activities of sensitivity analysis, uncertainty quantification, data assimilation, model calibration, and model validation) have been illustrated by numerous works by Cacuci, the most important of which are summarized in his book (Cacuci, 2025).

The *n*th-CASAM-N and the *n*th-CASAM-L methodologies enable the computation of arbitrarily high-order sensitivities of model responses *directly* with respect to the model's primary parameters. However, many models comprise not just individual primary model parameters but also functions of such primary model parameters. Such functions customarily appear in models in the form of correlations that describe *features* of the system under consideration, such as material properties and flow regimes. The number of such *feature* functions is considerably smaller than the total number of model parameters. However, a methodology for computing exact expressions of response sensitivities to functions/features of the primary model parameters has not been available until Cacuci (2023b) introduced the First- and Second-Order Features Adjoint Sensitivity Analysis Methodologies for Nonlinear Systems

(1st-FASAM-N and 2nd-FASAM-N), showing that the number of large-scale computations is significantly reduced by computing exact expressions of sensitivities of model responses with respect to functions/features of model parameters rather than the sensitivities with respect to the model parameters themselves. Cacuci (2024a, 2024b) has subsequently generalized the 1st-FASAM-N and 2nd-FASAM-N methodologies to compute the exact expressions of arbitrarily high-order sensitivities of model responses with respect to functions/features of model parameters, including uncertain boundaries and internal interfaces, for linear and/or nonlinear models. These methodologies are called the *n*th-Order Features Adjoint Sensitivity Analysis Methodologies for Nonlinear Systems (*n*th-FASAM-N) and the *n*th-Order Features Adjoint Sensitivity Analysis Methodologies for Response-Coupled Forward/Adjoint Linear Systems (*n*th-FASAM-L). The *n*th-FASAM-N methodology enables the most efficient computation of exact expressions of arbitrarily high-order response sensitivities to parameter features in nonlinear systems.

The aim of this book is to detail the general formulation of the *n*th-FASAM-N and *n*th-FASAM-L methodologies. It is important to note that although some linear systems can be treated as particular cases of nonlinear systems, response-coupled forward/adjoint systems do not fall into this category (since nonlinear operators do not admit adjoints) and should be treated in their own right (hence the development of the *n*th-FASAM-L). This book also presents applications of the *n*th-FASAM-N and *n*th-FASAM-L methodologies and is organized as detailed below.

Chapter 1 briefly reviews the basic motivations for computing high-order sensitivities and illustrates their importance by means of an OECD/NEA reactor physics benchmark, which is representative of a "large-scale system" involving many (21,976) uncertain parameters. Chapter 2 presents the 1st-FASAM-N, which enables the computation, with unparalleled efficiency, of first-order sensitivities of model responses to *functions* (*features*) of uncertain model parameters, including uncertain boundaries and internal interfaces, for linear and/or nonlinear models. Such functions of primary model parameters customarily describe characteristic *features* of the system under consideration, such as correlations that model material properties and flow regimes. The number of such *features* is considerably smaller than the total number of primary model parameters. The subsequent computation of the first-order response sensitivities with respect to the primary model parameters can be performed analytically using the corresponding response sensitivities to the features. This process involves practically no computational costs, as it only requires the respective functions/features of parameters rather than the entire model. The applicability and unparalleled efficiency of the 1st-FASAM-N for computing first-order sensitivities are illustrated by using a paradigm particle transport model that involves functions/features of parameters while admitting closed-form analytic solutions.

Chapter 3 presents the 2nd-FASAM-N, which enables the most efficient computation of exactly obtained expressions of the second-order sensitivities of a model response with respect to the functions (*features*) of primary model parameters. To compute all of the second-order sensitivities, the 2nd-FASAM-N requires only as many large-scale (adjoint) computations as there are non-zero first-order sensitivities with respect to the components of the *feature* function. The second-order sensitivities with respect to the primary model parameters are computed analytically, at virtually no computational cost, using the second-order sensitivities with respect to the feature functions. The 2nd-FASAM-N computes the mixed second-order sensitivities twice, involving distinct 2nd-level adjoint sensitivity functions. This property is essential for the mutual verification of the computational accuracy of the respective adjoint functions. These considerations are demonstrated by using a simple particle transport/shielding model, which admits closed-form expressions for all state functions and sensitivities.

Chapter 4 presents the mathematical framework of the *n*th-FASAM-N, which is shown to be the most efficient methodology for computing exact expressions of sensitivities, of any order, of model responses with respect to features of model parameters and, subsequently, with respect to the model's uncertain parameters, boundaries, and internal interfaces. The unparalleled efficiency and accuracy of the *n*th-FASAM-N methodology stems from the maximal reduction of the number of adjoint computations (which are considered to be *large-scale* computations) for computing high-order sensitivities. When applying the *n*th-FASAM-N methodology to compute the second- and higher-order sensitivities, the number of large-scale computations is proportional to the number of *model features* as opposed to being proportional to the number of model parameters (which are considerably more numerous than the number of features). When a model has no feature functions of parameters but only comprises primary parameters, the *n*th-FASAM-N methodology becomes identical to the *n*th-CASAM-N (*n*th-Order Comprehensive Adjoint Sensitivity Analysis Methodology for Nonlinear Systems) methodology. Both the *n*th-FASAM-N and the *n*th-CASAM-N methodologies are formulated in linearly increasing higher-dimensional dependent-variable spaces—as opposed to exponentially increasing parameter-dimensional spaces—thus overcoming the curse of dimensionality in sensitivity analysis of nonlinear systems.

Chapter 5 highlights the unparalleled efficiency of the *n*th-FASAM-N by considering the well-known Nordheim-Fuchs reactor dynamics/safety model. This model describes the time-dependent neutron flux, temperature distribution, and energy released during a short-time self-limiting power burst in a nuclear reactor system with a negative temperature coefficient. In such a system, a large amount of reactivity is suddenly inserted, either intentionally or by accident. For the Nordheim-Fuchs model, the following characteristics become evident: (i) the 1st-FASAM-N and the 1st-CASAM-N methodologies are equally efficient for computing the first-order sensitivities, each methodology requiring a single large-scale computation for solving the First-Level Adjoint Sensitivity System; (ii) the 2nd-FASAM-N methodology requires two large-scale computations to obtain all of the exact expressions of the 28 distinct second-order response sensitivities with respect to the model parameters, while the 2nd-CASAM-N methodology requires seven large-scale computations for obtaining these 28 second-order sensitivities; (iii) the 3rd-FASAM-N methodology is even more efficient than the 3rd-CASAM-N methodology; only two large-scale computations are needed to obtain the exact expressions of the 84 distinct third-order response sensitivities with respect to the Nordheim-Fuchs model's parameters when applying the 3rd-FASAM-N methodology, while the application of the 3rd-CASAM-N methodology requires at least 22 large-scale computations for computing the same 84 distinct third-order sensitivities.

Chapter 6 presents the mathematical/theoretical framework of the *n*th-FASAM-L, which enables the most efficient computation of exactly obtained mathematical expressions of arbitrarily high-order (*n*th-order) sensitivities of a generic system response that simultaneously depends on the system's forward and adjoint state functions. The mathematical framework underlying the *n*th-FASAM-L is developed in linearly increasing higher-dimensional Hilbert spaces and is established by mathematical induction. The unparalleled efficiency and accuracy of the *n*th-FASAM-L methodology stem from the maximal reduction of the number of *large-scale* computations for computing high-order sensitivities (with respect to the user-defined features of model parameters) of a generic system response that depends on both the forward and adjoint state functions.

Chapter 7 presents an illustrative application of the *n*th-FASAM-L to an analytically solvable model of the energy distribution of the *contributon-flux* of neutrons in a mixture of materials. The *contributon-flux* is a functional of both the forward and the adjoint neutron fluxes. This illustrative paradigm model highlights the efficiency of the *n*th-FASAM-L methodology: the higher the order of the computed sensitivities, the more efficient the *n*th-FASAM-L becomes by comparison to any other methodology, noting, in particular, that the probability of encountering identically vanishing high-order sensitivities is much larger when using the *n*th-FASAM-L than otherwise.

## REFERENCES


Bellman, R.E. 1957. *Dynamic Programming*. Rand Corporation, Princeton University Press, Princeton, NJ.

Cacuci, D.G. 1981a. Sensitivity Theory for Nonlinear Systems: I. Nonlinear Functional Analysis Approach. *J. Math. Phys. 22*, 2794–2802.

Cacuci, D.G. 1981b. Sensitivity Theory for Nonlinear Systems: II. Extensions to Additional Classes of Responses. *J. Math. Phys. 22*, 2803–2812.

Cacuci, D.G. 2003. *Sensitivity and Uncertainty Analysis: Theory*. Chapman & Hall/CRC, Boca Raton, FL, Vol. 1, p. 285.

Cacuci, D.G. 2015. Second-Order Adjoint Sensitivity Analysis Methodology (2nd-ASAM) for Computing Exactly and Efficiently First- and Second-Order Sensitivities in Large-Scale Linear Systems: I. Computational Methodology. *J. Comp. Phys. 284*, 687–699.

Cacuci, D.G. 2016. Second-Order Adjoint Sensitivity Analysis Methodology (2nd-ASAM) for Large-Scale Nonlinear Systems: I. Theory. *Nucl. Sci. Eng. 184*, 16–30.

Cacuci, D.G. 2018. *The Second-Order Adjoint Sensitivity Analysis Methodology*. Taylor & Francis/CRC Press, Boca Raton, FL, p. 305.

Cacuci, D.G. 2019. *BERRU Predictive Modeling: Best Estimate Results with Reduced Uncertainties*, 451 pages. Springer Verlag GmbH, Berlin, Germany. https://doi.org/10.1007/978-3-662-58395-1

Cacuci, D.G. 2022. *The nth-Order Comprehensive Adjoint Sensitivity Analysis Methodology (nth-CASAM): Overcoming the Curse of Dimensionality in Sensitivity and Uncertainty Analysis,* Volume I*: Linear Systems*. Springer Nature Switzerland, Cham, p. 362. https://doi.org/10.1007/978-3-030-96364-4

Cacuci, D.G. 2023a. *The nth-Order Comprehensive Adjoint Sensitivity Analysis Methodology (nth-CASAM): Overcoming the Curse of Dimensionality in Sensitivity and Uncertainty Analysis,* Volume III*: Nonlinear Systems*. Springer Nature Switzerland, Cham, p. 369. https://doi.org/10.1007/978-3-031-22757-8

Cacuci, D.G. 2023b. Computation of High-Order Sensitivities of Model Responses to Model Parameters. II: Introducing the Second-Order Adjoint Sensitivity Analysis Methodology for Computing Response Sensitivities to Functions/Features of Parameters. *Energies 16*, 6356. https://doi.org/10.3390/en16176356

Cacuci, D.G. 2024a. Introducing the nth-Order Features Adjoint Sensitivity Analysis Methodology for Nonlinear Systems (nth-FASAM-N): I. Mathematical Framework. *Am. J. Comp. Math. 14*, 11–42. https://doi.org/10.4236/ajcm.2024.141002

Cacuci, D.G. 2024b. The nth-Order Features Adjoint Sensitivity Analysis Methodology for Response-Coupled Forward/Adjoint Linear Systems (nth-FASAM-L): I. Mathematical Framework. *Front Energy Res. 12*, 1417594. https://doi.org/10.3389/fenrg.2024.1417594

Cacuci, D.G. 2025. *Advances in High-Order Predictive Modeling: Methodologies and Illustrative Problems*. CRC Press/Taylor & Francis, Boca Raton, FL, p. 288.

Cacuci, D.G., Ionescu-Bujor, M., Navon, M.I. 2005. *Sensitivity and Uncertainty Analysis: Applications to Large Scale Systems*. Chapman & Hall/CRC, Boca Raton, FL, Vol. 2, p. 352.

Cacuci, D.G., Ionescu-Bujor, M., Navon, M.I. 2014. *Computational Methods for Data Evaluation and Assimilation*. Chapman & Hall/CRC, Boca Raton, FL, p. 337.

# 1 Motivation for Computing High-Order Sensitivities of Model Responses to Model Parameters

## 1.1 INTRODUCTION

Scientific progress stems from the judicious combination of experimental information with results produced by computational models, none of which are perfectly accurate. On the one hand, computations are afflicted by errors stemming from numerical procedures, uncertain model parameters, boundary/initial conditions, and/or imperfectly known physical processes. On the other hand, results of measurements are afflicted by experimental errors, which means that around any reported experimental value, there always exists a range of values that may also be plausibly representative of the true but unknown value of the measured quantity. Extracting "best estimate" values for model parameters and predicted results, together with "best estimate" uncertainties for these parameters and results, requires reasoning from incomplete, error-afflicted, and occasionally discrepant information while combining experimental and computational data and their uncertainties. Under ideal circumstances, the result of predictive modeling is a probabilistic description of possible future outcomes based on all recognized errors and uncertainties.

The mathematical/computational models of physical systems comprise parameters, independent and dependent variables. The system's independent variables and parameters are related to the system's state (i.e., dependent) variables through a well-posed system of equations, which is usually nonlinear in all of its components. The minimum amount of information needed to use a model is the availability of nominal or mean values for the system parameters. With the known nominal parameter values, a model can be used to compute results of interest, which are called model/system "responses" or "objective functions" or "indices of performance." Since the physical processes themselves are seldom known precisely and since most of the model's parameters stem from experimental procedures that are also subject to imprecision and/or uncertainties, the results predicted by these models are also imprecise, being affected by the uncertainties underlying the

DOI: 10.1201/9781003478171-1 

respective model. Quantifying the reliability and accuracy of results (responses) computed using models is achieved by performing activities known as "sensitivity analysis," "uncertainty quantification," "model verification," "model validation," "data assimilation," "model calibration," and "model extrapolation." All of these activities are included in the overall activity that will be called "predictive modeling." *Sensitivity analysis* aims at quantifying the changes in the computed response that would be induced by changes in the model parameters along with understanding the model by ranking the importance of the various parameters in contributing to the response. The availability of such rankings makes it possible to prioritize improvements in the model and also streamline it by eliminating unimportant parameters and/or processes to develop faster-running yet accurate "reduced-order models" for routine computations. *Uncertainty quantification* aims at determining quantitatively the uncertainties induced in a model response due to model parameter uncertainties. *Model verification* deals with the question: "are the equations underlying the model solved correctly?" *Model validation* compares the model's computational results to experimental information in order to address the question "does the model represent reality?" *Data assimilation* combines experimental and computational information to improve model responses, while *model calibration* uses experimental information to adjust/calibrate the parameters to improve the accuracy of future computations. *Model extrapolation* refers to the application of the validated and calibrated model to explore novel applications. *Predictive modeling* subsumes all of the aforementioned activities, combining measured and computational information to obtain optimal best-estimate predictions of model responses and parameters, with reduced predicted uncertainties. All of these activities, as well as optimizing the system and using models for "inverse problems," require the availability of the functional derivatives – called "sensitivities" – of model responses with respect to the model parameters.

The definitions of the exact expressions of high-order sensitivities of model responses with respect to model parameters, or features/functions thereof, and the means for computing these exact expressions most efficiently are briefly reviewed in Section 1.2. The mathematical framework underlying the deterministic quantification of the effects of the high-order sensitivities on propagating uncertainties in the model parameters to create uncertainties in the model's response of interest is briefly summarized in Section 1.3. The use and impact of high-order sensitivities on combining computational and experimental information aimed at obtaining optimal best-estimate predicted responses and calibrated parameters, all with reduced predicted uncertainties, are briefly summarized in Section 1.4. Section 1.5 illustrates the typical applications and impacts of high-order sensitivities by using the polyethylene-reflected plutonium (acronym "PERP") reactor physics benchmark (Valentine, 2006), which is included in the OECD, Nuclear Energy Agency (NEA), and International Criticality Safety Benchmark Evaluation Project (ICSBEP). Section 1.6 presents a synopsis of the material covered in this chapter.

## 1.2 MATHEMATICAL DEFINITION OF HIGH-ORDER RESPONSE SENSITIVITIES WITH RESPECT TO FEATURE FUNCTIONS OF PARAMETERS

The generic modeling of a physical system and/or modeling the result of an indirect experimental measurement requires consideration of the following modeling components:

1. a mathematical model comprising linear and/or nonlinear equations that relate the system's independent variables and parameters to the system's state (i.e., dependent) variables;
2. statistical information (e.g., means, covariances, triple-correlations, quadruple correlations), inequality and/or equality constraints that characterize the ranges of possible variations in the system's parameters, since these parameters are never known exactly because they are afflicted by uncertainties;
3. one or several computational results, customarily referred to as system or model *responses* (or "objective functions," or "indices of performance"), which are computed using the mathematical model; and
4. experimentally measured responses, characterized by corresponding nominal (mean) values and uncertainties (variances, covariances, skewness, kurtosis, etc.).

Mathematically, a model response, denoted as $r_k(\boldsymbol{\alpha})$, is considered to be a function of $TP$ scalar parameters denoted as $\alpha_j$, $j = 1,\ldots,TP$. Without loss of generality, these parameters can be considered to be real-valued scalars. Since these parameters stem from experimental or computational procedures, they are never known exactly; only their nominal (or mean) values are known. These nominal values will be denoted as $\alpha_j^0$; the superscript "zero" will be used in this chapter to denote "nominal" or "mean" values. Historically, the concept of "sensitivity analysis" arose from the need to predict and quantify the changes induced in the model's responses by changes in the model's uncertain parameters and to rank the importance of the model's parameters in affecting the model's responses. Initially, such predictions were performed by simply recomputing the result of interest using finite-difference schemes in conjunction with "judiciously chosen" variations in the model parameters. Evidently, such methods can at best compute approximate values of a very limited number of sensitivities; re-computations become unfeasible for large-scale models with many uncertain parameters.

The only mathematically rigorous concept for defining the "sensitivity of a function with respect to its underlying parameters" is in terms of the partial derivatives of the respective function with respect to its underlying parameters. Thus, the "1st-order sensitivities" of a model response are the 1st-order partial functional derivatives of the response with respect to the parameters; the "2nd-order sensitivities" are the 2nd-order partial functional derivatives of the response with respect to the parameters; and so on; the "*n*th-order sensitivities" are the *n*th-order partial functional derivatives of the model response with respect to the model parameters. The scope of "sensitivity analysis" is to compute, as accurately and efficiently as possible, the sensitivities of various orders of model responses with respect to the model's parameters.

The multivariate Taylor-series expansion of a model response, $r_k(\boldsymbol{\alpha})$, $\boldsymbol{\alpha} \triangleq (\alpha_1,\ldots,\alpha_{TP})^\dagger \in \mathbb{R}^{TP}$, around the known nominal parameter values, $\boldsymbol{\alpha}^0 \triangleq \left(\alpha_1^0,\ldots,\alpha_{TP}^0\right)^\dagger$, has the following well-known formal expression:

$$r_k(\boldsymbol{\alpha}) = r_k(\boldsymbol{\alpha}^0) + \sum_{j_1=1}^{TP}\left\{\frac{\partial r_k(\boldsymbol{\alpha})}{\partial \alpha_{j_1}}\right\}_{\boldsymbol{\alpha}^0}\delta\alpha_{j_1} + \frac{1}{2}\sum_{j_1=1}^{TP}\sum_{j_2=1}^{TP}\left\{\frac{\partial^2 r_k(\boldsymbol{\alpha})}{\partial \alpha_{j_1}\partial \alpha_{j_2}}\right\}_{\boldsymbol{\alpha}^0}\delta\alpha_{j_1}\delta\alpha_{j_2}$$
$$+\frac{1}{3!}\sum_{j_1=1}^{TP}\sum_{j_2=1}^{TP}\sum_{j_3=1}^{TP}\left\{\frac{\partial^3 r_k(\boldsymbol{\alpha})}{\partial \alpha_{j_1}\partial \alpha_{j_2}\partial \alpha_{j_3}}\right\}_{\boldsymbol{\alpha}^0}\delta\alpha_{j_1}\delta\alpha_{j_2}\delta\alpha_{j_3} \tag{1.1}$$
$$+\frac{1}{4!}\sum_{j_1=1}^{TP}\sum_{j_2=1}^{TP}\sum_{j_3=1}^{TP}\sum_{j_4=1}^{TP}\left\{\frac{\partial^4 r_k(\boldsymbol{\alpha})}{\partial \alpha_{j_1}\partial \alpha_{j_2}\partial \alpha_{j_3}\partial \alpha_{j_4}}\right\}_{\boldsymbol{\alpha}^0}\delta\alpha_{j_1}\delta\alpha_{j_2}\delta\alpha_{j_3}\delta\alpha_{j_4}+\cdots,$$

where $\delta\alpha_j \triangleq \left(\alpha_j - \alpha_j^0\right)$, $j = 1,\ldots,TP$. The range of *validity* of the Taylor series shown in Eq. (1.1) is defined by its radius of convergence. Within its radius of convergence, the *accuracy* – as opposed to the "validity" – of the Taylor series in predicting the value of the response at an arbitrary point in the phase space of model parameters depends on the order of sensitivities retained in the Taylor expansion: the higher the respective order, the more accurate the respective response value predicted by the Taylor series. In the particular cases when the response happens to be a polynomial function of the parameters $\alpha_j$, the Taylor series provides an exact expression of the response.

As the Taylor expansion in Eq. (1.1) indicates, the sensitivities of a model response with respect to the model's parameters can be used for many activities, including: (i) understanding the model by ranking the importance of the various parameters; (ii) performing "reduced-order modeling" by eliminating unimportant parameters and/or processes; (iii) quantifying the uncertainties induced in a model response due to model parameter uncertainties; (iv) performing "model validation" by comparing computations to experiments to address the question "does the model represent reality?" (v) prioritizing improvements in the model; (vi) performing data assimilation and model calibration as part of forward "predictive modeling" to obtain best-estimate predicted results with reduced predicted uncertainties; (vii) performing inverse "predictive modeling" to calibrate/adjust the parameter values for enabling subsequent predictions with improved accuracy; (viii) designing and optimizing the system.

Since the explicit expression of $r_k(\boldsymbol{\alpha})$ as a function of the model parameters $\boldsymbol{\alpha} \triangleq (\alpha_1,\ldots,\alpha_{TP})^\dagger \in \mathbb{R}^{TP}$ is not available, the explicit expressions of the sensitivities of $r_k(\boldsymbol{\alpha})$ are not available, either, for numerical evaluation. It is evident from Eq. (1.1) that there are $TP$ 1st-order sensitivities, $TP\frac{(TP+1)}{2}$ distinct 2nd-order sensitivities, $TP\frac{(TP+1)(TP+2)}{6}$ distinct 3rd-order sensitivities, and so on. Thus, in the phase space of model parameters, the number of sensitivities increases exponentially with the order of the respective sensitivities. A computation using the model is considered to be a "large-scale" computation. Hence, the number of large-scale computations required by methods that involve direct differentiation of the original model equations and/or methods that require re-computations using the original model with altered parameter values (e.g., finite difference schemes, statistical assessments) increases exponentially, in the phase space of parameters, with the order of sensitivities, i.e., the computation by conventional methods of the $n$ th-order functional derivatives of a response with respect to the $TP$ parameters would require at least $O\left(TP^n\right)$ large-scale computations. The exponential increase – with the order of response sensitivities – of the number of large-scale computations needed to determine higher-order sensitivities is the manifestation of the "curse of dimensionality in sensitivity analysis," by analogy to the expression coined by Bellman (1957) to express the difficulty of using "brute-force" grid search when optimizing a function with many input variables.

Currently, the only method that overcomes, to the largest possible extent, the curse of dimensionality while providing exact expressions for computing most efficiently arbitrarily high-order sensitivities of responses with respect to model parameters (including imprecisely known domain boundaries and interfaces) is the "**nth**-Order **C**omprehensive **A**djoint **S**ensitivity **A**nalysis **M**ethodology (*n*th-CASAM)" developed by Cacuci (2021b, 2022a, 2022b, 2023a). The *n*th-CASAM comprises two mathematical frameworks, as follows: (a) the "*n*th-order Comprehensive Adjoint Sensitivity Analysis Methodology for Response-Coupled Forward/Adjoint Linear Systems" (*n*th-CASAM-L) conceived by Cacuci (2021b, 2022b) specifically for this unique class of systems, which often occur in nuclear engineering, and (b) the "*n*th-order Comprehensive Adjoint Sensitivity Analysis Methodology for Nonlinear Systems" (*n*th-CASAM-N) conceived by Cacuci (2022a, 2023a), which also subsumes the case of linear systems with responses that do not simultaneously depend on both the forward and the adjoint state/dependent variables. The *n*th-CASAM generalizes the original "1st-order adjoint sensitivity analysis methodology for nonlinear systems" conceived by Cacuci (1981a, 1981b), who has also demonstrated the efficacy and accuracy of this methodology by applying it to sensitivity analysis of large-scale systems encountered in the earth, atmospheric, and other sciences, as credited, e.g., by Práger and Kelemen (2014); Luo, Wang, and Liu (2020). The *n*th-CASAM involves large-scale computations in *linearly increasing higher-dimensional Hilbert spaces in the space of forward and adjoint state functions (dependent variables)*, as opposed to requiring large-scale computations in exponentially increasing parameter-dimensional spaces, as is the case when using automatic differentiation, finite differences, and/or statistical methods. In particular, for a scalar-valued response associated with a model comprising $TP$ model parameters, the *n*th-CASAM requires 1 adjoint computation for computing exactly all of the 1st-order response sensitivities, as opposed to at least $TP$ forward computations, as required by all other methods to obtain (usually approximate values for) these sensitivities. All of the (mixed) 2nd-order sensitivities are computed exactly by the *n*th-CASAM in at most $TP$ computations, as opposed to needing at least $TP\frac{(TP+1)}{2}$ computations, as required by all other methods, and so on. For every lower-order response-sensitivity of interest, the $n$th-CASAM computes the "*TP* next-higher-order" response-sensitivities in *one adjoint computation performed in a linearly increasing higher-dimensional Hilbert space*. The $n$th-CASAM applies to any model (deterministic, statistical, etc.) and is also applicable to the computation of sensitivities of operator-valued responses. The larger the number of model parameters, the more efficient the $n$th-CASAM becomes for computing arbitrarily high-order response sensitivities.

As an alternative to deterministic methods, "statistical methods" have been employed to compute so-called "measures of sensitivities" and/or "sensitivity indicators" constructed in terms of parameter variances to play the role of surrogates for (mostly 1st-order) response sensitivities. These "sensitivity indicators" are computed by initially constructing an approximate response distribution (often called "response surface") in the phase space of model parameters by performing many "forward" computations using the model with altered parameter values and subsequently using scatter plots, regression, rank transformation, correlations, and/or so-called "partial correlation analysis" in order to identify approximate expectation values, variances, and covariances for the responses. These statistical quantities are subsequently used to construct quantities that play the role of (approximate) 1st-order response sensitivities. Thus, statistical methods commence with "uncertainty analysis" and subsequently attempt to perform an approximate "sensitivity analysis" of the approximately computed "response surface" in the phase space of a subset of the model's parameters. Various variants of the statistical methods for uncertainty and sensitivity analysis are reviewed by Saltelli et al. (2008). The conclusions that emerge from examining these statistical methods are as follows:

1. The main advantage of using statistical methods for uncertainty and sensitivity analysis is that they are conceptually easy to implement.
2. Even 1st-order sensitivities cannot be computed exactly.
3. Since many thousands of simulations are needed to obtain reliable results, statistical methods are at best expensive (for small systems) or, at worst, impracticable (e.g., for large time-dependent systems), being subject to the curse of dimensionality.
4. Response sensitivities cannot be computed exactly; sensitivities beyond 1st-order cannot be estimated.
5. Since the response sensitivities and parameter uncertainties are amalgamated, inherently and inseparably, within the results produced by statistical methods, improvements in parameter uncertainties cannot be directly propagated to improve response uncertainties; rather, the entire set of simulations and statistical post-processing must be repeated anew.
6. A "fool-proof" statistical method for correctly analyzing models involving correlated parameters does not seem to exist currently, so particular care must be used when interpreting regression results obtained using such models in the presence of parameter correlations.

The practitioners of statistical methods often claim that these methods are "global" in the parameter space, contrasting them with the Taylor series of the response, which is labeled "local." In reality, the "sensitivity indicators" obtained using statistical methods are neither local nor global; they are approximate values, with an unknown range of validity, since such a range is never actually provided by these methods. On the other hand, the range of validity of the Taylor series is mathematically provided by its radius of convergence. The accuracy – as opposed to the "validity" – of the Taylor series in predicting the value of the response at an arbitrary point in the phase space of model parameters depends on the order of sensitivities retained in the Taylor expansion: the higher the respective order, the more accurate the respective response value predicted by the Taylor series. In the particular cases when the response happens to be a polynomial function of the model parameters, the Taylor series is actually exact (not just "non-local"). In contradistinction to the *n*th-CASAM and *n*th-FASAM methodologies, no statistical method is ever "exact." Many statistical sensitivity analyses are wrong, as noticed by the practitioners themselves (Saltelli et al. 2019), who "used results from an extensive systematic literature review to show that many highly cited papers (42% in the present analysis) fail the elementary requirement to properly explore the space of the input factors."

In many practical situations, the physical system and, hence, its computational model and its response depend not only on individual parameters but also on functions (e.g., correlations, material properties, flow regimes, etc.) of such parameters. In such cases, the model response can be considered to depend on the primary model parameters through "feature" functions of parameters. Formally, the model response $r_k[\mathbf{f}(\alpha)]$ can be considered to depend on a $TF$-dimensional vector-valued function, $\mathbf{f}(\alpha) \triangleq [f_1(\alpha),\ldots,f_{TF}(\alpha)]^\dagger$, having as components "feature functions" $f_i(\alpha), i = 1,\ldots,TF$, which can be considered without loss of generality to be real-valued functions of (some of) the primary model parameters $\alpha \in \mathbb{R}^{TP}$. The model response $r_k[\mathbf{f}(\alpha)]$ would therefore admit a Taylor series expansion around the nominal value $\mathbf{f}^0 \triangleq \mathbf{f}(\alpha^0)$, having the following form:

$$r_k[\mathbf{f}(\alpha)] = r_k(\mathbf{f}^0) + \sum_{j_1=1}^{TF}\left\{\frac{\partial r_k(\mathbf{f})}{\partial f_{j_1}}\right\}_{\mathbf{f}^0}\delta f_{j_1} + \frac{1}{2}\sum_{j_1=1}^{TF}\sum_{j_2=1}^{TF}\left\{\frac{\partial^2 r_k(\mathbf{f})}{\partial f_{j_1}\partial f_{j_2}}\right\}_{\mathbf{f}^0}\delta f_{j_1}\delta f_{j_2} + \cdots, \tag{1.2}$$

where $\delta f_j \triangleq [f_j(\alpha) - f_j^0]$; $f_j^0 \triangleq f_j(\alpha^0)$; $j = 1,\ldots,TF$. The "sensitivities of the model response with respect to the (feature) functions" are naturally defined as being the *functional derivatives* of $r_k[\mathbf{f}(\alpha)]$ with respect to the components ("features") $f_j(\alpha)$ of $\mathbf{f}(\alpha)$. By construction, the number of functions $f_j(\alpha)$ is smaller than the total number of model parameters, i.e., $TF \le TP$; the equality $TF = TP$ holds when no feature functions can be constructed. The construction/definition of the feature functions is model-dependent and is at the discretion of the modeler. Since $TF < TP$, the computations of the functional derivatives of $r_k[\mathbf{f}(\alpha)]$ with respect to the functions $f_j(\alpha)$, which appear in Eq. (1.2), will be considerably less expensive computationally than the computation of the functional derivatives involved in the Taylor series, provided in Eq. (1.1), of the response with respect to the model parameters. The functional derivatives of the response with respect to the parameters can be obtained from the functional derivatives of the response with respect to the "feature" functions $f_j(\boldsymbol{\alpha})$ by simply using the chain rule, i.e.,

$$\left\{\frac{\partial r_k(\boldsymbol{\alpha})}{\partial \alpha_{j1}}\right\}_{\boldsymbol{\alpha}^0}=\sum_{i_1=1}^{TF}\left\{\frac{\partial r_k(\mathbf{f})}{\partial f_{i_1}}\frac{\partial f_{i_1}(\boldsymbol{\alpha})}{\partial \alpha_{j1}}\right\}_{\boldsymbol{\alpha}^0};\quad \left\{\frac{\partial^2 r_k(\boldsymbol{\alpha})}{\partial \alpha_{j1}\partial \alpha_{j2}}\right\}_{\boldsymbol{\alpha}^0}=\frac{\partial}{\partial \alpha_{j2}}\sum_{i_1=1}^{TF}\left\{\frac{\partial r_k(\mathbf{f})}{\partial f_{i_1}}\frac{\partial f_{i_1}(\boldsymbol{\alpha})}{\partial \alpha_{j1}}\right\}_{\boldsymbol{\alpha}^0}; \tag{1.3}$$

and so on. Since the exact expressions of the feature functions $f_i(\boldsymbol{\alpha})$ in terms of the model parameters are known, the evaluation/computation of the functional derivatives $\frac{\partial f_{i_1}(\boldsymbol{\alpha})}{\partial \alpha_{j1}}$, $\frac{\partial^2 f_{i_1}(\boldsymbol{\alpha})}{\partial \alpha_{j1}\partial \alpha_{j2}}$, etc., does not require computations involving the model. Thus, the derivatives shown in Eq. (1,3) are trivial to compute by comparison to the evaluation of the functional derivatives ("sensitivities") of the response with respect to either the functions $f_j(\boldsymbol{\alpha})$ or the model parameters $\alpha_i, i=1,\ldots,TP$.

The range of *validity* of the Taylor series shown in Eq. (1.2) is defined by its radius of convergence. Within its radius of convergence, the *accuracy* of the Taylor series in predicting the value of the response at an arbitrary point in the phase space of model parameters depends on the order of sensitivities retained in the Taylor expansion: the higher the respective order, the more accurate the respective response value predicted by the Taylor series. In the particular case when the response happens to be a polynomial function of the "feature" functions $f_j(\boldsymbol{\alpha})$ and/or the model parameters, the Taylor-series presented in Eq. (1.2) provides an exact representation of the model response. The most efficient methodology for computing arbitrarily high-order sensitivities of a model response with respect to feature functions of parameters is the "**n**^th^-Order **F**eatures **A**djoint **S**ensitivity **A**nalysis **M**ethodology (*n*th-FASAM)" conceived by Cacuci (2024a, 2024b), which computes exactly determined expressions of sensitivities while overcoming, to the largest possible extent, the curse of dimensionality. This methodology will be presented in this book's Chapters 2 and 3. If the model does not admit construction of features/functions of model parameters, then $f_j(\boldsymbol{\alpha})=\alpha_j, j=1,\ldots,TP$, and consequently the *n*th-FASAM methodology becomes identical to the *n*th-CASAM methodology. Notably, there is no statistical method for estimating response sensitivities with respect to features/functions of parameters. On the other hand, the price for using the *n*th-FASAM or the *n*th-CASAM methodologies is the need for constructing and solving "adjoint sensitivity systems," as will be detailed in the remainder of this book.

## 1.3 USE OF SENSITIVITIES IN HIGH-ORDER UNCERTAINTY QUANTIFICATION

As has been mentioned in Section 1.2, the values of the model parameters $\boldsymbol{\alpha}\triangleq(\alpha_1,\ldots,\alpha_{TP})^\dagger\in\mathbb{R}^{TP}$ are most often determined experimentally and are therefore not known precisely. In practice, the following information is expected to be known about the model parameters: (i) the expected or nominal value, denoted as $\alpha_i^0$, of a model parameter $\alpha_i$; (ii) the covariance, denoted as $\mathrm{cov}(\alpha_i,\alpha_j)$, of two parameters,

$\alpha_i$ and $\alpha_j$, including, in particular, the standard deviation $\sigma_i$ of a parameter $\alpha_i$; (iii) the 3rd-order correlation, denoted as $t_{ijk}$, of three parameters, $\alpha_i$, $\alpha_j$, and $\alpha_k$, including, in particular, the skewness, denoted as $\mu_i^{(3)}$, of a parameter $\alpha_i$; (iv) the 4th-order correlation, denoted as $q_{ijkl}$, of four parameters, $\alpha_i$, $\alpha_j$, $\alpha_k$, and $\alpha_l$, including, in particular, the kurtosis, denoted as $\mu_i^{(4)}$, of a parameter $\alpha_i$.

Uncertainties in the model's parameters will give rise to uncertainties in the model responses $r_k(\boldsymbol{\alpha})$. The approximate moments of the unknown distribution of $r_k(\boldsymbol{\alpha})$ are obtained by using the so-called "propagation of errors" methodology pioneered by Tukey (1957), which involves the formal integration over the unknown distribution of model parameters of various expressions involving the truncated Taylor-series expansion of the response provided in Eq. (1.1). Tukey (1957) has presented formulas for the mean, variance, skewness, and kurtosis of the response distribution, which include, albeit inconsistently, terms up to 4th-order response sensitivities. Cacuci (2022b) has generalized Tukey's (1957) results, presenting formulas that are consistent up to and including the 6th-order parameter correlations. The uses and impacts of the sensitivities of various orders on the moments of the response distribution are illustrated below by presenting simplified (consistent up to 4th-order in the parameters' standard deviations) versions of the 6th-order expressions obtained by Cacuci (2022b), as follows:

i. The 4th-order expression of the expectation value, $E(r_k)$, of a response $r_k(\boldsymbol{\alpha})$, is provided below:

$$E(r_k)=r_k(\boldsymbol{\alpha}^0)+\frac{1}{2}\sum_{i=1}^{TP}\sum_{j=1}^{TP}\frac{\partial^2 r_k(\boldsymbol{\alpha}^0)}{\partial\alpha_i\,\partial\alpha_j}cov(\alpha_i,\alpha_j)$$

$$+\frac{1}{6}\sum_{i=1}^{TP}\sum_{j=1}^{TP}\sum_{\mu=1}^{TP}\frac{\partial^3 r_k(\boldsymbol{\alpha}^0)}{\partial\alpha_i\,\partial\alpha_j\,\partial\alpha_\mu}t_{ij\mu}+\frac{1}{4!}\sum_{i=1}^{TP}\sum_{j=1}^{TP}\sum_{\mu=1}^{TP}\sum_{\nu=1}^{TP}\frac{\partial^4 r_k(\boldsymbol{\alpha}^0)}{\partial\alpha_i\,\partial\alpha_j\,\partial\alpha_\mu\,\partial\alpha_\nu}q_{ij\mu\nu}+\cdots. \tag{1.4}$$

ii. The 4th-order expression of the covariance, $cov(r_k,r_l)$, between two responses $r_k(\boldsymbol{\alpha})$ and $r_l(\boldsymbol{\alpha})$, is provided below:

$$cov(r_k,r_l)=\sum_{i=1}^{TP}\sum_{j=1}^{TP}\frac{\partial r_k(\boldsymbol{\alpha}^0)}{\partial\alpha_i}\frac{\partial r_l(\boldsymbol{\alpha}^0)}{\partial\alpha_j}cov(\alpha_i,\alpha_j)$$

$$+\frac{1}{2}\sum_{i=1}^{TP}\sum_{j=1}^{TP}\sum_{\mu=1}^{TP}\left(\frac{\partial^2 r_k(\alpha^0)}{\partial\alpha_i\,\partial\alpha_j}\frac{\partial r_l(\alpha^0)}{\partial\alpha_\mu}+\frac{\partial r_k(\alpha^0)}{\partial\alpha_i}\frac{\partial^2 r_l(\alpha^0)}{\partial\alpha_j\,\partial\alpha_\mu}\right)t_{ij\mu}$$

$$+\frac{1}{4}\sum_{i=1}^{TP}\sum_{j=1}^{TP}\sum_{\mu=1}^{TP}\sum_{\nu=1}^{TP}\frac{\partial^2 r_k(\alpha^0)}{\partial\alpha_i\,\partial\alpha_j}\frac{\partial^2 r_l(\alpha^0)}{\partial\alpha_\mu\,\partial\alpha_\nu}\left[q_{ij\mu\nu}-cov(\alpha_i,\alpha_j)cov(\alpha_\mu,\alpha_\nu)\right]$$

$$+\frac{1}{6}\sum_{i=1}^{TP}\sum_{j=1}^{TP}\sum_{\mu=1}^{TP}\sum_{\nu=1}^{TP}\left(\frac{\partial^3 r_k(\alpha^0)}{\partial\alpha_i\,\partial\alpha_j\,\partial\alpha_\mu}\frac{\partial r_l(\alpha^0)}{\partial\alpha_\nu}+\frac{\partial r_k(\alpha^0)}{\partial\alpha_i}\frac{\partial^3 r_l(\alpha^0)}{\partial\alpha_j\,\partial\alpha_\mu\,\partial\alpha_\nu}\right)q_{ij\mu\nu}+\cdots. \tag{1.5}$$

iii. The 4th-order expression of the correlation, $cor(\alpha_i, r_k)$, between a parameter $\alpha_i$ and a response $r_k(\boldsymbol{\alpha})$, is provided below:

$$cor(\alpha_i, r_k) = \sum_{j=1}^{TP} \frac{\partial r_k(\boldsymbol{\alpha}^0)}{\partial \alpha_j} cov(\alpha_i, \alpha_j) + \frac{1}{2} \sum_{j=1}^{TP} \sum_{\mu=1}^{TP} \frac{\partial^2 r_k(\boldsymbol{\alpha}^0)}{\partial \alpha_j \partial \alpha_\mu} t_{ij\mu} + \frac{1}{6} \sum_{j=1}^{TP} \sum_{\mu=1}^{TP} \sum_{\nu=1}^{TP} \frac{\partial^3 r_k(\boldsymbol{\alpha}^0)}{\partial \alpha_j \partial \alpha_\mu \partial \alpha_\nu} q_{ij\mu\nu} + \cdots. \tag{1.6}$$

iv. The 4th-order expression of the triple correlation, $\mu_3(r_k, r_l, r_m)$, among three responses, $r_k, r_l, r_m$, is provided below:

$$\begin{aligned} \mu_3(r_k, r_l, r_m) = & \sum_{i=1}^{TP} \sum_{j=1}^{TP} \sum_{\mu=1}^{TP} \frac{\partial r_k(\boldsymbol{\alpha}^0)}{\partial \alpha_i} \frac{\partial r_l(\boldsymbol{\alpha}^0)}{\partial \alpha_j} \frac{\partial r_m(\boldsymbol{\alpha}^0)}{\partial \alpha_\mu} t_{ij\mu} \\ & + \frac{1}{2} \sum_{i=1}^{TP} \sum_{j=1}^{TP} \sum_{\mu=1}^{TP} \sum_{\nu=1}^{TP} \frac{\partial r_k(\boldsymbol{\alpha}^0)}{\partial \alpha_i} \frac{\partial r_l(\boldsymbol{\alpha}^0)}{\partial \alpha_j} \frac{\partial^2 r_m(\boldsymbol{\alpha}^0)}{\partial \alpha_\mu \partial \alpha_\nu} \left[ q_{ij\mu\nu} - cov(\alpha_i, \alpha_j) cov(\alpha_\mu, \alpha_\nu) \right] \\ & + \frac{1}{2} \sum_{i=1}^{TP} \sum_{j=1}^{TP} \sum_{\mu=1}^{TP} \sum_{\nu=1}^{TP} \frac{\partial r_k(\boldsymbol{\alpha}^0)}{\partial \alpha_i} \frac{\partial^2 r_l(\boldsymbol{\alpha}^0)}{\partial \alpha_j \partial \alpha_\mu} \frac{\partial r_m(\boldsymbol{\alpha}^0)}{\partial \alpha_\nu} \left[ q_{ij\mu\nu} - cov(\alpha_i, \alpha_\nu) cov(\alpha_j, \alpha_\mu) \right] \\ & + \frac{1}{2} \sum_{i=1}^{TP} \sum_{j=1}^{TP} \sum_{\mu=1}^{TP} \sum_{\nu=1}^{TP} \frac{\partial^2 r_k(\boldsymbol{\alpha}^0)}{\partial \alpha_i \partial \alpha_j} \frac{\partial r_l(\boldsymbol{\alpha}^0)}{\partial \alpha_\mu} \frac{\partial r_m(\boldsymbol{\alpha}^0)}{\partial \alpha_\nu} \left[ q_{ij\mu\nu} - cov(\alpha_i, \alpha_j) cov(\alpha_\mu, \alpha_\nu) \right] + \cdots. \end{aligned} \tag{1.7}$$

The expressions for the triple correlations, $cor(\alpha_i, \alpha_j, r_k)$ and $cor(\alpha_i, r_j, r_k)$, among parameters and responses, are provided by Cacuci (2022b).

v. The 4th-order expression of the quadruple-correlation, $\mu_4(r_k, r_l, r_m, r_n)$, among four responses, $r_k, r_l, r_m, r_n$, is provided below:

$$\mu_4(r_k, r_l, r_m, r_n) = \sum_{i=1}^{TP} \sum_{j=1}^{TP} \sum_{\mu=1}^{TP} \sum_{\nu=1}^{TP} \frac{\partial r_k(\boldsymbol{\alpha}^0)}{\partial \alpha_i} \frac{\partial r_l(\boldsymbol{\alpha}^0)}{\partial \alpha_j} \frac{\partial r_m(\boldsymbol{\alpha}^0)}{\partial \alpha_\mu} \frac{\partial r_n(\boldsymbol{\alpha}^0)}{\partial \alpha_\nu} q_{ij\mu\nu} + \cdots. \tag{1.8}$$

The expressions for the quadruple correlations, $cor(\alpha_i, r_j, r_k, r_l)$, $cor(\alpha_i, \alpha_j, r_k, r_l)$ and $cor(\alpha_i, \alpha_j, \alpha_k, r_l)$, among parameters and responses, are provided by Cacuci (2022b).

The expressions provided in Eqs. (1.4)–(1.8) highlight the major roles played by the sensitivities of various orders, as follows:

i. The 1st-order sensitivities contribute the leading terms to the 2nd-, 3rd-, and 4th-order moments of the response distribution, thus providing the leading contributions to the response variance/covariances, skewness, and kurtosis. Obtaining the exact and complete set of 1st-order sensitivities of responses to model parameters is of paramount importance for any analysis of a computational model.
ii. The 2nd-order sensitivities contribute the leading correction terms to the response's expected value, causing it to differ from the response's computed value. The 2nd-order sensitivities also contribute to the response variances and covariances. If the parameters follow a normal (Gaussian) multivariate distribution, the 2nd-order sensitivities contribute the leading terms to the response's 3rd-order moment. Thus, for normally distributed parameters, neglecting the 2nd-order response sensitivities would nullify the 3rd-order response correlations and hence would nullify the skewness of a response. Hence, a "1st-order sensitivity and uncertainty quantification" will always produce an erroneous $3^{rd}$-order moment (and hence skewness) of the predicted response distribution, unless the unknown response distribution happens to be symmetrical. At least 2nd-order sensitivities must be used in order to estimate the 3rd-order moment (and hence the skewness) of the response distribution. Skewness indicates the direction and relative magnitude of a distribution's deviation from the normal distribution while kurtosis indicates the propensity of the predicted response distribution to have heavy tails and/or outliers. With pronounced skewness, standard statistical inference procedures such as constructing a confidence interval for the mean (expectation) of a computed/predicted model response will be not only incorrect, in the sense that the true coverage level will differ from the nominal (e.g., 95%) level, but also the error probabilities will be unequal on each side of the predicted mean.

## 1.4 HIGH-ORDER PREDICTIVE MODELING: MAIN FEATURES OF THE 4TH-ORDER 4TH-BERRU-PM METHODOLOGY

The BERRU-PM methodology was originally conceived by Cacuci (2019a); the acronym "BERRU" stands for "Best-Estimate Results with Reduced Uncertainties" while "PM" stands for "predictive modeling." This methodology has significantly extended the conventional "data assimilation" methodologies (see, e.g., Lewis et al. 2006; Cacuci et al. 2014). Subsequently, Cacuci (2023b, 2023c) has extended his BERRU-PM methodology to incorporate consistently the 1st- and 2nd-order experimental and computational information, which was subsequently extended further by Cacuci (2023d, 2023e, 2025) to incorporate consistently all of the experimental and computational information up to, and including, 4th-order parameter and response correlations. This predictive modeling methodology, called the "4th-Order Best-Estimate Results with Reduced Uncertainties Predictive Modeling (4th-BERRU-PM)" methodology is the only methodology currently available that includes information (sensitivities, moments of the measured and experimental response distributions, and moments of the parameter distribution) higher than 2nd-order. The 4th-BERRU-PM methodology (Cacuci 2023d, 2023e, 2025) constructs the best-estimate 4th-order moments-constrained Maximum Entropy (MaxEnt) distribution, denoted as $p_4^{be}(\boldsymbol{\alpha},\mathbf{r})$, comprising all of the available knowledge in the combined phase space of the experimental and computational information, up to and including 4th-order, as described below. The MaxEnt distribution (Shannon 1948) is known to provide the optimal compatibility with the available information while simultaneously minimizing the introduction of spurious information. Specifically, the $p_4^{be}(\boldsymbol{\alpha},\mathbf{r})$ distribution incorporates the following experimental and computational a priori information:

1. the vector $\boldsymbol{\alpha}^0 \triangleq \left(\alpha_1^0,\dots,\alpha_{TP}^0\right)^\dagger$ of expected/nominal values of the model parameters, where $\alpha_i^0$ denotes the nominal/expected value of parameter $\alpha_i$, $i=1,\dots,TP$;
2. the parameter covariance matrix, $\mathbf{C}_{\alpha\alpha} \triangleq \left[c_{ij}^{\alpha}\right]_{TP\times TP}$, where the quantity $c_{ij}^{\alpha}$ denotes the covariance of two model parameters, $\alpha_i$ and $\alpha_j$, where $i,j=1,\dots,TP$;
3. the triple correlations of three model parameters $\alpha_i$, $\alpha_j$, and $\alpha_k$, denoted as $t_{ijk}^{\alpha}$, where $i,j,k=1,\dots,TP$;
4. the quadruple correlations of four model parameters $\alpha_i$, $\alpha_j$, $\alpha_k$, and $\alpha_\ell$, denoted as $q_{ijk\ell}^{\alpha}$, where $i,j,k,\ell=1,\dots,TP$;
5. the vector of expected values of the computed responses, denoted as $\mathbf{E}_c(\mathbf{r}) \triangleq \left[E_c(r_1),\dots,E_c(r_k),\dots,E_c(r_{TR})\right]^\dagger$, where $\mathbf{r} \triangleq \left[r_1,\dots,r_k,\dots,r_{TR}\right]^\dagger$; $E_c(r_k)$ denotes the expectation value of a computed response $r_k(\boldsymbol{\alpha})$, for $k=1,\dots,TR$; and where $TR$ denotes the total number of computed responses; see Eq. (1.4);
6. the parameter-response computed correlation matrix, $\mathbf{C}_{\alpha r}^{c} \triangleq \left[cor(\alpha_i,r_k)\right]_{TP\times TR}$, where $cor(\alpha_i,r_k)$ denotes the correlation between a parameter $\alpha_i$ and a computed response $r_k$, for $i=1,\dots,TP$ and $k=1,\dots,TR$; see Eq. (1.6);
7. the covariance matrix of computed responses, $\mathbf{C}_{rr}^{c} \triangleq \left[cov(r_k,r_l)\right]_{TR\times TR}$, where $cov(r_k,r_l)$ denotes the covariance between two computed responses $r_k$ and $r_l$, for $k,l=1,\dots,TR$; see Eq. (1.5);
8. the triple correlations, $\mu_3(r_k,r_l,r_m)$, among three computed responses, $r_k,r_l,r_m$, for $k,l,m=1,\dots,TR$; see Eq. (1.7); the skewness of the computed response $r_k$ is denoted as $\mu_3(r_k)$;
9. the quadruple correlations, $\mu_4(r_k,r_l,r_m,r_n)$, among four computed responses, $r_k,r_l,r_m,r_n$, for $k,l,m,n=1,\dots,TR$; see Eq. (1.8); the kurtosis of the computed response $r_k$ is denoted as $\mu_4(r_k)$;
10. the expectation values, denoted as $r_i^e$, of the experimentally measured system responses $r_i$;
11. the experimentally measured covariances, denoted as $c_{ij}^{e}$, for two responses $r_i$ and $r_j$, where $i,j=1,\dots,TR$; these covariances are considered to be components of the $TR\times TR$-dimensional experimentally measured covariance matrix of system responses $\mathbf{C}_{rr}^{e} \triangleq \left[c_{ij}^{e}\right]_{TR\times TR}$.
12. the experimentally measured triple correlations, denoted as $t_{ijk}^{e}$, among three system responses $r_i$, $r_j$, $r_k$, for $i,j,k=1,\dots,TR$; the skewness of the measured response $r_k$ is denoted as $t_k^e$;

13. the experimentally measured quadruple correlations, denoted as $q^e_{ijk\ell}$, among four system responses $r_i$, $r_j$, $r_k$, $r_\ell$, for $i, j, k, \ell = 1,\dots,TR$; the kurtosis of the measured response $r_k$ is denoted as $q^e_k$.

The moments of the distribution $p_4^{be}(\boldsymbol{\alpha},\mathbf{r})$ were derived by Cacuci (2023d, 2023e, 2025) using the saddle-point expansion (Shun and McCullagh 1995; Tang and Reid 2023) to obtain explicit closed-form expressions for the 1st-, 2nd-, 3rd-, and 4th-order moments of the best-estimate predicted responses and calibrated model parameters. The complete expressions for these moments are provided by Cacuci (2023d, 2023e, 2025). For illustrative purposes, the simplified expressions for these moments will be provided below for the lowest-order approximation, denoted as $\left(\mathbf{r}_s^{(0)},\boldsymbol{\alpha}_s^{(0)}\right)$, of the saddle point $(\mathbf{r},\boldsymbol{\alpha})=(\mathbf{r}_s,\boldsymbol{\alpha}_s)$ in the combined space of all computational and experimental information. The vectors $\mathbf{r}_s^{(0)} \triangleq \left[r_1^{(0)},\dots,r_k^{(0)},\dots,r_{TR}^{(0)}\right]^\dagger$ and $\boldsymbol{\alpha}_s^{(0)} \triangleq \left[\alpha_1^{(0)},\dots,\alpha_k^{(0)},\dots,\alpha_{TP}^{(0)}\right]^\dagger$ are defined as follows:

$$\mathbf{r}_s^{(0)} = \mathbf{r}^e + \mathbf{C}^e\left(\mathbf{C}^e + \mathbf{C}_{rr}^c\right)^{-1}\left[\mathbf{E}_c(\mathbf{r}) - \mathbf{r}^e\right], \tag{1.9}$$

$$\boldsymbol{\alpha}_s^{(0)} = \boldsymbol{\alpha}^0 - \mathbf{C}_{\alpha r}^c\left(\mathbf{C}^e + \mathbf{C}_{rr}^c\right)^{-1}\left[\mathbf{E}_c(\mathbf{r}) - \mathbf{r}^e\right]. \tag{1.10}$$

The superscript "(0)" in Eqs. (1.9) and (1.10) indicates that the quantities $\mathbf{r}_s^{(0)}$ and $\boldsymbol{\alpha}_s^{(0)}$ do *not* account for the triple and quadruple correlations among computed model responses but do take into account the triple and quadruple correlations among model parameters that enter in the expressions of the vector of mean values of the computed responses, $\mathbf{E}_c(\mathbf{r})$, and the matrices $\mathbf{C}_{rr}^c$ and $\mathbf{C}_{\alpha r}^c$. The vectors $\mathbf{r}_s^{(0)}$ and $\boldsymbol{\alpha}_s^{(0)}$ fully account for all 1st- and 2nd-order correlations among computed and measured responses and model parameters.

The simplified expressions of the predicted moments of the distribution $p_4^{be}(\boldsymbol{\alpha},\mathbf{r})$, of best-estimate responses and calibrated parameters in the combined space of all computational and experimental information are as follows:

1. The vector $\mathbf{r}^{be} \triangleq \left(r_1^{be},\dots,r_{TR}^{be}\right)^\dagger$ of mean values of the predicted best-estimate responses:

$$\mathbf{r}^{be} = \mathbf{r}_s^{(0)} + \mathbf{p}_1\left(\mathbf{r}_s^{(0)},\boldsymbol{\alpha}_s^{(0)}\right) + HOT, \tag{1.11}$$

where "*HOT*" denotes "higher order terms" and where:

$$\mathbf{p}_1\left(\mathbf{r}_s^{(0)},\boldsymbol{\alpha}_s^{(0)}\right) = \mathbf{C}^e\left(\mathbf{C}^e + \mathbf{C}_{rr}^c\right)^{-1}\left[\mathbf{C}_{rr}^c\mathbf{v}\left(\mathbf{r}_s^{(0)}\right) + \mathbf{C}_{r\alpha}^c\mathbf{w}\left(\boldsymbol{\alpha}_s^{(0)}\right)\right]; \tag{1.12}$$

$$\mathbf{v}\left(\mathbf{r}_s^{(0)}\right) \triangleq \left[v_1\left(\mathbf{r}_s^{(0)}\right),\dots,v_{TR}\left(\mathbf{r}_s^{(0)}\right)\right]^\dagger; \quad \mathbf{w}\left(\boldsymbol{\alpha}_s^{(0)}\right) \triangleq \left[w_1\left(\boldsymbol{\alpha}_s^{(0)}\right),\dots,w_{TP}\left(\boldsymbol{\alpha}_s^{(0)}\right)\right]^\dagger; \tag{1.13}$$

$$v_k\left(\mathbf{r}_s^{(0)}\right) \triangleq -\frac{\mu_3(r_k)}{2\left[\operatorname{var}(r_k,r_k)\right]^3}\left[r_k^{(0)} - E_c(r_k)\right]^2 - \frac{t_k^e}{2\left(c_{kk}^e\right)^3}\left(r_k^{(0)} - r_k^e\right)^2$$
$$+\frac{3\left[\operatorname{var}(r_k,r_k)\right]^2 - \mu_4(r_k)}{6\left[\operatorname{var}(r_k,r_k)\right]^4}\left[r_k^{(0)} - E_c(r_k)\right]^3 + \frac{3\left(c_{kk}^e\right)^2 - q_k^e}{6\left(c_{kk}^e\right)^4}\left(r_k^{(0)} - r_k^e\right)^3; \tag{1.14}$$

$$w_i\left(\boldsymbol{\alpha}_s^{(0)}\right) \triangleq -\frac{t_i^\alpha}{2\left(c_{ii}^\alpha\right)^3}\left(\alpha_i^{(0)} - \alpha_i^0\right)^2 + \frac{3\left(c_{ii}^\alpha\right)^2 - q_i^\alpha}{6\left(c_{ii}^\alpha\right)^4}\left(\alpha_i^{(0)} - \alpha_i^0\right)^3; \quad i = 1,\ldots,TP. \tag{1.15}$$

2. The vector of mean values of the predicted best-estimate calibrated model parameters:

$$\boldsymbol{\alpha}^{be} = \boldsymbol{\alpha}_s^{(0)} + \mathbf{p}_2\left(\mathbf{r}_s^{(0)},\boldsymbol{\alpha}_s^{(0)}\right) + HOT, \tag{1.16}$$

where:

$$\mathbf{p}_2\left(\mathbf{r}_s^{(0)},\boldsymbol{\alpha}_s^{(0)}\right) = \mathbf{C}_{\alpha r}^c\left[\mathbf{I}_r - \left(\mathbf{C}^e + \mathbf{C}_{rr}^c\right)^{-1}\mathbf{C}_{rr}^c\right]\mathbf{v}\left(\mathbf{r}_s^{(0)}\right)$$
$$+\left[\mathbf{C}_{\alpha\alpha} - \mathbf{C}_{\alpha r}^c\left(\mathbf{C}^e + \mathbf{C}_{rr}^c\right)^{-1}\mathbf{C}_{r\alpha}^c\right]\mathbf{w}\left(\boldsymbol{\alpha}_s^{(0)}\right). \tag{1.17}$$

3. The covariance matrix of the predicted best-estimate responses:

$$\mathbf{C}_{rr}^{be} = \mathbf{C}_{rr}^e - \mathbf{C}_{rr}^e\left(\mathbf{C}_{rr}^e + \mathbf{C}_{rr}^c\right)^{-1}\mathbf{C}_{rr}^e + HOT, \tag{1.18}$$

4. The covariance matrix of the best-estimate calibrated model parameters:

$$\mathbf{C}_{\alpha\alpha}^{be} = \mathbf{C}_{\alpha\alpha} - \mathbf{C}_{\alpha r}^c\left(\mathbf{C}_{rr}^e + \mathbf{C}_{rr}^c\right)^{-1}\mathbf{C}_{r\alpha}^c + HOT, \tag{1.19}$$

5. The correlation matrix of the predicted responses and calibrated model parameters

$$\mathbf{C}_{\alpha r}^{be} = -\mathbf{C}_{\alpha r}^c\left(\mathbf{C}_{rr}^e + \mathbf{C}_{rr}^c\right)^{-1}\mathbf{C}_{rr}^e + HOT. \tag{1.20}$$

6. The triple correlations among best-estimate predicted responses and calibrated model parameters, having the following expressions:
   a. the best-estimate predicted triple correlation, denoted as $\mu_3^{be}\left(r_k^{be},r_\ell^{be},r_m^{be}\right)$, among three best-estimate predicted responses, $r_k^{be},r_\ell^{be},r_m^{be}$:

$$\mu_3^{be}\left(r_k^{be},r_\ell^{be},r_m^{be}\right) = t_{ijk}^e - \rho_\ell c_{km}^e - \rho_m c_{k\ell}^e - \rho_k c_{\ell m}^e - \rho_k\rho_\ell\rho_m + HOT; \quad k,\ell,m = 1,\ldots,TR; \tag{1.21}$$

where:

$$\rho_k \triangleq \left\{ \mathbf{C}^e \left( \mathbf{C}^e + \mathbf{C}_{rr}^c \right)^{-1} \left[ \mathbf{E}_c(\mathbf{r}) - \mathbf{r}^e \right] + \mathbf{p}_1 \left( \mathbf{r}_s^{(0)}, \boldsymbol{\alpha}_s^{(0)} \right) \right\}_k . \tag{1.22}$$

b. The best-estimate predicted triple correlations, denoted as $\mu_3^{be}\left(\alpha_k^{be}, r_\ell^{be}, r_m^{be}\right)$, among the best-estimate calibrated model parameters, $\alpha_k^{be}$, and two best-estimate predicted responses, $r_\ell^{be}, r_m^{be}$:

$$\mu_3^{be}\left(\alpha_k^{be}, r_\ell^{be}, r_m^{be}\right) = \beta_k \left( c_{\ell m}^e + \rho_\ell \rho_m \right) + HOT; \tag{1.23}$$

where:

$$\beta_k \triangleq \left\{ \mathbf{C}_{\alpha r}^c \left( \mathbf{C}^e + \mathbf{C}_{rr}^c \right)^{-1} \left[ \mathbf{E}_c(\mathbf{r}) - \mathbf{r}^e \right] + \mathbf{p}_2 \left( \mathbf{r}_s^{(0)}, \boldsymbol{\alpha}_s^{(0)} \right) \right\}_k . \tag{1.24}$$

c. The best-estimate predicted triple correlations, denoted as $\mu_3^{be}\left(\alpha_k^{be}, \alpha_\ell^{be}, r_m^{be}\right)$, among the two best-estimate calibrated model parameters, $\alpha_k^{be}$, $\alpha_\ell^{be}$, and a best-estimate predicted responses, $r_m^{be}$:

$$\mu_3^{be}\left(\alpha_k^{be}, \alpha_\ell^{be}, r_m^{be}\right) = -\rho_m \left( c_{k\ell}^\alpha + \beta_k \beta_\ell \right) + HOT; \tag{1.25}$$

d. The best-estimate predicted triple correlations, denoted as $\mu_3^{be}\left(\alpha_k^{be}, \alpha_\ell^{be}, \alpha_m^{be}\right)$, among the three best-estimate calibrated model parameters:

$$\mu_3^{be}\left(\alpha_k^{be}, \alpha_\ell^{be}, \alpha_m^{be}\right) = t_{k\ell m}^\alpha + \beta_\ell c_{km}^\alpha + \beta_m c_{k\ell}^\alpha + \beta_k c_{\ell m}^\alpha + \beta_k \beta_\ell \beta_m + HOT; \tag{1.26}$$

7. The quadruple correlations among the best-estimate predicted responses and calibrated model parameters, as follows:
   a. The best-estimate predicted quadruple correlations, denoted as $\mu_4^{be}\left(r_k^{be}, r_\ell^{be}, r_m^{be}, r_n^{be}\right)$, among the four best-estimate predicted responses, $r_k^{be}$, $r_\ell^{be}$, $r_m^{be}$, $r_n^{be}$, as follows:

$$\begin{aligned} \mu_4^{be}\left(r_k^{be}, r_\ell^{be}, r_m^{be}, r_n^{be}\right) = {} & q_{k\ell mn}^e - \rho_\ell t_{kmn}^e - \rho_m t_{k\ell n}^e - \rho_n t_{k\ell m}^e - \rho_k t_{\ell mn}^e + \rho_\ell \rho_m c_{kn}^e \\ & + \rho_k \rho_\ell c_{mn}^e + \rho_k \rho_m c_{\ell n}^e + \rho_n \rho_\ell c_{km}^e + \rho_n \rho_m c_{k\ell}^e + \rho_k \rho_n c_{\ell m}^e \\ & + \rho_k \rho_\ell \rho_m \rho_n + HOT. \end{aligned} \tag{1.27}$$

   b. The best-estimate predicted quadruple correlations, denoted as $\mu_4^{be}\left(\alpha_k^{be}, r_\ell^{be}, r_m^{be}, r_n^{be}\right)$, among the best-estimate calibrated model parameters, $\alpha_k^{be}$, and three best-estimate predicted responses, $r_\ell^{be}$, $r_m^{be}$, $r_n^{be}$, as follows:

$$\begin{aligned} \mu_4^{be}\left(\alpha_k^{be}, r_\ell^{be}, r_m^{be}, r_n^{be}\right) = {} & -\rho_\ell t_{kmn}^e - \rho_m t_{k\ell n}^e + \beta_k t_{\ell mn}^e + c_{kn}^e \rho_\ell \rho_m \\ & + \beta_k \left( \rho_n c_{\ell m}^e - \rho_\ell c_{mn}^e - \rho_m c_{\ell n}^e - \rho_n \rho_\ell \rho_m \right) + HOT. \end{aligned} \tag{1.28}$$

c. The best-estimate predicted quadruple correlations, denoted as $\mu_4^{be}\left(\alpha_k^{be},\alpha_n^{be},r_\ell^{be},r_m^{be}\right)$, among the two best-estimate calibrated model parameters, $\alpha_k^{be}$, $\alpha_\ell^{be}$, and two best-estimate predicted responses, $r_m^{be}$, $r_n^{be}$, as follows:

$$\mu_4^{be}\left(\alpha_k^{be},\alpha_\ell^{be},r_m^{be},r_n^{be}\right)=c_{kn}^{e}\rho_\ell\rho_m+\beta_k\beta_n\left(c_{\ell m}^{e}+\rho_\ell\rho_m\right)+HOT. \tag{1.29}$$

d. The best-estimate predicted quadruple correlations, denoted as $\mu_4^{be}\left(\alpha_k^{be},\alpha_\ell^{be},\alpha_m^{be},r_n^{be}\right)$, among the three best-estimate calibrated model parameters, $\alpha_k^{be}$, $\alpha_\ell^{be}$, $\alpha_m^{be}$, and one best-estimate predicted response, $r_n^{be}$, as follows:

$$\mu_4^{be}\left(\alpha_k^{be},\alpha_\ell^{be},\alpha_m^{be},r_n^{be}\right)=\rho_n\left(c_{km}^{\alpha}\beta_\ell-\beta_k c_{\ell m}^{\alpha}-\beta_m c_{k\ell}^{\alpha}+\beta_k\beta_m\beta_\ell\right)+HOT. \tag{1.30}$$

e. The best-estimate predicted quadruple correlations, denoted as $\mu_4^{be}\left(\alpha_k^{be},\alpha_\ell^{be},\alpha_m^{be},\alpha_n^{be}\right)$, among the four best-estimate calibrated model parameters, $\alpha_k^{be}$, $\alpha_\ell^{be}$, $\alpha_m^{be}$, $\alpha_n^{be}$, defined as follows:

$$\begin{aligned}\mu_4^{be}\left(\alpha_k^{be},\alpha_\ell^{be},\alpha_m^{be},\alpha_n^{be}\right)&=q_{k\ell mn}^{\alpha}+\beta_\ell t_{kmn}^{\alpha}+\beta_m t_{k\ell n}^{\alpha}+\beta_k t_{\ell mn}^{\alpha}+\beta_n t_{k\ell m}^{\alpha}\\&+\beta_k\beta_\ell c_{mn}^{\alpha}+\beta_k\beta_m c_{\ell n}^{\alpha}+\beta_k\beta_n c_{\ell m}^{\alpha}+\beta_\ell\beta_m c_{kn}^{\alpha}+\beta_\ell\beta_n c_{km}^{\alpha}\\&+\beta_m\beta_n c_{k\ell}^{\alpha}+\beta_k\beta_\ell\beta_m\beta_n+HOT.\end{aligned} \tag{1.31}$$

8. Indicator of consistency among parameters and responses:

$$Q_{\min}^{(0)}=\left[\mathbf{E}_c(\mathbf{r})-\mathbf{r}^e\right]^{\dagger}\left(\mathbf{C}_{rr}^{e}+\mathbf{C}_{rr}^{c}\right)^{-1}\left[\mathbf{E}_c(\mathbf{r})-\mathbf{r}^e\right]. \tag{1.32}$$

The quantity $Q_{\min}^{(0)}$ represents the square of the length of the vector $\left[\mathbf{E}_c(\mathbf{r})-\mathbf{r}^e\right]$, measuring (in the corresponding metric) the deviations between the experimental and nominally computed responses. The quantity $Q_{\min}^{(0)}$ is independent of calibrating (or adjusting) the original data, so it can be evaluated directly from the given data (i.e., model parameters and computed and measured responses, together with their original uncertainties) after having computed the matrix $\left(\mathbf{C}_{rr}^{e}+\mathbf{C}_{rr}^{c}\right)^{-1}$. As the dimension of the vector $\left[\mathbf{E}_c(\mathbf{r})-\mathbf{r}^e\right]$ indicates, the number of degrees of freedom characteristic of the calibration under consideration is equal to the number *TR* of experimental responses. The quantity $Q_{\min}^{(0)}$ plays the role of a " $\chi^2$-like consistency indicator," which can be used in the course of quantitative model validation since it quantifies the degree of agreement between the computed and the experimentally measured responses before actually combining the computational and experimental information. Agreement between experimental and computational results is indicated when $Q_{\min}^{(0)}$ (per degree of freedom) is close to unity; values of $Q_{min}^{(0)}$ that differ greatly from unity indicate "inconsistency" (i.e., lack of validation) or perhaps even gross errors. In such cases, the individual contributions to $Q_{\min}^{(0)}$ must be examined, and the outliers (i.e., very large or very small individual contributions) need to be investigated as possible sources of inconsistencies that could invalidate either parts of the model or parts of the data (or both).

The results provided in Eqs. (1.11)–(1.31) highlight the fact that they can be utilized both for "forward/direct predictive modeling" and for "inverse predictive modeling." The "forward" or "direct problem" solves the "parameter-to-output" mapping that describes the "cause-to-effect" relationship in the physical process being modeled. The "inverse problem" attempts to solve the "output-to-parameters" mapping. Since the framework of the 4th-BERRU-PM methodology comprises the combined phase space of parameters and responses, it can be used for solving both forward/direct and inverse problems. The solution of the "forward/direct problem" is provided by the expressions for the predicted best-estimate responses together with the corresponding reduced predicted uncertainties. Conversely, the solution of the "inverse problem" is provided by the expressions for the predicted best-estimate calibrated parameters and their reduced predicted uncertainties.

## 1.5 ILLUSTRATIVE HIGH-ORDER SENSITIVITY ANALYSIS, UNCERTAINTY QUANTIFICATION, AND PREDICTIVE MODELING FOR THE PERP OECD/NEA/ICSBEP REACTOR PHYSICS BENCHMARK

The need for computing of higher-order sensitivities and their importance in uncertainty analysis and predictive modeling has been amply illustrated by Cacuci and Fang (2023) using, as a paradigm model, the polyethylene-reflected plutonium (acronym "PERP") reactor physics benchmark (Valentine 2006), which is included in the OECD/NEA, and ICSBEP documents. This subsection presents a synopsis of the salient results obtained by Cacuci and Fang (2023).

### 1.5.1 Computational Modeling of the PERP Reactor Physics Benchmark

The PERP reactor physics benchmark is a one-dimensional spherical subcritical nuclear system comprising an inner sphere (designated as "material 1") that is surrounded by a spherical shell (designated as "material 2"). The inner sphere of the PERP benchmark contains $\alpha$-phase plutonium, which acts as the source of spontaneous fission neutron particles; it has a radius of $r_1 = 3.794$ cm. This inner sphere is surrounded by a spherical shell reflector made of polyethylene of thickness 3.81 cm; the radius of the outer shell containing polyethylene is $r_s = 7.064$ cm. Table 1.1, comprises specifications regarding the constitutive materials of the PERP benchmark.

The distribution of neutrons within the radially symmentric benchmark is modeled by the linear time-independent inhomogeneous neutron transport (Boltzmann) equation in 4 independent variables, which has the following standard form in spherical geometry:

$$B(\alpha)\varphi(r,\Omega,E) = Q(r,E,\Omega), \tag{1.33}$$

**TABLE 1.1**
**Dimensions and Composition of the PERP Benchmark**

| Materials | Isotopes | Weight Fraction | Density (g/cm³) | Zones |
|---|---|---|---|---|
| Material 1 (plutonium metal) | Isotope 1 ($^{239}$Pu) | $9.3804\times10^{-1}$ | 19.6 | Material 1 is assigned to zone 1, which has a radius of 3.794 cm. |
| | Isotope 2 ($^{240}$Pu) | $5.9411\times10^{-2}$ | | |
| | Isotope 3 ($^{69}$Ga) | $1.5152\times10^{-3}$ | | |
| | Isotope 4 ($^{71}$Ga) | $1.0346\times10^{-3}$ | | |
| Material 2 (polyethylene) | Isotope 5 ($^{12}$C) | $8.5630\times10^{-1}$ | 0.95 | Material 2 is assigned to zone 2, which has an inner radius of 3.794 cm and an outer radius of 7.604 cm. |
| | Isotope 6 ($^{1}$H) | $1.4370\times10^{-1}$ | | |

where

$$B(\alpha)\varphi(r,\Omega,E) \triangleq \Omega\bullet\nabla\varphi(r,E,\Omega)+\Sigma_t(r,E)\varphi(r,E,\Omega)$$
$$-\int_0^{E_f} dE'\int_{4\pi} d\Omega'\Sigma_s(r,E'\to E,\Omega'\bullet\Omega)\varphi(r,E',\Omega') \tag{1.34}$$
$$-\int_0^{E_f} dE'\int_{4\pi} \chi(r,E'\to E)\nu\Sigma_f(r,E')\varphi(r,E',\Omega')\,d\Omega'.$$

The boundary condition appropriate for the PERP benchmark is that of "no incoming flux," which is mathematically expressed as follows:

$$\varphi(r_s,E,\Omega)=0,\quad r_s\in S_b,\quad \Omega\bullet\mathbf{n}<0, \tag{1.35}$$

where $r_s$ denotes the external radius of the PERP benchmark and where the vector $\mathbf{n}$ denotes the outward unit normal vector at each point on the sphere's outer boundary, denoted as $S_b$. The physical quantities that appear in the standard notation used in Eq. (1.34) are as follows:

i. $\varphi(r,E,\Omega)$ denotes the flux of particles (i.e., particle number density multiplied by the particle speed) in the energy range $dE$ about $E$, in the volume element $dr$ about the radial independent variable $r$, with directions of motion in the solid angle element $d\Omega$ about $\Omega$.
ii. $Q(r,E,\Omega)$ denotes the rate at which particles are produced in the same element of phase space from sources that are independent of the flux.
iii. $\Sigma_t(r,E)$ denotes the macroscopic total cross section.
iv. Assuming isotropic media, $\Sigma_s(r,E'\to E,\Omega'\bullet\Omega)$ denotes the macroscopic scattering transfer cross section from energy $E'$ to energy $E$ through a scattering angle $\Omega'\bullet\Omega$.

v. $\nu$ denotes the number of particles emitted isotropically $\left(\frac{1}{4\pi}\right)$ per fission.
vi. $\Sigma_f(r,E)$ denotes the macroscopic fission cross section.
vii. $\chi(r,E' \to E)$ denotes the fraction of fission particles appearing in energy $dE$ about $E$ from fissions in $dE'$ about $E'$.

The neutron flux distribution within the PERP benchmark was computed by using the deterministic software package PARTISN (Alcouffe et al. 2008), which solves the multigroup approximation of the transport equation. This approximation is obtained by partitioning the domain of the energy variable into G intervals, called "energy groups," and integrating Eqs. (1.34) and (1.35) formally over a generic energy group, labeled as "$g$," to obtain the following "multigroup transport equation" to be solved for the group-fluxes $\varphi^g(r,\Omega)$:

$$B^g(\alpha)\varphi^g(r,\Omega) = Q^g(\alpha;r), \quad g = 1,\ldots,G, \tag{1.36}$$

$$\varphi^g(r_s,\Omega) = 0, \quad r_s \in S_b, \quad \Omega \cdot \mathbf{n} < 0, \quad g = 1,\ldots,G, \tag{1.37}$$

where:

$$\begin{aligned} B^g(\alpha)\varphi^g(r,\Omega) \triangleq\ & \Omega \cdot \nabla\varphi^g(r,\Omega) + \Sigma_t^g(r)\varphi^g(r,\Omega) \\ & -\sum_{g'=1}^{G}\int_{4\pi}\Sigma_s^{g'\to g}(r,\Omega'\cdot\Omega)\varphi^{g'}(r,\Omega')\,d\Omega' \\ & -\chi^g(r)\sum_{g'=1}^{G}\int_{4\pi}(\nu\Sigma)_f^{g'}(r)\varphi^{g'}(r,\Omega')\,d\Omega', \end{aligned} \tag{1.38}$$

$$Q^g(\alpha;r) \triangleq \sum_{i=1}^{N_f}\lambda_i N_{i,1} F_i^{SF}\nu_i^{SF}\frac{1}{I_0}\int_{E^{g+1}}^{E^g} dE\, e^{-E/a_i}\sinh\sqrt{b_i E}, \quad g = 1,\ldots,G; \tag{1.39}$$

with:

$$I_0 \triangleq \frac{\sqrt{\pi a_i^3 b_i}}{2}e^{\frac{a_i b_i}{4}}. \tag{1.40}$$

In Eqs. (1.39) and (1.40), the subscript "$i$" denotes the number of nuclides within the spontaneous fission source.

The PARTISN (Alcouffe et al. 2008) computations used the MENDF71X library (Conlin et al. 2013), which comprises 618-group cross sections. These cross sections were collapsed to $G = 30$ energy groups, with group boundaries, $E^g$, as presented in Table 1.2. The MENDF71X library (Conlin et al. 2013) uses ENDF/B-VII.1 nuclear data (Chadwick et al. 2011). The group boundaries, $E^g$, are user-defined and are therefore considered to be perfectly well-known parameters.

**TABLE 1.2**
**Energy Group Structure, in [MeV], for PERP Benchmark Computations**

| $g$ | 1 | 2 | 3 | 4 | 5 | 6 |
|---|---|---|---|---|---|---|
| $E^g$ | $1.50\times10^1$ | $1.35\times10^1$ | $1.20\times10^1$ | $1.00\times10^1$ | $7.79\times10^0$ | $6.07\times10^0$ |
| $E^{g-1}$ | $1.70\times10^1$ | $1.50\times10^1$ | $1.35\times10^1$ | $1.20\times10^1$ | $1.00\times10^1$ | $7.79\times10^0$ |
| $g$ | 7 | 8 | 9 | 10 | 11 | 12 |
| $E^g$ | $3.68\times10^0$ | $2.87\times10^0$ | $2.23\times10^0$ | $1.74\times10^0$ | $1.35\times10^0$ | $8.23\times10^{-1}$ |
| $E^{g-1}$ | $6.07\times10^0$ | $3.68\times10^0$ | $2.87\times10^0$ | $2.23\times10^0$ | $1.74\times10^0$ | $1.35\times10^0$ |
| $g$ | 13 | 14 | 15 | 16 | 17 | 18 |
| $E^g$ | $5.00\times10^{-1}$ | $3.03\times10^{-1}$ | $1.84\times10^{-1}$ | $6.76\times10^{-2}$ | $2.48\times10^{-2}$ | $9.12\times10^{-3}$ |
| $E^{g-1}$ | $8.23\times10^{-1}$ | $5.00\times10^{-1}$ | $3.03\times10^{-1}$ | $1.84\times10^{-1}$ | $6.76\times10^{-2}$ | $2.48\times10^{-2}$ |
| $g$ | 19 | 20 | 21 | 22 | 23 | 24 |
| $E^g$ | $3.35\times10^{-3}$ | $1.24\times10^{-3}$ | $4.54\times10^{-4}$ | $1.67\times10^{-4}$ | $6.14\times10^{-5}$ | $2.26\times10^{-5}$ |
| $E^{g-1}$ | $9.12\times10^{-3}$ | $3.35\times10^{-3}$ | $1.24\times10^{-3}$ | $4.54\times10^{-4}$ | $1.67\times10^{-4}$ | $6.14\times10^{-5}$ |
| $g$ | 25 | 26 | 27 | 28 | 29 | 30 |
| $E^g$ | $8.32\times10^{-6}$ | $3.06\times10^{-6}$ | $1.13\times10^{-6}$ | $4.14\times10^{-7}$ | $1.52\times10^{-7}$ | $1.39\times10^{-10}$ |
| $E^{g-1}$ | $2.26\times10^{-5}$ | $8.32\times10^{-6}$ | $3.06\times10^{-6}$ | $1.13\times10^{-6}$ | $4.14\times10^{-7}$ | $1.52\times10^{-7}$ |

The source of neutrons in the PERP benchmark is provided by spontaneous fissions stemming from $^{239}Pu$ (Isotope 1) and $^{240}Pu$ (Isotope 2); there are no delayed neutrons or $(\alpha,n)$ sources. The spontaneous fission source has been computed using the code SOURCES4C (Wilson et al. 2002). For an actinide nuclide $k$, the spontaneous source depends on the following 6 model parameters: the decay constant $\lambda_k$, the atom density $N_{k,m}$, the average number of neutrons per spontaneous fission $\nu_k^{SF}$, the spontaneous fission branching ratio $F_k^{SF}$, and the two parameters $a_k$ and $b_k$ used in a Watt's fission spectrum to approximate the spontaneous fission neutron spectrum. The PERP benchmark contains two actinide nuclides, so that $k = 1,2$. The nominal values of these parameters (except for $N_{k,m}$) are available from a library file contained in SOURCES4C (Wilson et al. 2002), while the nominal values for $N_{k,m}$ are specified from the PERP benchmark. These imprecisely known source parameters also affect the accuracy of the neutron transport calculation.

The PARTISN (Alcouffe et al. 2008) software system uses the discrete-ordinates approximation to discretize the angular variable in the 1st and 2nd terms on the right side of Eq. (1.38) and a finite-moments expansion in spherical harmonics to approximate the angular variable in the 3rd and 4th terms on the right side of Eq. (1.38). The computations of the 1st-order sensitivities presented in this chapter were performed using a $P_3$ Legendre expansion of the scattering cross section, an angular quadrature of $S_{256}$, and a fine-mesh spacing of 0.005 cm, comprising 759 meshes for the plutonium sphere of radius 3.794 cm and 762 meshes for the polyethylene shell of thickness 3.81 cm. It is convenient to retain the continuous representation in the angular and radial variables since the spatial and angular discretization parameters are considered to be perfectly well known. The various quantities in Eqs. (1.36)–(1.40) have their usual meanings for the standard form of the multigroup neutron transport equation, as follows:

1. Using the notation employed in PARTISN (Alcouffe et al. 2008), the quantity $\varphi^g(r,\Omega) \triangleq \int_{E^{g-\frac{1}{2}}}^{E^{g+\frac{1}{2}}} \varphi(r,E,\Omega)dE$ denotes the "group-flux" for group $g$, and is the unknown state-function obtained by solving Eqs. (1.36) and (1.37).
2. The spontaneous-fission isotopes in the PERP benchmark are "isotope 1" ($^{239}Pu$) and "isotope 2" ($^{240}Pu$). The quantity $N_f$ denotes the total number of spontaneous-fission isotopes; for the PERP benchmark, $N_f = 2$. The spontaneous fission neutron spectra of $^{239}Pu$ and, respectively, $^{240}Pu$, are approximated by Watt's fission spectra, each spectrum using two evaluated parameters, denoted as $a_k$ and $b_k$, respectively. The decay constant for actinide nuclide $k$ is denoted as $\lambda_k$, while $F_k^{SF}$ denotes the fraction of decays that are spontaneous fission (the "spontaneous fission branching fraction"). There are 10 source parameters, as follows: $\lambda_1, \lambda_2$; $F_1^{SF}, F_2^{SF}$; $a_1, a_2$; $b_1, b_2$; $\nu_1^{SF}, \nu_2^{SF}$.
3. The quantity $N_{i,m}$ denotes the atom density of isotope $i$ in material $m$; $i = 1,\ldots,I$, $m = 1,\ldots,M$, where $I$ denotes the total number of isotopes, and $M$ denotes the total number of materials. The atom density $N_{i,m}$ is defined as follows:

$$N_{i,m} \triangleq \frac{\rho_m w_{i,m} N_A}{A_i}, \tag{1.41}$$

where $\rho_m$ denotes the mass density of material $m$, $m = 1,\ldots,M$; $w_{i,m}$ denotes the weight fraction of isotope $i$ in material $m$; $A_i$ denotes the atomic weight of isotope $i$, $i = 1,\ldots,I$; $N_A$ denotes Avogadro's number. For the PERP benchmark, $I = 6$ and $M = 2$, but since the respective isotopes are all distinct (i.e., are not repeated) in the PERP benchmark's distinct materials, as specified in Table 1.1, it follows that there are 6 isotopic number densities for this benchmark, as follows: $N_{1,1}$, $N_{2,1}$, $N_{3,1}$, $N_{4,1}$, $N_{5,2}$, $N_{6,2}$.
4. The quantity $\Sigma_s^{g'\to g}(r,\Omega'\bullet\Omega)$ represents the scattering transfer cross section from energy group $g'$, $g' = 1,\ldots,G$ into energy group $g$, $g = 1,\ldots,G$. The transfer cross sections are computed in terms of the $l$th-order Legendre coefficients $\sigma_{s,l,i}^{g'\to g}$ (of the Legendre-expanded microscopic scattering cross section from energy group $g'$ into energy group $g$, for isotope $i$), which are tabulated parameters, using the following expansion:

$$\begin{aligned} \Sigma_s^{g'\to g}(r,\Omega'\bullet\Omega) &= \sum_{m=1}^{M=2} \Sigma_{s,m}^{g'\to g}(r,\Omega'\bullet\Omega), \\ \Sigma_{s,m}^{g'\to g}(r,\Omega'\bullet\Omega) &\cong \sum_{i=1}^{I=6} N_{i,m} \sum_{l=0}^{ISCT=3} (2l+1)\sigma_{s,l,i}^{g'\to g}(r) P_l(\Omega'\bullet\Omega), \quad m = 1,2, \end{aligned} \tag{1.42}$$

where $ISCT = 3$ denotes the order of the respective finite expansion in the Legendre polynomial. The spatial variable $r$ will henceforth no longer appear in the arguments of the various cross sections since the cross sections for every material are treated in the PARTISN (Alcouffe et al. 2008) computations as being space-independent within the respective material. There are 21,600 multigroup microscopic scattering cross sections $\sigma_{s,l,i}^{g'\to g}$, for $i = 1,\ldots,6$; $l = 0,\ldots,3$; $g, g' = 1,\ldots,30$.

5. The total cross section $\Sigma_t^g$ for energy group $g, g = 1,\ldots,G$, and material $m$, is computed for the PERP benchmark using the following expression:

$$\Sigma_t^g = \sum_{m=1}^{M=2} \Sigma_{t,m}^g; \; \Sigma_{t,m}^g = \sum_{i=1}^{I} N_{i,m}\sigma_{t,i}^g = \sum_{i=1}^{I} N_{i,m}\left[\sigma_{f,i}^g + \sigma_{c,i}^g + \sum_{g'=1}^{G} \sigma_{s,l=0,i}^{g\to g'}\right], m = 1,2, \tag{1.43}$$

where $\sigma_{f,i}^g$ and $\sigma_{c,i}^g$ denote, respectively, the tabulated group microscopic fission and neutron capture cross sections for group $g, g = 1,\ldots,G$. Other nuclear reactions, including $(n,2n)$ and $(n,3n)$ reactions, are not present in this benchmark. The expressions in Eqs. (1.42) and (1.43) indicate that the zeroth-order (i.e., $l = 0$) scattering cross sections must be separately considered from the higher-order (i.e., $l \geq 1$) scattering cross sections, since the $l = 0$ term contributes to the total cross sections, while the $l \geq 1$ terms do not contribute. There are 180 multigroup microscopic total cross sections $\sigma_{t,i}^g$, for $i = 1,\ldots,6$; $g = 1,\ldots,30$.

6. PARTISN (Alcouffe et al. 2008) computes the quantity $(\nu\Sigma_f)^g$ using the microscopic quantities $(\nu\sigma)_{f,i}^g$, which are provided in data files for each isotope $i$, and energy group $g$, as follows:

$$(\nu\Sigma_f)^g = \sum_{m=1}^{M=2} (\nu\Sigma_f)_m^g; (\nu\Sigma_f)_m^g = \sum_{i=1}^{I=6} N_{i,m}(\nu\sigma_f)_i^g, \quad m = 1,2. \tag{1.44}$$

For the purposes of sensitivity analysis, the quantity $\nu_i^g$, which denotes the number of neutrons that are produced per fission by isotope $i$ and energy group $g$, can be obtained by using the relation $\nu_i^g = \frac{(\nu\sigma_f)_i^g}{\sigma_{f,i}^g}$, where the isotopic fission cross sections $\sigma_{f,i}^g$ are available in data files for computing reaction rates. There are 60 parameters $\nu_i^g$ and 60 multigroup microscopic fission cross sections $\sigma_{f,i}^g$, for $i = 1,2$; $g = 1,\ldots,30$.

7. The quantity $\chi^g$ denotes the fission spectrum in the energy group $g$; $\chi^g$ is defined in PARTISN (Alcouffe et al. 2008) as a space-independent quantity, as follows:

$$\chi^g \triangleq \frac{\sum_{i=1}^{N_f} \chi_i^g N_{i,m} \sum_{g'=1}^{G} (\nu\sigma_f)_i^{g'} f_i^{g'}}{\sum_{i=1}^{N_f} N_{i,m} \sum_{g'=1}^{G} (\nu\sigma_f)_i^{g'} f_i^{g'}}, \quad \sum_{g=1}^{G} \chi_i^g = 1, \tag{1.45}$$

where $\chi_i^g$ denotes the isotopic fission spectrum in group $g$, while $f_i^g$ denotes the corresponding spectrum weighting function. There are 60 fission spectrum parameters $\chi_i^g$, for $i = 1, 2$; $g = 1, \ldots, 30$.

8. The summary of all of the model parameters described in the foregoing is as follows:
   i. The total cross section $\Sigma_t^g \to \Sigma_t^g(\mathbf{t})$ is characterized by the vector of parameters $\boldsymbol{t}$, which is defined as follows:

$$\mathbf{t} \triangleq \left[t_1, \ldots, t_{J_t}\right]^\dagger \triangleq \left[t_1, \ldots, t_{J_{\sigma t}}; n_1, \ldots, n_{J_n}\right]^\dagger \triangleq \left[\boldsymbol{\sigma}_t; \mathbf{N}\right]^\dagger, \quad J_t \triangleq J_{\sigma t} + J_n, \tag{1.46}$$

$$\mathbf{N} \triangleq \left[n_1, \ldots, n_{J_n}\right]^\dagger \triangleq \left[N_{1,1}, N_{2,1}, N_{3,1}, N_{4,1}, N_{5,2}, N_{6,2}\right]^\dagger, \quad J_n = 6, \tag{1.47}$$

$$\begin{aligned} \boldsymbol{\sigma}_t &\triangleq \left[t_1, \ldots, t_{J_{\sigma t}}\right]^\dagger \triangleq \left[\sigma_{t,i=1}^1, \sigma_{t,i=1}^2, \ldots, \sigma_{t,i=1}^G, \ldots, \sigma_{t,i}^g, \ldots, \sigma_{t,i=I}^1, \ldots, \sigma_{t,i=I}^G\right]^\dagger, \\ & i = 1, \ldots, I; \quad g = 1, \ldots, G; \quad J_{\sigma t} = I \times G; \end{aligned} \tag{1.48}$$

   In Eqs. (1.46)–(1.48), $\sigma_{t,i}^g$ denotes the microscopic total cross section for isotope $i$ and energy group $g$, $N_{i,m}$ denotes the respective isotopic number density, and $J_n$ denotes the total number of isotopic number densities in the model. Thus, the vector $\boldsymbol{t}$ comprises a total of $J_t = 30 \times 6 + 6 = 186$ imprecisely known components ("model parameters").

   ii. The scattering cross section $\Sigma_s^{g' \to g}(\boldsymbol{\Omega}' \bullet \boldsymbol{\Omega}) \to \Sigma_s^{g' \to g}(\mathbf{s}; \boldsymbol{\Omega}' \bullet \boldsymbol{\Omega})$ is characterized by the vector of parameters $\boldsymbol{s}$, which is defined as follows:

$$\mathbf{s} \triangleq \left[s_1, \ldots, s_{J_s}\right]^\dagger \triangleq \left[s_1, \ldots, s_{J_{\sigma s}}; n_1, \ldots, n_{J_n}\right]^\dagger \triangleq \left[\boldsymbol{\sigma}_s; \mathbf{N}\right]^\dagger, \quad J_s \triangleq J_{\sigma s} + J_n, \tag{1.49}$$

   where

$$\begin{aligned} \boldsymbol{\sigma}_s &\triangleq \left[s_1, \ldots, s_{J_{\sigma s}}\right]^\dagger \\ &\triangleq \left[\sigma_{s,l=0,i=1}^{g'=1\to g=1}, \sigma_{s,l=0,i=1}^{g'=2\to g=1}, \ldots, \sigma_{s,l=0,i=1}^{g'=G\to g=1}, \sigma_{s,l=0,i=1}^{g'=1\to g=2}, \sigma_{s,l=0,i=1}^{g'=2\to g=2}, \ldots, \sigma_{s,l,i}^{g'\to g}, \ldots, \sigma_{s,ISCT,i=I}^{G\to G}\right]^\dagger, \\ & l = 0, \ldots, ISCT; \quad i = 1, \ldots, I; \quad g, g' = 1, \ldots, G; \quad J_{\sigma s} = (G \times G) \times I \times (ISCT + 1). \end{aligned} \tag{1.50}$$

iii. The quantity $(\nu\Sigma_f)^g \to [\nu\Sigma_f(\mathbf{f})]^g$ in the fission integral $\int_{4\pi} (\nu\Sigma)_f^{g'} \varphi^{g'}(r,\Omega')d\Omega'$ depends on the vector of parameters $\mathbf{f}$, which is defined as follows:

$$\mathbf{f} \triangleq \left[f_1,\ldots,f_{J_{\sigma f}};f_{J_{\sigma f}+1},\ldots,f_{J_{\sigma f}+J_\nu};f_{J_{\sigma f}+J_\nu+1},\ldots,f_{J_f}\right]^\dagger \triangleq \left[\boldsymbol{\sigma}_f;\boldsymbol{\nu};\mathbf{N}\right]^\dagger, \\ J_f = J_{\sigma f} + J_\nu + J_n, \tag{1.51}$$

with

$$\boldsymbol{\sigma}_f \triangleq \left[\sigma_{f,i=1}^1,\sigma_{f,i=1}^2,\ldots,\sigma_{f,i=1}^G,\ldots,\sigma_{f,i}^g,\ldots,\sigma_{f,i=N_f}^1,\ldots,\sigma_{f,i=N_f}^G\right]^\dagger \\ \triangleq \left[f_1,\ldots,f_{J_{\sigma f}}\right]^\dagger, \quad i=1,\ldots,N_f; \quad g=1,\ldots,G; \quad J_{\sigma f} = G\times N_f, \tag{1.52}$$

$$\boldsymbol{\nu} \triangleq \left[\nu_{i=1}^1,\nu_{i=1}^2,\ldots,\nu_{i=1}^G,\ldots,\nu_i^g,\ldots,\nu_{i=N_f}^1,\ldots,\nu_{i=N_f}^G\right]^\dagger \\ \triangleq \left[f_{J_{\sigma f}+1},\ldots,f_{J_{\sigma f}+J_\nu}\right]^\dagger, \quad i=1,\ldots,N_f; \quad g=1,\ldots,G; \quad J_\nu = G\times N_f, \tag{1.53}$$

where $\sigma_{f,i}^g$ denotes the microscopic fission cross section for the isotope $i$ and energy group $g$; $\nu_i^g$ denotes the average number of neutrons per fission for the isotope $i$ and energy group $g$; and $N_f$ denotes the total number of fissionable isotopes.

iv. The fission spectrum is considered to depend on the vector of parameters $\mathbf{p}$, which is defined as follows:

$$\mathbf{p} \triangleq \left[p_1,\ldots,p_{J_p}\right]^\dagger \triangleq \left[\chi_{i=1}^{g=1},\chi_{i=1}^{g=2},\ldots,\chi_{i=1}^G,\ldots,\chi_i^g,\ldots,\chi_{N_f}^G\right]^\dagger, \\ i=1,\ldots,N_f; \quad g=1,\ldots,G; \quad J_p = G\times N_f. \tag{1.54}$$

v. The quantities $\chi^g$ further depend on the parameters $\chi_i^g$, $N_{i,m}$, $f_i^g$, $(\nu\sigma_f)_i^g$, but these latter dependences can be taken into account by applying the chain rule to the 1st-order sensitivities $\dfrac{\partial L}{\partial \chi^g}$, once these sensitivities have been obtained.

vi. The source $Q^g(r) \to Q^g(\mathbf{q};\mathbf{N})$ depends on the vector of model parameters $\mathbf{q}$, which is defined as follows:

$$\mathbf{q} \triangleq \left[q_1,\ldots,q_{J_q}\right]^\dagger \triangleq \left[\lambda_1,\lambda_2;F_1^{SF},F_2^{SF};a_1,a_2;b_1,b_2;\nu_1^{SF},\nu_2^{SF}\right]^\dagger, \quad J_q = 10. \tag{1.55}$$

Summarizing the information presented in Eqs. (1.46)–(1.55) indicates that the model parameters characterizing the PERP benchmark can all be considered to be the components of the "vector of model parameters" $\boldsymbol{\alpha}$, which is defined below:

$$\boldsymbol{\alpha} \triangleq [\alpha_1,\ldots,\alpha_{TP}]^\dagger \triangleq [\boldsymbol{\sigma}_t;\boldsymbol{\sigma}_s;\boldsymbol{\sigma}_f;\boldsymbol{\nu};\mathbf{p};\mathbf{q};\mathbf{N}]^\dagger,$$
$$TP = J_{\sigma t} + J_{\sigma s} + J_{\sigma f} + J_\nu + J_p + J_q + J_n. \tag{1.56}$$

Thus, the "vector of imprecisely known model parameters" $\boldsymbol{\alpha} \triangleq [\alpha_1,\ldots,\alpha_{TP}]^\dagger$, which appears in the argument of the Boltzmann operator $B^g(\boldsymbol{\alpha})$, comprises a total of $TP = (I \times G) + (G \times G) \times I \times (ISCT + 1) + 2(G \times N_f) + G \times N_f + 10 + 6 = 21{,}976$ imprecisely known (i.e., uncertain) model parameters. However, only 7,477 parameters have nonzero nominal values, as follows: (a) 180 group-averaged total microscopic cross sections; (b) 7,101 non-zero group-averaged scattering microscopic cross sections; the remaining group-averaged scattering cross sections, of which there are 21,600 in total, are zero due to the choice of group boundaries and the scattering laws of neutrons with matter; (c) 120 fission process parameters; (d) 60 fission spectrum parameters; (e) 10 parameters describing the experiment's nuclear sources; and (f) six isotopic number densities.

The *total leakage* of neutrons out of the PERP benchmark is considered to be the representative *model response* of interest for sensitivity analysis, uncertainty quantification, and predictive modeling, since it does not depend on the detector configuration (as opposed to count rates) and has been measured reliably (Valentine 2006). Mathematically, the total neutron leakage from the PERP sphere, which is denoted below as $L(\boldsymbol{\alpha})$, depends on all model parameters (indirectly, through the neutron flux) and is defined as follows:

$$L(\boldsymbol{\alpha}) \triangleq \int_V dV \int_0^\infty dE \int_{\boldsymbol{\Omega}\cdot\mathbf{n}>0} d\Omega \boldsymbol{\Omega}\bullet\mathbf{n}\,\delta(r - r_s)\varphi(r,E,\boldsymbol{\Omega}) = \int_{S_b} dS \sum_{g=1}^{G} \int_{\boldsymbol{\Omega}\cdot\mathbf{n}>0} d\Omega \boldsymbol{\Omega}\bullet\mathbf{n}\,\varphi^g(r,\boldsymbol{\Omega}). \tag{1.57}$$

The distribution of neutrons within the benchmark is computed by numerically solving the multigroup Boltzmann equation, namely Eqs. (1.36) and (1.37), comprising the above-mentioned 21,976 parameters. Solving the Boltzmann equation is representative of a "large-scale" computation. The nominal value of total leakage, computed by using Eq. (1.57) at the nominal parameter values (which are denoted using the usual notation $\boldsymbol{\alpha}^0$) is $L(\boldsymbol{\alpha}^0) = 1.7648 \times 10^6$ neutrons/sec.

### 1.5.2 Computation of the Largest High-Order Sensitivities of the PERP Leakage Response with Respect to PERP Parameters

The 1st-order sensitivities of the leakage response with respect to the PERP benchmark's model parameters were derived and computed (Cacuci and Fang 2023) using the 1st-Order Comprehensive Sensitivity Analysis Methodology for Linear Systems (1st-CASAM-L) developed by Cacuci (2021b, 2022b). The mathematical expressions of the 1st-order sensitivities of the leakage response with respect to the PERP benchmark's parameters were obtained from the following expression of the 1st-order total differential of the leakage response with respect to arbitrary variations in the PERP benchmark's parameters, denoted below as $\delta L(\boldsymbol{\alpha},\varphi;\delta\boldsymbol{\alpha})$:

$$\delta L(\boldsymbol{\alpha},\varphi;\delta\boldsymbol{\alpha}) = \int dV \int_{4\pi} d\boldsymbol{\Omega} \int_0^\infty dE \, \psi^{(1)}(r,\boldsymbol{\Omega},E) Q^{(1)}(\boldsymbol{\alpha},\varphi;\delta\boldsymbol{\alpha}) \triangleq \sum_{m=1}^{TP} \frac{\partial L(\boldsymbol{\alpha},\varphi;\psi^{(1)})}{\partial \alpha_m} \delta\alpha_m . \tag{1.58}$$

where

$$\begin{aligned} Q^{(1)}(\boldsymbol{\alpha},\varphi;\delta\boldsymbol{\alpha}) \triangleq{}& \delta Q(\mathbf{q};r,\boldsymbol{\Omega},E) - \delta\Sigma_t(\mathbf{t};r,E)\varphi(r,\boldsymbol{\Omega},E) \\ &+ \int_{4\pi} d\boldsymbol{\Omega}' \int_0^\infty dE' \varphi(r,\boldsymbol{\Omega}',E')\left[\delta\Sigma_s(s;r,E'\to E,\boldsymbol{\Omega}'\bullet\boldsymbol{\Omega})\right] \\ &+ \int_{4\pi} d\boldsymbol{\Omega}' \int_0^\infty dE' \delta\chi(\mathbf{p};r,E'\to E)\varphi(r,\boldsymbol{\Omega}',E')\left[\nu\Sigma_f(\mathbf{f};r,E')\right] \\ &+ \int_{4\pi} d\boldsymbol{\Omega}' \int_0^\infty dE' \chi(\mathbf{p};r,E'\to E)\varphi(r,\boldsymbol{\Omega}',E')\delta\left[\nu\Sigma_f(\mathbf{f};r,E')\right], \end{aligned} \tag{1.59}$$

and where the 1st-level adjoint sensitivity function $\psi^{(1)}(r,\boldsymbol{\Omega},E)$ is the solution of the following 1st-Level Adjoint Sensitivity System (1st-LASS):

$$A^{(1)}(\boldsymbol{\alpha})\psi^{(1)}(r,\boldsymbol{\Omega},E) = \boldsymbol{\Omega}\bullet\mathbf{n}\delta(r-r_s), \tag{1.60}$$

together with the adjoint boundary condition provided below:

$$\psi^{(1)}(r_s,\boldsymbol{\Omega},E) = 0, \quad r_s \in S_b, \quad \boldsymbol{\Omega}\bullet\mathbf{n} > 0. \tag{1.61}$$

The 1st-level adjoint operator $A^{(1)}(\boldsymbol{\alpha})$ on the left side of Eq. (1.60) is defined as follows:

$$\begin{aligned} A^{(1)}(\boldsymbol{\alpha})\psi^{(1)} \triangleq{}& -\boldsymbol{\Omega}\bullet\nabla\psi^{(1)}(r,\boldsymbol{\Omega},E) + \Sigma_t(\mathbf{t};r,E)\psi^{(1)}(r,\boldsymbol{\Omega},E) \\ &- \int_{4\pi} d\boldsymbol{\Omega}' \int_0^\infty dE' \Sigma_s(\mathbf{s};r,E\to E',\boldsymbol{\Omega}\bullet\boldsymbol{\Omega}')\psi^{(1)}(r,\boldsymbol{\Omega}',E') \\ &- \nu\Sigma_f(\mathbf{f};r,E) \int_{4\pi} d\boldsymbol{\Omega}' \int_0^\infty dE' \chi(\mathbf{p};r,E\to E')\psi^{(1)}(r,\boldsymbol{\Omega}',E'), \end{aligned} \tag{1.62}$$

In Eq. (1.58), the quantity $\frac{\partial L\left(\alpha,\varphi;\psi^{(1)}\right)}{\partial \alpha_m}$ denotes the 1st-*order sensitivity of the leakage response with respect to a generic parameter* $\alpha_m$, $m = 1,\ldots,TP$. These partial 1st-order sensitivities are obtained by substituting the expression of $Q^{(1)}\left(\alpha,\varphi;\delta\alpha\right)$ from Eq. (1.59) into Eq. (1.58) and subsequently identifying the quantities that formally multiply the various parameter variations $\delta\alpha_m$. The quantities appearing in Eqs. (1.58)–(1.62) are evaluated at the nominal values for the parameters, but, in order to simplify the notation, the superscript "zero" denoting "nominal values" has been omitted and will be omitted henceforth.

The 1st-LASS defined by Eqs. (1.60) and (1.61) is also solved using PARTISN (Alcouffe et al. 2008), using the same 30-group structure as for solving the forward Boltzmann transport equation. Solving the 1st-LASS yields the multigroup adjoint fluxes $\psi^{(1),g}\left(r,\Omega\right)$, $g = 1,\ldots,G$. The multigroup form of the 1st-LASS is as follows:

$$A^{(1),g}\left(\alpha\right)\psi^{(1),g}\left(r,\Omega\right) = \Omega \bullet \mathbf{n}\delta\left(r - r_s\right), \quad g = 1,\ldots,G, \tag{1.63}$$

where

$$\begin{aligned} A^{(1),g}\left(\alpha\right)\psi^{(1),g}\left(r,\Omega\right) &\triangleq -\Omega \bullet \nabla\psi^{(1),g}\left(r,\Omega\right) + \Sigma_t^g\left(\mathbf{t}^g\right)\psi^{(1),g}\left(r,\Omega\right) \\ &- \sum_{g'=1}^{G}\int_{4\pi} d\Omega' \Sigma_s^{g\to g'}\left(\mathbf{s}^{gg'};\Omega \bullet \Omega'\right)\psi^{(1),g'}\left(r,\Omega'\right) \\ &- \nu\Sigma_f^g\left(\mathbf{f}^g\right)\sum_{g'=1}^{G}\int_{4\pi} d\Omega' \chi^{g\to g'}\left(\mathbf{p}^{gg'}\right)\psi^{(1),g'}\left(r,\Omega'\right), \quad g = 1,\ldots,G, \end{aligned} \tag{1.64}$$

subject to the following adjoint boundary condition describing "zero outgoing adjoint flux:"

$$\psi^{(1),g}\left(r_s,\Omega\right) = 0, \quad r_s \in S_b, \quad \Omega \bullet \mathbf{n} > 0. \tag{1.65}$$

The detailed expressions of the 1st-order sensitivities are listed below (Cacuci and Fang 2023):

1. The 1st-order sensitivities of the PERP leakage response with respect to the total group-averaged microscopic cross sections are given by the following integral involving the components $\psi^{(1),g}\left(r,\Omega\right)$, $g = 1,\ldots,G$, of the 1st-level adjoint function:

$$\frac{\partial L(\boldsymbol{\alpha})}{\partial \sigma_{t,i}^{g}} = -N_{i,m} \int_{V} dV \int_{4\pi} d\Omega \psi^{(1),g}(r,\Omega) \varphi^{g}(r,\Omega), \tag{1.66}$$

$$i = 1,\ldots,I; \quad g = 1,\ldots,G; \quad m = 1,\ldots,M.$$

2. The 1st-order sensitivities of the PERP leakage response with respect to the zeroth-order ($l = 0$) moment of the group-averaged microscopic scattering cross sections are given by the following integral involving the components $\psi^{(1),g}(r,\Omega)$, $g = 1,\ldots,G$, of the 1st-level adjoint function:

$$\left(\frac{\partial L(\boldsymbol{\alpha})}{\partial s_j}\right)_{(s=\sigma_{s,l=0})} = N_{i_j,m_j} \int_{V} dV \varphi_0^{g_j'}(r) \xi_0^{(1),g_j}(r)$$

$$-N_{i_j,m_j} \int_{V} dV \int_{4\pi} d\Omega\, \psi^{(1),g_j'}(r,\Omega) \varphi^{g_j'}(r,\Omega), \quad \text{for } j = 1,\ldots,J_{\sigma s,l=0}, \tag{1.67}$$

where the forward and adjoint flux moments $\varphi_0^{g_j'}(r)$ and $\xi_0^{(1),g_j}(r)$ are defined as follows:

$$\varphi_0^{g}(r) \triangleq \int_{4\pi} d\Omega \varphi^{g}(r,\Omega), \tag{1.68}$$

$$\xi_0^{(1),g}(r) \triangleq \int_{4\pi} d\Omega \psi^{(1),g}(r,\Omega). \tag{1.69}$$

3. The 1st-order sensitivities of the PERP leakage response with respect to the higher-order ($l \geq 1$) moments of the group-averaged microscopic scattering cross sections are given by the following integral involving the components $\psi^{(1),g}(r,\Omega)$, $g = 1,\ldots,G$, of the 1st-level adjoint function:

$$\left(\frac{\partial L(\boldsymbol{\alpha})}{\partial s_j}\right)_{(s=\sigma_{s,l\geq 1})} = N_{i_j,m_j}(2l_j + 1) \int_{V} dV \varphi_{l_j}^{g_j'}(r) \xi_{l_j}^{(1),g_j}(r), \quad j = 1,\ldots,J_{\sigma s,l\geq 1}, \tag{1.70}$$

where the forward and adjoint flux moments $\varphi_{l_j}^{g_j'}(r)$ and $\xi_{l_j}^{(1),g_j}(r)$ are defined as follows:

$$\varphi_l^{g}(r) \triangleq \int_{4\pi} d\Omega P_l(\Omega) \varphi^{g}(r,\Omega), \tag{1.71}$$

$$\xi_l^{(1),g}(r) \triangleq \int_{4\pi} d\Omega P_l(\Omega) \psi^{(1),g}(r,\Omega). \tag{1.72}$$

4. The 1st-order sensitivities of the PERP leakage response with respect to the group-averaged microscopic fission cross sections are given by the following integral involving the components $\psi^{(1),g}(r,\Omega)$, $g = 1,\dots,G$, of the 1st-level adjoint function:

$$\frac{\partial L(\alpha)}{\partial \sigma_{f,i}^{g}} = N_{i,m}\int_V dV \nu_i^g \varphi_0^g(r)\sum_{g'=1}^{G}\chi^{g'}\xi_0^{(1),g'}(r) - N_{i,m}\int_V dV\int_{4\pi} d\Omega \psi^{(1),g}(r,\Omega)\varphi^g(r,\Omega),$$

$$\text{for } i = 1,\dots,I;\quad g = 1,\dots,G;\quad m = 1,\dots,M, \tag{1.73}$$

where the flux moments $\varphi_0^g(r)$ and $\xi_0^{(1),g'}(r)$ are as defined in (1.68) and (1.69),

5. The 1st-order sensitivities of the PERP leakage response with respect to the group-averaged number of neutrons per fission are given by the following integral involving the components $\psi^{(1),g}(r,\Omega)$, $g = 1,\dots,G$, of the 1st-level adjoint function:

$$\frac{\partial L(\alpha)}{\partial \nu_i^g} = N_{i,m}\int_V dV \sigma_{f,i}^g \varphi_0^g(r)\sum_{g'=1}^{G}\chi^{g'}\xi_0^{(1),g'}(r), \tag{1.74}$$

$$\text{for } i = 1,\dots,I;\ g = 1,\dots,G;\ m = 1,\dots,M,$$

where the flux moments $\varphi_0^g(r)$ and $\xi_0^{(1),g'}(r)$ are as defined in (1.68) and (1.69),

6. The 1st-order sensitivities of the PERP leakage response with respect to the zeroth-order moment of the group-averaged fractions of the neutron spectrum are given by the following integral involving the components $\psi^{(1),g}(r,\Omega)$, $g = 1,\dots,G$ of the 1st-level adjoint function:

$$\frac{\partial L(\alpha)}{\partial \chi_i^g} = \int_V dV\left(\frac{N_{i,1}\sum_{g'=1}^{G}\nu_i^{g'}\sigma_{f,i}^{g'}f_i^{g'}}{\sum_{i=1}^{N_f}N_{i,1}\sum_{g'=1}^{G}\nu_i^{g'}\sigma_{f,i}^{g'}f_i^{g'}}\right)\xi_0^{(1),g}(r)\sum_{g'=1}^{G}\left(\nu\Sigma_f\right)^{g'}\varphi_0^{g'}(r), \tag{1.75}$$

where the flux moments $\varphi_0^g(r)$ and $\xi_0^{(1),g}(r)$ are as defined in (1.68) and (1.69),

Notably, in addition to solving the forward Boltzmann equation to compute the leakage response, the only "large-scale" computation needed for evaluating the sensitivities defined through the integrals presented in Eqs. (1.66)–(1.75) is the computation of the 1st-level multigroup adjoint fluxes $\psi^{(1),g}(r,\Omega)$, $g = 1,\dots,G$, by solving the 1st-LASS. The evaluation of the integrals, using high-order quadrature formulas, is trivial by comparison to solving either the forward Boltzmann equation or the

1st-LASS. For example, Cacuci and Fang (2023) reported that, using a DELL desktop computer (AMD FX-8350) with an 8-core processor, the CPU time for a forward computation for solving the multigroup Boltzmann equation with an $S_{256}$ angular quadrature is ca. 118 seconds; the CPU time for a typical adjoint computation of the 1st-level adjoint sensitivity function $\psi^{(1)}(r,E,\Omega)$ using an $S_{256}$ angular quadrature is ca. 85 seconds; and the typical CPU time needed to perform the integration (quadrature) over the forward or adjoint functions to compute one sensitivity is ca. 0.0012 seconds. Solving the 1st-LASS, comprising Eqs. (1.60) and (1.61), is faster (i.e., takes less CPU time) than solving the original forward system, comprising Eqs. (1.33) and (1.35). This is because "inverting" the adjoint Boltzmann operator on the left side of Eq. (1.60) subject to the adjoint boundary condition provided in Eq. (1.61) is computationally equivalent to inverting the forward Boltzmann operator on the left side of Eq. (1.33) subject to the forward boundary condition provided in Eq. (1.35), but the expression of the source term for the 1st-LASS is considerably simpler than that of the source term $Q(r,E,\Omega)$ for the forward Boltzmann equation.

The vast majority of the 1st-order sensitivities were found to have relative values below 10%, but 16 of them were larger than 100%. The largest of the 1st-order relative sensitivities was the sensitivity of the leakage response with respect to the total group cross section, denoted as $\sigma_{t,6}^{30}$, of hydrogen (labeled "isotope #6") in group 30 (thermal neutrons), which had the relative value $S^{(1)}\left(\sigma_{t,6}^{30}\right) = -9.366$. Among the benchmark's parameters, the total group-averaged microscopic cross sections had the largest sensitivities (and hence the largest impact on the leakage response). The following sensitivities of the leakage response with respect to the microscopic total cross sections of hydrogen also had large values for energy groups 17 through 20: $S^{(1)}\left(\sigma_{t,6}^{16}\right) = -1.164$; $S^{(1)}\left(\sigma_{t,6}^{17}\right) = -1.173$; $S^{(1)}\left(\sigma_{t,6}^{18}\right) = -1.141$; $S^{(1)}\left(\sigma_{t,6}^{19}\right) = -1.094$; $S^{(1)}\left(\sigma_{t,6}^{20}\right) = -1.033$. Also included in the group of relative sensitivities larger than 155% were the sensitivities of the leakage response with respect to the total microscopic cross sections of $^{239}$Pu (labeled "isotope #1") in energy groups 12 and 13, which had the values $S^{(1)}\left(\sigma_{t,1}^{12}\right) = -1.320$ and $S^{(1)}\left(\sigma_{t,1}^{13}\right) = -1.154$, respectively. Notably, all of the 1st-order sensitivities of the leakage response with respect to the group-averaged total microscopic cross sections have negative values.

The 2nd-order sensitivities also must be computed in order to compare them with the values of the largest 1st-order sensitivities and consequently decide which, if any, of the 2nd-order sensitivities must be retained for subsequent uses (e.g., for uncertainty quantification and/or data assimilation and predictive modeling). Therefore, all of the distinct $(7{,}477 \times 7{,}478/2)$ 2nd-order sensitivities of the PERP benchmark's leakage response with respect to the benchmark's parameters were computed (Cacuci and Fang 2023) by applying the 2nd-Order Comprehensive Sensitivity Analysis Methodology (2nd-CASAM) conceived by Cacuci (2015, 2016, 2018, 2021b, 2022b). It was found (Cacuci and Fang 2023) that many 2nd-order sensitivities of the OECD benchmark's response to the benchmark's uncertain parameters were much larger than the largest 1st-order ones. It was thus established that ca. 1,000 relative 2nd-order sensitivities had values larger than unity, with over 50 of them having values between 10.0 and 100.0. The overall largest 2nd-order relative sensitivity was the unmixed 2nd-order sensitivity of the leakage response with respect to the group-averaged total microscopic cross section of hydrogen in energy group 30, which had the value

$S^{(2)}\left(\sigma_{t,6}^{30},\sigma_{t,6}^{30}\right) = 429.6$. In particular, all of the 2nd-order sensitivities of the PERP benchmark's leakage response with respect to group-averaged total microscopic cross sections have positive values.

The large values of the 2nd-order sensitivities of the leakage response with respect to the benchmark's group-averaged total microscopic cross sections have motivated the computation (Cacuci and Fang 2023) of the 3rd-order sensitivities of the leakage response with respect to these cross sections by applying the "3rd-Order Comprehensive Adjoint Sensitivity Analysis Methodology" (3rd-CASAM) conceived by Cacuci (2019b, 2021b, 2022b). Recall that the number of sensitivities of the PERP leakage response that involve solely the group-averaged total microscopic cross sections are as follows: (i) 180 1st-order sensitivities; (ii) 32,400 2nd-order sensitivities, of which 16,290 are distinct; and (iii) 5,832,000 3rd-order sensitivities, of which 988,260 are distinct. In order to reduce the computational effort, the angular quadrature for solving the forward Boltzmann equation, namely Eqs. (1.36) and (1.37), was reduced from $S_{256}$ to $S_{32}$ ($ISN$=32), with a loss of accuracy of less than 0.1% in the computation of the leakage response. The corresponding adjoint Boltzmann equation (1st-LASS) was also solved using an $S_{32}$ ($ISN$=32) angular quadrature. Using the DELL AMD FX-8350 computer with an 8-core processor, the CPU time for a typical forward neutron transport computation using the neutron transport solver PARTISN with an angular quadrature of $S_{32}$ is ca. 45 seconds. The CPU time for a typical adjoint neutron transport computation with an angular quadrature of $S_{32}$ is ca. 24 seconds. The CPU time for computing the integrals over the various adjoint functions that appear in the definition of the respective sensitivity is ca. 0.0012 seconds. Thus, as detailed by Cacuci and Fang (2023), the CPU time requirements for computing the 1st-, 2nd-, and 3rd-order sensitivities of the leakage response with respect to the 180 microscopic total cross sections are as follows:

1. To compute the 180 1st-order sensitivities, one adjoint large-scale computation is needed in order to obtain the 1st-level adjoint sensitivities function (which requires ca. 24 seconds CPU time) plus ca. 1 second for computing the 180 integrals over this adjoint function. By comparison, ca. 270 minutes would be needed to compute these 1st-order sensitivities using a 2nd-order accurate finite-difference (FD) formula, assuming that the optimal step-size would be guessed ab initio.
2. To compute the 180 1st-order sensitivities together with the $\frac{180(180+1)}{2} = 16,290$ distinct 2nd-order sensitivities, $180 \times 2 = 360$ adjoint large-scale computations are needed to obtain the 2nd-level adjoint functions, requiring ca. 2.4 hours CPU time. An additional CPU time of ca. 3 minutes is needed for computing the integrals over these adjoint functions, which provide the values of both the 1st- and 2nd-order sensitivities. In addition to obtaining the 1st-order sensitivities, the 2nd-order mixed sensitivities are computed twice, using two distinct adjoint functions, for verification purposes. By comparison, ca. 810 CPU-hours would be needed to directly compute the 2nd-order sensitivities (only) using a 2nd-order accurate FD-formula, assuming that the optimal step-sizes would be guessed ab initio.

3. To compute the 180 1st-order sensitivities together with the $\frac{180(180+1)}{2}=$ 16,290 distinct 2nd-order sensitivities and the $\frac{180(180+1)(180+2)}{3!}=988,260$ distinct 3rd-order sensitivities, 32,940 adjoint large-scale computations are needed to obtain the 3rd-level adjoint functions, which require ca. 220 hours CPU time. An additional CPU time of ca. 0.6 hours is required for computing the integrals over these adjoint functions to obtain all of the 1st-order, 2nd-order, and 3rd-order sensitivities. As before, the mixed 2nd-order sensitivities would be computed twice, while the mixed sensitivities would be computed three times, using distinct adjoint functions in all cases.

The computational time that would be required to compute the 3rd-order sensitivities using the 2nd-order accurate FD-formulas, assuming that the optimal step-sizes would be guessed ab initio, can be estimated as follows:

a. The 3rd-order *unmixed* sensitivities, $\frac{\partial^3 L(\alpha)}{\partial t_j^3}$, of the leakage response, $L(\alpha)$, can be *approximately* computed by re-computations using the 2nd-order accurate finite-difference formula presented below:

$$\frac{\partial^3 L(\alpha)}{\partial t_j^3} \approx \frac{1}{2h_j^3}\left(L_{j+2}-2L_{j+1}+2L_{j-1}-L_{j-2}\right)+O\left(h_j^2\right), \quad j=1,\ldots,J_\alpha, \qquad (1.76)$$

where $L_{j+2} \triangleq L\left(t_j+2h_j\right)$, $L_{j+1} \triangleq L\left(t_j+h_j\right)$, $L_{j-1} \triangleq L\left(t_j-h_j\right)$, $L_{j-2} \triangleq L\left(t_j-2h_j\right)$, and where $h_j$ denotes a "judiciously chosen" variation in the parameter $t_j$ around its nominal value $t_j^0$.

b. The 3rd-order *mixed* sensitivities, $\frac{\partial^3 L}{\partial t_{j3}\,\partial t_{j2}\,\partial t_{j1}}$, can be calculated to 2nd-order accuracy by using the following finite-difference formula:

$$\begin{aligned}\frac{\partial^3 L(\alpha)}{\partial t_{j3}\,\partial t_{j2}\,\partial t_{j1}} \approx \frac{1}{8h_{j1}h_{j2}h_{j3}}&\left(L_{j1+1,j2+1,j3+1}-L_{j1+1,j2+1,j3-1}-L_{j1+1,j2-1,j3+1}\right.\\ &\left.+L_{j1+1,j2-1,j3-1}-L_{j1-1,j2+1,j3+1}+L_{j1-1,j2+1,j3-1}+L_{j1-1,j2-1,j3+1}-L_{j1-1,j2-1,j3-1}\right)\\ &+O\left(h_{j1}^2,h_{j2}^2,h_{j3}^2\right),\end{aligned} \qquad (1.77)$$

with $L_{j1+1,j2+1,j3+1} \triangleq L\left(t_{j1}+h_{j1},t_{j2}+h_{j2},t_{j3}+h_{j3}\right)$, $L_{j1-1,j2+1,j3+1} \triangleq L\left(t_{j1}-h_{j1},t_{j2}+h_{j2},t_{j3}+h_{j3}\right)$, etc. The values of the quantities on the rights-sides of the expressions shown in (1.76) and (1.77) are obtained by re-solving the forward Boltzmann equation repeatedly, using the changed parameter values $\left(t_{j1}\pm h_{j1}\right)$, $\left(t_{j1}\pm h_{j2}\right)$ and $\left(t_{j3}\pm h_{j3}\right)$. Of course, the values of $h_{j1}$, $h_{j2}$ and $h_{j3}$, respectively, must be chosen "judiciously" by "trial and error" for each of the parameters $t_{j1}$, $t_{j2}$ and $t_{j3}$. Of course, these "trial and error" computations would necessitate additional CPU time. The finite difference formulas introduce their intrinsic "methodological errors" of order $O\left(h_{j1}^2, h_{j2}^2, h_{j3}^2\right)$ which are in addition to, and independent of, the errors that might be incurred in the computation of $\frac{\partial^3 L}{\partial t_{j3}\,\partial t_{j2}\,\partial t_{j1}}$. Using (1.76) and (1.77), the number of forward computations needed to obtain all of the *distinct* 3rd-order sensitivities by the FD-approximation method is 7,905,360, comprising $180 \times 4$ forward PARTISN computations for the 180 unmixed 3rd-order sensitivities and $(988{,}260 - 180) \times 8$ forward PARTISN computations for the 988,080 mixed 3rd-order sensitivities. Assuming that the optimal step sizes would be guessed ab initio, Table 1.3, presents the comparison of the CPU times for all of the *distinct* 3rd-order sensitivities $\frac{\partial^3 L}{\partial t_{j3}\,\partial t_{j2}\,\partial t_{j1}}$ for $j_1 = 1,\ldots,J_{\sigma t}$; $j_2 = 1,\ldots,j_1$; $j_3 = 1,\ldots,j_2$, required by applying the 3rd-CASAM versus using the FD-approximations shown in Eqs. (1.76) and (1.77).

The comparisons presented in Table 1.3 evidently indicate that the 3rd-CASAM-L is the only practically applicable methodology capable of computing 3rd-order sensitivities without introducing methodological errors. Finite-difference schemes, on the other hand, are evidently rendered impractical by the "curse of dimensionality."

The largest 1st-, 2nd-, and 3rd-order relative sensitivities of the PERP benchmark's leakage response with respect to the microscopic total cross sections are

**TABLE 1.3**

**Comparison of CPU Times for Computing $\frac{\partial^3 L}{\partial t_{j3}\,\partial t_{j2}\,\partial t_{j1}}$**

| FD-Approximation | 3rd-CASAM |
|---|---|
| Nr. Forward comp. = 7,905,360 | Nr. Adjoint comp. = 33,301 |
| CPU time ≈ 98,817 hours | CPU time ≈ 175 hours |

**TABLE 1.4**

**Summary of the Large Relative Sensitivities for the PERP Benchmark**

| | 1st-Order | 2nd-Order | 3rd-Order |
|---|---|---|---|
| Number of sensitivities with absolute values between 1.0 and 10.0 | 8 | 665 | 45,970 |
| Number of sensitivities with absolute values between 10.0 and 100.0 | 0 | 54 | 11,861 |
| Number of sensitivities with absolute values > 100.0 | 0 | 1 | 1,199 |
| Largest relative sensitivity | $S^{(1)}\left(\sigma_{t,6}^{30}\right) = -9.366$ | $S^{(2)}\left(\sigma_{t,6}^{30}, \sigma_{t,6}^{30}\right) = 429.6$ | $S^{(3)}\left(\sigma_{t,1}^{30}, \sigma_{t,6}^{30}, \sigma_{t,6}^{30}\right) = -1.88 \times 10^5$ |

summarized in Table 1.4, underscoring the finding that the total number of 3rd-order relative sensitivities that have large values (greater than 1.0) is significantly higher than the number of large 1st- and 2nd-order sensitivities. All of the 1st-order and 3rd-order sensitivities of the leakage response with respect to the microscopic total cross sections have negative values. The absolute value of the largest 3rd-order relative sensitivity is ca. 437 times larger than the largest 2nd-order sensitivity and is ca. 20,000 times larger than the largest 1st-order sensitivity. All of the largest 1st-, 2nd-, and 3rd-order sensitivities involve the microscopic total cross section for the lowest (30th) energy group of the isotope $^{1}$H (i.e., $\sigma_{t,6}^{30}$). The largest overall 3rd-order sensitivity is the mixed 3rd-order sensitivity $S^{(3)}\left(\sigma_{t,1}^{g=30},\sigma_{t,6}^{g'=30},\sigma_{t,6}^{g''=30}\right)=-1.88\times10^{5}$, which also involves the microscopic total cross section for the 30th energy group of the isotope $^{239}$Pu (i.e., $\sigma_{t,1}^{g=30}$).

The above results for the 3rd-order sensitivities have motivated the computation (Cacuci and Fang 2023) of the 4th-order sensitivities of the leakage response with respect to the total microscopic cross section of hydrogen by applying the "4th-Order Comprehensive Adjoint Sensitivity Analysis Methodology" (4th-CASAM) conceived by Cacuci (2021a, 2021b, 2022b). As has already been discussed, the number of sensitivities of the PERP leakage response that involve solely the group-averaged total microscopic cross sections are as follows: (i) 180 1st-order sensitivities; (ii) 32,400 2nd-order sensitivities, of which 16,290 are distinct; (iii) 5,832,000 3rd-order sensitivities, of which 988,260 are distinct; and (iv) 1,049,760,000 4th-order sensitivities, of which 45,212,895 are distinct.

As detailed by Cacuci and Fang (2023), a total number of 2,075,341 adjoint-like or forward-like large-scale computations are needed to obtain all of the *distinct* 4th-order sensitivities $\dfrac{\partial^4 L(\alpha)}{\partial t_{j4}\,\partial t_{j3}\,\partial t_{j2}\,\partial t_{j1}}$ using the 4th-CASAM. Therefore, the computation of the 4th-order sensitivities $\dfrac{\partial^4 L(\alpha)}{\partial t_{j4}\,\partial t_{j3}\,\partial t_{j2}\,\partial t_{j1}}$ must be prioritized by computing the largest (and, hence, the most important) sensitivities with the highest priority. Based on the magnitudes of the 1st-, 2nd-, and 3rd-order sensitivities, which were previously computed, the most important 4th-order sensitivities are expected to include the 180 unmixed 4th-order sensitivities $\dfrac{\partial^4 L(\alpha)}{\left(\partial t_j\right)^4}$, $j=1,\ldots,J_{\sigma t}$. As detailed by Cacuci and Fang (2023), the computation of the 180 unmixed 4th-order sensitivities requires 2,521 large-scale adjoint computations.

The 1st-order through 4th-order sensitivities of the PERP leakage response with respect to the PERP model parameters are summarized in the book by Cacuci and Fang (2023), where it is shown that a significant number of 2nd-order sensitivities are larger than the 1st-order sensitivities, several 3rd-order sensitivities are much larger than the 2nd-order ones, and several 4th-order sensitivities are much larger than the 3rd-order ones. Table 1.5, presents an illustrative comparison of the values of the unmixed relative sensitivities, from 1st-order through 4th-order, for isotope 6 ($^{1}$H). Sensitivities that have absolute values larger than unity are presented in bold characters. All of the 4th-order sensitivities with respect to the group-averaged

**TABLE 1.5**
**Comparison of the Unmixed Relative Sensitivities, $S^{(1)}\left(\sigma^g_{t,6}\right)$, $S^{(2)}\left(\sigma^g_{t,6},\sigma^g_{t,6}\right)$, $S^{(3)}\left(\sigma^g_{t,6},\sigma^g_{t,6},\sigma^g_{t,6}\right)$, and $S^{(4)}\left(\sigma^g_{t,6},\sigma^g_{t,6},\sigma^g_{t,6},\sigma^g_{t,6}\right)$, $g=1,\ldots,30$, for Isotope 6 ($^1$H)**

| *G* | 1st-Order | 2nd-Order | 3rd-Order | 4th-Order |
|---|---|---|---|---|
| 1 | $-8.471\times10^{-6}$ | $7.636\times10^{-7}$ | $6.322\times10^{-8}$ | $1.460\times10^{-7}$ |
| 2 | $-2.060\times10^{-5}$ | $2.280\times10^{-6}$ | $4.516\times10^{-8}$ | $4.956\times10^{-7}$ |
| 3 | $-6.810\times10^{-5}$ | $9.021\times10^{-6}$ | $-4.677\times10^{-7}$ | $2.245\times10^{-6}$ |
| 4 | $-3.932\times10^{-4}$ | $6.673\times10^{-5}$ | $-8.758\times10^{-6}$ | $2.039\times10^{-5}$ |
| 5 | $-2.449\times10^{-3}$ | $5.549\times10^{-4}$ | $-1.216\times10^{-4}$ | $2.142\times10^{-4}$ |
| 6 | $-9.342\times10^{-3}$ | $2.935\times10^{-3}$ | $-1.123\times10^{-3}$ | $1.553\times10^{-3}$ |
| 7 | $-7.589\times10^{-2}$ | $3.949\times10^{-2}$ | $-2.690\times10^{-2}$ | $3.513\times10^{-2}$ |
| 8 | $-9.115\times10^{-2}$ | $5.604\times10^{-2}$ | $-4.380\times10^{-2}$ | $5.536\times10^{-2}$ |
| 9 | $-1.358\times10^{-1}$ | $1.014\times10^{-1}$ | $-9.758\times10^{-2}$ | $1.416\times10^{-1}$ |
| 10 | $-1.659\times10^{-1}$ | $1.428\times10^{-1}$ | $-1.604\times10^{-1}$ | $2.582\times10^{-1}$ |
| 11 | $-1.899\times10^{-1}$ | $1.849\times10^{-1}$ | $-2.385\times10^{-1}$ | $4.233\times10^{-1}$ |
| 12 | $-4.446\times10^{-1}$ | $6.620\times10^{-1}$ | **$-1.373\times10^{0}$** | **$3.815\times10^{0}$** |
| 13 | $-5.266\times10^{-1}$ | $9.782\times10^{-1}$ | **$-2.590\times10^{0}$** | **$9.015\times10^{0}$** |
| 14 | $-5.772\times10^{-1}$ | **$1.262\times10^{0}$** | **$-3.991\times10^{0}$** | **$1.650\times10^{1}$** |
| 15 | $-5.820\times10^{-1}$ | **$1.391\times10^{0}$** | **$-4.581\times10^{0}$** | **$2.208\times10^{1}$** |
| 16 | **$-1.164\times10^{0}$** | **$4.460\times10^{0}$** | **$-2.530\times10^{1}$** | **$1.890\times10^{2}$** |
| 17 | **$-1.173\times10^{0}$** | **$4.853\times10^{0}$** | **$-2.991\times10^{1}$** | **$2.432\times10^{2}$** |
| 18 | **$-1.141\times10^{0}$** | **$4.828\times10^{0}$** | **$-3.049\times10^{1}$** | **$2.543\times10^{2}$** |
| 19 | **$-1.094\times10^{0}$** | **$4.619\times10^{0}$** | **$-2.913\times10^{1}$** | **$2.428\times10^{2}$** |
| 20 | **$-1.033\times10^{0}$** | **$4.284\times10^{0}$** | **$-2.655\times10^{1}$** | **$2.175\times10^{2}$** |
| 21 | $-9.692\times10^{-1}$ | **$3.937\times10^{0}$** | **$-2.388\times10^{1}$** | **$1.915\times10^{2}$** |
| 22 | $-8.917\times10^{-1}$ | **$3.515\times10^{0}$** | **$-2.069\times10^{1}$** | **$1.609\times10^{2}$** |
| 23 | $-8.262\times10^{-1}$ | **$3.177\times10^{0}$** | **$-1.823\times10^{1}$** | **$1.382\times10^{2}$** |
| 24 | $-7.495\times10^{-1}$ | **$2.792\times10^{0}$** | **$-1.552\times10^{1}$** | **$1.140\times10^{2}$** |
| 25 | $-7.087\times10^{-1}$ | **$2.604\times10^{0}$** | **$-1.427\times10^{1}$** | **$1.033\times10^{2}$** |
| 26 | $-6.529\times10^{-1}$ | **$2.349\times10^{0}$** | **$-1.260\times10^{1}$** | **$8.932\times10^{1}$** |
| 27 | $-5.845\times10^{-1}$ | **$2.039\times10^{0}$** | **$-1.061\times10^{1}$** | **$7.288\times10^{1}$** |
| 28 | $-5.474\times10^{-1}$ | **$1.885\times10^{0}$** | **$-9.678\times10^{0}$** | **$6.565\times10^{1}$** |
| 29 | $-5.439\times10^{-1}$ | **$1.891\times10^{0}$** | **$-9.800\times10^{0}$** | **$6.705\times10^{1}$** |
| 30 | **$-9.366\times10^{0}$** | **$4.296\times10^{2}$** | **$-2.966\times10^{4}$** | **$2.720\times10^{6}$** |

total microscopic cross sections have positive values. As shown in Table 1.5, the largest absolute values for the 1st-, 2nd-, 3rd-, and 4th-order unmixed relative sensitivities all occur for the lowest-energy group ($g=30$; thermal neutrons), which are significantly larger than the values of the sensitivities in other energy groups. Notably, the largest 4th-order unmixed relative sensitivity attains a very large value: $S^{(4)}\left(\sigma^{g=30}_{t,6},\sigma^{g=30}_{t,6},\sigma^{g=30}_{t,6},\sigma^{g=30}_{t,6}\right)=2.720\times10^{6}$.

For comparison with the values obtained by applying the 4th-CASAM methodology, several of the largest 4th-order sensitivities were also computed using 2nd-order accurate finite-difference formulas. In particular, the 4th-order *unmixed* sensitivities, $\frac{\partial^4 L(\alpha)}{\partial \alpha_j^4}$, of the leakage response, $L(\alpha)$, with respect to the model parameters, can be estimated using the 2nd-order accurate finite-difference formula presented below:

$$\frac{\partial^4 L(\alpha)}{(\partial t_j)^4} \approx \frac{1}{h_j^4}\left(L_{j+2} - 4L_{j+1} + 6L_j - 4L_{j-1} + L_{j-2}\right) + O\left(h_j^2\right), \tag{1.78}$$

where $L_{j+2} \triangleq L\left(t_j + 2h_j\right)$, $L_{j+1} \triangleq L\left(t_j + h_j\right)$, $L_{j-1} \triangleq L\left(t_j - h_j\right)$, $L_{j-2} \triangleq L\left(t_j - 2h_j\right)$ and $h_j$ denotes a "judiciously chosen" variation in the parameter $t_j$ around its nominal value $t_j^0$. Several computation of were performed using Eq. (1.78) with various step sizes $h_j$, ranging from 0.125% to 2% of $\sigma_{t,6}^{g=30}$. The results of these computations are summarized in Table 1.6, which indicates that using a very small step size (e.g., a 0.125% change in the microscopic total cross section $\sigma_{t,6}^{g=30}$) within the finite difference method causes a very large error of ca. –6,010% by comparison to the exact value $S^{(4)}\left(\sigma_{t,6}^{g=30}, \sigma_{t,6}^{g=30}, \sigma_{t,6}^{g=30}, \sigma_{t,6}^{g=30}\right) = 2.720 \times 10^6$ obtained using the 4th-CASAM-L methodology. Increasing the step size to a 0.5% change in $\sigma_{t,6}^{g=30}$ reduces dramatically the error (to –17.7%). The smallest error (of just –0.45%) between the approximate result produced using the FD-formula and the exact result produced by using the 4th-CASAM-L was attained for a 0.60% change in $\sigma_{t,6}^{g=30}$. Increasing further the step size, however, worsened the results produced by the FD-formula: thus, a 1.0% change in $\sigma_{t,6}^{g=30}$ increased the error to 31.4%, while a 2.0% change in $\sigma_{t,6}^{g=30}$ further increased the error of the FD-formula (by comparison to the exact result produced by the 4th-CASAM) to 698%. For $h_j > 2.5\% \times \sigma_{t,6}^{30}$, the forward neutron transport (re-) computations using PARTISAN failed to converge.

**TABLE 1.6**
**FD-Computations of $S^{(4)}\left(\sigma_{t,6}^{g=30}, \sigma_{t,6}^{g=30}, \sigma_{t,6}^{g=30}, \sigma_{t,6}^{g=30}\right)$ Using Eq. (1.78)**

| Step-Size $h_j$ | FD-Method | $\frac{FD - 4th\,CASAM}{4th\,CASAM}$ |
|---|---|---|
| $0.125\% \times \sigma_{t,6}^{g=30}$ | $-1.607 \times 10^8$ | –6,010% |
| $0.25\% \times \sigma_{t,6}^{g=30}$ | $-7.488 \times 10^6$ | –375% |
| $0.50\% \times \sigma_{t,6}^{g=30}$ | $2.239 \times 10^6$ | –17.7% |
| $0.60\% \times \sigma_{t,6}^{g=30}$ | $2.708 \times 10^6$ | –0.45% |
| $0.65\% \times \sigma_{t,6}^{g=30}$ | $2.838 \times 10^6$ | 4.3% |
| $0.75\% \times \sigma_{t,6}^{g=30}$ | $3.063 \times 10^6$ | 12.6% |
| $1.00\% \times \sigma_{t,6}^{g=30}$ | $3.574 \times 10^6$ | 31.4% |
| $2.00\% \times \sigma_{t,6}^{g=30}$ | $2.171 \times 10^7$ | 698% |
| $> 2.5\% \times \sigma_{t,6}^{g=30}$ | – | – |

The detailed computations of the sensitivities, from 1st- through 4th-order, of the PERP benchmark's leakage response with respect to all of the PERP benchmark's parameters are presented in the book by (Cacuci and Fang 2023).

### 1.5.3 High-Order Uncertainty Quantification

The effects of high-order sensitivities on the moments (i.e., mean, variance, skewness) of the response distribution will be illustrated in this section by considering the leakage response of the PERP reactor physics benchmark discussed in Section 1.5.1, in the phase space of total microscopic cross sections, since the largest sensitivities of the leakage response are with respect to these parameters. A complete uncertainty analysis of the PERP leakage response can be found in the book by Cacuci and Fang (2023).

In order to simplify the algebraic computations, the group-averaged microscopic total cross sections will be considered to be uncorrelated and normally distributed. Under these conditions, the first three moments of the distribution of the leakage response are provided by the following expressions (Cacuci 2022b; Cacuci and Fang 2023), which are particular cases of Eqs. (1.4), (1.5), and (1.7):

i. The expected value of the leakage response has the following particular expression:

$$[E(L)]_t^{(U,N)} = L(\alpha^0) + [E(L)]_t^{(2,U,N)} + [E(L)]_t^{(4,U,N)}, \tag{1.79}$$

where the superscript "*U,N*" indicates contributions from *uncorrelated* and *normally distributed* parameters, the subscript "*t*" indicates group-averaged microscopic "total" cross section, and the letter "*L*" denotes "leakage response." In Eq. (1.79), the contributions from the 3rd-order (and all odd-order) sensitivities vanish because the parameters are uncorrelated. Furthermore, the quantity $L(\alpha^0)$ represents the leakage response computed using the nominal cross section values, and the quantities $[E(L)]_t^{(2,U,N)}$ and $[E(L)]_t^{(4,U,N)}$ denote the contributions from the 2nd-order and 4th-order response sensitivities, respectively, which are provided by the following expressions:

$$[E(L)]_t^{(2,U,N)} = \frac{1}{2}\sum_{j_1=1}^{J_{\sigma t}} \frac{\partial^2 L(\alpha)}{\partial^2 t_{j_1}} \sigma_{j_1}^2, \tag{1.80}$$

$$[E(L)]_t^{(4,U,N)} = \frac{1}{8}\sum_{j_1=1}^{J_{\sigma t}} \frac{\partial^4 L(\alpha)}{\partial^4 t_{j_1}} \sigma_{j_1}^4. \tag{1.81}$$

In Eqs. (1.80) and (1.81), the quantity $t_{j_1}$ represents the "$j_1$-th microscopic total cross section," while the quantity $J_{\sigma t} = G \times I = 180$ denotes the total number of microscopic total cross sections for $G = 30$ groups and $I = 6$

isotopes contained in the PERP benchmark. The traditional notation "$\sigma$" for "microscopic cross sections" is not used in the formulas to be presented in this section since this notation will be used to denote "standard deviation."

ii. The variance of the leakage response for the PERP benchmark takes on the following particular form:

$$\left[\mathrm{var}(L)\right]_t^{(U,N)} = \sum_{i=1}^{4} \left[\mathrm{var}(L)\right]_t^{(i,U,N)}, \tag{1.82}$$

where $\left[\mathrm{var}(L)\right]_t^{(1,U,N)}$, $\left[\mathrm{var}(L)\right]_t^{(2,U,N)}$, $\left[\mathrm{var}(L)\right]_t^{(3,U,N)}$ and $\left[\mathrm{var}(L)\right]_t^{(4,U,N)}$ denote the contributions of the terms involving the 1st-order through 4th-order sensitivities, respectively, to the variance $\left[\mathrm{var}(L)\right]_t^{(U,N)}$ and are defined by the following expressions:

$$\left[\mathrm{var}(L)\right]_t^{(1,U,N)} \triangleq \sum_{j_1=1}^{J_{\sigma t}} \left[\frac{\partial L(\alpha)}{\partial t_{j_1}}\right]^2 \left(\sigma_{j_1}\right)^2, \tag{1.83}$$

$$\left[\mathrm{var}(L)\right]_t^{(2,U,N)} \triangleq \frac{1}{2} \sum_{j_1=1}^{J_{\sigma t}} \sum_{j_2=1}^{J_{\sigma t}} \left[\frac{\partial^2 L(\alpha)}{\partial t_{j_1} \partial t_{j_2}} \sigma_{j_1} \sigma_{j_2}\right]^2, \tag{1.84}$$

$$\left[\mathrm{var}(L)\right]_t^{(3,U,N)} = \sum_{j_1=1}^{J_{\sigma t}} \sum_{j_2=1}^{J_{\sigma t}} \left[\frac{\partial^3 L(\alpha)}{\partial t_{j_1} \partial t_{j_1} \partial t_{j_2}} \frac{\partial L(\alpha)}{\partial t_{j_2}}\right] \sigma_{j_1}^2 \sigma_{j_2}^2 + \frac{15}{36} \sum_{j_1=1}^{J_{\sigma t}} \left[\frac{\partial^3 L(\alpha)}{\partial t_{j_1} \partial t_{j_1} \partial t_{j_1}}\right]^2 \sigma_{j_1}^6, \tag{1.85}$$

$$\left[\mathrm{var}(L)\right]_t^{(4,U,N)} = \frac{1}{2} \sum_{j_1=1}^{J_{\sigma t}} \left[\frac{\partial^4 L(\alpha)}{\left(\partial t_{j_1}\right)^4} \frac{\partial^2 L(\alpha)}{\left(\partial t_{j_1}\right)^2}\right] \sigma_{j_1}^6. \tag{1.86}$$

iii. The 3rd-order moment of the leakage response for the PERP benchmark takes on the following particular form:

$$\left[\mu_3(L)\right]_t^{(U,N)} = \sum_{i=1}^{4} \left[\mu_3(L)\right]_t^{(i,U,N)}, \tag{1.87}$$

where $\left[\mu_3(L)\right]_t^{(1,U,N)}$, $\left[\mu_3(L)\right]_t^{(2,U,N)}$, $\left[\mu_3(L)\right]_t^{(3,U,N)}$ and $\left[\mu_3(L)\right]_t^{(4,U,N)}$ denote the contributions to $\left[\mu_3(L)\right]_t^{(U,N)}$ of the terms involving the 1st-order through 4th-order sensitivities, respectively; these quantities have the following expressions:

$$\left[\mu_3(L)\right]_t^{(2,U,N)} = 3 \sum_{j_1=1}^{J_{\sigma t}} \sum_{j_2=1}^{J_{\sigma t}} \frac{\partial L(\alpha)}{\partial t_{j_1}} \frac{\partial L(\alpha)}{\partial t_{j_2}} \frac{\partial^2 L(\alpha)}{\partial t_{j_1} \partial t_{j_2}} \left(\sigma_{j_1} \sigma_{j_2}\right)^2 + \sum_{j_1=1}^{J_{\sigma t}} \left[\frac{\partial^2 L(\alpha)}{\left(\partial t_{j_1}\right)^2}\right]^3 \sigma_{j_1}^6, \tag{1.88}$$

$$\left[\mu_3(L)\right]_t^{(3,U,N)} = 6\sum_{j_1=1}^{J_{\sigma t}} \frac{\partial L(\alpha)}{\partial t_{j_1}} \frac{\partial^2 L(\alpha)}{\left(\partial t_{j_1}\right)^2} \frac{\partial^3 L(\alpha)}{\left(\partial t_{j_1}\right)^3} \sigma_{j_1}^6, \tag{1.89}$$

$$\left[\mu_3(L)\right]_t^{(4,U,N)} = \frac{3}{2}\sum_{j_1=1}^{J_{\sigma t}} \left[\frac{\partial L(\alpha)}{\partial t_{j_1}}\right]^2 \frac{\partial^4 L(\alpha)}{\left(\partial t_{j_1}\right)^4} \sigma_{j_1}^6. \tag{1.90}$$

The skewness, denoted as $\gamma_1(L)$, of the response $L(\alpha)$ indicates the degree of the distribution's asymmetry with respect to its mean and is defined as follows:

$$\left[\gamma_1(L)\right]_t^{(U,N)} = \frac{\left[\mu_3(L)\right]_t^{(U,N)}}{\left\{\left[\text{var}(L)\right]_t^{(U,N)}\right\}^{\frac{3}{2}}}. \tag{1.91}$$

Using Eqs. (1.79)–(1.91), the effects of the sensitivities of various orders on the leakage response's expectation, variance, and skewness have been quantified by considering uniform standard deviations of 1% (small) and 5% (moderate), respectively, for the microscopic total cross sections. These results are presented in Tables 1.7 and 1.8, respectively.

The results presented in Table 1.7 were obtained by considering a small relative standard deviation of 1% for each of the uncorrelated microscopic total cross sections of the isotopes included in the PERP benchmark. The effects of the 2nd-order and 4th-order sensitivities on the expected response value $\left[E(L)\right]_t^{(U,N)}$ are both negligibly small, since $\left[E(L)\right]_t^{(2,U,N)} \cong 2.5\% \times \left[E(L)\right]_t^{(U,N)}$ and $\left[E(L)\right]_t^{(4,U,N)} \cong 0.3\% \times \left[E(L)\right]_t^{(U,N)}$. The results presented in the 2nd column in Table 1.7 imply that $\left[\text{var}(L)\right]_t^{(1,U,N)} \cong 70\% \times \left[\text{var}(L)\right]_t^{(U,N)}$, $\left[\text{var}(L)\right]_t^{(2,U,N)} \cong 6\% \times \left[\text{var}(L)\right]_t^{(U,N)}$, $\left[\text{var}(L)\right]_t^{(3,U,N)} \cong 20\% \times \left[\text{var}(L)\right]_t^{(U,N)}$,

**TABLE 1.7**
**Expected Response Value, Variance, and Skewness for 1% Relative Standard Deviations (RSD) of the Normally Distributed and Uncorrelated Microscopic Total Cross Sections**

| Expected Value | Variance | 3rd-Order Moment and Skewness |
|---|---|---|
| $L(\alpha^0) = 1.765 \times 10^6$ | $\left[\text{var}(L)\right]_t^{(1,U,N)} = 3.419 \times 10^{10}$ | $\left[\mu_3(L)\right]_t^{(2,U,N)} = 6.663 \times 10^{15}$ |
| $\left[E(L)\right]_t^{(2,U,N)} = 4.598 \times 10^4$ | $\left[\text{var}(L)\right]_t^{(2,U,N)} = 2.879 \times 10^9$ | $\left[\mu_3(L)\right]_t^{(3,U,N)} = 3.948 \times 10^{15}$ |
| | $\left[\text{var}(L)\right]_t^{(3,U,N)} = 9.841 \times 10^9$ | $\left[\mu_3(L)\right]_t^{(4,U,N)} = 1.973 \times 10^{15}$ |
| $\left[E(L)\right]_t^{(4,U,N)} = 6.026 \times 10^3$ | $\left[\text{var}(L)\right]_t^{(4,U,N)} = 1.825 \times 10^9$ | $\left[\mu_3(L)\right]_t^{(U,N)} = 1.258 \times 10^{16}$ |
| $\left[E(L)\right]_t^{(U,N)} = 1.817 \times 10^6$ | $\left[\text{var}(L)\right]_t^{(U,N)} = 4.874 \times 10^{10}$ | $\left[\gamma_1(L)\right]_t^{(U,N)} = 1.169$ |

**TABLE 1.8**
**Expected Response Value, Variance, and Skewness for 5% Relative Standard Deviations (RSD) of the Normally Distributed and Uncorrelated Microscopic Total Cross Sections**

| Expected Value | Variance | 3rd-Order Moment and Skewness |
|---|---|---|
| $L(\alpha^0) = 1.765 \times 10^6$ | $[\mathrm{var}(L)]_t^{(1,U,N)} = 8.549 \times 10^{11}$ | $[\mu_3(L)]_t^{(2,U,N)} = 1.070 \times 10^{19}$ |
| $[E(L)]_t^{(2,U,N)} = 1.149 \times 10^6$ | $[\mathrm{var}(L)]_t^{(2,U,N)} = 1.799 \times 10^{12}$ | $[\mu_3(L)]_t^{(3,U,N)} = 6.169 \times 10^{19}$ |
| | $[\mathrm{var}(L)]_t^{(3,U,N)} = 2.338 \times 10^{13}$ | $[\mu_3(L)]_t^{(4,U,N)} = 3.083 \times 10^{19}$ |
| $[E(L)]_t^{(4,U,N)} = 3.766 \times 10^6$ | $[\mathrm{var}(L)]_t^{(4,U,N)} = 2.852 \times 10^{13}$ | $[\mu_3(L)]_t^{(U,N)} = 1.032 \times 10^{20}$ |
| $[E(L)]_t^{(U,N)} = 6.681 \times 10^6$ | $[\mathrm{var}(L)]_t^{(U,N)} = 5.456 \times 10^{13}$ | $[\gamma_1(L)]_t^{(U,N)} = 0.256$ |

and $[\mathrm{var}(L)]_t^{(4,U,N)} \cong 4\% \times [\mathrm{var}(L)]_t^{(U,N)}$, indicating that the contributions stemming from the 1st-order sensitivities to the response variance are significantly larger (by ca. 70%) than those stemming from higher-order sensitivities. By comparison, the 2nd-order sensitivities contribute about 6% to the response variance, the 3rd-order sensitivities contribute about 20% to the response variance, while the 4th-order ones only contribute about 4% to the response variance. The results presented in Table 1.7 also indicate that $[\mu_3(L)]_t^{(2,U,N)} \cong 53\% \times [\mu_3(L)]_t^{(U,N)}$, $[\mu_3(L)]_t^{(3,U,N)} \cong 31\% \times [\mu_3(L)]_t^{(U,N)}$ and $[\mu_3(L)]_t^{(4,U,N)} \cong 16\% \times [\mu_3(L)]_t^{(U,N)}$; thus, the contributions to the 3rd-order response moment $[\mu_3(L)]_t^{(U,N)}$ stemming from the 2nd-order sensitivities are the largest (e.g., around 53% in this case), followed by the contributions stemming from the 3rd-order sensitivities, while the contributions stemming from the 4th-order sensitivities are the smallest. The response skewness, $[\gamma_1(L)]_t^{(U,N)}$, is positive, causing the leakage response distribution to be skewed toward the positive direction from its expected value $[E(L)]_t^{(U,N)}$.

Table 1.8, presents results obtained by considering a relative standard deviation of 5% (which is a typical measurement precision) for each of the uncorrelated microscopic total cross sections. These results show that $[E(L)]_t^{(2,U,N)} \cong 65\% \times L(\alpha^0) \cong 17\% \times [E(L)]_t^{(U,N)}$, thereby indicating that the contributions from the 2nd-order sensitivities to the expected response are around 65% of the computed leakage value $L(\alpha^0)$, and contribute around 17% to the expected value $[E(L)]_t^{(U,N)}$ of the leakage response. Furthermore, the results presented in Table 1.8 also imply that $[E(L)]_t^{(4,U,N)} \cong 213\% \times L(\alpha^0) \cong 56\% \times [E(L)]_t^{(U,N)}$, indicating that the contributions from the 4th-order sensitivities to the expected value of the leakage response are about 2.1 times larger than the computed leakage value $L(\alpha^0)$, and contribute around 56% to the expected value $[E(L)]_t^{(U,N)}$. Therefore, if the computed value, $L(\alpha^0)$, is considered to be the actual expected value of the leakage response, neglecting the 4th-order sensitivities would produce an error of ca. 210%.

For a typical relative standard deviation of 5% for the uncorrelated microscopic total cross sections, the results presented in Table 1.8 indicate that $[\text{var}(L)]_t^{(1,U,N)} \cong 2\% \times [\text{var}(L)]_t^{(U,N)}$, $[\text{var}(L)]_t^{(2,U,N)} \cong 3\% \times [\text{var}(L)]_t^{(U,N)}$, $[\text{var}(L)]_t^{(3,U,N)} \cong 43\% \times [\text{var}(L)]_t^{(U,N)}$, and $[\text{var}(L)]_t^{(4,U,N)} \cong 52\% \times [\text{var}(L)]_t^{(U,N)}$, which means the contributions stemming from the 3rd- and 4th-order sensitivities to the response variance are remarkably larger than those from the 1st- and 2nd-order ones. The results in Table 1.8 also show that $[\mu_3(L)]_t^{(2,U,N)} \cong 10\% \times [\mu_3(L)]_t^{(U,N)}$, $[\mu_3(L)]_t^{(3,U,N)} \cong 60\% \times [\mu_3(L)]_t^{(U,N)}$ and $[\mu_3(L)]_t^{(4,U,N)} \cong 30\% \times [\mu_3(L)]_t^{(U,N)}$; thus, the contributions from the 3rd-order sensitivities are the largest (e.g., around 60%), followed by the contributions from the 4th-order sensitivities, which contribute about 30%; the smallest contributions stem from the 2nd-order sensitivities.

The results shown in Tables 1.7 and 1.8 highlight the fact that the successively higher-order terms become increasingly less important for small (1%) parameter uncertainties but become increasingly more important for larger (e.g., 5%) parameter uncertainties. Recall that the radius/domain of convergence of the Taylor series in Eq. (1.1) determines the largest values of the parameter variations $\delta\alpha_j$ which are admissible before the respective series becomes divergent. In turn, these maximum admissible parameter variations limit the largest parameter covariances/standard deviations that can be considered for using the Taylor-series expansion for the subsequent purposes of computing moments of the distribution of computed responses, namely Eqs. (1.4)–(1.8). Applying the ratio test for convergence of infinite series to the 3rd- and 4th-order sensitivities indicates that a uniform relative parameter variation of ca. 5% (which would correspond to a 5% uniform relative standard deviation for the model parameters) would yield a ratio slightly larger than unity, which is just outside the boundary of the domain of convergence of the Taylor series expansion shown in Eq. (1.1). Consequently, in the case of the PERP benchmark, if the relative standard deviations of the parameters (such as the total cross sections of Hydrogen $^{1}$H and/or Plutonium $^{239}$Pu isotopes) that have large sensitivities are 5% or larger, then these standard deviations would need to be reduced by recalibrating the respective parameters in order to keep the Taylor series expansion shown in Eq. (1.1) within its radius of convergence. Such a recalibration could be performed by using measurements of parameters and/or responses in conjunction with the high-order predictive modeling methodology developed by Cacuci (2023e, 2025), which will be briefly reviewed in the next section.

### 1.5.4 Fourth-Order BERRU (Best Estimate Results with Reduced Uncertainties) Predictive Modeling Applied to the PERP Benchmark

The impact of combining computational and experimental information (such as measurements of the leakage response) within the 4th-BERRU-PM methodology summarized in Section 1.4 will be briefly illustrated in this subsection by considering a measurement of the leakage response having the value nominal measured value $L^e = 1.941 \times 10^6$ neutrons/sec, with a measured experimental relative deviation $RSD^{(e)} = 5\%$; the superscript "$e$" is used to denote quantities stemming from "experiments." Specifically, the application of the 4th-BERRU-PM methodology will be illustrated by using Eqs. (1.11) and (1.18) to obtain the best-estimate results predicted for the leakage response and for the reduced predicted response standard deviation, respectively. Recall that the computed value of the leakage response is $L(\alpha^0) = 1.7648 \times 10^6$ neutrons/ sec. As has been discussed in the foregoing, the largest sensitivities of the leakage response are with respect to the group-averaged microscopic total cross sections. The vast majority of the sensitivities with respect to the remaining parameters are very small. In particular, their combined impact on the expectation value and variance of the computed leakage response is of the order of 5% (Cacuci and Fang 2023). Therefore, these parameters can be safely neglected, without affecting the conclusions that will be reached by considering just the microscopic total cross sections. Hence, only the 180 microscopic total cross sections will be considered as the benchmark's imprecisely known parameters for the illustrative application of the 4th-Order BERRU Predictive Modeling (4th-BERRU-PM) methodology to be presented in this section.

Table 1.9, presents the results predicted by the 4th-BERRU-PM methodology for the best-estimate mean value $L^{(be,i)}$ predicted for the leakage response value and the predicted response standard deviation $SD^{(be,i)}$, $i = 1,\ldots,4$, when considering parameter $RSD = 2\%$, which are within the radius of convergence of the Taylor series, cf. Eq. (1.1), of the leakage response.

For parameter relative standard deviations $RSD = 5\%$, which are just outside the radius of convergence of the Taylor series shown in Eq. (1.1) for the leakage response, the results predicted by Eqs. (1.11) and (1.18) are presented in Table 1.10.

**TABLE 1.9**
**Results for $L^{(be,i)} \pm SD^{(be,i)}$, $i = 1,\ldots,4$, for Parameters $RSD = 2\%$**

| Order ($i$) of Sensitivities | $L^e \pm SD^{(e)}$ (Neutrons/Sec) | $L^{(be,i)} \pm SD^{(be,i)}$ (Neutrons/Sec) |
|---|---|---|
| $i = 1$ | $1.941 \times 10^6 \pm 9.706 \times 10^4$ | $1.930 \times 10^6 \pm 9.388 \times 10^4$ |
| $i = 2$ | | $1.942 \times 10^6 \pm 9.466 \times 10^4$ |
| $i = 3$ | | $1.942 \times 10^6 \pm 9.593 \times 10^4$ |
| $i = 4$ | | $1.942 \times 10^6 \pm 9.596 \times 10^4$ |

**TABLE 1.10**
**Results for $L^{(be,i)} \pm SD^{(be,i)}$, $i = 1,\ldots,4$, for Parameters $RSD = 5\%$**

| Order ($i$) of Sensitivities | $L^e \pm SD^{(e)}$ (Neutrons/Sec) | $L^{(be,i)} \pm SD^{(be,i)}$ (Neutrons/Sec) |
|---|---|---|
| $i = 1$ | $1.941 \times 10^6 \pm 9.706 \times 10^4$ | $1.939 \times 10^6 \pm 9.653 \times 10^4$ |
| $i = 2$ | | $1.945 \times 10^6 \pm 9.689 \times 10^4$ |
| $i = 3$ | | $1.942 \times 10^6 \pm 9.705 \times 10^4$ |
| $i = 4$ | | $1.942 \times 10^6 \pm 9.705 \times 10^4$ |

As indicated by the results presented in Tables 1.9 and 1.10, the application of the 4th-BERRU-PM methodology yields a best-estimate value for the predicted leakage response that falls in-between the experimentally measured and computed values of this response, with an accompanying predicted standard deviation that is smaller than either the measured or the computed standard deviations. The predicted leakage response is closer to the experimentally measured response rather than to the computed response because the standard deviation of the experimentally measured response is smaller than (which means that it is "more accurately known") the standard deviation of the computed leakage response. The results obtained in Table 1.10, for parameter-$RSD=5\%$, indicate that the application of the 4th-BERRU-PM methodology can produce physically useful/meaningful results even when the Taylor-series expansion of the response in terms of parameter standard deviations diverges slightly.

The predicted skewness of the (predicted) leakage response is computed within the 4th-BERRU-PM methodology as a particular case of Eq. (1.21). Table 1.11 presents results obtained by using Eq. (1.21) in conjunction with a measured response having the nominal value $L^e = 1.941\times10^6$ neutrons/sec and a relative standard deviation $SD^{(e)} = 5\%$, for relative standard deviations for the parameters of 2% (inside the convergence region of the Taylor series of the computed response) and, respectively, 5% (just outside the convergence region of the Taylor series of the computed response). The notation $Skew^{(be,i)}\left(L^{(be,i)}\right)$, $i = 1,\ldots,4$, indicates the highest-order of sensitivities retained for computing the corresponding result: $Skew^{(be,1)}\left(L^{(be,1)}\right)$ indicates that only the 1st-order sensitivities were retained; $Skew^{(be,2)}\left(L^{(be,2)}\right)$ indicates that the 1st- and 2nd-order sensitivities were retained; $Skew^{(be,3)}\left(L^{(be,3)}\right)$ indicates that all of the sensitivities up to and including 3rd-order were retained; $Skew^{(be,4)}\left(L^{(be,4)}\right)$ indicates that all of the sensitivities up to and including 4th-order were retained. The results presented in Table 1.11 indicate that the 2nd- and higher-order sensitivities significantly impact the skewness of the predicted response, since the skewness is positive if the 2nd- and higher-order sensitivities are neglected but becomes negative when the higher-order sensitivities are included.

The predicted kurtosis of the (predicted) leakage response is computed within the 4th-BERRU-PM methodology as a particular case of Eq. (1.27). Table 1.12, presents the results thus obtained in conjunction with a measured response having the nominal value $L^e = 1.941\times10^6$ neutrons/sec and a relative standard deviation $SD^{(e)} = 5\%$, for relative standard deviations for the parameters of 2% (inside the convergence region of the Taylor series of the computed response) and, respectively, 5% (just outside

**TABLE 1.11**
**Best-Estimate Skewness of Predicted Leakage Response**

| | $Skew^{(be,1)}\left(L^{(be,1)}\right)$ | $Skew^{(be,2)}\left(L^{(be,2)}\right)$ | $Skew^{(be,3)}\left(L^{(be,3)}\right)$ | $Skew^{(be,4)}\left(L^{(be,4)}\right)$ |
|---|---|---|---|---|
| $SD = 2\%$ | $3.902\times10^{-1}$ | $-1.216\times10^{-2}$ | $-5.546\times10^{-3}$ | $-5.959\times10^{-2}$ |
| $SD = 5\%$ | $6.045\times10^{-2}$ | $-1.070\times10^{-2}$ | $-1.088\times10^{-2}$ | $-2.530\times10^{-2}$ |

**TABLE 1.12**
**Best-Estimate Kurtosis of Predicted Leakage Response**

| | $Kurt^{(be,1)}\left(L^{(be,1)}\right)$ | $Kurt^{(be,2)}\left(L^{(be,2)}\right)$ | $Kurt^{(be,3)}\left(L^{(be,3)}\right)$ | $Kurt^{(be,4)}\left(L^{(be,4)}\right)$ |
|---|---|---|---|---|
| $SD = 2\%$ | 3.522 | 3.317 | 3.145 | 3.114 |
| $SD = 5\%$ | 3.069 | 3.029 | 3.002 | 3.001 |

the convergence region of the Taylor series of the computed response). The notation $Kurt^{(be,i)}\left(L^{(be,i)}\right)$, $i = 1,\ldots,4$, indicates the highest-order of sensitivities retained for computing the corresponding result: $Kurt^{(be,1)}\left(L^{(be,1)}\right)$ indicates that only the 1st-order sensitivities were retained; $Kurt^{(be,2)}\left(L^{(be,2)}\right)$ indicates that the 1st- and 2nd-order sensitivities were retained; $Kurt^{(be,3)}\left(L^{(be,3)}\right)$ indicates that all of the sensitivities up to and including 3rd-order were retained; $Kurt^{(be,4)}\left(L^{(be,4)}\right)$ indicates that all of the sensitivities up to and including 4th-order were retained. The results presented in Table 1.12 for parameter standard deviations of 2% (for which the Taylor-series expansion of the computed response in terms of parameter deviations converges) indicate that when only the 1st-order sensitivities are considered, the kurtosis of the MaxEnt distribution of the predicted best-estimate leakage response is larger than 3.00, thus being somewhat "more peaked" than the normal distribution. However, when the 2nd- and higher-order sensitivities are also considered, the MaxEnt distribution of the predicted best-estimate leakage response approaches the normal distribution (for which "kurtosis = 3.00"). The same trend is also observed when considering parameter relative standard deviations of 5%.

## 1.6 CHAPTER SUMMARY

Sensitivities are used for the following purposes: (i) understanding the model by ranking the importance of the various parameters; (ii) performing "reduced-order modeling" by eliminating unimportant parameters and/or processes; (iii) quantifying the uncertainties induced in a model response due to model parameter uncertainties; (iv) performing "model validation" by comparing computations to experiments to address the question "does the model represent reality?" (v) prioritizing improvements in the model; (vi) performing data assimilation and model calibration as part of forward "predictive modeling" to obtain best-estimate predicted results with reduced predicted uncertainties; (vii) performing inverse "predictive modeling"; (viii) designing and optimizing the system. The applications of high-order sensitivities for quantifying the uncertainties induced in a model's response by uncertainties in the model's parameters and for predictive modeling have been illustrated by considering the Polyethylene-Reflected Plutonium Metal Sphere (acronym: PERP) OECD/NEA reactor physics benchmark (Valentine 2006), which is a one-dimensional spherical subcritical nuclear system driven by a source of spontaneous fission neutrons. The computational model of the PERP benchmark entails solving the neutron transport equation (which is representative of a "large-scale" computation) comprising 21,976 model parameters. It has been shown that the higher-order sensitivities can become considerably larger than the lower-order ones. The considerable impacts of the 2nd-, 3rd-, and 4th-order sensitivities on the expected value, standard deviation, skewness, and kurtosis of both the computed response and the predicted response have also been illustrated.

This chapter has also illustrated the practical inability of finite-difference schemes to compute higher-order sensitivities. This is because finding the proper step size varies from sensitivity to sensitivity and must therefore be determined by trial and error. Furthermore, the result produced by finite-difference schemes cannot be verified unless the correct value of the respective sensitivity is known ahead of time. Thus, finite-difference methods are imprecise and are subject to the "curse of dimensionality." These shortcomings are also characteristic of statistical methods, which are also limited by the curse of dimensionality and cannot compute any sensitivity exactly. Notably, the practitioners of the statistical methods have themselves drawn an alarm signal against the indiscriminate use of these methods by pointing out the erroneous results that their use can bring about in the suggestively entitled work: "Why So Many Published Sensitivity Analyses Are False. A Systematic Review of Sensitivity Analysis Practices" by Saltelli et al. (2019). One uses finite-difference and statistical methods at one's own peril.

Clearly, only the *n*th-CASAM-L (Cacuci 2021b) and the *n*th-CASAM-N (Cacuci 2022a) methodologies produce reliable results that are based on obtaining and evaluating exact deterministic expressions for sensitivities of arbitrarily high-order by solving exact deterministic equations. However, implementing this methodology entails developing/constructing and solving corresponding *n*th-Level Adjoint Sensitivity Systems (*n*th-LASS), which is a non-trivial undertaking. Cacuci (2024a) has generalized and maximized the computational efficiency of these methodologies by conceiving the "*n*th-Order Features Adjoint Sensitivity Analysis Methodology for Nonlinear Systems (*n*th-FASAM-N)," as will be detailed in subsequent chapters.

# 2 The 1st-FASAM-N Methodology for Nonlinear Systems

## 2.1 INTRODUCTION

Chapter 1 has highlighted the need for computing high-order sensitivities and their subsequent uses for performing sensitivity analysis of model responses to model parameters, quantification of uncertainties induced in responses by uncertainties in parameters, and their essential contributions to high-order predictive modeling – which combines computational with experimental information to obtain best-estimate predicted responses and calibrated model parameters with reduced predicted uncertainties. For large-scale models involving many parameters, even the 1st-order sensitivities are computationally very expensive to determine accurately by conventional methods. Furthermore, the "curse of dimensionality" (Bellman 1957) prohibits the accurate computation of higher-order sensitivities by conventional methods. Cacuci (2021b, 2022b) has developed the "*n*th-Order Comprehensive Adjoint Sensitivity Analysis Methodology for Response-Coupled Forward/Adjoint Linear Systems" (abbreviated as "*n*th-CASAM-L"), which overcomes the curse of dimensionality in sensitivity analysis for such systems. The *n*th-CASAM-L methodology was developed specifically for linear systems because the most important model responses produced by such systems are various Lagrangian functionals which depend simultaneously on both the forward and adjoint state functions governing the respective linear system. Since nonlinear operators do not admit adjoint operators, responses that simultaneously depend on both the forward and the adjoint state functions can occur only for linear systems. The *n*th-CASAM-L was extended by Cacuci (2022a, 2023a) to nonlinear systems by developing the "*n*th-Order Comprehensive Adjoint Sensitivity Analysis Methodology for Nonlinear Systems" (abbreviated as: *n*th-CASAM-N). Just like the *n*th-CASAM-L, the *n*th-CASAM-N is formulated in linearly increasing higher-dimensional Hilbert spaces (as opposed to exponentially increasing parameter-dimensional spaces), thus overcoming the curse of dimensionality in sensitivity analysis of nonlinear systems. The *n*th-CASAM-L and the *n*th-CASAM-N methodologies enable the most efficient computation of exactly determined expressions of arbitrarily high-order sensitivities of generic linear and/or nonlinear system responses with respect to model parameters, uncertain boundaries, and uncertain internal interfaces in the model's phase-space.

The *n*th-CASAM-L and the *n*th-CASAM-N methodologies compute the sensitivities of model responses directly with respect to the model's primary parameters. However, many models comprise not just disparate primary model parameters but


DOI: 10.1201/9781003478171-2

also functions of such primary model parameters. Such functions customarily appear in models in the form of correlations that describe "features" of the system under consideration, such as material properties, flow regimes, etc. Usually, the number of such "feature" functions is considerably smaller than the total number of model parameters. For example, the numerical model of the OECD/NEA reactor physics benchmark analyzed by Cacuci and Fang (2023) comprises 21,976 uncertain primary model parameters. However, in the large-scale numerical computation which solves the Boltzmann neutron transport equation to compute the neutron flux distribution within this benchmark, the computational model does not use the primary parameters directly but uses various functions of these parameters. High-order sensitivity analyses and uncertainty quantification investigations by Cacuci and Fang (2023) have shown, as discussed in Chapter 1, that the most important of these functions are the 180 group-averaged total macroscopic cross sections, which are functions/features of the microscopic cross sections and isotopic number densities, which in turn are uncertain primary model parameters. The methodology for computing exact expressions of response sensitivities to functions/features of the primary model parameters was conceived by Cacuci (2023f, 2024a, 2024b, 2024c, 2024d, 2024e). This chapter will present the "1st-Order Functions/Features Adjoint Sensitivity Analysis Methodology for Nonlinear Systems" (1st-FASAM-N), which enables the most efficient computation of exactly obtained expressions of first-order sensitivities of model responses to functions/features of model parameters.

This chapter is structured as follows: Section 2.2 presents the general mathematical modeling of a generic nonlinear system comprising functions/features of uncertain parameters and boundaries. Section 2.3 introduces the mathematical framework of the "First-Order Function/Feature Adjoint Sensitivity Analysis Methodology for Nonlinear Systems" (1st-FASAM-N). Section 2.4 illustrates the application of the 1st-FASAM-N to determine the 1st-order sensitivities of a "dosimetry" detector response for a paradigm model of transport of particles (neutrons and/or gamma rays) through a shielding material. This benchmark is representative of large-scale computations, such as those performed for the OECD/NEA reactor physics benchmark mentioned above. This paradigm particle transport model admits closed-form analytical solutions for the sensitivities of the model's response with respect to the model's "feature" functions, thereby highlighting the efficiency of the 1st-FASAM-N to obtain exact expressions for these sensitivities and compute the respective expressions most efficiently. Comparisons with the 1st-CASAM-N (Cacuci 2022a, 2023a) are also provided, illustrating the transition/connection from the 1st-FASAM-N (which provides exact sensitivities to functions/features of parameters) to the 1st-CASAM-N (which provides response sensitivities directly with respect to the primary parameters).

## 2.2 MATHEMATICAL MODELING OF A GENERIC NONLINEAR SYSTEM COMPRISING FUNCTIONS ("FEATURES") OF UNCERTAIN PARAMETERS AND BOUNDARIES

The mathematical model that underlies the numerical evaluation of a process and/or state of a physical system comprises equations that relate the system's independent variables and parameters to the system's state/dependent variables. These coupled equations, which are in general nonlinear, can be represented generically in operator form as follows:

$$\mathbf{N}[\mathbf{u}(\mathbf{x});\mathbf{x};\mathbf{g}(\alpha)] = \mathbf{Q}[\mathbf{x};\mathbf{g}(\alpha)], \quad \mathbf{x} \in \Omega_x(\mathbf{x}), \tag{2.1}$$

$$\mathbf{B}[\mathbf{u}(\mathbf{x});\mathbf{g}(\alpha);\ \lambda(\alpha);\omega(\alpha)] = \mathbf{C}[\mathbf{g}(\alpha);\ \lambda(\alpha);\omega(\alpha)], \quad \mathbf{x} \in \partial\Omega_x[\lambda(\alpha);\omega(\alpha)]. \tag{2.2}$$

The results computed using a mathematical model are customarily called "model responses" (or "system responses" or "objective functions" or "indices of performance"). As will be discussed under "item 9" in the list below, the sensitivity analysis of any model response can be performed by considering the following generic integral representation:

$$R[\mathbf{u}(\mathbf{x});\mathbf{f}(\alpha)] \triangleq \int_{\lambda_1(\alpha)}^{\omega_1(\alpha)} \cdots \int_{\lambda_{TI}(\alpha)}^{\omega_{TI}(\alpha)} S[\mathbf{u}(\mathbf{x});\mathbf{g}(\alpha);\mathbf{h}(\alpha);\mathbf{x}]dx_1 \ldots dx_{TI}, \tag{2.3}$$

where $S[\mathbf{u}(\mathbf{x});\mathbf{g}(\alpha);\mathbf{h}(\alpha);\mathbf{x}]$ is a suitably differentiable nonlinear function of the vectors $\mathbf{u}(\mathbf{x})$ and $\alpha$.

Without loss of generality, the quantities which appear in Eqs. (2.1)–(2.3) can be considered to be real-valued, having the following meanings:

1. Matrices are denoted using capital bold letters, while vectors will be denoted using either capital or lower-case bold letters. The symbol "$\triangleq$" will be used to denote "is defined as" or "is by definition equal to." Transposition will be indicated by a dagger $(\dagger)$ superscript. The equalities are considered to hold in the weak ("distributional") sense. The expressions in Eqs. (2.1)–(2.3), respectively, may contain "generalized functions/functionals," particularly Dirac-distributions and derivatives thereof.
2. The $TP$-dimensional column-vector $\alpha \triangleq (\alpha_1,\ldots,\alpha_{TP})^\dagger \in \mathbb{R}^{TP}$ represents the "vector of primary model parameters" and has components denoted as $\alpha_1,\ldots,\alpha_{TP}$, where $TP$ denotes the "total number of parameters" involved in the model under consideration. These model parameters usually stem from processes that are external to the system under consideration and are seldom, if ever, known precisely. The known characteristics of the model parameters usually include their nominal (expected/mean) values and, possibly, higher-order moments or cumulants (i.e., variance/covariances, skewness, kurtosis), which are usually determined from experimental data and/or processes external to the physical system under consideration. Occasionally, just the lower and the upper bounds may be known for some model parameters, expressed by inequality and/or equality constraints that delimit the ranges of the system's parameters. Without loss of generality, the imprecisely known model parameters can be considered to be real-valued scalar quantities. It is important to note that the components of the vector

$\boldsymbol{\alpha}$ not only include parameters that appear in Eqs. (2.1) and (2.2) but also include parameters that may specifically occur just in the definition of the model's response provided in Eq. (2.3). The nominal parameter values will be denoted as $\boldsymbol{\alpha}^0 \triangleq \left[\alpha_1^0,\ldots,\alpha_i^0,\ldots,\alpha_{TP}^0\right]^\dagger$; the superscript "0" will be used throughout this book to denote "nominal values."

3. The $TI$-dimensional column vector $\mathbf{x} \triangleq (x_1,\ldots,x_{TI})^\dagger \in \mathbb{R}^{TI}$ comprises the model's independent variables, denoted as $x_i, i = 1,\ldots,TI$; the sub/superscript "$TI$" denotes the "total number of independent variables." The vector $\mathbf{x} \in \mathbb{R}^{TI}$ is considered to be defined on a phase-space domain denoted as $\Omega_x(\mathbf{x}) \triangleq \{\lambda_i(\boldsymbol{\alpha}) \le x_i \le \omega_i(\boldsymbol{\alpha}); i = 1,\ldots,TI\}$, including the particular cases when $\lambda_i(\boldsymbol{\alpha}) = -\infty, \omega_i(\boldsymbol{\alpha}) = \infty$ for some independent variables $x_i$, $i = 1,\ldots,TI$. The domain boundary $\partial\Omega_x[\boldsymbol{\lambda}(\boldsymbol{\alpha});\boldsymbol{\omega}(\boldsymbol{\alpha})] \triangleq \{\lambda_i(\boldsymbol{\alpha}) \cup \omega_i(\boldsymbol{\alpha}), i = 1,\ldots,TI\}$ of $\Omega_x[\mathbf{f}(\boldsymbol{\alpha})]$ comprises the set of all endpoints $\lambda_i(\boldsymbol{\alpha}), \omega_i(\boldsymbol{\alpha}), i = 1,\ldots,TI$. For subsequent mathematical developments, it is convenient to consider that the endpoints, $\lambda_i(\boldsymbol{\alpha}), \omega_i(\boldsymbol{\alpha}), i = 1,\ldots,TI$, are components of column vectors $\boldsymbol{\lambda}(\boldsymbol{\alpha}) \triangleq [\lambda_1(\boldsymbol{\alpha}),\ldots,\lambda_{TI}(\boldsymbol{\alpha})]^\dagger$ and $\boldsymbol{\omega}(\boldsymbol{\alpha}) \triangleq [\omega_1(\boldsymbol{\alpha}),\ldots,\omega_{TI}(\boldsymbol{\alpha})]^\dagger$, respectively. These endpoints depend on the physical system's geometrical dimensions, which may be imprecisely known because of manufacturing tolerances, and are considered therefore to be functions of the vector $\boldsymbol{\alpha} \triangleq (\alpha_1,\ldots,\alpha_{TP})^\dagger \in \mathbb{R}^{TP}$ of primary model parameters. Furthermore, the boundary-endpoints, $\lambda_i(\boldsymbol{\alpha}), \omega_i(\boldsymbol{\alpha}), i = 1,\ldots,TI$, may also depend on the parameters that define the material properties of the respective medium. For example, in models based on diffusion theory, the boundary conditions for materials facing air/vacuum are imposed on a physics-based mathematical construct called the "extrapolated boundary" of the respective spatial domain. The "extrapolated boundary" depends both on the imprecisely known physical dimensions of the system's materials and also on the materials' properties, such as atomic number densities and microscopic transport cross sections. Therefore, the boundary end-points can be considered, in general, to be functions of (some of) the primary model parameters.
4. The $TD$-dimensional column vector $\mathbf{u}(\mathbf{x}) \triangleq [u_1(\mathbf{x}),\ldots,u_{TD}(\mathbf{x})]^\dagger$ comprises the model's dependent variables (also called "state functions") $u_i(\mathbf{x}), i = 1,\ldots,TD$; the abbreviation "$TD$" denotes "total number of dependent variables."
5. The vector $\mathbf{g}(\boldsymbol{\alpha}) \triangleq [g_1(\boldsymbol{\alpha}),\ldots,g_{TG}(\boldsymbol{\alpha})]$ is a $TG$-dimensional vector having components $g_i(\boldsymbol{\alpha}), i = 1,\ldots,TG$, which are real-valued functions of (some of) the primary model parameters $\boldsymbol{\alpha} \in \mathbb{R}^{TP}$. Such functions customarily appear in models in the form of correlations that describe "features" of the system under consideration, such as material properties, flow regimes, etc. Usually, the number of functions $g_i(\boldsymbol{\alpha})$ is considerably smaller than the total number of model parameters, i.e., $TG \ll TP$. For example, the numerical model (Cacuci and Fang 2023) of the OECD/NEA reactor physics benchmark discussed in Chapter 1 comprises 21,976 uncertain primary model parameters (including microscopic cross sections and isotopic

number densities) but the neutron transport equation, which is solved to determine the neutron flux distribution within the benchmark, does not use these parameters directly but instead uses "group-averaged macroscopic cross sections" – which are functions/features of the microscopic cross sections and isotopic number densities (which in turn are uncertain quantities that would be components of the vector of primary model parameters). In particular, a component $g_j(\boldsymbol{\alpha})$ may simply be one of the primary model parameters $\alpha_j$, i.e., $g_j(\boldsymbol{\alpha}) \equiv \alpha_j$.

6. The $TD$-dimensional column vector $\mathbf{N}[\mathbf{u}(\mathbf{x});\mathbf{x};\mathbf{g}(\boldsymbol{\alpha})] \triangleq (N_1,\ldots,N_{TD})^\dagger$ comprises components $N_i[\mathbf{u}(\mathbf{x});\mathbf{x};\mathbf{g}(\boldsymbol{\alpha})], i=1,\ldots,TD$, which are operators (including differential, difference, integral, distributions, and/or finite or infinite matrices) acting nonlinearly (in general) on the dependent variables $\mathbf{u}(\mathbf{x})$, the independent variables $\mathbf{x}$ and on the functions $\mathbf{g}(\boldsymbol{\alpha})$ of model parameters $\boldsymbol{\alpha}$.
7. The $TD$-dimensional column vector $\mathbf{Q}[\mathbf{u}(\mathbf{x});\mathbf{x};\mathbf{g}(\boldsymbol{\alpha})] \triangleq (q_1,\ldots,q_{TD})^\dagger$, having components $q_i[\mathbf{u}(\mathbf{x});\mathbf{x};\mathbf{g}(\boldsymbol{\alpha})], i=1,\ldots,TD$, denotes inhomogeneous source terms, which usually depend nonlinearly on the uncertain parameters $\boldsymbol{\alpha}$.
8. The components of $\mathbf{B}[\mathbf{u}(\mathbf{x});\mathbf{g}(\boldsymbol{\alpha});\mathbf{x}]$ are nonlinear operators defined on the system's boundary $\partial\Omega_x$, while the components of $\mathbf{C}[\mathbf{x},\mathbf{g}(\boldsymbol{\alpha})]$ represent inhomogeneous boundary sources defined on the boundary $\partial\Omega_x$.
9. The integral representation of the response provided in Eq. (2.3) can represent "averaged" as well as "point-valued" quantities in the phase-space of independent variables. For example, if $R[\mathbf{u}(\mathbf{x});\mathbf{f}(\boldsymbol{\alpha})]$ represents the computation or the measurement (which would be a "detector-response") of a quantity of interest at a point $\mathbf{x}_d$ in the phase-space of independent variables, then $S[\mathbf{u}(\mathbf{x});\mathbf{g}(\boldsymbol{\alpha});\mathbf{h}(\boldsymbol{\alpha});\mathbf{x}]$ would contain a Dirac-delta functional of the form $\delta(\mathbf{x}-\mathbf{x}_d)$. Responses that represent "differentials/derivatives of quantities" would contain derivatives of Dirac-delta functionals in the definition of $S[\mathbf{u}(\mathbf{x});\mathbf{g}(\boldsymbol{\alpha});\mathbf{h}(\boldsymbol{\alpha});\mathbf{x}]$. The vector $\mathbf{h}(\boldsymbol{\alpha}) \triangleq [h_1(\boldsymbol{\alpha}),\ldots,h_{TH}(\boldsymbol{\alpha})]$, having components $h_i(\boldsymbol{\alpha}), i=1,\ldots,TH$, which appears among the arguments of the function $S[\mathbf{u}(\mathbf{x});\mathbf{g}(\boldsymbol{\alpha});\mathbf{h}(\boldsymbol{\alpha});\mathbf{x}]$, represents functions of primary parameters that often appear solely in the definition of the response but do not appear in the mathematical definition of the model represented by Eqs. (2.1) and (2.2). The quantity $TH$ denotes the total number of such functions that appear exclusively in the definition of the model's response $R[\mathbf{u}(\mathbf{x});\mathbf{f}(\boldsymbol{\alpha})]$. Evidently, the response will depend directly and/or indirectly (through the "feature"-functions) on all of the primary model parameters. This fact has been indicated in Eq. (2.3) by using the vector-valued function $\mathbf{f}(\boldsymbol{\alpha})$ as an argument in the definition of the response $R[\mathbf{u}(\mathbf{x});\mathbf{f}(\boldsymbol{\alpha})]$ to represent the concatenation of all of the "features" of the model and response under consideration. The vector $\mathbf{f}(\boldsymbol{\alpha})$ is thus defined as follows:

$$\mathbf{f}(\boldsymbol{\alpha}) \triangleq [\mathbf{g}(\boldsymbol{\alpha});\mathbf{h}(\boldsymbol{\alpha});\boldsymbol{\lambda}(\boldsymbol{\alpha});\boldsymbol{\omega}(\boldsymbol{\alpha})]^\dagger \triangleq [f_1(\boldsymbol{\alpha}),\ldots,f_{TF}(\boldsymbol{\alpha})]^\dagger; \quad TF \triangleq TG+TH+2TI. \tag{2.4}$$

The response $R[\mathbf{u}(\mathbf{x});\mathbf{f}(\boldsymbol{\alpha})]$ defined in Eq. (2.3) is a functional (i.e., a scalar-valued function) of the dependent variables $\mathbf{u}(\mathbf{x})$. The reason for considering model responses that are functionals of $\mathbf{u}(\mathbf{x})$ is because the differentials of these response will be expressed in terms of adjoint functions to be introduced via an "inner product" which is itself a functional (i.e., a scalar-valued function) of the variation $\delta\mathbf{u}(\mathbf{x})$, as will be shown in the remainder of this chapter. Reponses that are functions (rather than functionals) of $\mathbf{u}(\mathbf{x})$ are treated by expanding them in a generalized Fourier series over the domain of definition $\Omega_x(\mathbf{x}) \triangleq \{\lambda_i(\boldsymbol{\alpha}) \le x_i \le \omega_i(\boldsymbol{\alpha});\, i = 1,\ldots,TI\}$ and subsequently treating the coefficients of this generalized Fourier series, which will be functionals of $\mathbf{u}(\mathbf{x})$, as "model responses."

Solving Eqs. (2.1) and (2.2) at the nominal parameter values, $\boldsymbol{\alpha}^0 \triangleq [\alpha_1^0,\ldots,\alpha_i^0,\ldots,\alpha_{TP}^0]^\dagger$, provides the "nominal solution" $\mathbf{u}^0(\mathbf{x})$, i.e., the vectors $\mathbf{u}^0(\mathbf{x})$ and $\boldsymbol{\alpha}^0$ satisfy the following equations:

$$\mathbf{N}\left[\mathbf{u}^0(\mathbf{x});\mathbf{g}(\boldsymbol{\alpha}^0)\right] = \mathbf{Q}\left[\mathbf{x},\mathbf{g}(\boldsymbol{\alpha}^0)\right], \quad \mathbf{x} \in \Omega_x(\mathbf{x}), \tag{2.5}$$

$$\begin{gathered}\mathbf{B}\left[\mathbf{u}^0(\mathbf{x});\mathbf{g}(\boldsymbol{\alpha}^0);\boldsymbol{\lambda}(\boldsymbol{\alpha}^0);\boldsymbol{\omega}(\boldsymbol{\alpha}^0)\right] = \mathbf{C}\left[\mathbf{g}(\boldsymbol{\alpha}^0);\boldsymbol{\lambda}(\boldsymbol{\alpha}^0);\boldsymbol{\omega}(\boldsymbol{\alpha}^0)\right],\\ \mathbf{x} \in \partial\Omega_x\left[\boldsymbol{\lambda}(\boldsymbol{\alpha}^0);\boldsymbol{\omega}(\boldsymbol{\alpha}^0)\right].\end{gathered} \tag{2.6}$$

Using the nominal parameter values $\boldsymbol{\alpha}^0$ together with the "nominal solution" $\mathbf{u}^0(\mathbf{x})$ in Eq. (2.3) yields the nominal value of the response, namely:

$$R^0 \triangleq R\left[\mathbf{u}^0(\mathbf{x});\mathbf{f}(\boldsymbol{\alpha}^0)\right] \triangleq \int_{\lambda_1(\boldsymbol{\alpha}^0)}^{\omega_1(\boldsymbol{\alpha}^0)} \cdots \int_{\lambda_{TI}(\boldsymbol{\alpha}^0)}^{\omega_{TI}(\boldsymbol{\alpha}^0)} S\left[\mathbf{u}^0(\mathbf{x});\mathbf{g}(\boldsymbol{\alpha}^0);\mathbf{h}(\boldsymbol{\alpha}^0);\mathbf{x}\right] dx_1 \ldots dx_{TI}. \tag{2.7}$$

As defined in Eq. (2.3), the model response $R[\mathbf{u}(\mathbf{x});\mathbf{f}(\boldsymbol{\alpha})]$ can depend on the components $f_j(\boldsymbol{\alpha})$, $j = 1,\ldots,TF$, of the feature function $\mathbf{f}(\boldsymbol{\alpha})$ both directly and indirectly, through the components of the function $\mathbf{u}(\mathbf{x})$, which is the solution of Eqs. (2.1) and (2.2). The model response is considered to be sufficiently differentiable to admit a Taylor-series expansion around the nominal value $\mathbf{f}^0 \triangleq \mathbf{f}(\boldsymbol{\alpha}^0)$, having the following form, where the implicit dependence of $\mathbf{u}(\mathbf{x})$ on $\mathbf{f}(\boldsymbol{\alpha})$ is not explicitly shown, for simplicity:

$$R[\mathbf{f}(\boldsymbol{\alpha})] = R(\mathbf{f}^0) + \sum_{j1=1}^{TF}\left\{\frac{\partial R(\mathbf{f})}{\partial f_{j1}}\right\}_{\mathbf{f}^0}\delta f_{j1} + \frac{1}{2}\sum_{j1=1}^{TF}\sum_{j2=1}^{TF}\left\{\frac{\partial^2 R(\mathbf{f})}{\partial f_{j1}\,\partial f_{j2}}\right\}_{\mathbf{f}^0}\delta f_{j1}\delta f_{j2} + \cdots, \tag{2.8}$$

where $\delta f_j \triangleq [f_j(\boldsymbol{\alpha}) - f_j^0]$; $f_j^0 \triangleq f_j(\boldsymbol{\alpha}^0)$; $j = 1,\ldots,TF$. The "sensitivities of the model response with respect to the feature functions" are naturally defined as being the *functional derivatives* of $R[\mathbf{f}(\boldsymbol{\alpha})]$ with respect to the components ("features") $f_j(\boldsymbol{\alpha})$ of $\mathbf{f}(\boldsymbol{\alpha})$. Since $TF \ll TP$, the computations of the functional derivatives of $R[\mathbf{f}(\boldsymbol{\alpha})]$

with respect to the functions $f_j(\boldsymbol{\alpha})$, which appear in Eq. (2.8), will be considerably less expensive computationally than the computation of the functional derivatives involved in the Taylor series of the response with respect to the model parameters. The functional derivatives of the response with respect to the parameters can be obtained from the functional derivatives of the response with respect to the "feature" functions $f_j(\boldsymbol{\alpha})$ by simply using the chain rule, i.e.:

$$\left\{\frac{\partial R(\boldsymbol{\alpha})}{\partial \alpha_{j1}}\right\}_{\boldsymbol{\alpha}^0} = \sum_{i_1=1}^{TF}\left\{\frac{\partial R(\mathbf{f})}{\partial f_{i1}}\frac{\partial f_{i_1}(\boldsymbol{\alpha})}{\partial \alpha_{j1}}\right\}_{\boldsymbol{\alpha}^0}; \quad \left\{\frac{\partial^2 R(\boldsymbol{\alpha})}{\partial \alpha_{j1}\partial \alpha_{j2}}\right\}_{\boldsymbol{\alpha}^0} = \frac{\partial}{\partial \alpha_{j2}}\sum_{i_1=1}^{TF}\left\{\frac{\partial R(\mathbf{f})}{\partial f_{i1}}\frac{\partial f_{i_1}(\boldsymbol{\alpha})}{\partial \alpha_{j1}}\right\}_{\boldsymbol{\alpha}^0}; \tag{2.9}$$

and so on. Since the functional forms of the functions $f_j(\boldsymbol{\alpha})$ are "user-chosen" and are therefore perfectly well-known in closed-form, the evaluation/computation of the functional derivatives $\partial f_{i1}(\boldsymbol{\alpha})/\partial \alpha_{j1}$, $\partial^2 f_{i1}(\boldsymbol{\alpha})/\partial \alpha_{j1}\partial \alpha_{j2}$, etc., does not require computations involving the model and is therefore trivial by comparison to the evaluation of the functional derivatives ("sensitivities") of the response with respect to either the functions ("features") $f_j(\boldsymbol{\alpha})$ or the model parameters $\alpha_i$, $i = 1,\ldots,TP$.

The range of *validity* of the Taylor series shown in Eq. (2.8) is defined by its radius of convergence. The *accuracy* – as opposed to the "validity"– of the Taylor series in predicting the value of the response at an arbitrary point in the phase-space of model parameters depends on the order of sensitivities retained in the Taylor-series expansion: the higher the respective order, the more accurate the respective response value predicted by the Taylor series. In the particular cases when the response happens to be a polynomial function of the "feature" functions $f_j(\boldsymbol{\alpha})$, the Taylor series is actually exact.

In turn, the functions $f_j(\boldsymbol{\alpha})$ can also be formally expanded in a multivariate Taylor series around the nominal (mean) parameter values $\boldsymbol{\alpha}^0$, namely:

$$\begin{aligned} f_i(\boldsymbol{\alpha}) &= f_i(\boldsymbol{\alpha}^0) + \sum_{j1=1}^{TP}\left\{\frac{\partial f_i(\boldsymbol{\alpha})}{\partial \alpha_{j1}}\right\}_{\boldsymbol{\alpha}^0}\delta\alpha_{j1} + \frac{1}{2}\sum_{j1=1}^{TP}\sum_{j2=1}^{TP}\left\{\frac{\partial^2 f_i(\boldsymbol{\alpha})}{\partial \alpha_{j1}\partial \alpha_{j2}}\right\}_{\boldsymbol{\alpha}^0}\delta\alpha_{j1}\delta\alpha_{j2} \\ &+ \frac{1}{3!}\sum_{j_1=1}^{TP}\sum_{j_2=1}^{TP}\sum_{j_3=1}^{TP}\left\{\frac{\partial^3 f_i(\boldsymbol{\alpha})}{\partial \alpha_{j_1}\partial \alpha_{j_2}\partial \alpha_{j_3}}\right\}_{\boldsymbol{\alpha}^0}\delta\alpha_{j1}\delta\alpha_{j2}\delta\alpha_{j3} + \ldots, \end{aligned} \tag{2.10}$$

The domain of validity of the Taylor series in Eq. (2.10) is defined by its own radius of convergence.

In practice, the model parameters are not known exactly. Even though these parameters are not bona fide random quantities, they are formally considered to be variates that obey a multivariate probability distribution function, denoted as $p_\alpha(\boldsymbol{\alpha})$, which is usually unknown. The moments of $p_\alpha(\boldsymbol{\alpha})$ are defined in a standard manner. The formal definition for the "expected (or mean) value" of a function, $u(\boldsymbol{\alpha})$, of the parameters $\boldsymbol{\alpha}$, which is defined over the domain of definition of $p_\alpha(\boldsymbol{\alpha})$, is as follows:

$$E[u(\boldsymbol{\alpha})] \triangleq \int u(\boldsymbol{\alpha})\, p_\alpha(\boldsymbol{\alpha})\, d\boldsymbol{\alpha}. \tag{2.11}$$

In particular, the vector of expected values, denoted as $\alpha_j^0$, of the model parameters $\alpha_j$ is defined as follows:

$$\boldsymbol{\alpha}^0 \triangleq \left(\alpha_1^0,\ldots,\alpha_{TP}^0\right)^\dagger, \quad \alpha_j^0 \triangleq E\left(\alpha_j\right), \quad j=1,\ldots,TP. \tag{2.12}$$

The covariances, $\operatorname{cov}\left(\alpha_i,\alpha_j\right)$, between two parameters, $\alpha_i$ and $\alpha_j$, are defined as follows:

$$\operatorname{cov}\left(\alpha_i,\alpha_j\right) \triangleq E\left(\delta\alpha_i\delta\alpha_j\right) \triangleq \rho_{ij}\sigma_i\sigma_j, \quad i,j=1,\ldots,TP. \tag{2.13}$$

In Eq. (2.13), the quantities $\sigma_i$ and $\sigma_j$ denote the standard deviations of $\alpha_i$ and $\alpha_j$, respectively, while $\rho_{ij}$ denotes the correlation between the respective parameters.

The third-order correlation, $t_{j_1 j_2 j_3}$, among three parameters is defined as follows:

$$t_{j_1 j_2 j_3}\sigma_{j_1}\sigma_{j_2}\sigma_{j_3} \triangleq E\left(\delta\alpha_{j_1}\delta\alpha_{j_2}\delta\alpha_{j_3}\right); \quad j_1,j_2,j_3=1,\ldots,TP. \tag{2.14}$$

The fourth-order correlation $q_{j_1 j_2 j_3 j_4}$ among four parameters is defined as follows:

$$q_{j_1 j_2 j_3 j_4}\sigma_{j_1}\sigma_{j_2}\sigma_{j_3}\sigma_{j_4} \triangleq E\left(\delta\alpha_{j_1}\delta\alpha_{j_2}\delta\alpha_{j_3}\delta\alpha_{j_4}\right); \quad j_1,j_2,j_3,j_4=1,\ldots,TP. \tag{2.15}$$

In practice, all of the statistical properties of the model parameters are deduced from measurements and/or computations performed to determine the respective parameters, prior to using them in a computational model.

In terms of the moments of the parameter distribution, the moments of the distribution of the function $f_i(\boldsymbol{\alpha})$ in the phase-space of model parameters are obtained by using the Taylor series provided in Eq. (2.10) in conjunction with the definitions provided in Eqs. (2.12)–(2.15) to obtain the following expressions:

i. The mean (or expected value) of $f_i(\boldsymbol{\alpha})$:

$$\begin{aligned} E\left[f_i(\boldsymbol{\alpha})\right] = {} & f_i\left(\boldsymbol{\alpha}^0\right) + \frac{1}{2}\sum_{j_1=1}^{TP}\sum_{j_2=1}^{TP}\left\{\frac{\partial^2 f_i(\boldsymbol{\alpha})}{\partial\alpha_{j_1}\,\partial\alpha_{j_2}}\right\}_{\boldsymbol{\alpha}^0}\rho_{j_1 j_2}\sigma_{j_1}\sigma_{j_2} \\ & + \frac{1}{3!}\sum_{j_1=1}^{TP}\sum_{j_2=1}^{TP}\sum_{j_3=1}^{TP}\left\{\frac{\partial^3 f_i(\boldsymbol{\alpha})}{\partial\alpha_{j_1}\,\partial\alpha_{j_2}\,\partial\alpha_{j_3}}\right\}_{\boldsymbol{\alpha}^0} t_{j_1 j_2 j_3}\sigma_{j_1}\sigma_{j_2}\sigma_{j_3} + \ldots, \end{aligned} \tag{2.16}$$

ii. the covariance $\operatorname{cov}\left(f_k,f_\ell\right)$ of two functions $f_k(\boldsymbol{\alpha})$ and $f_\ell(\boldsymbol{\alpha})$:

$$\operatorname{cov}\left(f_k,f_\ell\right) = \sum_{j_1=1}^{TP}\sum_{j_2=1}^{TP}\left\{\frac{\partial f_k(\boldsymbol{\alpha})}{\partial\alpha_{j_1}}\frac{\partial f_\ell(\boldsymbol{\alpha})}{\partial\alpha_{j_2}}\right\}_{\boldsymbol{\alpha}^0}\rho_{j_1 j_2}\sigma_{j_1}\sigma_{j_2} + \cdots. \tag{2.17}$$

Higher-order moments of the distribution of $\mathbf{f}(\boldsymbol{\alpha})$ in terms of the moments of the distribution of the model parameters $\boldsymbol{\alpha}$ are obtained similarly.

Since the parameters $\boldsymbol{\alpha}$ are affected by uncertainties, these uncertainties will induce uncertainties in the components of the feature function $\mathbf{f}(\boldsymbol{\alpha})$, which in turn will induce uncertainties in the function $\mathbf{u}(\mathbf{x})$, which will ultimately induce, collectively, uncertainties in the response $R[\mathbf{u}(\mathbf{x});\mathbf{f}(\boldsymbol{\alpha})]$. Thus, in terms of the moments of the distribution of $\mathbf{f}(\boldsymbol{\alpha})$, the moments of $R[\mathbf{f}(\boldsymbol{\alpha})]$ are obtained by using the Taylor series shown in Eq. (2.8). Thus, the expectation value $E[R(\mathbf{f})]$ will have the following expression, where the implicit dependence of $\mathbf{u}(\mathbf{x})$ on $\mathbf{f}(\boldsymbol{\alpha})$ is not explicitly shown, for simplicity:

$$E[R(\mathbf{f})] = R(\mathbf{f}^0) + \frac{1}{2}\sum_{j_1=1}^{TF}\sum_{j_2=1}^{TF}\left\{\frac{\partial^2 R(\mathbf{f})}{\partial f_{j_1}\,\partial f_{j_2}}\right\}_{\mathbf{f}^0} \operatorname{cov}(f_{j_1}, f_{j_2}) + \cdots. \tag{2.18}$$

Using the Taylor expansions of two responses, denoted as $R_k[\mathbf{f}(\boldsymbol{\alpha})]$ and $R_\ell(\mathbf{f})$, will yield the following expression for the covariance, $\operatorname{cov}[R_k(\mathbf{f}), R_\ell(\mathbf{f})]$, of these two responses:

$$\operatorname{cov}[R_k(\mathbf{f}), R_\ell(\mathbf{f})] = \sum_{j_1=1}^{TF}\sum_{j_2=1}^{TF}\left\{\frac{\partial R_k(\mathbf{f})}{\partial f_{j_1}}\frac{\partial R_\ell(\mathbf{f})}{\partial f_{j_2}}\right\}_{\boldsymbol{\alpha}^0} \operatorname{cov}(f_{j_1}, f_{j_2}) + \cdots. \tag{2.19}$$

Higher moments of the distribution of responses are also obtained using their corresponding Taylor expansions, as discussed in Chapter 1. The remainder of this chapter will present the most efficient methodology, called the "First-Order Function/Feature Adjoint Sensitivity Analysis Methodology for Nonlinear Systems (1st-FASAM-N)", for obtaining and computing exact expressions of the first-order sensitivities of a model response with respect to the components $f_i(\boldsymbol{\alpha})$, $i = 1,\dots,TF$, of the feature function $\mathbf{f}(\boldsymbol{\alpha})$.

## 2.3 INTRODUCING THE 1ST-FASAM-N: FIRST-ORDER FUNCTION/FEATURE ADJOINT SENSITIVITY ANALYSIS METHODOLOGY FOR NONLINEAR SYSTEMS

The "First-Order Function/Feature Adjoint Sensitivity Analysis Methodology for Nonlinear Systems" (1st-FASAM-N), which will be developed in this section, will enable the most efficient computation of model response sensitivities to the components of the "features" function $\mathbf{f}(\boldsymbol{\alpha})$ of model parameters. The nominal (or mean) parameter values, $\boldsymbol{\alpha}^0$, are considered to be known, but these values will differ from the true but unknown values, $\boldsymbol{\alpha}$, by variations $\delta\boldsymbol{\alpha} \triangleq (\delta\alpha_1,\dots,\delta\alpha_{TP})^\dagger$, where $\delta\alpha_i \triangleq \alpha_i - \alpha_i^0$. The parameter variations $\delta\boldsymbol{\alpha}$ will induce variations $\delta\mathbf{f}(\boldsymbol{\alpha}) \triangleq [\delta f_1(\boldsymbol{\alpha}),\dots,\delta f_{TF}(\boldsymbol{\alpha})]^\dagger$ in the vector-valued function $\mathbf{f}(\boldsymbol{\alpha})$ and will also induce variations $\mathbf{v}^{(1)}(\mathbf{x}) \triangleq [\delta u_1(\mathbf{x}),\dots,\delta u_{TD}(\mathbf{x})]^\dagger$ around the nominal solution $\mathbf{u}^0(\mathbf{x})$ in the forward state functions, since the forward state functions $\mathbf{u}(\mathbf{x})$ are related to $\mathbf{f}(\boldsymbol{\alpha})$ through Eqs. (2.1) and (2.2). In turn, the variations $\delta\mathbf{f}(\boldsymbol{\alpha})$ and $\mathbf{v}^{(1)}(\mathbf{x})$ will induce variations $\delta R[\mathbf{u}^0(\mathbf{x});\mathbf{f}^0(\boldsymbol{\alpha});\mathbf{v}^{(1)}(\mathbf{x});\delta\mathbf{f}(\boldsymbol{\alpha})]$ in the system's response $R[\mathbf{u}(\mathbf{x});\mathbf{f}(\boldsymbol{\alpha})]$.

Mathematically, the quantity $\delta R\left[\mathbf{u}^0(\mathbf{x});\mathbf{f}^0(\alpha);\mathbf{v}^{(1)}(\mathbf{x});\delta\mathbf{f}(\alpha)\right]$ is the first-order Gateaux ($G$)-variation of the response $R[\mathbf{u}(\mathbf{x});\mathbf{f}(\alpha)]$, induced by arbitrary variations $\left[\mathbf{v}^{(1)}(\mathbf{x});\delta\mathbf{f}(\alpha)\right]$ in a neighborhood around the nominal functions and parameter values $\left[\mathbf{u}^0(\mathbf{x});\mathbf{f}^0(\alpha)\right]$ which is defined as follows:

$$\delta R\left[\mathbf{u}^0(\mathbf{x});\mathbf{f}^0(\alpha);\mathbf{v}^{(1)}(\mathbf{x});\delta\mathbf{f}(\alpha)\right]\triangleq\left\{\frac{d}{d\varepsilon}\mathbf{R}\left[\mathbf{u}^0+\varepsilon\mathbf{v}^{(1)};\mathbf{f}^0+\varepsilon(\delta\mathbf{f})\right]\right\}_{\varepsilon=0}, \quad (2.20)$$

where $\varepsilon$ is a scalar quantity. The G-variation $\delta R\left[\mathbf{u}^0(\mathbf{x});\mathbf{f}^0(\alpha);\mathbf{v}^{(1)}(\mathbf{x});\delta\mathbf{f}(\alpha)\right]$ is an operator defined on the same domain as $R[\mathbf{u}(\mathbf{x});\mathbf{f}(\alpha)]$, and having the same range as $R[\mathbf{u}(\mathbf{x});\mathbf{f}(\alpha)]$. The existence of the G-variation $\delta R\left[\mathbf{u}^0(\mathbf{x});\mathbf{f}^0(\alpha);\mathbf{v}^{(1)}(\mathbf{x});\delta\mathbf{f}(\alpha)\right]$ does not guarantee its numerical computability. Numerical methods most often require that $\delta R\left[\mathbf{u}^0(\mathbf{x});\mathbf{f}^0(\alpha);\mathbf{v}^{(1)}(\mathbf{x});\delta\mathbf{f}(\alpha)\right]$ be linear in $\left[\mathbf{v}^{(1)}(\mathbf{x});\delta\mathbf{f}(\alpha)\right]$ in a neighborhood $\left[\mathbf{u}^0+\varepsilon\mathbf{v}^{(1)};\mathbf{f}^0+\varepsilon(\delta\mathbf{f})\right]$ around $\left[\mathbf{u}^0(\mathbf{x});\mathbf{f}^0(\alpha)\right]$, thereby admitting a total first-order *G-differential*, which will be denoted as $dR\left[\mathbf{u}^0(\mathbf{x});\mathbf{f}^0(\alpha);\mathbf{v}^{(1)}(\mathbf{x});\delta\mathbf{f}(\alpha)\right]$. The necessary and sufficient conditions for a nonlinear operator $R[\mathbf{u}(\mathbf{x});\mathbf{f}(\alpha)]$ to admit a total G-differential in a neighborhood $\left[\mathbf{u}^0+\varepsilon\mathbf{v}^{(1)};\mathbf{f}^0+\varepsilon(\delta\mathbf{f})\right]$ around $\left[\mathbf{u}^0(\mathbf{x});\mathbf{f}^0(\alpha)\right]$ are as follows:

i. $R[\mathbf{u}(\mathbf{x});\mathbf{f}(\alpha)]$ satisfies a weak Lipschitz condition at $\left[\mathbf{u}^0(\mathbf{x});\mathbf{f}^0(\alpha)\right]$, namely:

$$\left\|R\left[\mathbf{u}^0+\varepsilon\mathbf{v}^{(1)};\mathbf{f}^0+\varepsilon(\delta\mathbf{f})\right]-R\left(\mathbf{u}^0;\mathbf{f}^0\right)\right\|\le k\left\|\varepsilon\left(\mathbf{u}^0;\mathbf{f}^0\right)\right\|,\quad k<\infty. \quad (2.21)$$

ii. $R[\mathbf{u}(\mathbf{x});\mathbf{f}(\alpha)]$ satisfies the following condition for a scalar $\varepsilon$ and vectors $\mathbf{v}_2$, $\mathbf{f}_2$:

$$\begin{aligned}&R\left[\mathbf{u}^0+\varepsilon\mathbf{v}^{(1)}+\varepsilon\mathbf{v}_2;\mathbf{f}^0+\varepsilon(\delta\mathbf{f})+\varepsilon\mathbf{f}_2\right]-R\left[\mathbf{u}^0+\varepsilon\mathbf{v}^{(1)};\mathbf{f}^0+\varepsilon(\delta\mathbf{f})\right]\\&-R\left[\mathbf{u}^0+\varepsilon\mathbf{v}_2;\mathbf{f}^0+\varepsilon\mathbf{f}_2\right]+R\left(\mathbf{u}^0;\mathbf{f}^0\right)=o(\varepsilon).\end{aligned} \quad (2.22)$$

In practice, the relations provided in Eqs. (2.21) and (2.22) are seldom used directly since the computation of the expression on the right side of Eq. (2.20) reveals immediately if the respective expression is linear (or not) in the vectors $\mathbf{v}^{(1)}(\mathbf{x})$ and/or $\delta\mathbf{f}(\alpha)$.

Numerical methods (e.g., Newton's method and variants thereof) for solving Eqs. (2.1) and (2.2) also require the existence of the first-order G-differential of the original model equations. Therefore, the conditions provided in Eqs. (2.21) and (2.22) are henceforth considered to be satisfied by the model responses and also by the operators underlying the physical system modeled by Eqs. (2.1)–(2.3), which implies that all of the operators/functions considered in this work admit G-differential.

When the 1st-order G-variation $\delta R\left[\mathbf{u}^0(\mathbf{x});\mathbf{f}^0(\alpha);\mathbf{v}^{(1)}(\mathbf{x});\delta\mathbf{f}(\alpha)\right]$ satisfies the conditions provided in Eqs. (2.21) and (2.22), it can be written as follows:

$$\delta R\left[\mathbf{u}^0(\mathbf{x});\mathbf{f}^0(\alpha);\mathbf{v}^{(1)}(\mathbf{x});\delta\mathbf{f}(\alpha)\right]\triangleq\left\{dR\left[\mathbf{u}(\mathbf{x});\mathbf{f}(\alpha);\delta\mathbf{f}\right]\right\}_{dir}+\left\{dR\left[\mathbf{u}(\mathbf{x});\mathbf{f}(\alpha);\mathbf{v}^{(1)}(\mathbf{x})\right]\right\}_{ind}$$

$$\triangleq\left\{\frac{d}{d\varepsilon}\int_{\lambda_1(\alpha^0)+\varepsilon(\delta\lambda_1)}^{\omega_1(\alpha^0)+\varepsilon(\delta\omega_1)}\cdots\int_{\lambda_{TI}(\alpha^0)+\varepsilon(\delta\lambda_{TI})}^{\omega_{TI}(\alpha^0)+\varepsilon(\delta\omega_{TI})}S\left[\mathbf{u}^0+\varepsilon\mathbf{v}^{(1)},\mathbf{g}^0+\varepsilon(\delta\mathbf{g});\mathbf{h}^0+\varepsilon(\delta\mathbf{h});\mathbf{x}\right]dx_1\ldots dx_{TI}\right\}_{\varepsilon=0}. \quad (2.23)$$

In Eq. (2.23), the "direct-effect" term, denoted as $\left\{dR\left[\mathbf{u}(\mathbf{x});\mathbf{f}(\alpha);\delta\mathbf{f}\right]\right\}_{dir}$, comprises only dependencies on $\delta\mathbf{f}(\alpha)$ and is defined as follows:

$$\left\{dR\left[\mathbf{u}(\mathbf{x});\mathbf{f};\delta\mathbf{f}\right]\right\}_{dir}\triangleq\left\{\frac{\partial\mathbf{R}(\mathbf{u};\mathbf{f})}{\partial\mathbf{f}}\right\}_{\alpha^0}\delta\mathbf{f}\triangleq\sum_{j_1=1}^{TF}\left\{R^{(1)}\left[j_1;\mathbf{u}(\mathbf{x});\mathbf{f}(\alpha)\right]\right\}_{dir}\delta f_{j_1}, \quad (2.24)$$

where:

$$\frac{\partial[\;]}{\partial\mathbf{f}}\delta\mathbf{f}\triangleq\sum_{i=1}^{TF}\frac{\partial[\;]}{\partial f_i}\delta f_i=\sum_{i=1}^{TG}\frac{\partial[\;]}{\partial g_i}\delta g_i+\sum_{i=1}^{TH}\frac{\partial[\;]}{\partial h_i}\delta h_i+\sum_{i=1}^{TI}\frac{\partial[\;]}{\partial\lambda_i}\delta\lambda_i+\sum_{i=1}^{TI}\frac{\partial[\;]}{\partial\omega_i}\delta\omega_i, \quad (2.25)$$

The components of the direct-effect term have the following expressions:

i. for $j_1=1,\ldots,TG$:

$$\left\{R^{(1)}\left[j_1;\mathbf{u}(\mathbf{x});\mathbf{f}(\alpha)\right]\right\}_{dir}\delta f_{j_1}\triangleq\left\{\int_{\lambda_1(\alpha)}^{\omega_1(\alpha)}\cdots\int_{\lambda_{TI}(\alpha)}^{\omega_{TI}(\alpha)}\frac{\partial S(\mathbf{u};\mathbf{g};\mathbf{h})}{\partial g_i}dx_1\ldots dx_{TI}\right\}_{\alpha^0}\delta g_i; \quad (2.26)$$

ii. for $j_1=TG+1,\ldots,TG+TH$:

$$\left\{R^{(1)}\left[j_1;\mathbf{u}(\mathbf{x});\mathbf{f}(\alpha)\right]\right\}_{dir}\delta f_{j_1}\triangleq\left\{\int_{\lambda_1(\alpha)}^{\omega_1(\alpha)}\cdots\int_{\lambda_{TI}(\alpha)}^{\omega_{TI}(\alpha)}\frac{\partial S(\mathbf{u};\mathbf{h})}{\partial h_i}dx_1\ldots dx_{TI}\right\}_{\alpha^0}\delta h_i, \quad (2.27)$$

*where* $i=j_1-TG$;

iii. for $j_1=TG+TH+1,\ldots,TG+TH+TI$:

$$\left\{R^{(1)}\left[j_1;\mathbf{u}(\mathbf{x});\mathbf{f}(\alpha)\right]\right\}_{dir}\delta f_{j_1}$$

$$\triangleq\left\{\int_{\lambda_1}^{\omega_1}dx_1\cdots\int_{\lambda_{i-1}}^{\omega_{i-1}}dx_{i-1}\int_{\lambda_{i+1}}^{\omega_{i+1}}dx_{i+1}\cdots\int_{\lambda_{TI}}^{\omega_{TI}}dx_{TI}S\left[\mathbf{u}(x_1,\ldots,\omega_i(\alpha),\ldots,x_{N_x});\mathbf{g};\mathbf{h}\right]\right\}_{\alpha^0}\delta\omega_i(\alpha),$$

*where* $i=j_1-TG-TH$;

(2.28)

iv. for $j_1 = TG + TH + TI + 1,\ldots,TG + TH + 2TI$:

$$\left\{R^{(1)}\left[j_1;\mathbf{u}(\mathbf{x});\mathbf{f}(\alpha)\right]\right\}_{dir}\delta f_{j_1}$$

$$\triangleq -\left\{\int_{\lambda_1}^{\omega_1} dx_1 \cdots \int_{\lambda_{i-1}}^{\omega_{i-1}} dx_{i-1} \int_{\lambda_{1+1}}^{\omega_{i+1}} dx_{i+1} \cdots \int_{\lambda_{TI}}^{\omega_{TI}} dx_{TI} S\left[\mathbf{u}\left(x_1,\ldots,\lambda_i(\alpha),\ldots,x_{N_x}\right);\mathbf{g};\mathbf{h}\right]\right\}_{\alpha^0} \delta\lambda_i,$$

$$where \quad i = j_1 - TG - TH - TI. \tag{2.29}$$

The notation on the left side of Eq. (2.25) represents an inner product between two vectors (so it comprises an implied multiplication of a vector with a transposed vector), but the "dagger (†), which indicates "transposition", has been omitted in order to keep the notation as simple as possible. "Daggers" indicating transposition will also be omitted in other inner products, whenever possible, while avoiding ambiguities.

The direct-effect term can be computed after having solved Eqs. (2.1) and (2.2) to obtain the nominal values, $\mathbf{u}^0(\mathbf{x})$, of the dependent variables. On the other hand, the quantity $\left\{dR\left[\mathbf{u}(\mathbf{x});\mathbf{f}(\alpha);\mathbf{v}^{(1)}(\mathbf{x})\right]\right\}_{ind}$ defined in Eq. (2.23) comprises only variations in the state functions and is therefore called the "indirect-effect term", being defined as follows:

$$\left\{dR\left[\mathbf{u}(\mathbf{x});\mathbf{f}(\alpha);\mathbf{v}^{(1)}(\mathbf{x})\right]\right\}_{ind} \triangleq \left\{\int_{\lambda_1(\alpha)}^{\omega_1(\alpha)} dx_1 \cdots \int_{\lambda_{TI}(\alpha)}^{\omega_{TI}(\alpha)} dx_{TI} \frac{\partial S(\mathbf{u};\mathbf{g};\mathbf{h})}{\partial \mathbf{u}} \mathbf{v}^{(1)}(\mathbf{x})\right\}_{\alpha^0}, \tag{2.30}$$

where

$$\frac{\partial[\ ]}{\partial \mathbf{u}}\mathbf{v}^{(1)}(\mathbf{x}) \triangleq \sum_{i=1}^{TD} \frac{\partial[\ ]}{\partial u_i(\mathbf{x})}\delta u_i(\mathbf{x}). \tag{2.31}$$

The "indirect-effect" term induces variations in the response through the variations in the state functions, which are, in turn, caused by the parameter variations through the equations underlying the model. The indirect-effect term can be quantified only after having determined the variations $\mathbf{v}^{(1)}(\mathbf{x})$ in terms of the variations $\delta\mathbf{g}$, $\delta\boldsymbol{\lambda}$, and $\delta\boldsymbol{\omega}$.

The first-order relationship between the vectors $\mathbf{v}^{(1)}(\mathbf{x})$ and the variations $\delta\mathbf{g}$, $\delta\boldsymbol{\lambda}$, and $\delta\boldsymbol{\omega}$ is determined by solving the equations obtained by applying the definition of the G-differential to Eqs. (2.1) and (2.2), which yields the following equations:

$$\left\{\frac{d}{d\varepsilon}\mathbf{N}\left[\mathbf{u}^0 + \varepsilon\mathbf{v}^{(1)}(\mathbf{x});\mathbf{g}(\alpha^0)+\varepsilon(\delta\mathbf{g})\right]\right\}_{\varepsilon=0} = \left\{\frac{d}{d\varepsilon}\mathbf{Q}\left[\mathbf{x},\mathbf{g}(\alpha^0)+\varepsilon(\delta\mathbf{g})\right]\right\}_{\varepsilon=0}, \tag{2.32}$$

$$\left\{\frac{d}{d\varepsilon}\mathbf{B}\left[\mathbf{u}^0+\varepsilon\mathbf{v}^{(1)}(\mathbf{x});\mathbf{g}\left(\boldsymbol{\alpha}^0\right)+\varepsilon(\delta\mathbf{g});\boldsymbol{\lambda}\left(\boldsymbol{\alpha}^0\right)+\varepsilon(\delta\boldsymbol{\lambda});\boldsymbol{\omega}\left(\boldsymbol{\alpha}^0\right)+\varepsilon(\delta\boldsymbol{\omega})\right]\right\}_{\varepsilon=0}$$

$$-\left\{\frac{d}{d\varepsilon}\mathbf{C}\left[\mathbf{g}\left(\boldsymbol{\alpha}^0\right)+\varepsilon(\delta\mathbf{g});\boldsymbol{\lambda}\left(\boldsymbol{\alpha}^0\right)+\varepsilon(\delta\boldsymbol{\lambda});\boldsymbol{\omega}\left(\boldsymbol{\alpha}^0\right)+\varepsilon(\delta\boldsymbol{\omega})\right]\right\}_{\varepsilon=0}=\mathbf{0}, \tag{2.33}$$

Carrying out the differentiations with respect to $\varepsilon$ in Eqs. (2.32) and (2.33), and setting $\varepsilon = 0$ in the resulting expressions yields the following equations:

$$\left\{\mathbf{N}^{(1)}(\mathbf{u};\mathbf{g})\mathbf{v}^{(1)}(\mathbf{x})\right\}_{\boldsymbol{\alpha}^0}=\left\{\mathbf{q}_V^{(1)}(\mathbf{u};\mathbf{g};\delta\mathbf{g})\right\}_{\boldsymbol{\alpha}^0}, \quad \mathbf{x}\in\Omega_x, \tag{2.34}$$

$$\left\{\mathbf{b}_V^{(1)}\left(\mathbf{u};\mathbf{g};\boldsymbol{\lambda};\boldsymbol{\omega};\mathbf{v}^{(1)};\delta\mathbf{g};\delta\boldsymbol{\lambda};\delta\boldsymbol{\omega}\right)\right\}_{\boldsymbol{\alpha}^0}=\mathbf{0}, \quad \mathbf{x}\in\partial\Omega_x. \tag{2.35}$$

In Eqs. (2.34) and (2.35), the superscript "(1)" indicates "1st-Level" and the various quantities which appear in these equations are defined as follows:

$$\mathbf{N}^{(1)}(\mathbf{u};\mathbf{g})\triangleq\left\{\frac{\partial\mathbf{N}(\mathbf{u};\mathbf{g})}{\partial\mathbf{u}}\right\}\triangleq\left\{\frac{\partial N_i}{\partial u_j}\right\}_{TD\times TD}; \tag{2.36}$$

$$\mathbf{q}_V^{(1)}(\mathbf{u};\mathbf{g};\delta\mathbf{g})\triangleq\frac{\partial\left[\mathbf{Q}(\mathbf{g})-\mathbf{N}(\mathbf{u};\mathbf{g})\right]}{\partial\mathbf{g}}\delta\mathbf{g}\triangleq\sum_{j_1=1}^{TG}\mathbf{s}_V^{(1)}(j_1;\mathbf{u};\mathbf{g})\delta g_{j_1}; \tag{2.37}$$

$$\begin{aligned}&\left\{\mathbf{b}_V^{(1)}\left(\mathbf{u};\mathbf{g};\boldsymbol{\lambda};\boldsymbol{\omega};\mathbf{v}^{(1)};\delta\mathbf{g};\delta\boldsymbol{\lambda};\delta\boldsymbol{\omega}\right)\right\}_{\boldsymbol{\alpha}^0}\\&\triangleq\left\{\frac{\partial\mathbf{B}(\mathbf{u};\mathbf{g};\boldsymbol{\lambda};\boldsymbol{\omega})}{\partial\mathbf{u}}\right\}_{\boldsymbol{\alpha}^0}\mathbf{v}^{(1)}+\left\{\frac{\partial\left[\mathbf{B}(\mathbf{u};\mathbf{g};\boldsymbol{\lambda};\boldsymbol{\omega})-\mathbf{C}(\mathbf{g};\boldsymbol{\lambda};\boldsymbol{\omega})\right]}{\partial\mathbf{g}}\delta\mathbf{g}\right.\\&\left.+\frac{\partial\left[\mathbf{B}(\mathbf{u};\mathbf{g};\boldsymbol{\lambda};\boldsymbol{\omega})-\mathbf{C}(\mathbf{g};\boldsymbol{\lambda};\boldsymbol{\omega})\right]}{\partial\boldsymbol{\lambda}}\delta\boldsymbol{\lambda}+\frac{\partial\left[\mathbf{B}(\mathbf{u};\mathbf{g};\boldsymbol{\lambda};\boldsymbol{\omega})-\mathbf{C}(\mathbf{g};\boldsymbol{\lambda};\boldsymbol{\omega})\right]}{\partial\boldsymbol{\omega}}\delta\boldsymbol{\omega}\right\}_{\boldsymbol{\alpha}^0}.\end{aligned} \tag{2.38}$$

The system of equations comprising Eqs. (2.34) and (2.35) is called the "1st-Level Variational Sensitivity System" (1st-LVSS). The solution, $\mathbf{v}^{(1)}(\mathbf{x})$, of the 1st-LVSS will be a function of the variations $\delta\mathbf{g}$, $\delta\boldsymbol{\lambda}$, and $\delta\boldsymbol{\omega}$. Hence, when the function $\mathbf{v}^{(1)}(\mathbf{x})$ is introduced into the expression of the indirect-effect term defined in Eq. (2.30), it will introduce dependencies of the response sensitivities on $\delta\mathbf{g}$, $\delta\boldsymbol{\lambda}$, and $\delta\boldsymbol{\omega}$, which will be in addition to the dependencies displayed by the direct-effect term defined in Eq. (2.24). As Eqs. (2.34) and (2.35) indicate, every parameter variation $\delta\alpha_{j_1}$, $j_1 = 1,\ldots,TP$, will induce, directly or indirectly, variations in the model's state variables. In principle, therefore, for every parameter variation $\delta\alpha_{j_1}$, $j_1 = 1,\ldots,TP$, there would correspond a solution $\mathbf{v}^{(1)}(j_1;\mathbf{x}),\ j_1 = 1,\dots,TP$, of the 1st-LVSS. Thus, if the effect of every parameter variation were of interest, then the 1st-LVSS would need to be solved $TP$ times, with distinct right sides and boundary conditions for each parameter variation $\delta\alpha_{j1}$, which would require at least $TP$ large-scale computations. If only variations induced by the functions $\delta\mathbf{g}$, $\delta\boldsymbol{\lambda}$, and/or $\delta\boldsymbol{\omega}$ were of interest, then fewer than $TP$ large-scale computations would be required.

However, solving the 1st-LVSS can be avoided altogether by using the ideas underlying the "adjoint sensitivity analysis methodology" as originally conceived by Cacuci (1981a) and subsequently generalized by Cacuci (2015, 2016, 2018, 2019b, 2021a, 2021b, 2022a, 2022b, 2023a) to enable the computation of arbitrarily high-order response sensitivities to model parameters. Thus, the need for computing the vectors, $\mathbf{v}^{(1)}(j_1;\mathbf{x}),\ j_1 = 1,\dots,TP$, is eliminated by expressing the indirect-effect term defined in Eq. (2.30) in terms of the solutions of the "1st-Level Adjoint Sensitivity System" (1st-LASS) that is constructed by introducing a (real) Hilbert space denoted as $\mathcal{H}_1(\Omega_x)$, endowed with an inner product of two vectors $\mathbf{w}^{(1)}(\mathbf{x}) \in \mathcal{H}_1$ and $\mathbf{w}^{(2)}(\mathbf{x}) \in \mathcal{H}_1$ denoted as $\left\langle \mathbf{w}^{(1)},\mathbf{w}^{(2)} \right\rangle_1$ and defined as follows:

$$\left\langle \mathbf{w}^{(1)},\mathbf{w}^{(2)} \right\rangle_1 \triangleq \left\{ \int_{\lambda_1(\alpha)}^{\omega_1(\alpha)} \cdots \int_{\lambda_{TI}(\alpha)}^{\omega_{TI}(\alpha)} \left[ \mathbf{w}^{(1)}(\mathbf{x}) \bullet \mathbf{w}^{(2)}(\mathbf{x}) \right] dx_1 \dots dx_{TI} \right\}_{\alpha^0} . \tag{2.39}$$

In Eq. (2.39), the "dagger" (†), which indicates "transposition," has been omitted to simplify the notation for the scalar product $\mathbf{w}^{(1)}(\mathbf{x}) \bullet \mathbf{w}^{(2)}(\mathbf{x}) \triangleq \sum_{i=1}^{TD} w_i^{(1)}(\mathbf{x})\, w_i^{(2)}(\mathbf{x})$.

The 1st-LASS is now constructed by considering a vector $\mathbf{a}^{(1)}(\mathbf{x}) \in \mathcal{H}_1$, which is an element in $\mathcal{H}_1(\Omega_x)$ but is otherwise arbitrary at this stage, and by using Eq. (2.39) to form the inner product of $\mathbf{a}^{(1)}(\mathbf{x}) \in \mathcal{H}_1$ with the relation provided in Eq. (2.34) to obtain:

$$\left\{ \left\langle \mathbf{a}^{(1)}, \mathbf{N}^{(1)}(\mathbf{u};\mathbf{g})\,\mathbf{v}^{(1)} \right\rangle_1 \right\}_{\alpha^0} = \left\{ \left\langle \mathbf{a}^{(1)}, \mathbf{q}_V^{(1)}(\mathbf{u};\mathbf{g};\delta\mathbf{g}) \right\rangle_1 \right\}_{\alpha^0}, \quad \mathbf{x} \in \Omega_x. \tag{2.40}$$

The left side of Eq. (2.40) is transformed by using the definition of the adjoint operator in $\mathcal{H}_1(\Omega_x)$, as follows:

$$\left\{ \left\langle \mathbf{a}^{(1)}, \mathbf{N}^{(1)}(\mathbf{u};\mathbf{g})\,\mathbf{v}^{(1)} \right\rangle_1 \right\}_{\alpha^0} = \left\{ \left\langle \mathbf{A}^{(1)}(\mathbf{u};\mathbf{g})\,\mathbf{a}^{(1)}, \mathbf{v}^{(1)} \right\rangle_1 \right\}_{\alpha^0} + \left\{ \left[ P^{(1)}\left(\mathbf{u};\mathbf{g};\boldsymbol{\lambda};\boldsymbol{\omega};\mathbf{v}^{(1)};\mathbf{a}^{(1)}\right) \right]_{\partial\Omega_x} \right\}_{\alpha^0}, \tag{2.41}$$

where $\left[ P^{(1)}\left(\mathbf{u};\mathbf{g};\boldsymbol{\lambda};\boldsymbol{\omega};\mathbf{a}^{(1)};\mathbf{v}^{(1)}\right) \right]_{\partial\Omega_x}$ denotes the associated bilinear concomitant evaluated on the domain's boundary $\partial\Omega_x$ and where $\mathbf{A}^{(1)}(\mathbf{u};\mathbf{g}) \triangleq \left[ \mathbf{N}^{(1)}(\mathbf{u};\mathbf{g}) \right]^*$ denotes the operator formally adjoint to $\mathbf{N}^{(1)}(\mathbf{u};\mathbf{g})$. The symbol $[\ ]^*$ indicates "formal adjoint" operator.

The first term on the right side of Eq. (2.41) is now required to represent the indirect-effect term defined in Eq. (2.30) by imposing the following relationship:

$$\left\{\mathbf{A}^{(1)}(\mathbf{u};\mathbf{g})\mathbf{a}^{(1)}(\mathbf{x})\right\}_{\alpha^0} = \left\{\partial S(\mathbf{u};\mathbf{g};\mathbf{h})/\partial\mathbf{u}\right\}_{\alpha^0} \triangleq \mathbf{q}_A^{(1)}\left[\mathbf{u}(\mathbf{x});\mathbf{g};\mathbf{h}\right], \quad \mathbf{x}\in\Omega_x. \tag{2.42}$$

The domain of $\mathbf{A}^{(1)}(\mathbf{u};\mathbf{g})$ is determined by selecting appropriate adjoint boundary and/or initial conditions, which will be denoted in operator form as:

$$\left\{\mathbf{b}_A^{(1)}\left(\mathbf{u};\mathbf{a}^{(1)};\mathbf{g}\right)\right\}_{\alpha^0} = \mathbf{0}, \quad \mathbf{x}\in\partial\Omega_x. \tag{2.43}$$

The above boundary conditions for the adjoint operator $\mathbf{A}^{(1)}(\mathbf{u};\mathbf{g})$ are obtained by imposing the following requirements:

i. they must be independent of unknown values of $\mathbf{v}^{(1)}(\mathbf{x})$ and $\delta\mathbf{g}$;
ii. the substitution of the boundary and initial conditions represented by Eqs. (2.35) and (2.43) into the expression of $\left\{\left[P^{(1)}\left(\mathbf{u};\mathbf{g};\mathbf{a}^{(1)};\mathbf{v}^{(1)}\right)\right]_{\partial\Omega_x}\right\}_{\alpha^0}$ must cause all terms containing unknown values of $\mathbf{v}^{(1)}(\mathbf{x})$ to vanish.

Using the adjoint and forward variational boundary conditions represented by Eqs. (2.43) and (2.35) into (2.41) reduces the bilinear concomitant $\left\{\left[P^{(1)}\left(\mathbf{u};\mathbf{g};\mathbf{a}^{(1)};\mathbf{v}^{(1)}\right)\right]_{\partial\Omega_x}\right\}_{\alpha^0}$ to a residual term denoted as $\left\{\left[\hat{P}^{(1)}\left(\mathbf{u};\mathbf{g};\boldsymbol{\lambda};\boldsymbol{\omega};\mathbf{a}^{(1)};\delta\mathbf{g};\delta\boldsymbol{\lambda};\delta\boldsymbol{\omega}\right)\right]_{\partial\Omega_x}\right\}_{\alpha^0}$, which will contain boundary terms involving only known values of $\delta\mathbf{g}$, $\delta\boldsymbol{\lambda};\delta\boldsymbol{\omega}$; $\mathbf{g}$, $\mathbf{u}$, and $\mathbf{a}^{(1)}$. The residual term $\left\{\left[\hat{P}^{(1)}\left(\mathbf{u};\mathbf{g};\boldsymbol{\lambda};\boldsymbol{\omega};\mathbf{a}^{(1)};\delta\mathbf{g};\delta\boldsymbol{\lambda};\delta\boldsymbol{\omega}\right)\right]_{\partial\Omega_x}\right\}_{\alpha^0}$ is linear in $\delta\mathbf{g}$, $\delta\boldsymbol{\lambda};\delta\boldsymbol{\omega}$ and can therefore be expressed in the following form:

$$\begin{aligned}\left\{\left[\hat{P}^{(1)}\left(\mathbf{u};\mathbf{g};\boldsymbol{\lambda};\boldsymbol{\omega};\mathbf{a}^{(1)};\delta\mathbf{g};\delta\boldsymbol{\lambda};\delta\boldsymbol{\omega}\right)\right]_{\partial\Omega_x}\right\}_{\alpha^0} &= \left\{\sum_{j_1=1}^{TG}\left[\partial\hat{P}^{(1)}/\partial g_{j_1}\right]\delta g_{j_1}\right\}_{\alpha^0} \\ &+\left\{\sum_{j_1=1}^{TI}\left[\partial\hat{P}^{(1)}/\partial\lambda_{j_1}\right]\delta\lambda_{j_1}\right\}_{\alpha^0} + \left\{\sum_{j_1=1}^{TI}\left[\partial\hat{P}^{(1)}/\partial\omega_{j_1}\right]\delta\omega_{j_1}\right\}_{\alpha^0}.\end{aligned} \tag{2.44}$$

The results obtained in (2.41) and (2.42) are now replaced in (2.30) to obtain the following expression of the indirect-effect term as a function of $\mathbf{a}^{(1)}(\mathbf{x})$:

$$\begin{aligned}\left\{\delta R\left[\mathbf{u}(\mathbf{x});\mathbf{f}(\boldsymbol{\alpha});\mathbf{v}^{(1)}(\mathbf{x})\right]\right\}_{ind} &= \left\{\left\langle\mathbf{a}^{(1)},\mathbf{q}_V^{(1)}(\mathbf{u};\mathbf{g};\delta\mathbf{g})\right\rangle_1\right\}_{\alpha^0} \\ &-\left\{\left[\hat{P}^{(1)}\left(\mathbf{u};\mathbf{g};\boldsymbol{\lambda};\boldsymbol{\omega};\mathbf{a}^{(1)};\delta\mathbf{g};\delta\boldsymbol{\lambda};\delta\boldsymbol{\omega}\right)\right]_{\partial\Omega_x}\right\}_{\alpha^0}.\end{aligned} \tag{2.45}$$

Replacing in Eq. (2.24) the result obtained in Eq. (2.45) together with the expression for the direct-effect term provided in Eq. (2.26) yields the following expression for the first-order G-differential of the response $R[\mathbf{u}(\mathbf{x});\mathbf{f}(\alpha)]$:

$$\left\{dR\left[\mathbf{u}(\mathbf{x});\mathbf{f}(\alpha);\mathbf{v}^{(1)}(\mathbf{x});\delta\mathbf{f}(\alpha)\right]\right\}_{\alpha^0} = \left\{dR\left[\mathbf{u}(\mathbf{x});\mathbf{f}(\alpha);\delta\mathbf{f}\right]\right\}_{dir}$$

$$+\left\{\left\langle \mathbf{a}^{(1)},\mathbf{q}_V^{(1)}(\mathbf{u};\mathbf{g};\delta\mathbf{g})\right\rangle_1\right\}_{\alpha^0} - \left\{\left[\hat{P}^{(1)}\left(\mathbf{u};\mathbf{g};\boldsymbol{\lambda};\boldsymbol{\omega};\mathbf{a}^{(1)};\delta\mathbf{g};\delta\boldsymbol{\lambda};\delta\boldsymbol{\omega}\right)\right]_{\partial\Omega_x}\right\}_{\alpha^0}$$

$$\triangleq \left\{\sum_{j_1=1}^{TF} R^{(1)}\left[j_1;\mathbf{u}(\mathbf{x});\mathbf{a}^{(1)}(\mathbf{x});\mathbf{f}(\alpha)\right]\delta f_{j_1}\right\}_{\alpha^0},$$

where $R^{(1)}\left[j_1;\mathbf{u}(\mathbf{x});\mathbf{a}^{(1)}(\mathbf{x});\mathbf{f}(\alpha)\right] \triangleq \partial R[\mathbf{u}(\mathbf{x});\mathbf{f}(\alpha)]/\partial f_{j_1}$ denotes the first-order sensitivity of the response $R[\mathbf{u}(\mathbf{x});\mathbf{f}(\alpha)]$ with respect to the components $f_{j_1}$ of the "feature" function $\mathbf{f}(\alpha) \triangleq [\mathbf{g}(\alpha);\mathbf{h}(\alpha);\boldsymbol{\lambda}(\alpha);\boldsymbol{\omega}(\alpha)]^{\dagger}$. Each sensitivity $R^{(1)}\left[j_1;\mathbf{u}(\mathbf{x});\mathbf{a}^{(1)}(\mathbf{x});\mathbf{f}(\alpha)\right]$ is obtained by identifying the expression that multiplies the corresponding component of the variations $\delta\mathbf{g}$, $\delta\mathbf{h}$, $\delta\boldsymbol{\lambda}$, and $\delta\boldsymbol{\omega}$, respectively, on the right side of Eq. (2.46). Each of the 1st-order sensitivities $R^{(1)}\left[j_1;\mathbf{u}(\mathbf{x});\mathbf{a}^{(1)}(\mathbf{x});\mathbf{f}(\alpha)\right]$ of the response $R[\mathbf{u}(\mathbf{x});\mathbf{f}(\alpha)]$ with respect to the components of the functions **g**, **h**, $\boldsymbol{\lambda}$, and $\boldsymbol{\omega}$ can be computed inexpensively after having obtained the 1st-level adjoint function $\mathbf{a}^{(1)}(\mathbf{x}) \in \mathcal{H}_1$, using just quadrature formulas to evaluate the inner products involving $\mathbf{a}^{(1)}(\mathbf{x}) \in \mathcal{H}_1$ in Eq. (2.46). The function $\mathbf{a}^{(1)}(\mathbf{x}) \in \mathcal{H}_1$ is obtained by solving numerically (2.42) and (2.43), which is the only large-scale computation needed for obtaining all of the first-order sensitivities.

Equations (2.42) and (2.43) are called the *1st-Level Adjoint Sensitivity System* (1st-LASS). The solution, $\mathbf{a}^{(1)}(\mathbf{x}) \in \mathcal{H}_1(\Omega_x)$, of the 1st-LASS is called the 1st-level adjoint function. It is very important to note that the 1st-LASS is independent of all parameter variations $\delta\mathbf{g}$, $\delta\mathbf{h}$, $\delta\boldsymbol{\lambda}$, and $\delta\boldsymbol{\omega}$ (or, equivalently, of all parameter variations $\delta\alpha_{j_1}, j_1 = 1,\ldots,TP$) and therefore needs to be solved only once, regardless of the number of model parameters under consideration. Furthermore, since the 1st-LASS is linear in $\mathbf{a}^{(1)}(\mathbf{x})$, solving it requires less computational effort than solving the original model, which is nonlinear in $\mathbf{u}(\mathbf{x})$. The first-order sensitivity $R^{(1)}\left[j_1;\mathbf{u}(\mathbf{x});\mathbf{a}^{(1)}(\mathbf{x});\mathbf{f}(\alpha)\right]$ can be represented formally in the following integral form, for $j_1 = 1,\ldots,TF$:

$$R^{(1)}\left[j_1;\mathbf{u}(\mathbf{x});\mathbf{a}^{(1)}(\mathbf{x});\mathbf{f}(\alpha)\right] \triangleq \int_{\lambda_1(\alpha)}^{\omega_1(\alpha)} \ldots \int_{\lambda_{TI}(\alpha)}^{\omega_{TI}(\alpha)} S^{(1)}\left[j_1;\mathbf{u}(\mathbf{x});\mathbf{a}^{(1)}(\mathbf{x});\mathbf{f}(\alpha)\right] dx_1 \ldots dx_{TI}. \tag{2.47}$$

The functions $S^{(1)}\left[j_1;\mathbf{u}(\mathbf{x});\mathbf{a}^{(1)}(\mathbf{x});\mathbf{f}(\alpha)\right]$ will be subsequently used for determining the exact expressions of the 2nd-order sensitivities of the response with respect to the components of the function $\mathbf{f}(\alpha)$ of model parameters.

### 2.3.1 Comparison between 1st-FASAM-N and 1st-CASAM-N

The general methodology for computing most efficiently the exactly derived expressions of the 1st-order sensitivities of a model response *directly* with respect to the model's parameters is the "1st-Order Comprehensive Adjoint Sensitivity Analysis Methodology" (1st-CASAM-N) developed by Cacuci (2016, 2022a). The conceptual difference between the 1st-CASAM-N and the 1st-FASAM presented in the foregoing stems from the fact that within the 1st-CASAM-N the possible functions ("features") of model parameters are not explicitly considered, so that all differentials are computed directly with respect to the basic model parameters. Consequently, the main distinctions between the 1st-CASAM-N and the 1st-FASAM-N are as follows:

i. Instead of the expressions provided in Eq. (2.23)–(2.29), the "direct-effect" term in the 1st-CASAM-N framework, denoted as $\left\{\delta R\left[\mathbf{u}(\mathbf{x});\boldsymbol{\alpha};\delta\boldsymbol{\alpha}\right]\right\}_{dir}$, comprises the direct dependencies on $\delta\boldsymbol{\alpha}$ and is defined as follows:

$$\left\{\delta R\left[\mathbf{u}(\mathbf{x});\boldsymbol{\alpha};\delta\boldsymbol{\alpha}\right]\right\}_{dir} \triangleq \left\{\frac{\partial \mathbf{R}(\mathbf{u};\boldsymbol{\alpha})}{\partial \boldsymbol{\alpha}}\right\}_{\boldsymbol{\alpha}^0}\delta\boldsymbol{\alpha} \triangleq \sum_{j_1=1}^{TP}\left\{R^{(1)}\left[j_1;\mathbf{u}(\mathbf{x});\boldsymbol{\alpha}\right]\right\}_{dir}\delta\alpha_{j_1}, \quad (2.48)$$

where $\partial\mathbf{R}(\mathbf{u};\boldsymbol{\alpha})/\partial\boldsymbol{\alpha}$ denotes the partial G-derivatives of $\mathbf{R}(\mathbf{e})$ with respect to $\boldsymbol{\alpha}$, evaluated at the nominal parameter values, and where the following definitions were used:

$$\frac{\partial[\ ]}{\partial\boldsymbol{\alpha}}\delta\boldsymbol{\alpha} \triangleq \sum_{i=1}^{TP}\frac{\partial[\ ]}{\partial\alpha_i}\delta\alpha_i, \quad (2.49)$$

$$\begin{aligned}
\left\{R^{(1)}\left[j_1;\mathbf{u}(\mathbf{x});\boldsymbol{\alpha}\right]\right\}_{dir} \triangleq & \left\{\int_{\lambda_1(\boldsymbol{\alpha})}^{\omega_1(\boldsymbol{\alpha})}\cdots\int_{\lambda_{TI}(\boldsymbol{\alpha})}^{\omega_{TI}(\boldsymbol{\alpha})}\frac{\partial S(\mathbf{u};\boldsymbol{\alpha};\boldsymbol{\alpha})}{\partial\alpha_{j_1}}dx_1\ldots dx_{TI}\right\}_{\boldsymbol{\alpha}^0} \\
& +\sum_{j=1}^{TI}\left\{\frac{\partial\omega_j(\boldsymbol{\alpha})}{\partial\alpha_{j_1}}\int_{\lambda_1}^{\omega_1}dx_1\ldots\int_{\lambda_{j-1}}^{\omega_{j-1}}dx_{i-1}\int_{\lambda_{j+1}}^{\omega_{j+1}}dx_{i+1}\ldots\int_{\lambda_{TI}}^{\omega_{TI}}dx_{TI}S\left[\mathbf{u}\left(x_1,\ldots,\omega_j(\boldsymbol{\alpha}),\ldots,x_{N_x}\right);\boldsymbol{\alpha}\right]\right\}_{\boldsymbol{\alpha}^0} \\
& -\sum_{j=1}^{TI}\left\{\frac{\partial\lambda_j(\boldsymbol{\alpha})}{\partial\alpha_{j_1}}\int_{\lambda_1}^{\omega_1}dx_1\ldots\int_{\lambda_{j-1}}^{\omega_{j-1}}dx_{i-1}\int_{\lambda_{j+1}}^{\omega_{j+1}}dx_{i+1}\ldots\int_{\lambda_{TI}}^{\omega_{TI}}dx_{TI}S\left[\mathbf{u}\left(x_1,\ldots,\lambda_j(\boldsymbol{\alpha}),\ldots,x_{N_x}\right);\boldsymbol{\alpha}\right]\right\}_{\boldsymbol{\alpha}^0}.
\end{aligned} \quad (2.50)$$

ii. Instead of Eqs. (2.34) and (2.35), the 1st-LVSS within the 1st-CASAM-N framework has the following expression:

$$\left\{\mathbf{N}^{(1)}(\mathbf{u};\boldsymbol{\alpha})\mathbf{v}^{(1)}(j_1;\mathbf{x})\right\}_{\boldsymbol{\alpha}^0} = \left\{\mathbf{s}_V^{(1)}(j_1;\mathbf{u};\boldsymbol{\alpha})\right\}_{\boldsymbol{\alpha}^0}, \quad j_1 = 1,\ldots,TP; \quad \mathbf{x}\in\Omega_x, \quad (2.51)$$

$$\left\{\mathbf{b}_V^{(1)}\left[\mathbf{u};\boldsymbol{\alpha};\mathbf{v}^{(1)}(j_1;\mathbf{x})\right]\right\}_{\boldsymbol{\alpha}^0}=\mathbf{0};\quad j_1=1,\ldots,TP;\quad \mathbf{x}\in\partial\Omega_x\left(\boldsymbol{\alpha}^0\right). \tag{2.52}$$

where:

$$\mathbf{s}_V^{(1)}(j_1;\mathbf{u};\boldsymbol{\alpha})\triangleq\frac{\partial[\mathbf{Q}(\boldsymbol{\alpha})-\mathbf{N}(\mathbf{u};\boldsymbol{\alpha})]}{\partial\alpha_{j_1}}, \tag{2.53}$$

$$\left\{\mathbf{b}_V^{(1)}\left[\mathbf{u};\boldsymbol{\alpha};\mathbf{v}^{(1)}(j_1;\mathbf{x})\right]\right\}_{\boldsymbol{\alpha}^0}\triangleq\left\{\frac{\partial\mathbf{B}(\mathbf{u};\boldsymbol{\alpha})}{\partial\mathbf{u}}\mathbf{v}^{(1)}(j_1;\mathbf{x})+\frac{\partial[\mathbf{B}(\mathbf{u};\boldsymbol{\alpha})-\mathbf{C}(\boldsymbol{\alpha})]}{\partial\alpha_{j_1}}\right\}_{\boldsymbol{\alpha}^0}. \tag{2.54}$$

To determine the solutions $\mathbf{v}^{(1)}(j_1;\mathbf{x})$ of the 1st-LVSS that would correspond to every parameter variation $\delta\alpha_{j_1}$, $j_1=1,\ldots,TP$, the 1st-LVSS would need to be solved $TP$ times, with distinct right sides for each $\delta\alpha_{j_1}$, thus requiring $TP$ large-scale computations.

iii. The first-order G-differential of the response induced by variations in the model parameters and, consequently, the first-order sensitivities $\partial R[\mathbf{u}(\mathbf{x});\boldsymbol{\alpha}]/\partial\alpha_{j_1}$ of the response with respect to the model parameters have the following expressions:

$$\left\{\delta R\left[\mathbf{u}(\mathbf{x});\boldsymbol{\alpha};\mathbf{v}^{(1)}(\mathbf{x});\delta\boldsymbol{\alpha}\right]\right\}_{\boldsymbol{\alpha}^0}=\left\{\delta R[\mathbf{u}(\mathbf{x});\boldsymbol{\alpha};\delta\boldsymbol{\alpha}]\right\}_{dir}+\left\{\left\langle\mathbf{a}^{(1)},\mathbf{q}_V^{(1)}(\mathbf{u};\boldsymbol{\alpha};\delta\boldsymbol{\alpha})\right\rangle_1\right\}_{\boldsymbol{\alpha}^0}$$
$$-\left\{\left[\hat{P}^{(1)}\left(\mathbf{u};\boldsymbol{\alpha};\mathbf{a}^{(1)};\delta\boldsymbol{\alpha}\right)\right]_{\partial\Omega_x}\right\}_{\boldsymbol{\alpha}^0}\triangleq\sum_{j_1=1}^{TP}\left\{R^{(1)}\left[j_1;\mathbf{u}(\mathbf{x});\mathbf{a}^{(1)}(\mathbf{x});\boldsymbol{\alpha}\right]\right\}_{\boldsymbol{\alpha}^0}\delta\alpha_{j_1}, \tag{2.55}$$

where, for each $j_1=1,\ldots,TP$, the quantity $R^{(1)}\left[j_1;\mathbf{u}(\mathbf{x});\mathbf{a}^{(1)}(\mathbf{x});\boldsymbol{\alpha}\right]\triangleq\partial R[\mathbf{u}(\mathbf{x});\boldsymbol{\alpha}]/\partial\alpha_{j_1}$ denotes the 1st-order sensitivities of the response $R[\mathbf{u}(\mathbf{x});\boldsymbol{\alpha}]$ with respect to the model parameters $\alpha_{j_1}$ and has the following expression, for $j_1=1,\ldots,TP$:

$$R^{(1)}\left[j_1;\mathbf{u}(\mathbf{x});\mathbf{a}^{(1)}(\mathbf{x});\boldsymbol{\alpha}\right]=-\frac{\partial\left[\hat{P}^{(1)}\left(\mathbf{u};\mathbf{a}^{(1)};\boldsymbol{\alpha};\right)\right]_{\partial\Omega_x}}{\partial\alpha_{j_1}}$$
$$+\left\{R^{(1)}[j_1;\mathbf{u}(\mathbf{x});\boldsymbol{\alpha}]\right\}_{dir}+\int_{\lambda_1(\boldsymbol{\alpha})}^{\omega_1(\boldsymbol{\alpha})}dx_1\ldots\int_{\lambda_{TI}(\boldsymbol{\alpha})}^{\omega_{TI}(\boldsymbol{\alpha})}\mathbf{a}^{(1)}(\mathbf{x})\frac{\partial[\mathbf{Q}(\boldsymbol{\alpha})-\mathbf{N}(\mathbf{u};\boldsymbol{\alpha})]}{\partial\alpha_{j_1}}dx_{TI} \tag{2.56}$$
$$\triangleq\int_{\lambda_1(\boldsymbol{\alpha})}^{\omega_1(\boldsymbol{\alpha})}\ldots\int_{\lambda_{TI}(\boldsymbol{\alpha})}^{\omega_{TI}(\boldsymbol{\alpha})}S^{(1)}\left[j_1;\mathbf{u}(\mathbf{x});\mathbf{a}^{(1)}(\mathbf{x});\boldsymbol{\alpha}\right]dx_1\ldots dx_{TI}\triangleq\frac{\partial R[\mathbf{u}(\mathbf{x});\boldsymbol{\alpha}]}{\partial\alpha_{j_1}}.$$

Comparing Eq. (2.56) to Eq. (2.46) indicates that the same first-level adjoint function $\mathbf{a}^{(1)}(\mathbf{x})\in\mathscr{H}_1$ is used in all of these expressions. Therefore, the 1st-LASS needs to be solved only once, to obtain $\mathbf{a}^{(1)}(\mathbf{x}) \in \mathscr{H}_1$, regardless of whether the response sensitivities are computed directly with respect to the model parameters or whether the sensitivities are first computed with respect to the components ("features") of the function $\mathbf{f}(\boldsymbol{\alpha})$ and subsequently computed with respect to the components of the vector $\boldsymbol{\alpha}$ of model parameters. This is as expected, since the 1st-LASS is independent of parameter variations $\delta\alpha_{j_1}$, $j_1 = 1,\ldots,TP$, and therefore needs to be solved only once, regardless of the number of "features" or model parameters under consideration.

It is more advantageous to use the 1st-FASAM-N to compute the sensitivities of the response to the "feature" functions provided in Eq. (2.46) since these expressions involve fewer integrals ("quadratures") over the 1st-level adjoint function $\mathbf{a}^{(1)}(\mathbf{x}) \in \mathscr{H}_1$ than using Eq. (2.56), since $TG < TP$. The transition from the sensitivities with respect to "features" provided in Eq. (2.46) to the sensitivities with respect to the model parameters provided in Eq. (2.56) can be done very inexpensively by using the "chain-rule" of differentiation of compound functions, namely $\delta g_i(\boldsymbol{\alpha}) = [\partial g_i(\boldsymbol{\alpha})/\partial \boldsymbol{\alpha}]\delta\boldsymbol{\alpha}$, $\delta h_i(\boldsymbol{\alpha}) = [\partial h_i(\boldsymbol{\alpha})/\partial \boldsymbol{\alpha}]\delta\boldsymbol{\alpha}$, $\delta\lambda_i(\boldsymbol{\alpha}) = [\partial\lambda_i(\boldsymbol{\alpha})/\partial \boldsymbol{\alpha}]\delta\boldsymbol{\alpha}$, $\delta\omega_i(\boldsymbol{\alpha}) = [\partial\omega_i(\boldsymbol{\alpha})/\partial \boldsymbol{\alpha}]\delta\boldsymbol{\alpha}$, which implies that:

$$\frac{\partial R[\mathbf{u}(\mathbf{x});\boldsymbol{\alpha}]}{\partial \alpha_{j_1}} = \sum_{i=1}^{TF} \frac{\partial R[\mathbf{u}(\mathbf{x});\boldsymbol{\alpha}]}{\partial f_i} \frac{\partial f_i(\boldsymbol{\alpha})}{\partial \alpha_{j_1}}.$$

The features of the 1st-FASAM-N will be illustrated in Section 2.4, by means of a paradigm neutron diffusion model which admits closed-form exact expressions for all quantities involved (i.e., forward dependent variable, model response, and all sensitivities). The considerable reduction in the number of large-scale computations when using the "function/feature adjoint sensitivity analysis methodology" will become apparent for the computation of second-order (and, subsequently, higher-order) sensitivities, as will be shown in Chapters 3 and 4.

## 2.4 ILLUSTRATIVE APPLICATION OF THE 1ST-FASAM-N TO A PARADIGM PARTICLE TRANSPORT MODEL

The application of the 1st-FASAM-N will be illustrated in this section by considering the sensitivity analysis of an illustrative computational model of a dosimetry response (e.g., energy deposition) stemming from a shielded volumetric source of directly and/or indirectly ionizing radiation (see, e.g., Shultis and Faw 2000). Such models typically describe the radiation and/or particle transport within shielded containers/drums holding vitrified radioactive waste material for deep underground disposal. In practice, the distribution of gamma-ray photons emitted by waste products within a shielded vitrified container is computed using sophisticated time-dependent three-dimensional energy-dependent radiation decay and transport models. To illustrate the application of the 1st-FASAM-N, however, the radiation transport model will be greatly simplified by considering that all processes are steady-state, the radiation is monoenergetic, the sources of radiation are additive, and the geometry is one-dimensional (slab) rather than three-dimensional. These drastic simplifications will enable the analytical solving of the transport equation, thereby minimizing cumbersome algebraic manipulations while highlighting the application of the fundamental concepts underlying the 1st-FASAM-N.

Consider that a total of "*MQ*" radioactive sources, each emitting $q_i, i = 1,\ldots,MQ$, particles per unit time [e.g., "second"] per unit area [e.g., "cm$^2$"] are homogeneously distributed (vitrified) within a slab of homogeneous material, which will be designated as "Material A." There are many radioactive sources (i.e., *MQ* is a large number); even after several years of "cooling" there remain long-lived radionuclides which require adequate shielding. Included among the most relevant for shielding of waste disposal containers are 11 fission products, 12 actinides, and five activation products (Bonin 2010). The vitrified containment material, designated as Material A, is considered to be homogeneous, comprising "*MA*" nuclides/elements. Each of these nuclides is characterized by a total microscopic cross section denoted as $\sigma_t^{(A,i)}$, $i = 1,\ldots,MA$, and an isotopic number density denoted as $N^{(A,i)}$, $i = 1,\ldots,MA$, where *MA* denotes the "total number of elements/nuclides within the glassy "Material A." Geometrically, consider that Material A is a slab of a thickness $2a$ [cm], which is imprecisely known due to manufacturing tolerances. The slab of Material A, which contains radioactive sources, is considered to be shielded on each of its sides by another slab, made of "Material B," of thickness $b$ [cm], which is also imprecisely known due to manufacturing tolerances. "Material B" is considered to comprise a total number of *MB* nuclides/elements, each of which is characterized by a total microscopic cross section denoted as $\sigma_t^{(B,i)}$, $i = 1,\ldots,MB$, and an isotopic number density denoted as $N^{(B,i)}$, $i = 1,\ldots,MB$. For example, "Material B" could be magnetite concrete (approximate density: 3.53 g/cm$^3$), composed of $MB = 12$ nuclides/elements, with approximate partial densities in [g/cm$^3$] provided in parentheses, as follows: 1. Hydrogen (0.011); 2: Oxygen (1.168); 3. Silicon (0.091); 4. Calcium (0.251); 5. Magnesium (0.033); 6. Aluminum (0.083); 7. Sulfur (0.005); 8. Iron (1.676); 9. Titanium (0.192); 10. Chromium (0.006); 11. Manganese (0.007); 12. Vanadium (0.011).

The distribution of monoenergetic particles within the aforementioned arrangement of slabs is symmetrical with respect to the midplane of the inner slab, so the center of the coordinate system can be chosen at this midplane. Within the inner slab, the uncollided angular flux of monoenergetic neutral particles (e.g., photons and/or neutrons), denoted as $\varphi_A(z,\omega)$, in the positive direction perpendicular to the slab, labeled as "direction $z > 0$", is modeled by the following form of the particle transport equation:

$$\omega \frac{d\varphi_A(z,\omega)}{dz} + \mu_A \varphi_A(z,\omega) = Q, \quad 0 < z < a, \quad 0 \le \omega \le 1. \tag{2.58}$$

In Eq. (2.58), the source term is defined as follows:

$$Q \triangleq \sum_{i=1}^{MQ} q_i. \tag{2.59}$$

Also in Eq. (2.58), the quantity $\omega$ denotes the cosine of the angle between the particle's direction and the $z$-axis, while $\mu_A$ denotes the interaction coefficient of the particles with the inner slab's homogenized material and is defined as follows:

$$\mu_A \triangleq \sum_{i=1}^{MA} N^{(A,i)} \sigma_t^{(A,i)}. \tag{2.60}$$

Due to the model's symmetry, choosing the origin of the coordinate system at the inner slab's midplane, the appropriate boundary condition for the inner slab is:

$$\frac{d\varphi_A(z,\omega)}{dz} = 0, \quad \text{at} \quad z = 0, \quad 0 \le \omega \le 1. \tag{2.61}$$

The solution of Eqs. (2.58) and (2.61) is the following constant function:

$$\varphi_A(z,\omega) = \frac{Q}{\mu_A}, \quad 0 < z < a, \quad 0 \le \omega \le 1. \tag{2.62}$$

The particle transport equation governing the uncollided angular flux of monoenergetic particles, denoted as $\varphi(z,\omega)$, through the outer slab ("Material B") in particle directions $0 \le \omega \le 1$ has the following form:

$$\omega \frac{d\varphi(z,\omega)}{dz} + \mu_B \varphi(z,\omega) = 0, \quad a < z < b, \quad 0 \le \omega \le 1, \tag{2.63}$$

where the interaction coefficient $\mu_B$ is defined as follows:

$$\mu_B \triangleq \sum_{i=1}^{MB} N^{(B,i)} \sigma_t^{(B,i)}. \tag{2.64}$$

The uncollided flux is continuous across the interface at $z = a$, i.e.:

$$\varphi(a,\omega) = \varphi_A(a,\omega) = \frac{Q}{\mu_A}. \tag{2.65}$$

For a subsequent verification of the expressions to be obtained for the response sensitivities to parameters, the closed-form expression of the uncollided flux $\varphi(z,\omega)$, which is the solution of (2.63) and (2.65), is provided below:

$$\varphi(z,\omega) = \frac{Q}{\mu_A} \exp\left[\frac{\mu_B}{\omega}(a - z)\right], \quad a \le z \le b, \quad 0 \le \omega \le 1. \tag{2.66}$$

The "model response" of interest for this illustrative paradigm model is the total flux of particles interacting with a "kerma response" detector placed on the outer slab's external surface, which can be represented mathematically by the following expression:

$$R[\varphi(z,\omega);\boldsymbol{\alpha}] \triangleq \mu_D \int_0^1 d\omega \int_a^b \varphi(z,\omega) \delta(z - b) dz. \tag{2.67}$$

The interaction coefficient, $\mu_D$, which characterizes the detector's material, is defined as follows:

$$\mu_D \triangleq \sum_{i=1}^{MD} N^{(D,i)} \sigma_t^{(D,i)}, \tag{2.68}$$

where the quantity $MD$ denotes the number of distinct nuclides/elements comprising the detector's response function, $\sigma_t^{(D,i)}$ denotes the total microscopic interaction cross section of the respective nuclide with the source particles, and $N^{(D,i)}$ denotes the respective nuclide's atomic number density, for $i = 1,\ldots, MD$.

Replacing the expression of $\varphi(z,\omega)$ obtained in Eq. (2.66) into Eq. (2.67) yields the following expression for the response $R[\varphi(z,\omega);\boldsymbol{\alpha}]$:

$$R(\varphi;\boldsymbol{\alpha}) = \frac{Q\mu_D}{\mu_A} E_2[(b-a)\mu_B], \tag{2.69}$$

where the exponential-integral function $E_n(z)$ is defined as follows:

$$E_n(z) \triangleq \int_0^1 y^{n-2} e^{-z/y} dy; \quad E_0(z) = \frac{e^{-z}}{z}; \quad E_1(z) \triangleq \int_0^1 y^{-1} e^{-z/y} dy = -Ei(-z); \quad E_2(0) = 1; \tag{2.70}$$

and possesses the following important properties:

$$E_n(z) = \frac{1}{n-1}\left[e^{-z} - zE_{n-1}(z)\right], \text{ for } n > 1; \quad \frac{dE_n(z)}{dz} = -E_{n-1}(z); \quad E_n(z) = \int_z^{\infty} E_{n-1}(y) dy. \tag{2.71}$$

The uncertain parameters that describe the properties (microscopic cross sections, atomic number densities, source, slab boundaries.) characterizing the two materials and the detector, will be generically denoted as $\alpha_i, i = 1,\ldots, TP$, where $TP$ denotes the "total number of model parameters." These parameters will be considered to be the components of the (column) "vector of imprecisely known model parameters" defined as follows:

$$\begin{aligned} \boldsymbol{\alpha} &\triangleq (\alpha_1,\ldots,\alpha_{TP})^{\dagger} \\ &\triangleq \left[\sigma_t^{(A,i)}, N^{(A,i)}; i = 1,\ldots, MA; \sigma_t^{(B,i)}, N^{(B,i)}; i = 1,\ldots, MB; \sigma_t^{(D,i)}, N^{(D,i)};\right. \\ &\quad \left. i = 1,\ldots, MD; q_1,\ldots,q_{MQ}; a, b\right]^{\dagger}; \\ TP &= 2MA + 2MB + 2MD + MQ + 2. \end{aligned} \tag{2.72}$$

From the definitions provided in Eqs. (2.59), (2.60), (2.64) and (2.68), it follows that the interaction coefficients and the source are functions of the vector $\boldsymbol{\alpha}$ of model parameters.

### 2.4.1 Application of the 1st-CASAM-N to Compute First-Order Response Sensitivities Directly with Respect to Model Parameters

The total first-order sensitivity of the response defined in Eq. (2.67) to the model's primary parameters (including interface and boundary locations) is provided by the Gateaux-(G-) differential $\{\delta R(\varphi;\alpha;\delta\varphi;\delta\alpha)\}_{\alpha^0}$ of the response $R(\varphi;\alpha)$ for arbitrary variations $(\delta\varphi,\delta\alpha)$ around the nominal parameter and state functions values, which is defined as follows:

$$\delta R(\varphi;\alpha;\delta\varphi;\delta\alpha)_{\alpha^0} \triangleq \frac{d}{d\varepsilon}\left\{\left[\sum_{i=1}^{MD}\left(N^{(D,i)}+\varepsilon\delta N^{(D,i)}\right)\left(\sigma_t^{(D,i)}+\varepsilon\delta\sigma_t^{(D,i)}\right)\right.\right.$$
$$\left.\left.\times\int_0^1 d\omega\int_{a+\varepsilon\delta a}^{b+\varepsilon\delta b}(\varphi+\varepsilon\delta\varphi)\delta(z-b-\varepsilon\delta b)\right]_{\alpha^0} dz\right\}_{\varepsilon=0} \tag{2.73}$$
$$=\sum_{i=1}^{TP}\frac{\partial R}{\partial\alpha_i}\delta\alpha_i\left\{\delta R(\varphi;\alpha;\delta\alpha)_{\alpha^0}\right\}_{dir}+\left\{\delta R(\varphi;\alpha;\delta\varphi)_{\alpha^0}\right\}_{ind},$$

where the "direct-effect" term $\left\{\delta R(\varphi;\alpha;\delta\alpha)_{\alpha^0}\right\}_{dir}$ depends only on variations $\delta\alpha$ in the parameters and is defined as follows:

$$\left\{\delta R(\varphi;\alpha;\delta\alpha)_{\alpha^0}\right\}_{dir} \triangleq \left\{\sum_{i=1}^{MD}\left(N^{(D,i)}\delta\sigma_t^{(D,i)}+\sigma_t^{(D,i)}\delta N^{(D,i)}\right)\times\int_0^1 d\omega\int_a^b dz\varphi(z,\omega)\delta(z-b)\right\}_{\alpha^0}$$
$$-(\delta b)\left\{\mu_D\int_0^1 d\omega\int_a^b\varphi(z,\omega)\delta'(z-b)dz\right\}_{\alpha^0}, \tag{2.74}$$

and where the "indirect-effect" term $\left\{\delta R(\varphi;\alpha;\delta\varphi)_{\alpha^0}\right\}_{ind}$ depends only on variations $\delta\varphi(z,\omega)$ in the dependent variable (state function) and is defined as follows:

$$\left\{\delta R(\varphi;\alpha;\delta\varphi)_{\alpha^0}\right\}_{ind} \triangleq \left\{\mu_D(\alpha)\int_0^1 d\omega\int_a^b\delta\varphi(z,\omega)\delta(z-b)dz\right\}_{\alpha^0}. \tag{2.75}$$

Since the flux $\varphi(z,\omega)$ is known, the direct-effect term in Eq. (2.74) can be computed immediately at this stage, for any parameter variations that appear in its expression. On the contrary, the "indirect-effect" terms in Eq. (2.75) cannot be computed at this stage since the 1st-order variation, $\delta\varphi(z,\omega)$, is not available at this stage. The variational function $\delta\varphi(z,\omega)$ could be determined by solving the 1st-Level Variational

Sensitivity System (1st-LVSS), which is obtained by G-differentiating Eqs. (2.63) and (2.65) at the nominal parameter values, to obtain the following system of equations:

$$\left\{\frac{d}{d\varepsilon}\left[\omega\frac{d(\varphi+\varepsilon\delta\varphi)}{dz}+\mu_B(\alpha+\varepsilon\delta\alpha)(\varphi+\varepsilon\delta\varphi)\right]_{\alpha^0}\right\}_{\varepsilon=0}=0,\quad a^0<z<b^0,\quad 0\le\omega\le 1; \tag{2.76}$$

$$\left\{\frac{d}{d\varepsilon}\left[\varphi\left(z=a^0+\varepsilon\delta a;\omega\right)+\varepsilon\delta\varphi\left(z=a^0+\varepsilon\delta a;\omega\right)\right]\right\}_{\varepsilon=0}=\left\{\frac{d}{d\varepsilon}\left[\frac{Q(\alpha+\varepsilon\delta\alpha)}{\mu_A(\alpha+\varepsilon\delta\alpha)}\right]_{\alpha^0}\right\}_{\varepsilon=0}. \tag{2.77}$$

Carrying out the operations with respect to $\varepsilon$ in (2.76) and (2.77) yields the following 1st-LVSS to be satisfied by the 1st-order forward variational function $\delta\varphi(z,\omega)$:

$$\left\{\omega\frac{d}{dz}[\delta\varphi(z,\omega)]+\mu_B(\alpha)[\delta\varphi(z,\omega)]\right\}_{\alpha^0}=-\left\{\varphi(z,\omega)\sum_{i=1}^{MB}\frac{\partial\mu_B(\alpha)}{\partial\alpha_i}(\delta\alpha_i)\right\}_{\alpha^0}; \tag{2.78}$$

$a^0<z<b^0,\quad 0\le\omega\le 1;$

$$\begin{aligned}\{\delta\varphi(z,\omega)\}_{z=a^0}&=\left\{\frac{1}{\mu_A(\alpha)}\sum_{i=1}^{MQ}\frac{\partial Q(\alpha)}{\partial\alpha_i}(\delta\alpha_i)-\frac{Q(\alpha)}{\mu_A^2(\alpha)}\sum_{i=1}^{MA}\frac{\partial\mu_A(\alpha)}{\partial\alpha_i}(\delta\alpha_i)\right\}_{\alpha^0}\\&\quad-(\delta a)\left\{\frac{d\varphi(z,\omega)}{dz}\right\}_{z=a^0}\\&=\left\{\frac{1}{\mu_A(\alpha)}\sum_{i=1}^{MQ}\frac{\partial Q(\alpha)}{\partial\alpha_i}(\delta\alpha_i)-\frac{Q(\alpha)}{\mu_A^2(\alpha)}\sum_{i=1}^{MA}\frac{\partial\mu_A(\alpha)}{\partial\alpha_i}(\delta\alpha_i)\right\}_{\alpha^0}\\&\quad+(\delta a)\left\{\frac{Q(\alpha)}{\mu_A(\alpha)}\frac{\mu_B(\alpha)}{\omega}\right\}_{\alpha^0};\quad z=a^0,\quad 0\le\omega\le 1.\end{aligned} \tag{2.79}$$

It is evident from the 1st-LVSS that a distinct function $\delta\varphi(z,\omega)$ would be obtained by solving Eqs. (2.78) and (2.79) for each parameter variation $\delta\alpha_i$, $i=1,\ldots,TP$. Performing so many large-scale computations would be impractical for large-scale systems involving many parameters.

The alternative to solving repeatedly the 1st-LVSS for every possible variation in the system's imprecisely known parameters is to use the 1st-Level Adjoint Sensitivity System (1st-LASS), which is constructed by applying the following sequence of steps:

1. Consider that the function $\delta\varphi(z,\omega)$ belongs to a Hilbert space, which will be denoted as $\mathcal{H}_1(\Omega_z)$, $\Omega_z\triangleq\{\omega\in[0,1]\otimes z\in[a,b]\}$. This Hilbert space

is endowed with an inner product of two functions $\chi(z,\omega)\in\mathcal{H}_1$ and $\psi(z,\omega)\in\mathcal{H}_1$, which is denoted as $\langle\chi(z,\omega),\psi(z,\omega)\rangle_1$ and is defined as follows:

$$\langle\chi(z,\omega),\psi(z,\omega)\rangle_1 \triangleq \left\{\int_0^1 d\omega \int_a^b \chi(z,\omega)\psi(z,\omega)dz\right\}_{\alpha^0}. \tag{2.80}$$

2. Using the definition provided in Eq. (2.80), construct the inner product of a function $a^{(1)}(z,\omega)\in\mathcal{H}_1$ with Eq. (2.78) to obtain the following relation:

$$\begin{aligned}&\left\{\int_0^1 d\omega \int_a^b a^{(1)}(z,\omega)\left[\omega\frac{d}{dz}\delta\varphi(z,\omega)+\mu_B(\alpha)\delta\varphi(z,\omega)\right]dz\right\}_{\alpha^0}\\&=-\left\{\sum_{i=1}^{MB}\frac{\partial\mu_B(\alpha)}{\partial\alpha_i}(\delta\alpha_i)\int_0^1 d\omega\int_a^b a^{(1)}(z,\omega)\varphi(z,\omega)dz\right\}_{\alpha^0}.\end{aligned} \tag{2.81}$$

3. Integrate by parts, over the independent variable $z$, the left side of Eq. (2.81) to obtain the following relation:

$$\begin{aligned}&\left\{\int_0^1 d\omega \int_a^b a^{(1)}(z,\omega)\left[\omega\frac{d}{dz}\delta\varphi(z,\omega)+\mu_B(\alpha)\delta\varphi(z,\omega)\right]dz\right\}_{\alpha^0}\\&=\int_0^1 \omega\, d\omega\left\{a^{(1)}(b,\omega)\delta\varphi(b,\omega)-a^{(1)}(a,\omega)\delta\varphi(a,\omega)\right\}_{\alpha^0}\\&+\left\{\int_0^1 d\omega\int_a^b \delta\varphi(z,\omega)\left[-\omega\frac{d}{dz}a^{(1)}(z,\omega)+\mu_B(\alpha)a^{(1)}(z,\omega)\right]dz\right\}_{\alpha^0}.\end{aligned} \tag{2.82}$$

4. Require the last term on the right side of Eq. (2.82) to represent the indirect-effect term defined in Eq. (2.75) by imposing the following relationship:

$$\begin{aligned}&\left\{\int_0^1 d\omega\int_a^b \delta\varphi(z,\omega)\left[-\omega\frac{d}{dz}a^{(1)}(z,\omega)+\mu_B(\alpha)a^{(1)}(z,\omega)\right]dz\right\}_{\alpha^0}\\&=\left\{\mu_D(\alpha)\int_0^1 d\omega\int_a^b \delta\varphi(z,\omega)\delta(z-b)dz\right\}_{\alpha^0}.\end{aligned} \tag{2.83}$$

5. The relation in Eq. (2.83) implies that the following equation holds, in the weak sense, at the nominal parameter values:

$$\left\{-\omega\frac{d}{dz}a^{(1)}(z,\omega)+\mu_B(\alpha)a^{(1)}(z,\omega)\right\}_{\alpha^0}=\left\{\mu_D(\alpha)\delta(z-b)\right\}_{\alpha^0},\quad a^0<z<b^0;\quad 0\le\omega\le 1. \tag{2.84}$$

6. Complete the definition of the function $a^{(1)}(z,\omega)$ by requiring that the unknown value of the function $\delta\varphi(b,\omega)$ be eliminated from appearing on the right side of Eq. (2.82). This requirement is met by imposing the following condition to be satisfied by the function $a^{(1)}(z,\omega)$ on the model's outer boundary:

$$a^{(1)}(z,\omega)=0,\quad at\quad z=b^0,\quad 0\le\omega\le 1. \tag{2.85}$$

Altogether, Eqs. (2.84) and (2.85) constitute a well-posed system, called the "1st-Level Adjoint Sensitivity System" (1st-LASS), for determining the function $a^{(1)}(z,\omega)\in\mathcal{H}_1$. The function $a^{(1)}(z,\omega)\in\mathcal{H}_1$ is called the "1st-level adjoint sensitivity function." Since the 1st-LASS does not depend on the parameter variations, it needs to be solved only once in order to obtain the 1st-level adjoint sensitivity function $\psi_1(z,\omega)$. For the paradigm model under consideration, the 1st-LASS given by Eqs. (2.84) and (2.85) can be solved exactly to obtain the following closed-form expression for the 1st-level adjoint function $a^{(1)}(z,\omega)$:

$$a^{(1)}(z,\omega)=\left\{\frac{\mu_D(\alpha)}{\omega}\left[1-H(z-b)\right]\exp\left[\frac{\mu_B(\alpha)(z-b)}{\omega}\right]\right\}_{\alpha^0}, \tag{2.86}$$

where $H(z-b)$ denotes the Heaviside functional defined as follows:

$$H(z-b)=\begin{cases}1, & z\ge b;\\ 0, & z<b.\end{cases} \tag{2.87}$$

Collecting the results obtained in Eqs. (2.81)–(2.85) leads to the following expression for the indirect-effect term:

$$\begin{aligned}\left\{\delta R(\varphi;\alpha;\delta\varphi)_{\alpha^0}\right\}_{ind}&=\left\{\int_0^1 d\omega\left[\omega a^{(1)}(a,\omega)\right]\left[\frac{1}{\mu_A(\alpha)}\sum_{i=1}^{MQ}\frac{\partial Q(\alpha)}{\partial\alpha_i}(\delta\alpha_i)\right.\right.\\&\left.\left.-\frac{Q(\alpha)}{\mu_A^2(\alpha)}\sum_{i=1}^{MA}\frac{\partial\mu_A(\alpha)}{\partial\alpha_i}(\delta\alpha_i)+(\delta a)\frac{Q(\alpha)}{\mu_A(\alpha)}\frac{\mu_B(\alpha)}{\omega}\right]\right\}_{\alpha^0}\\&-\left\{\sum_{i=1}^{MB}\frac{\partial\mu_B(\alpha)}{\partial\alpha_i}(\delta\alpha_i)\int_0^1 d\omega\int_{b_1}^{b_2}a^{(1)}(z,\omega)\varphi(z,\omega)dz\right\}_{\alpha^0}\\&\triangleq\left\{\delta R\left(\varphi;a^{(1)};\alpha;\delta\alpha\right)\right\}_{ind}.\end{aligned} \tag{2.88}$$

The appearance of the 1st-level adjoint function $a^{(1)}(z,\omega)$ in the list of arguments of the indirect-effect term $\left\{\delta R\left(\varphi;a^{(1)};\alpha;\delta\alpha\right)\right\}_{ind}$ on the rightmost-side of Eq. (2.88) emphasizes the fact the function $\delta\varphi(z,\omega)$, which depends implicitly on parameter variations, has been eliminated from appearing in the final expression of $\left\{\delta R\left(\varphi;a^{(1)};\alpha;\delta\alpha\right)\right\}_{ind}$. Instead of depending on $\delta\varphi(z,\omega)$, the indirect-effect term $\left\{\delta R\left(\varphi;a^{(1)};\alpha;\delta\alpha\right)\right\}_{ind}$ now depends on the 1st-level adjoint function $a^{(1)}(z,\omega)$, which is independent of the model's parameter variations. Hence, the 1st-LASS needs to be solved just once (which means a single "large-scale computation" comparable to solving the original equations underlying the model) to obtain the 1st-level adjoint function $a^{(1)}(z,\omega)$. After obtaining $a^{(1)}(z,\omega)$, all of the partial sensitivities included in the indirect-effect term, cf. Eq. (2.88), are obtained by quadrature formulas for evaluating the integrals involving $a^{(1)}(z,\omega)$.

Adding the results obtained in Eqs. (2.88) and (2.74) yields the following complete expression for the total 1st-order response sensitivity $\left\{\delta R\left(\varphi;\alpha;a^{(1)};\delta\alpha\right)\right\}_{\alpha^0}$:

$$\begin{aligned}\left\{\delta R\left(\varphi;\alpha;a^{(1)};\delta\alpha\right)\right\}_{\alpha^0} &= -(\delta b)\left\{\mu_D\int_0^1 d\omega\int_a^b \varphi(z,\omega)\delta'(z-b)dz\right\}_{\alpha^0}\\ &+\left\{\sum_{i=1}^{M_D}\left(N^{(D,i)}\delta\sigma_t^{(D,i)}+\sigma_t^{(D,i)}\delta N^{(D,i)}\right)\int_0^1 d\omega\int_a^b dz\varphi(z,\omega)\delta(z-b)\right\}_{\alpha^0}\\ &+\left\{\int_0^1 \omega\, d\omega a^{(1)}(a,\omega)\left[\frac{1}{\mu_A(\alpha)}\sum_{i=1}^{MQ}\frac{\partial Q(\alpha)}{\partial\alpha_i}(\delta\alpha_i)-\frac{Q(\alpha)}{\mu_A^2(\alpha)}\sum_{i=1}^{MA}\frac{\partial\mu_A(\alpha)}{\partial\alpha_i}(\delta\alpha_i)\right.\right.\\ &\left.\left.+(\delta a)\frac{Q(\alpha)}{\mu_A(\alpha)}\frac{\mu_B(\alpha)}{\omega}\right]\right\}_{\alpha^0}-\left\{\sum_{i=1}^{MB}\frac{\partial\mu_B(\alpha)}{\partial\alpha_i}(\delta\alpha_i)\int_0^1 d\omega\int_{b_1}^{b_2} a^{(1)}(z,\omega)\varphi(z,\omega)dz\right\}_{\alpha^0}.\end{aligned}\tag{2.89}$$

The specific expression of each 1st-order partial sensitivity of the response $R(\varphi;\alpha)$ to each uncertain parameter is obtained by identifying the expression that multiplies the respective parameter variation, which yields the following results:

i. Sensitivities stemming solely from the direct-effect term:

$$\begin{aligned}\left\{\frac{\partial R(\varphi;\alpha)}{\partial b}\right\}_{\alpha^0} &= -\left\{\mu_D(\alpha)\int_0^1 d\omega\int_a^b \varphi(z,\omega)\delta'(z-b)dz\right\}_{\alpha^0}\\ &= -\left\{\frac{Q(\alpha)\mu_B(\alpha)\mu_D(\alpha)}{\mu_A(\alpha)}E_1\left[(b-a)\mu_B(\alpha)\right]\right\}_{\alpha^0};\end{aligned}\tag{2.90}$$

$$\left\{\frac{\partial R(\varphi;\alpha)}{\partial \sigma_t^{(D,i)}}\right\}_{\alpha^0} = \left\{N^{(D,i)}\int_0^1 d\omega \int_a^b dz\varphi(z,\omega)\delta(z-b)\right\}_{\alpha^0}$$
$$= \left\{\frac{Q(\alpha)N^{(D,i)}}{\mu_A}E_2\left[(b-a)\mu_B\right]\right\}_{\alpha^0}; \quad i=1,\dots,MD; \tag{2.91}$$

$$\left\{\frac{\partial R(\varphi;\alpha)}{\partial N^{(D,i)}}\right\}_{\alpha^0} = \left\{\sigma_t^{(D,i)}\int_0^1 d\omega \int_a^b dz\varphi(z,\omega)\delta(z-b)\right\}_{\alpha^0}$$
$$= \left\{\frac{Q(\alpha)\sigma_t^{(D,i)}}{\mu_A}E_2\left[(b-a)\mu_B\right]\right\}_{\alpha^0} \quad i=1,\dots,MD; \tag{2.92}$$

ii. Sensitivities stemming solely from the indirect-effect term:

$$\left\{\frac{\partial R(\varphi;\alpha)}{\partial N^{(A,i)}}\right\}_{\alpha^0} = -\left\{\sigma_t^{(A,i)}\frac{Q(\alpha)}{\mu_A^2(\alpha)}\int_0^1 a^{(1)}(a,\omega)\omega d\omega\right\}_{\alpha^0}$$
$$= -\left\{\sigma_t^{(A,i)}\frac{Q(\alpha)\mu_D(\alpha)}{\mu_A^2(\alpha)}E_2\left[(b-a)\mu_B(\alpha)\right]\right\}_{\alpha^0}; \quad i=1,\dots,MA; \tag{2.93}$$

$$\left\{\frac{\partial R(\varphi;\alpha)}{\partial \sigma_t^{(A,i)}}\right\}_{\alpha^0} = -\left\{N^{(A,i)}\frac{Q(\alpha)}{\mu_A^2(\alpha)}\int_0^1 a^{(1)}(a,\omega)\omega d\omega\right\}_{\alpha^0}$$
$$= -\left\{N^{(A,i)}\frac{Q(\alpha)\mu_D(\alpha)}{\mu_A^2(\alpha)}E_2\left[(b-a)\mu_B(\alpha)\right]\right\}_{\alpha^0}; \quad i=1,\dots,MA; \tag{2.94}$$

$$\left\{\frac{\partial R(\varphi;\alpha)}{\partial N^{(B,i)}}\right\}_{\alpha^0} = -\left\{\sigma_t^{(B,i)}\int_0^1 d\omega \int_a^b a^{(1)}(z,\omega)\varphi(z,\omega)dz\right\}_{\alpha^0}$$
$$= \left\{\sigma_t^{(B,i)}\frac{Q(\alpha)\mu_D(\alpha)}{\mu_A(\alpha)}(a-b)E_1\left[(b-a)\mu_B(\alpha)\right]\right\}_{\alpha^0}; \tag{2.95}$$
$$i=1,\dots,MB;$$

$$\left\{\frac{\partial R(\varphi;\alpha)}{\partial \sigma_t^{(B,i)}}\right\}_{\alpha^0} = -\left\{N^{(B,i)}\int_0^1 d\omega \int_a^b a^{(1)}(z,\omega)\varphi(z,\omega)dz\right\}_{\alpha^0}$$

$$= \left\{N^{(B,i)}\frac{Q(\alpha)\mu_D(\alpha)}{\mu_A(\alpha)}(a-b)E_1\left[(b-a)\mu_B(\alpha)\right]\right\}_{\alpha^0}; \tag{2.96}$$

$$i=1,\dots,MB;$$

$$\left\{\frac{\partial R(\varphi;\alpha)}{\partial q_i}\right\}_{\alpha^0} = \left\{\frac{1}{\mu_A(\alpha)}\int_0^1 a^{(1)}(a,\omega)\omega d\omega\right\}_{\alpha^0}$$
$$= \left\{\frac{\mu_D(\alpha)}{\mu_A(\alpha)} E_2\left[(b-a)\mu_B(\alpha)\right]\right\}_{\alpha^0}; \quad i=1,\dots,MQ; \tag{2.97}$$

$$\left\{\frac{\partial R(\varphi;\alpha)}{\partial a}\right\}_{\alpha^0} = \left\{\frac{Q(\alpha)\mu_B(\alpha)}{\mu_A(\alpha)}\int_0^1 a^{(1)}(a,\omega)d\omega\right\}_{\alpha^0}$$
$$= \left\{\frac{Q(\alpha)\mu_B(\alpha)\mu_D(\alpha)}{\mu_A(\alpha)} E_1\left[(b-a)\mu_B(\alpha)\right]\right\}_{\alpha^0}. \tag{2.98}$$

It is noteworthy that the absolute sensitivities $\partial R(\varphi;\alpha)/\partial q_i$ all have equal values, i.e., $\partial R(\varphi;\alpha)/\partial q_1 = \cdots = \partial R(\varphi;\alpha)/\partial q_{MQ}$. Also noteworthy are the results that the relative sensitivities of the response with respect to the nuclide densities have the same values as the relative sensitivities of the response with respect to the microscopic cross sections, i.e.,

$$\left\{\frac{\partial R(\varphi;\alpha)}{\partial N^{(A,i)}}\frac{N^{(A,i)}}{R(\varphi;\alpha)}\right\}_{\alpha^0} = \left\{\frac{\partial R(\varphi;\alpha)}{\partial \sigma_t^{(A,i)}}\frac{\sigma_t^{(A,i)}}{R(\varphi;\alpha)}\right\}_{\alpha^0}$$
$$= -\left\{\frac{\sigma_t^{(A,i)}N^{(A,i)}}{\mu_A}\right\}_{\alpha^0}; \quad i=1,\dots,MA; \tag{2.99}$$

$$\left\{\frac{\partial R(\varphi;\alpha)}{\partial N^{(B,i)}}\frac{N^{(B,i)}}{R(\varphi;\alpha)}\right\}_{\alpha^0} = \left\{\frac{\partial R(\varphi;\alpha)}{\partial \sigma_t^{(B,i)}}\frac{\sigma_t^{(B,i)}}{R(\varphi;\alpha)}\right\}_{\alpha^0}$$
$$= \left\{\sigma_t^{(B,i)}N^{(B,i)}(a-b)\frac{E_1\left[(b-a)\mu_B(\alpha)\right]}{E_2\left[(b-a)\mu_B\right]}\right\}_{\alpha^0}; \tag{2.100}$$

$$i=1,\dots,MB;$$

$$\left\{\frac{\partial R(\varphi;\alpha)}{\partial N^{(D,i)}}\frac{N^{(D,i)}}{R(\varphi;\alpha)}\right\}_{\alpha^0} = \left\{\frac{\partial R(\varphi;\alpha)}{\partial \sigma_t^{(D,i)}}\frac{\sigma_t^{(D,i)}}{R(\varphi;\alpha)}\right\}_{\alpha^0}$$
$$= \left\{\frac{N^{(D,i)}\sigma_t^{(D,i)}}{\mu_D(\alpha)}\right\}_{\alpha^0}; \quad i=1,\dots,N_D. \tag{2.101}$$

Notably, obtaining all of the 1st-order partial response sensitivities to the primary model parameters has necessitated a single "large-scale" computation for determining the 1st-level adjoint sensitivity function $a^{(1)}(z,\omega)$. This is in contradistinction with the use of the 1st-LVSS, which would require at least *TP* large-scale computations to obtain the corresponding *TP* 1st-order sensitivities. Furthermore, the expressions obtained for these sensitivities are exact, as opposed to being approximate, as would have been the case if these sensitivities had been computed by finite-difference or statistical procedures. Also, many more large-scale computations would have been necessary to compute these 1st-order partial response sensitivities by any other (finite-difference or statistical) procedure. Thus, the 1st-CASAM-N is the most efficient computational method for obtaining the exact expressions of the 1st-order partial response sensitivities to the primary model parameters.

The fact that so many relative sensitivities have the same values, as indicated in Eqs. (2.99)–(2.101), respectively, would render quasi-useless the statistical methods that attempt to rank the sensitivities by their relative values, because such statistical methods break down in situations when many relative sensitivities have equal values. On the other hand, if statistical methods attempted to rank sensitivities by their absolute values, these statistical methods would fail when attempting to determine the response sensitivities to the material properties of the detector. In all cases, the only reliable method for computing response sensitivities accurately (and also most efficiently, from a computational standpoint) is the 1st-CASAM-N methodology.

### 2.4.2 Most Efficient Alternative Procedure: Applying the 1st-FASAM-N to Compute First-Order Response Sensitivities to Functions/Features of Model Parameters

The particle transport equation governing the uncollided angular flux $\varphi(z,\omega)$, i.e., Eqs. (2.63) and (2.65), can be written in the following form which highlights the "feature" functions of the primary model parameters:

$$\omega\frac{d\varphi(z,\omega)}{dz}+g_2(\boldsymbol{\alpha})\varphi(z,\omega)=0,\quad a<z<b,\quad 0\le\omega\le 1; \tag{2.102}$$

$$\varphi(a,\omega)=g_1(\boldsymbol{\alpha}); \tag{2.103}$$

$$R[\varphi(z,\omega);\mathbf{f}(\boldsymbol{\alpha})]\triangleq h(\boldsymbol{\alpha})\int_0^1 d\omega\int_a^b \varphi(z,\omega)\delta(z-b)dz. \tag{2.104}$$

where:

$$\mathbf{f}(\boldsymbol{\alpha})\triangleq[\mathbf{g}(\boldsymbol{\alpha});\mathbf{h}(\boldsymbol{\alpha});\boldsymbol{\lambda}(\boldsymbol{\alpha});\boldsymbol{\omega}(\boldsymbol{\alpha})]^{\dagger}\triangleq[f_1(\boldsymbol{\alpha}),\dots,f_{TF}(\boldsymbol{\alpha})]^{\dagger};\quad TF\triangleq 2+1+2; \tag{2.105}$$

$$\mathbf{g}(\alpha) \triangleq [g_1(\alpha), g_2(\alpha)]^\dagger; \quad g_1(\alpha) \triangleq Q(\alpha)\left[\sum_{i=1}^{MA} N^{(A,i)}\sigma_t^{(A,i)}\right]^{-1};$$

$$g_2(\alpha) \triangleq \mu_B(\alpha) = \sum_{i=1}^{MB} N^{(B,i)}\sigma_t^{(B,i)}; \tag{2.106}$$

$$h(\alpha) \triangleq \mu_D(\alpha) = \sum_{i=1}^{MD} N^{(D,i)}\sigma_t^{(D,i)}; \quad \lambda(\alpha) \triangleq a; \quad \omega(\alpha) \triangleq b.$$

In terms of the components of the "feature" function $\mathbf{f}(\alpha)$, the uncollided flux and the model response have the following expressions:

$$\varphi(z,\omega) = g_1(\alpha)\exp\left[\frac{g_2(\alpha)}{\omega}(a-z)\right], \quad a \le z \le b, \quad 0 \le \omega \le 1. \tag{2.107}$$

$$R(\varphi;\alpha) = g_1(\alpha)h(\alpha)E_2[(b-a)g_2(\alpha)]. \tag{2.108}$$

The 1st-order total sensitivity of $R(\varphi;\mathbf{f})$ is obtained by determining the first-order G-differential $\delta R(\varphi;\mathbf{f};\delta\varphi;\delta\mathbf{f})_{\alpha^0}$ of $R(\varphi;\mathbf{f})$ at the nominal parameter values, which is determined by applying the definition of the G-differential to Eq. (2.104). This operation yields the following expression:

$$\begin{aligned}\delta R\left(\varphi;\mathbf{f};v^{(1)};\delta\mathbf{f}\right)_{\alpha^0} &\triangleq \left\{\frac{d}{d\varepsilon}\left[(h+\varepsilon\delta h)\int_0^1 d\omega \int_{a+\varepsilon\delta a}^{b+\varepsilon\delta b}\left(\varphi+\varepsilon v^{(1)}\right)\delta(z-b-\varepsilon\delta b)\,dz\right]_{\alpha^0}\right\}_{\varepsilon=0} \\ &= \left\{\delta R(\varphi;\mathbf{f};\delta\mathbf{f})_{\alpha^0}\right\}_{dir} + \left\{\delta R\left(\varphi;\mathbf{f};v^{(1)}\right)_{\alpha^0}\right\}_{ind},\end{aligned} \tag{2.109}$$

where the direct-effect term $\left\{\delta R(\varphi;\mathbf{f};\delta\mathbf{f})_{\alpha^0}\right\}_{dir}$ depends only on the parameter variations $\delta\mathbf{f}$ and is defined as follows:

$$\begin{aligned}\left\{\delta R(\varphi;\mathbf{f};\delta\mathbf{f})_{\alpha^0}\right\}_{dir} &\triangleq (\delta h)\left\{\int_0^1 d\omega \int_a^b \varphi(z,\omega)\delta(z-b)\,dz\right\}_{\alpha^0} \\ &\quad -(\delta b)\left\{h(\alpha)\int_0^1 d\omega\,\varphi(z,\omega)\delta'(z-b)\right\}_{\alpha^0},\end{aligned} \tag{2.110}$$

and where the indirect-effect term $\left\{\delta R\left(\varphi;\mathbf{f};v^{(1)}\right)_{\alpha^0}\right\}_{ind}$ depends only on the variational function $v^{(1)}(z,\omega)$ and is defined as follows:

$$\left\{\delta R\left(\varphi;\mathbf{f};v^{(1)}\right)_{\alpha^0}\right\}_{ind} \triangleq \left\{h(\alpha)\int_0^1 d\omega\int_a^b v^{(1)}(z,\omega)\delta(z-b)dz\right\}_{\alpha^0}. \tag{2.111}$$

The 1st-LVSS to be satisfied by the 1st-order forward variational function $v^{(1)}(z,\omega)$ is obtained by G-differentiating Eqs. (2.102) and (2.103), to obtain the following system of equations:

$$\left\{\omega\frac{dv^{(1)}(z,\omega)}{dz}+g_2(\alpha)v^{(1)}(z,\omega)\right\}_{\alpha^0} = -\left\{(\delta g_2)\varphi(z,\omega)\right\}_{\alpha^0};\quad a^0<z<b^0,\quad 0\le\omega\le 1; \tag{2.112}$$

$$\begin{aligned}\left\{v^{(1)}(z,\omega)\right\}_{z=a^0} &= \left\{\delta g_1(\alpha)-(\delta a)g_1(\alpha)\left[\frac{d\varphi(z,\omega)}{dz}\right]_{z=a}\right\}_{\alpha^0}\\ &= \left\{\delta g_1(\alpha)\right\}_{\alpha^0}+(\delta a)\left\{\frac{g_1(\alpha)g_2(\alpha)}{\omega}\right\}_{\alpha^0};\quad z=a^0,\quad 0\le\omega\le 1.\end{aligned} \tag{2.113}$$

It is noteworthy that the above 1st-LVSS would need to be solved just three times to obtain the three variational functions $v^{(1)}(i;z,\omega)$, $i=1,2,3$, which would correspond to the variations $\delta g_1$, $\delta g_2$, and $\delta a$; the function $v^{(1)}(1;z,\omega)$ would be obtained by setting $\delta g_1=1$, $\delta g_2=0$, and $\delta a=0$; the function $v^{(1)}(2;z,\omega)$ would be obtained by setting $\delta g_1=0$, $\delta g_2=1$, and $\delta a=0$; and the function $v^{(1)}(3;z,\omega)$ would be obtained by setting $\delta g_1=0$, $\delta g_2=0$, and $\delta a=1$.

The need for computing the functions, $v^{(1)}(i;z,\omega)$, $i=1,2,3$, is circumvented by expressing the indirect-effect term $\left\{\delta R\left(\varphi;\mathbf{f};v^{(1)}\right)_{\alpha^0}\right\}_{ind}$ defined in Eq. (2.111) in terms of a 1st-level adjoint function, which will be denoted as $a^{(1)}(z,\omega)$, and which is the solution of a 1st-Level Adjoint Sensitivity System (1st-LASS) constructed by performing the same sequence of operations as indicated in Eqs. (2.80)–(2.82). The resulting 1st-LASS for the 1st-level adjoint function $a^{(1)}(z,\omega)$ is the same as was obtained in Eqs. (2.84) and (2.85), namely:

$$\left\{-\omega\frac{d}{dz}a^{(1)}(z,\omega)+g_2(\alpha)a^{(1)}(z,\omega)\right\}_{\alpha^0} = \left\{h(\alpha)\delta(z-b)\right\}_{\alpha^0},\quad a^0<z<b^0;\quad 0\le\omega\le 1. \tag{2.114}$$

$$a^{(1)}(z,\omega)=0,\quad at\quad z=b^0,\quad 0\le\omega\le 1. \tag{2.115}$$

The solution of Eqs. (2.114) and (2.115) has the following expression:

$$a^{(1)}(z,\omega)=\left\{\frac{h(\alpha)}{\omega}\left[1-H(z-b)\right]\exp\left[\frac{g_2(\alpha)(z-b)}{\omega}\right]\right\}_{\alpha^0}. \tag{2.116}$$

In terms of the 1st-level adjoint function $a^{(1)}(z,\omega)$, the indirect-effect term $\left\{\delta R\left(\varphi;\mathbf{f};v^{(1)}\right)_{\alpha^0}\right\}_{ind}$ will have the following expression:

$$\begin{aligned}\left\{\delta R\left(\varphi;\mathbf{f};v^{(1)}\right)_{\alpha^0}\right\}_{ind}&=\left\{\int_0^1 a^{(1)}(a,\omega)\left[(\delta g_1)+(\delta a)\frac{g_1(\alpha)g_2(\alpha)}{\omega}\right]\omega d\omega\right\}_{\alpha^0}\\&-\left\{(\delta g_2)\int_0^1 d\omega\int_a^b a^{(1)}(z,\omega)\varphi(z,\omega)dz\right\}_{\alpha^0}.\end{aligned} \tag{2.117}$$

Adding the expression for the indirect-effect term obtained in Eq. (2.117) with the expression of the direct-effect term obtained in Eq. (2.110) yields the following expression for the total 1st-order G-differential $\delta R(\varphi;\mathbf{f};\delta\mathbf{f})_{\alpha^0}$:

$$\begin{aligned}\delta R(\varphi;\mathbf{f};\delta\mathbf{f})_{\alpha^0}&=(\delta h)\left\{\int_0^1 d\omega\int_a^b \varphi(z,\omega)\delta(z-b)dz\right\}_{\alpha^0}\\&-(\delta b)\left\{h(\alpha)\int_0^1 d\omega\varphi(z,\omega)\delta'(z-b)\right\}_{\alpha^0}\\&-\left\{(\delta g_2)\int_0^1 d\omega\int_a^b a^{(1)}(z,\omega)\varphi(z,\omega)dz\right\}_{\alpha^0}\\&+\left\{\int_0^1 a^{(1)}(a,\omega)\left[(\delta g_1)+(\delta a)\frac{g_1(\alpha)g_2(\alpha)}{\omega}\right]\omega d\omega\right\}_{\alpha^0}.\end{aligned} \tag{2.118}$$

It follows from Eq. (2.118) that the expressions of the partial sensitivities of the response $R(\varphi;\alpha)$ with respect to the components of the "feature" function $\mathbf{f}(\alpha)$ are as follows:

$$\left\{\frac{\partial R(\varphi;\mathbf{f})}{\partial g_1(\alpha)}\right\}_{\alpha^0}=\left\{\int_0^1 a^{(1)}(a,\omega)\omega d\omega\right\}_{\alpha^0}=\left\{h(\alpha)E_2\left[(b-a)g_2(\alpha)\right]\right\}_{\alpha^0}; \tag{2.119}$$

$$\left\{\frac{\partial R(\varphi;\mathbf{f})}{\partial g_2(\boldsymbol{\alpha})}\right\}_{\boldsymbol{\alpha}^0} = -\left\{\int_0^1 d\omega \int_a^b a^{(1)}(z,\omega)\varphi(z,\omega)dz\right\}_{\boldsymbol{\alpha}^0} = \left\{g_1(\boldsymbol{\alpha})h(\boldsymbol{\alpha})(a-b)E_1\left[(b-a)g_2(\boldsymbol{\alpha})\right]\right\}_{\boldsymbol{\alpha}^0}; \tag{2.120}$$

$$\left\{\frac{\partial R(\varphi;\mathbf{f})}{\partial h(\boldsymbol{\alpha})}\right\}_{\boldsymbol{\alpha}^0} = \left\{\int_0^1 d\omega \int_a^b \varphi(z,\omega)\delta(z-b)dz\right\}_{\boldsymbol{\alpha}^0} = \left\{g_1(\boldsymbol{\alpha})E_2\left[(b-a)g_2(\boldsymbol{\alpha})\right]\right\}_{\boldsymbol{\alpha}^0} \tag{2.121}$$

$$\left\{\frac{\partial R(\varphi;\mathbf{f})}{\partial a}\right\}_{\boldsymbol{\alpha}^0} = \left\{g_1(\boldsymbol{\alpha})g_2(\boldsymbol{\alpha})\int_0^1 a^{(1)}(a,\omega)d\omega\right\}_{\boldsymbol{\alpha}^0} = \left\{g_1(\boldsymbol{\alpha})g_2(\boldsymbol{\alpha})h(\boldsymbol{\alpha})E_1\left[(b-a)g_2(\boldsymbol{\alpha})\right]\right\}_{\boldsymbol{\alpha}^0}; \tag{2.122}$$

$$\left\{\frac{\partial R(\varphi;\mathbf{f})}{\partial b}\right\}_{\boldsymbol{\alpha}^0} = -\left\{h(\boldsymbol{\alpha})\int_0^1 d\omega \int_a^b \varphi(z,\omega)\delta'(z-b)dz\right\}_{\boldsymbol{\alpha}^0} = -\left\{g_1(\boldsymbol{\alpha})g_2(\boldsymbol{\alpha})h(\boldsymbol{\alpha})E_1\left[(b-a)g_2(\boldsymbol{\alpha})\right]\right\}_{\boldsymbol{\alpha}^0}. \tag{2.123}$$

The above expressions can be used to obtain efficiently the sensitivities of the response with respect to the primary model parameters by using the chain-rule of differentiation, as follows:

$$\frac{\partial R(\varphi;\boldsymbol{\alpha})}{\partial q_i} = \frac{\partial R(\varphi;\mathbf{f})}{\partial g_1(\boldsymbol{\alpha})}\frac{\partial g_1(\boldsymbol{\alpha})}{\partial q_i} = \frac{1}{\mu_A(\boldsymbol{\alpha})}\frac{\partial R(\varphi;\mathbf{f})}{\partial g_1(\boldsymbol{\alpha})}; \quad i=1,\ldots,MQ; \tag{2.124}$$

$$\frac{\partial R(\varphi;\boldsymbol{\alpha})}{\partial N^{(A,i)}} = \frac{\partial R(\varphi;\mathbf{f})}{\partial g_1(\boldsymbol{\alpha})}\frac{\partial g_1(\boldsymbol{\alpha})}{\partial N^{(A,i)}} = -\frac{q\sigma_t^{(A,i)}}{\mu_A^2(\boldsymbol{\alpha})}\frac{\partial R(\varphi;\mathbf{f})}{\partial g_1(\boldsymbol{\alpha})}; \quad i=1,\ldots,M_A; \tag{2.125}$$

$$\frac{\partial R(\varphi;\boldsymbol{\alpha})}{\partial \sigma_t^{(A,i)}} = \frac{\partial R(\varphi;\mathbf{f})}{\partial g_1(\boldsymbol{\alpha})}\frac{\partial g_1(\boldsymbol{\alpha})}{\partial \sigma_t^{(A,i)}} = -\frac{qN^{(A,i)}}{\mu_A^2(\boldsymbol{\alpha})}\frac{\partial R(\varphi;\mathbf{f})}{\partial g_1(\boldsymbol{\alpha})}; \quad i=1,\ldots,M_A; \tag{2.126}$$

$$\frac{\partial R(\varphi;\boldsymbol{\alpha})}{\partial N^{(B,i)}} = \frac{\partial R(\varphi;\mathbf{f})}{\partial g_2(\boldsymbol{\alpha})}\frac{\partial g_2(\boldsymbol{\alpha})}{\partial N^{(B,i)}} = \sigma_t^{(B,i)}\frac{\partial R(\varphi;\mathbf{f})}{\partial g_2(\boldsymbol{\alpha})}; \quad i=1,\ldots,M_B; \tag{2.127}$$

$$\frac{\partial R(\varphi;\boldsymbol{\alpha})}{\partial \sigma_t^{(B,i)}} = \frac{\partial R(\varphi;\mathbf{f})}{\partial g_2(\boldsymbol{\alpha})}\frac{\partial g_2(\boldsymbol{\alpha})}{\partial \sigma_t^{(B,i)}} = N^{(B,i)}\frac{\partial R(\varphi;\mathbf{f})}{\partial g_2(\boldsymbol{\alpha})}; \quad i=1,\ldots,M_B; \tag{2.128}$$

$$\frac{\partial R(\varphi;\boldsymbol{\alpha})}{\partial N^{(D,i)}} = \frac{\partial R(\varphi;\mathbf{f})}{\partial h(\boldsymbol{\alpha})}\frac{\partial h(\boldsymbol{\alpha})}{\partial N^{(D,i)}} = \sigma_t^{(D,i)}\frac{\partial R(\varphi;\mathbf{f})}{\partial h(\boldsymbol{\alpha})}; \quad i=1,\ldots,M_D; \tag{2.129}$$

$$\frac{\partial R(\varphi;\boldsymbol{\alpha})}{\partial \sigma_t^{(D,i)}} = \frac{\partial R(\varphi;\mathbf{f})}{\partial h(\boldsymbol{\alpha})}\frac{\partial h(\boldsymbol{\alpha})}{\partial \sigma_t^{(D,i)}} = N^{(D,i)}\frac{\partial R(\varphi;\mathbf{f})}{\partial h(\boldsymbol{\alpha})}; \quad i=1,\ldots,M_D. \tag{2.130}$$

It can be readily verified that the expressions shown in Eqs. (2.122)–(2.130) are identical to the expressions of the corresponding sensitivities to the model parameters obtained using the 1st-CASAM-N framework in Eqs. (2.90)–(2.98).

It is important to note that while there are a total number of $TP = MQ + 2MA + 2MB + 2MD + 2$ 1st-order sensitivities of the response with respect to the primary model parameters, there are only $TF \triangleq 3+2=5$ 1st-order sensitivities of the response with respect to the five components of the "feature" function $\mathbf{f}(\boldsymbol{\alpha}) \triangleq \left[g_1(\boldsymbol{\alpha}), g_2(\boldsymbol{\alpha}); h(\boldsymbol{\alpha}); a; b\right]^{\dagger}$, of which three of these sensitivities are with respect to the bona-fide feature functions $g_1(\boldsymbol{\alpha}), g_2(\boldsymbol{\alpha}); h(\boldsymbol{\alpha})$, while the other two sensitivities are with respect to the boundary parameters $(a,b)$, which are "primary parameters" in this illustrative example. The decisive advantages of the 1st-FASAM-N framework over the 1st-CASAM-N framework will become apparent when computing the second-order sensitivities, since, as will be shown in the next Chapter, only $TF = 5$ 2nd-level adjoint sensitivity functions would be needed within the 2nd-FASAM-N framework, versus the much larger number of $TP = MQ + 2MA + 2MB + 2MD + 2$ 2nd-level adjoint sensitivity functions that would be needed within the 2nd-CASAM-N framework.

The illustrative paradigm model analyzed in the foregoing admitted feature functions which depend on mutually exclusive groups of primary parameters, so that none of the components of the "feature" function $\mathbf{f}(\boldsymbol{\alpha}) \triangleq \left[g_1(\boldsymbol{\alpha}), g_2(\boldsymbol{\alpha}); h(\boldsymbol{\alpha}); a; b\right]^{\dagger}$ had any primary parameters in common. It is advantageous to choose components of the "feature" function in this way (i.e., to not depend on any common primary parameters) since it is subsequently easier to compute the sensitivities with respect to the primary parameters from the sensitivities with respect to the components of the "feature" function. Of course, it may not always be possible to choose the components of the "feature" function to not share dependencies on the mutually independent primary model parameters. Hence, it may happen that some components of the "feature" function $\mathbf{f}(\boldsymbol{\alpha}) \triangleq \left[g_1(\boldsymbol{\alpha}), g_2(\boldsymbol{\alpha}); h(\boldsymbol{\alpha}); a; b\right]^{\dagger}$ would share dependencies on some common primary parameters. In such instances, particular attention would need to be given to ensuring that the sensitivities of the response with respect to the primary parameters are correctly extracted from the sensitivities of the response with respect to the components of the "feature" function, to ensure that all contributions are properly considered.

## 2.5 CHAPTER SUMMARY

This chapter has presented the "1st-Order Function/Feature Adjoint Sensitivity Analysis Methodology for Nonlinear Systems" (1st-FASAM-N). This methodology enables the most efficient computation of exactly obtained expressions of the first-order sensitivities of a model response with respect to a *function* ("feature") of primary model parameters. For the computation of the first-order such sensitivities, the 1st-FASAM-N requires a single large-scale computation, which is needed for solving the "1st-level adjoint sensitivity system" (1st-LASS). This is similar to the computational requirement needed by the extant 1st-CASAM-N (Cacuci 2016, 2022a) for obtaining the sensitivities to the primary model parameters. All of these considerations were demonstrated in this chapter by using a simple particle transport/shielding model which admits closed-forms analytic expressions for all of the quantities of interest.

# 3 The 2nd-FASAM-N Methodology for Nonlinear Systems

## 3.1 INTRODUCTION

Chapter 2 has presented the "1st-Order Functions/Features Adjoint Sensitivity Analysis Methodology for Nonlinear systems" (1st-FASAM-N), which enables the most efficient computation of exactly obtained expressions of 1st-order sensitivities of model responses to functions/features of model parameters. This chapter presents the "2nd-Order Function/Feature Adjoint Sensitivity Analysis Methodology" (2nd-FASAM-N), which enables the most efficient computation of exactly obtained expressions for the 2nd-order sensitivities of a model response with respect to a multi-component *function* ("feature") of primary model parameters. This chapter is structured as follows: Section 3.2 presents the mathematical framework of the "Second-Order Function/Feature Adjoint Sensitivity Analysis Methodology for Nonlinear Systems" (2nd-FASAM-N). Section 3.3 illustrates the application of the 2nd-FASAM-N to determine the 2nd-order sensitivities of a "dosimetry" detector response for the paradigm model considered in Chapter 2, which models the transport of particles (neutrons and/or gamma rays) through a shielding material and admits closed-form analytical expressions for the 1st- and 2nd-order sensitivities of the model's response with respect to the model's "feature" functions and/or parameters. This paradigm illustrative model highlights the efficiency of the 2nd-FASAM-N for obtaining and computing exact expressions for the 2nd-order sensitivities. Comparisons with the 2nd-CASAM-N (Cacuci 2016, 2022a, 2023a) are also provided, illustrating the connection between the 2nd-FASAM-N – which provides exact sensitivities with respect to functions/features of parameters – and the 2nd-CASAM-N – which provides 2nd-order response sensitivities directly with respect to the primary parameters. Section 3.4 summarizes the material presented in this chapter.

## 3.2 MATHEMATICAL FRAMEWORK OF THE 2ND-FASAM-N METHODOLOGY

The "Second-Order Function/Feature Adjoint Sensitivity Analysis Methodology" (2nd-FASAM-N) determines the exact expressions of the 2nd-order sensitivities $\partial^2 R[\mathbf{u}(\mathbf{x});\mathbf{f}(\alpha)]/\partial f_{j2}\,\partial f_{j1}$ of the response with respect to the components of the "feature" function $\mathbf{f}(\alpha) \triangleq [\mathbf{g}(\alpha);\mathbf{h}(\alpha);\boldsymbol{\lambda}(\alpha);\boldsymbol{\omega}(\alpha)]^{\dagger}$ and computes them with unparalleled efficiency. Conceptually, the 2nd-FASAM-N considers the 1st-order


DOI: 10.1201/9781003478171-3

sensitivities $R^{(1)}\left[j_1;\mathbf{u}(\mathbf{x});\mathbf{a}^{(1)}(\mathbf{x});\mathbf{f}(\boldsymbol{\alpha})\right]\triangleq\partial R\left[\mathbf{u}(\mathbf{x});\mathbf{f}(\boldsymbol{\alpha})\right]/\partial f_{j_1}$, which were obtained in Chapter 2, to be the equivalent of "model responses" and determines the 2nd-order sensitivities by applying the concepts underlying 1st-FASAM-N to the "response" $R^{(1)}\left[j_1;\mathbf{u}(\mathbf{x});\mathbf{a}^{(1)}(\mathbf{x});\mathbf{f}(\boldsymbol{\alpha})\right]$. In other words, the 2nd-FASAM-N determines the 2nd-order sensitivities as the "1st-order sensitivities of the 1st-order sensitivities."

The 1st-order G-differential of a generic 1st-order sensitivity $R^{(1)}\left[j_1;\mathbf{u}(\mathbf{x});\mathbf{a}^{(1)}(\mathbf{x});\mathbf{f}\right]$, $j_1=1,\ldots,TF$, is obtained, by definition, as follows:

$$
\begin{aligned}
&\left\{\delta R^{(1)}\left[j_1;\mathbf{u}(\mathbf{x});\mathbf{a}^{(1)}(\mathbf{x});\mathbf{f};\mathbf{v}^{(1)}(\mathbf{x});\delta\mathbf{a}^{(1)}(\mathbf{x});\delta\mathbf{f}\right]\right\}_{\boldsymbol{\alpha}^0}\\
&\triangleq\left\{\frac{d}{d\varepsilon}\delta R^{(1)}\left[j_1;\mathbf{u}(\mathbf{x})+\varepsilon\mathbf{v}^{(1)}(\mathbf{x});\mathbf{a}^{(1)}(\mathbf{x})+\varepsilon\delta\mathbf{a}^{(1)}(\mathbf{x});\mathbf{f}+\varepsilon\delta\mathbf{f}\right]\right\}_{\varepsilon=0}\\
&=\sum_{j_2=1}^{TF}\left\{\frac{\partial R^{(1)}\left[j_1;\mathbf{u};\mathbf{a}^{(1)};\mathbf{f}\right]}{\partial f_{j_2}}\right\}_{\boldsymbol{\alpha}^0}\delta f_{j_2}+\left\{\delta R^{(1)}\left[j_1;\mathbf{u};\mathbf{a}^{(1)};\mathbf{f};\mathbf{v}^{(1)}(\mathbf{x});\delta\mathbf{a}^{(1)}(\mathbf{x})\right]\right\}_{ind},
\end{aligned}
\tag{3.1}
$$

where the indirect-effect term $\left\{\delta R^{(1)}\left[j_1;\mathbf{u};\mathbf{a}^{(1)};\mathbf{f};\mathbf{v}^{(1)}(\mathbf{x});\delta\mathbf{a}^{(1)}(\mathbf{x})\right]\right\}_{ind}$ comprises all dependencies on the vectors $\mathbf{v}^{(1)}(\mathbf{x})$ and $\delta\mathbf{a}^{(1)}(\mathbf{x})$ of variations in the state functions $\mathbf{u}(\mathbf{x})$ and $\mathbf{a}^{(1)}(\mathbf{x})$, respectively, and is defined as follows:

$$
\begin{aligned}
&\left\{\delta R^{(1)}\left[j_1;\mathbf{u};\mathbf{a}^{(1)};\mathbf{f};\mathbf{v}^{(1)};\delta\mathbf{a}^{(1)}\right]\right\}_{ind}\\
&\triangleq\int_{\lambda_1(\boldsymbol{\alpha})}^{\omega_1(\boldsymbol{\alpha})}dx_1\ldots\int_{\lambda_{TI}(\boldsymbol{\alpha})}^{\omega_{TI}(\boldsymbol{\alpha})}dx_{TI}\left\{\frac{\partial S^{(1)}\left[j_1;\mathbf{u}(\mathbf{x});\mathbf{a}^{(1)}(\mathbf{x});\mathbf{f}\right]}{\partial\mathbf{u}}\mathbf{v}^{(1)}(\mathbf{x})\right.\\
&\left.+\frac{\partial S^{(1)}\left[j_1;\mathbf{u}(\mathbf{x});\mathbf{a}^{(1)}(\mathbf{x});\mathbf{f}\right]}{\partial\mathbf{a}^{(1)}}\delta\mathbf{a}^{(1)}(\mathbf{x})\right\}_{\boldsymbol{\alpha}^0},
\end{aligned}
\tag{3.2}
$$

The functions $\mathbf{v}^{(1)}(\mathbf{x})$ and $\delta\mathbf{a}^{(1)}(\mathbf{x})$ are obtained by solving the 2nd-Level Variational Sensitivity System (2nd-LVSS), which consists of concatenating the 1st-LVSS with the G-differentiated 1st-LASS to obtain the following system (2nd-LVSS):

$$
\begin{aligned}
&\left\{\mathbf{VM}^{(2)}\left[2\times 2;\mathbf{U}^{(2)}(2;\mathbf{x});\mathbf{f}\right]\mathbf{V}^{(2)}(2;\mathbf{x})\right\}_{\boldsymbol{\alpha}^0}\\
&=\left\{\mathbf{Q}_V^{(2)}\left[2;\mathbf{U}^{(2)}(2;\mathbf{x});\mathbf{f};\delta\boldsymbol{\alpha}\right]\right\}_{\boldsymbol{\alpha}^0},\quad \mathbf{x}\in\Omega_x,
\end{aligned}
\tag{3.3}
$$

$$
\left\{\mathbf{B}_V^{(2)}\left[2;\mathbf{U}^{(2)}(2;\mathbf{x});\mathbf{V}^{(2)}(2;\mathbf{x});\mathbf{f};\delta\mathbf{f}\right]\right\}_{\boldsymbol{\alpha}^0}=\mathbf{0}[2]\triangleq[\mathbf{0},\mathbf{0}]^{\dagger},\quad \mathbf{x}\in\partial\Omega_x\left(\boldsymbol{\alpha}^0\right).
\tag{3.4}
$$

The argument "2," which appears in the list of arguments of the vector $\mathbf{U}^{(2)}(2;\mathbf{x})$ and of the "variational vector" $\mathbf{V}^{(2)}(2;\mathbf{x})$ in Eq. (3.3) indicates that each of these vectors is a two-block column vector, each block comprising a column vector of dimension $TD$, defined as follows:

$$\mathbf{U}^{(2)}(2;\mathbf{x}) \triangleq \begin{pmatrix} \mathbf{u}^{(1)}(\mathbf{x}) \\ \mathbf{a}^{(1)}(\mathbf{x}) \end{pmatrix}; \quad \mathbf{V}^{(2)}(2;\mathbf{x}) \triangleq \delta\mathbf{U}^{(2)}(2;\mathbf{x}) \triangleq \begin{pmatrix} \mathbf{v}^{(2)}(1;\mathbf{x}) \\ \mathbf{v}^{(2)}(2;\mathbf{x}) \end{pmatrix}$$

$$\triangleq \begin{pmatrix} \mathbf{v}^{(1)}(\mathbf{x}) \\ \delta\mathbf{a}^{(1)}(\mathbf{x}) \end{pmatrix}. \tag{3.5}$$

To distinguish block vectors from block matrices, two bold capital letters have been used (and will henceforth be used) to denote block matrices, as in the case of "the 2nd-level variational matrix" $\mathbf{VM}^{(2)}\left[2\times 2;\mathbf{u}^{(2)}(\mathbf{x});\boldsymbol{\alpha}\right]$. The "2nd-level" is indicated by the superscript "(2)." The argument "$2\times 2$," which appears in the list of arguments of $\mathbf{VM}^{(2)}\left[2\times 2;\mathbf{u}^{(2)}(\mathbf{x});\mathbf{f}\right]$, indicates that this matrix is a $2\times 2$-dimensional block matrix comprising four matrices, each of dimensions $TD\times TD$. In particular, the structure of this block matrix is provided below:

$$\mathbf{VM}^{(2)}\left[2\times 2;\mathbf{U}^{(2)}(2;\mathbf{x});\mathbf{f}\right] \triangleq \begin{pmatrix} \mathbf{N}^{(1)} & \mathbf{0} \\ \mathbf{V}_{21}^{(2)} & \mathbf{V}_{22}^{(2)} \end{pmatrix}, \tag{3.6}$$

where:

$$\mathbf{V}_{21}^{(2)}\left(\mathbf{u};\mathbf{a}^{(1)};\mathbf{f}\right) \triangleq \frac{\partial\left\{\left[\mathbf{N}^{(1)}(\mathbf{u};\mathbf{f})\right]^{*}\mathbf{a}^{(1)}\right\}}{\partial\mathbf{u}} - \frac{\mathbf{q}_{A}^{(1)}\left[\mathbf{u}(\mathbf{x});\mathbf{f}\right]}{\partial\mathbf{u}}; \tag{3.7}$$

$$\mathbf{V}_{22}^{(2)}(\mathbf{u};\mathbf{f}) \triangleq \left[\mathbf{N}^{(1)}(\mathbf{u};\mathbf{f})\right]^{*}. \tag{3.8}$$

The other quantities that appear in Eqs. (3.3) and (3.4) are two-block vectors having the same structure as $\mathbf{V}^{(2)}(2;\mathbf{x})$, and are defined as follows:

$$\mathbf{Q}_{V}^{(2)}\left[2;\mathbf{U}^{(2)}(2;\mathbf{x});\mathbf{f};\delta\mathbf{f}\right] \triangleq \begin{pmatrix} \mathbf{q}_{V}^{(2)}\left(1;\mathbf{U}^{(2)}(2;\mathbf{x});\mathbf{f};\delta\mathbf{f}\right) \\ \mathbf{q}_{V}^{(2)}\left(2;\mathbf{U}^{(2)}(2;\mathbf{x});\mathbf{f};\delta\mathbf{f}\right) \end{pmatrix}; \tag{3.9}$$

$$\mathbf{q}_{V}^{(2)}\left[i;\mathbf{U}^{(2)}(2;\mathbf{x});\mathbf{f};\delta\mathbf{f}\right] \triangleq \sum_{j_2=1}^{TP} \mathbf{s}_{V}^{(2)}\left[i;j_2;\mathbf{U}^{(2)}(2;\mathbf{x});\boldsymbol{\alpha}\right]\delta f_{j_2}, \quad i=1,2; \tag{3.10}$$

$$\mathbf{s}_{V}^{(2)}\left[1;j_2;\mathbf{U}^{(2)}(2;\mathbf{x});\mathbf{f}\right] \triangleq \frac{\partial\left[\mathbf{Q}(\mathbf{f})-\mathbf{N}(\mathbf{u};\mathbf{f})\right]}{\partial f_{j_2}}; \tag{3.11}$$

$$\mathbf{s}_V^{(2)}\left[2;j_2;\mathbf{U}^{(2)}(2;\mathbf{x});\mathbf{f}\right]\triangleq\frac{\partial\mathbf{q}_A^{(1)}[\mathbf{u}(\mathbf{x});\mathbf{f}]}{\partial f_{j2}}-\frac{\partial\left\{\left[\mathbf{N}^{(1)}(\mathbf{u};\mathbf{f})\right]^*\mathbf{a}^{(1)}(\mathbf{x})\right\}}{\partial f_{j2}}; \quad (3.12)$$

$$\mathbf{B}_V^{(2)}\left[2;\mathbf{U}^{(2)}(2;\mathbf{x});\mathbf{V}^{(2)}(2;\mathbf{x});\mathbf{f};\delta\mathbf{f}\right]\triangleq\begin{pmatrix}\mathbf{b}_V^{(1)}\left(\mathbf{u}^{(1)};\mathbf{f};\delta\mathbf{u}^{(1)};\delta\mathbf{f}\right)\\ \delta\mathbf{b}_A^{(1)}\left[\mathbf{U}^{(2)}(2;\mathbf{x});\mathbf{V}^{(2)}(2;\mathbf{x});\mathbf{f};\delta\mathbf{f}\right]\end{pmatrix}$$
$$\triangleq\begin{pmatrix}\mathbf{b}_V^{(2)}\left[1;\mathbf{U}^{(2)}(2;\mathbf{x});\mathbf{V}^{(2)}(2;\mathbf{x});\mathbf{f};\delta\mathbf{f}\right]\\ \mathbf{b}_V^{(2)}\left[2;\mathbf{U}^{(2)}(2;\mathbf{x});\mathbf{V}^{(2)}(2;\mathbf{x});\mathbf{f};\delta\mathbf{f}\right]\end{pmatrix}; \quad (3.13)$$

$$\delta\mathbf{b}_A^{(1)}\left(\mathbf{u};\mathbf{a}^{(1)};\mathbf{f}\right)\triangleq\frac{\partial\mathbf{b}_A^{(1)}}{\partial\mathbf{u}}\mathbf{v}^{(1)}(\mathbf{x})+\frac{\partial\mathbf{b}_A^{(1)}}{\partial\mathbf{a}^{(1)}}\delta\mathbf{a}^{(1)}(\mathbf{x})+\frac{\partial\mathbf{b}_A^{(1)}}{\partial\mathbf{f}}\delta\mathbf{f}. \quad (3.14)$$

The argument "2" in the expression $\mathbf{0}[2]\triangleq[\mathbf{0},\mathbf{0}]^\dagger$ in Eq. (3.4) indicates that this expression is a two-block column vector comprising two vectors, each having $TD$ -components, all of which are zero-valued.

The need for solving the 2nd-LVSS is circumvented by deriving an alternative expression for the indirect-effect term $\left\{\delta R^{(1)}\left[j_1;\mathbf{u};\mathbf{a}^{(1)};\mathbf{f};\mathbf{V}^{(2)}(2;\mathbf{x})\right]\right\}_{ind}$ defined in Eq. (3.2), in which the function $\mathbf{V}^{(2)}(2;\mathbf{x})$ will be replaced by a 2nd-level adjoint function that is independent of variations in the model parameter and state functions. This 2nd-level adjoint function is the solution of a 2nd-Level Adjoint Sensitivity System (2nd-LASS) that will be constructed by using the same principles as employed for deriving the 1st-LASS. The 2nd-LASS is constructed in a Hilbert space, denoted as $\mathcal{H}_2(\Omega_x)$, comprising as elements block-vectors of the same form as $\mathbf{V}^{(2)}(2;\mathbf{x})$. The Hilbert space $\mathcal{H}_2(\Omega_x)$ is endowed with the following inner product, denoted as $\left\langle\mathbf{\Psi}^{(2)}(2;\mathbf{x}),\mathbf{\Phi}^{(2)}(\mathbf{x})\right\rangle_2$, between two vectors $\mathbf{\Psi}^{(2)}(2;\mathbf{x})\triangleq\left[\boldsymbol{\psi}^{(2)}(1;\mathbf{x}),\boldsymbol{\psi}^{(2)}(2;\mathbf{x})\right]^\dagger\in\mathcal{H}_2(\Omega_x)$ and $\mathbf{\Phi}^{(2)}(\mathbf{x})\triangleq\left[\boldsymbol{\varphi}^{(2)}(1;\mathbf{x}),\boldsymbol{\varphi}^{(2)}(2;\mathbf{x})\right]^\dagger\in\mathcal{H}_2(\Omega_x)$:

$$\left\langle\mathbf{\Psi}^{(2)}(2;\mathbf{x}),\mathbf{\Phi}^{(2)}(\mathbf{x})\right\rangle_2\triangleq\sum_{i=1}^{2}\left\langle\boldsymbol{\psi}^{(2)}(i;\mathbf{x}),\boldsymbol{\varphi}^{(2)}(i;\mathbf{x})\right\rangle_1 \quad (3.15)$$

The inner product defined in Eq. (3.15) is used to construct the 2nd-Level Adjoint Sensitivity System (2nd-LASS) for the 2nd-level adjoint function $\mathbf{A}^{(2)}(2;j_1;\mathbf{x})\triangleq\left[\mathbf{a}^{(2)}(1;j_1;\mathbf{x}),\mathbf{a}^{(2)}(2;j_1;\mathbf{x})\right]^\dagger\in\mathcal{H}_2(\Omega_x)$, having two components denoted as $\mathbf{a}^{(2)}(1;j_1;\mathbf{x})$ and $\mathbf{a}^{(2)}(2;j_1;\mathbf{x})$ that are distinguished from each other by the use of the numbers "1" and "2" in the respective list of arguments for each $j_1 = 1,\ldots,TF$. This 2nd-LASS is constructed by following the same steps as those for constructing the 1st-LASS in Chapter 2, as follows:

1. Form the inner product of the 2nd-level adjoint function of the form $\mathbf{A}^{(2)}(2;j_1;\mathbf{x}) \triangleq \left[\mathbf{a}^{(2)}(1;j_1;\mathbf{x}),\mathbf{a}^{(2)}(2;j_1;\mathbf{x})\right]^{\dagger}$ with Eq. (3.3) to obtain the following relation:

$$\left\{\left\langle \mathbf{A}^{(2)}(2;j_1;\mathbf{x}),\mathbf{VM}^{(2)}\mathbf{V}^{(2)}(2;\mathbf{x})\right\rangle_2\right\}_{\alpha^0} = \left\{\left\langle \mathbf{A}^{(2)}(2;j_1;\mathbf{x}),\mathbf{Q}_V^{(2)}\right\rangle_2\right\}_{\alpha^0}. \quad (3.16)$$

2. Use the definition of the adjoint operator to transform the left side of Eq. (3.16) as follows:

$$\left\{\left\langle \mathbf{A}^{(2)}(2;\mathbf{x}),\mathbf{VM}^{(2)}\mathbf{V}^{(2)}(2;\mathbf{x})\right\rangle_2\right\}_{\alpha^0} = \left\{\left[P^{(2)}\left(\mathbf{U}^{(2)};\mathbf{A}^{(2)};\mathbf{V}^{(2)};\mathbf{f}\right)\right]_{\partial\Omega_x}\right\}_{\alpha^0}$$
$$+\left\{\left\langle \mathbf{V}^{(2)}(2;\mathbf{x}),\mathbf{AM}^{(2)}\left[2\times 2;\mathbf{U}^{(2)}(2;\mathbf{x});\boldsymbol{\alpha}\right]\mathbf{A}^{(2)}(2;\mathbf{x})\right\rangle_2\right\}_{\alpha^0}, \quad (3.17)$$

where the quantity $\left\{\left[P^{(2)}\left(\mathbf{U}^{(2)};\mathbf{A}^{(2)};\mathbf{V}^{(2)};\mathbf{f}\right)\right]_{\partial\Omega_x}\right\}_{\alpha^0}$ denotes the corresponding bilinear concomitant on the domain's boundary, evaluated at the nominal values for the parameters and respective state functions, and where the block matrix operator $\mathbf{AM}^{(2)}\left[2\times 2;\mathbf{U}^{(2)}(2;\mathbf{x});\mathbf{f}\right]$ is the formal adjoint of $\mathbf{VM}^{(2)}\left(2\times 2;\mathbf{U}^{(2)}(2;\mathbf{x});\mathbf{f}\right)$, i.e., $\mathbf{AM}^{(2)}\left[2\times 2;\mathbf{U}^{(2)}(2;\mathbf{x});\mathbf{f}\right] \triangleq \left[\mathbf{VM}^{(2)}\left(2\times 2;\mathbf{U}^{(2)}(2;\mathbf{x});\mathbf{f}\right)\right]^{*}$, comprising $(2\times 2)$ block matrices, each of dimensions $TD^2$ (thus comprising a total of $(2\times 2)TD^2$ elements), and is defined as follows:

$$\mathbf{AM}^{(2)}\left[2\times 2;\mathbf{U}^{(2)}(2;\mathbf{x});\mathbf{f}\right] \triangleq \begin{pmatrix} \left\{\left[\mathbf{N}^{(1)}\right]^{*}\right\}^{\dagger} & \left[\mathbf{V}_{21}^{(2)*}\right]^{\dagger} \\ \mathbf{0} & \left[\mathbf{V}_{22}^{(2)*}\right]^{\dagger} \end{pmatrix} \quad (3.18)$$

3. The second term on the right side of Eq. (3.17) is now required to represent the indirect-effect term defined in Eq. (3.2) by imposing the relation

$$\left\{\mathbf{AM}^{(2)}\left[2\times 2;\mathbf{U}^{(2)}(2;\mathbf{x});\mathbf{f}\right]\mathbf{A}^{(2)}(2;j_1;\mathbf{x})\right\}_{\alpha^0}$$
$$= \left\{\mathbf{Q}_A^{(2)}\left[2;j_1;\mathbf{U}^{(2)}(2;\mathbf{x});\mathbf{f}\right]\right\}_{\alpha^0}, \quad j_1 = 1,\ldots,TF; \quad \mathbf{x}\in\Omega_x, \quad (3.19)$$

where:

$$\mathbf{Q}_A^{(2)}\left[2; j_1; \mathbf{U}^{(2)}(2;\mathbf{x}); \mathbf{f}\right] \triangleq \begin{pmatrix} \mathbf{q}_A^{(2)}\left(1; j_1; \mathbf{U}^{(2)}; \mathbf{f}\right) \\ \mathbf{q}_A^{(2)}\left(2; j_1; \mathbf{U}^{(2)}; \mathbf{f}\right) \end{pmatrix}$$

$$\triangleq \begin{pmatrix} \partial S^{(1)}\left[j_1; \mathbf{u}(\mathbf{x}); \mathbf{a}^{(1)}(\mathbf{x}); \mathbf{f}; \mathbf{v}^{(1)}(\mathbf{x})\right] \big/ \partial \mathbf{u} \\ \partial S^{(1)}\left[j_1; \mathbf{u}(\mathbf{x}); \mathbf{a}^{(1)}(\mathbf{x}); \mathbf{f}; \mathbf{v}^{(1)}(\mathbf{x})\right] \big/ \partial \mathbf{a}^{(1)} \end{pmatrix}, \quad (3.20)$$

$$j_1 = 1, \ldots, TF.$$

4. The 2nd-level adjoint boundary/initial conditions for the 2nd-level adjoint sensitivity function $\mathbf{A}^{(2)}(2; j_1; \mathbf{x}) \triangleq \left[\mathbf{a}^{(2)}(1; j_1; \mathbf{x}), \mathbf{a}^{(2)}(2; j_1; \mathbf{x})\right]^{\dagger}$ are now selected by requiring that they satisfy the following conditions:
   a. they must be independent of unknown values of $\mathbf{V}^{(2)}(2;\mathbf{x})$;
   b. the substitution of the boundary and/or initial conditions represented by Eqs. (3.21) and (3.4) into the expression of $\left\{\left[P^{(2)}\left(\mathbf{U}^{(2)}; \mathbf{A}^{(2)}; \mathbf{V}^{(2)}; \mathbf{f}\right)\right]_{\partial\Omega_x}\right\}_{\alpha^0}$ must cause all terms containing unknown values of $\mathbf{V}^{(2)}(2;\mathbf{x})$ to vanish.

The 2nd-level adjoint boundary/initial conditions thus determined are represented in operator form as follows:

$$\begin{gathered} \left\{\mathbf{B}_A^{(2)}\left[2; \mathbf{U}^{(2)}(2;\mathbf{x}); \mathbf{A}^{(2)}(2; j_1; \mathbf{x}); \boldsymbol{\alpha}\right]\right\}_{\alpha^0} \\ = \mathbf{0}[2]; \quad j_1 = 1, \ldots, TF; \quad \mathbf{x} \in \partial\Omega_x\left(\alpha^0\right), \end{gathered} \quad (3.21)$$

The 2nd-LASS for the 2nd-level adjoint sensitivity function $\mathbf{A}^{(2)}(2; j_1; \mathbf{x}) \triangleq \left[\mathbf{a}^{(2)}(1; j_1; \mathbf{x}), \mathbf{a}^{(2)}(2; j_1; \mathbf{x})\right]^{\dagger}$ comprises Eqs. (3.19) and (3.21). Using in Eq. (3.2) the relations defining 2nd-LASS together with the 2nd-LVSS and the relation provided in Eq. (3.17) yields the following alternative expression for the indirect-effect term in terms of the 2nd-level adjoint sensitivity function $\mathbf{A}^{(2)}(2; j_1; \mathbf{x}) \triangleq \left[\mathbf{a}^{(2)}(1; j_1; \mathbf{x}), \mathbf{a}^{(2)}(2; j_1; \mathbf{x})\right]^{\dagger}$:

$$\begin{gathered} \left\{\delta R^{(1)}\left[j_1; \mathbf{u}; \mathbf{a}^{(1)}; \mathbf{f}; \mathbf{V}^{(2)}(2;\mathbf{x})\right]\right\}_{ind} = -\left\{\left[\hat{P}^{(2)}\left(\mathbf{U}^{(2)}; \mathbf{A}^{(2)}; \mathbf{f}; \delta\mathbf{f}\right)\right]_{\partial\Omega_x}\right\}_{\alpha^0} \\ + \left\{\left\langle \mathbf{A}^{(2)}(2; j_1; \mathbf{x}), \mathbf{Q}_V^{(2)}\left[2; \mathbf{U}^{(2)}(2;\mathbf{x}); \mathbf{f}; \delta\mathbf{f}\right]\right\rangle_2\right\}_{\alpha^0}, \end{gathered} \quad (3.22)$$

where $\left\{\left[\hat{P}^{(2)}\left(\mathbf{U}^{(2)}; \mathbf{A}^{(2)}; \mathbf{f}; \delta\mathbf{f}\right)\right]_{\partial\Omega_x}\right\}_{\alpha^0}$ denotes residual boundary terms that may not have vanished after having used the boundary and/or initial conditions represented by Eqs. (3.4) and (3.21).

Replacing the expression obtained in Eq. (3.22) into Eq. (3.1) yields the following expression:

$$\left\{\delta R^{(1)}\left[j_1;\mathbf{U}^{(2)}(2;\mathbf{x});\mathbf{A}^{(2)}(2;j_1;\mathbf{x});\mathbf{f};\delta\mathbf{f}\right]\right\}_{\alpha^0}$$
$$=\sum_{j_2=1}^{TP}\left\{R^{(2)}\left[j_2;j_1;\mathbf{U}^{(2)}(2;\mathbf{x});\mathbf{A}^{(2)}(2;j_1;\mathbf{x});\mathbf{f}\right]\right\}_{\alpha^0}\delta f_{j_2},\quad j_1=1,\ldots,TF, \tag{3.23}$$

where the quantity $R^{(2)}\left[j_2;j_1;\mathbf{U}^{(2)}(2;\mathbf{x});\mathbf{A}^{(2)}(2;j_1;\mathbf{x});\mathbf{f}\right]$ denotes the second-order sensitivity of the generic scalar-valued response $R[\mathbf{u}(\mathbf{x});\boldsymbol{\alpha}]$ with respect to the components $f_{j_1}$ and $f_{j_2}$ of the "feature" function $\mathbf{f}$, and has the following expression for $j_1,j_2=1,\ldots,TF$:

$$R^{(2)}\left[j_2;j_1;\mathbf{U}^{(2)}(2;\mathbf{x});\mathbf{A}^{(2)}(2;j_1;\mathbf{x});\mathbf{f}\right]$$
$$\triangleq\frac{\partial R^{(1)}\left[j_1;\mathbf{u}(\mathbf{x});\mathbf{a}^{(1)}(\mathbf{x});\mathbf{f}\right]}{\partial j_{j_2}}+\sum_{i=1}^{2}\left\langle\mathbf{a}^{(2)}(i;j_1;\mathbf{x}),\mathbf{s}_V^{(2)}\left[i;j_2;\mathbf{U}^{(2)}(2;\mathbf{x});\mathbf{f}\right]\right\rangle_1 \tag{3.24}$$
$$-\frac{\left\{\partial\hat{P}^{(2)}\left[\mathbf{U}^{(2)}(2;\mathbf{x});\mathbf{A}^{(2)}(2;j_1;\mathbf{x});\mathbf{f}\right]\right\}_{\partial\Omega_x}}{\partial f_{j_2}}\triangleq\frac{\partial^2R[\mathbf{u}(\mathbf{x});\alpha]}{\partial f_{j_2}\,\partial f_{j_1}}.$$

The quantities in Eq. (3.24) are to be computed at the nominal values of the parameters and respective state functions, but the superscript "zero," which is used to denote this fact, has been omitted in order to keep the notation as simple as possible.

Notably, the 2nd-LASS is independent of variations $\delta\mathbf{f}$ and, hence, of parameter variations $\delta\alpha$) and variations $\mathbf{V}^{(2)}(2;\mathbf{x})$ in the respective state functions. It is also important to note that the $(2\times TD)^2$-dimensional matrix $\mathbf{AM}^{(2)}\left[2\times2;\mathbf{U}^{(2)}(2;\mathbf{x});\mathbf{f}\right]$ is independent of the index $j_1$. Only the source term $\mathbf{Q}_A^{(2)}\left[2;j_1;\mathbf{U}^{(2)}(2;\mathbf{x});\mathbf{f}\right]$ depends on the index $j_1$. Therefore, the same solver can be used to invert the matrix $\mathbf{AM}^{(2)}\left[2\times2;\mathbf{U}^{(2)}(2;\mathbf{x});\mathbf{f}\right]$ when solving numerically the 2nd-LASS repeatedly, for each $j_1$-dependent source $\mathbf{Q}_A^{(2)}\left[2;j_1;\mathbf{U}^{(2)}(2;\mathbf{x});\mathbf{f}\right]$, in order to obtain the corresponding $j_1$-dependent $2\times TD$-dimensional 2nd-level adjoint function $\mathbf{A}^{(2)}(2;j_1;\mathbf{x})\triangleq\left[\mathbf{a}^{(2)}(1;j_1;\mathbf{x}),\mathbf{a}^{(2)}(2;j_1;\mathbf{x})\right]^\dagger$. Computationally, it would be most efficient to store, if possible, the inverse matrix $\left\{\mathbf{AM}^{(2)}\left[2\times2;\mathbf{U}^{(2)}(2;\mathbf{x});\mathbf{f}\right]\right\}^{-1}$, in order to multiply directly the inverse matrix $\left\{\mathbf{AM}^{(2)}\left[2\times2;\mathbf{U}^{(2)}(2;\mathbf{x});\mathbf{f}\right]\right\}^{-1}$ with the corresponding source term $\mathbf{Q}_A^{(2)}\left[2;j_1;\mathbf{U}^{(2)}(2;\mathbf{x});\mathbf{f}\right]$, for each index $j_1$, in order to obtain the corresponding $j_1$-dependent $2\times TD$-dimensional 2nd-level adjoint function $\mathbf{A}^{(2)}(2;j_1;\mathbf{x})\triangleq\left[\mathbf{a}^{(2)}(1;j_1;\mathbf{x}),\mathbf{a}^{(2)}(2;j_1;\mathbf{x})\right]^\dagger$.

Since the adjoint matrix $\mathbf{AM}^{(2)}\left[2\times 2;\mathbf{U}^{(2)}(2;\mathbf{x});\mathbf{f}\right]$ is block-diagonal, solving the 2nd-LASS is equivalent to solving two 1st-LASS with two different source terms. Thus, the "solvers" and the computer program used for solving the 1st-LASS can also be used for solving the 2nd-LASS. The 2nd-LASS was designated as the "2nd-level" rather than the "2nd-order" adjoint sensitivity system, since the 2nd-LASS does not involve any explicit 2nd-order G-derivatives of the operators underlying the original system but involves the inversion of the same operators that need to be inverted for solving the 1st-LASS.

If the 2nd-LASS is solved $TF$-times, the 2nd-order mixed sensitivities $R^{(2)}\left[j_2;j_1;\mathbf{U}^{(2)}(2;\mathbf{x});\mathbf{A}^{(2)}(2;j_1;\mathbf{x});\mathbf{f}\right]\equiv \partial^2 R/\partial f_{j2}\,\partial f_{j1}$ will be computed twice, in two different ways, in terms of two distinct 2nd-level adjoint functions. Consequently, the symmetry property $\partial^2 R[\mathbf{u}(\mathbf{x});\mathbf{f}]/\partial f_{j2}\,\partial f_{j1} = \partial^2 R[\mathbf{u}(\mathbf{x});\mathbf{f}]/\partial f_{j1}\,\partial f_{j2}$ provides an intrinsic (numerical) verification that the 1st-level adjoint function $\mathbf{a}^{(1)}(\mathbf{x})$ and the components of the 2nd-level adjoint function $\mathbf{A}^{(2)}(2;j_1;\mathbf{x})$ are computed accurately.

### 3.2.1 Similarities and Differences between the 2nd-FASAM-N and the 2nd-CASAM-N

As has been discussed in Chapter 2, the "1st-Order Comprehensive Adjoint Sensitivity Analysis Methodology" (1st-CASAM-N) computes the exactly-derived expressions of the 1st-order sensitivities of a model response *directly* with respect to the model's parameters, while the "1st-Order Features Adjoint Sensitivity Analysis Methodology" (1st-FASAM-N) computes the exactly-derived expressions of the 1st-order sensitivities of a model response with respect to functions/features of the model. Mathematically, the formal correspondence between the 1st-FASAM-N and the 1st-CASAM-N is obtained by replacing the components of the "features" vector-valued function $\mathbf{f}(\boldsymbol{\alpha})\triangleq[\mathbf{g}(\boldsymbol{\alpha});\mathbf{h}(\boldsymbol{\alpha});\boldsymbol{\lambda}(\boldsymbol{\alpha});\boldsymbol{\omega}(\boldsymbol{\alpha})]^{\dagger}\triangleq[f_1(\boldsymbol{\alpha}),\ldots,f_{TF}(\boldsymbol{\alpha})]^{\dagger}$ within the 1st-FASAM-N formalism with the components of the vector of primary parameters $\boldsymbol{\alpha}\triangleq(\alpha_1,\ldots,\alpha_{TP})^{\dagger}$, which would yield the 1st-CASAM-N mathematical formalism. The formal correspondence described in Chapter 2 between the 1st-FASAM-N and the 1st-CASAM-N also translates, for each of the respective 1st-order sensitivities, to the formal correspondence between the 2nd-FASAM-N and the 2nd-CASAM-N (Cacuci 2016, 2023a) frameworks, i.e., replacing formally the components of $\mathbf{f}(\boldsymbol{\alpha})\triangleq[f_1(\boldsymbol{\alpha}),\ldots,f_{TF}(\boldsymbol{\alpha})]^{\dagger}$ in the 2nd-FASAM-N framework with the components of $\boldsymbol{\alpha}\triangleq(\alpha_1,\ldots,\alpha_{TP})^{\dagger}$ yields the 2nd-CASAM-N framework. It is very important to note, however, that while there is a total number of $TP$ primary model parameters (and, hence, $TP$ 1st-order sensitivities) within the 1st-CASAM-N, there are only $TF$ feature functions (and, hence, $TF$ 1st-order sensitivities) within the 1st-FASAM-N, where $TF \ll TP$. Each of the respective 1st-order sensitivities becomes a "response" that generates a corresponding "2nd-level adjoint sensitivity system" (2nd-LASS) for computing the corresponding 2nd-order sensitivities. Therefore, there is a total number of $TP$ "2nd-level adjoint sensitivity systems" to be solved within the 2nd-CASAM-N framework, but there are only $TF$ "2nd-level adjoint sensitivity systems" to be solved within the 2nd-FASAM-N formalism. Since the 2nd-LASS within either the 2nd-CASAM-N or the 2nd-CASAM-N frameworks involves the same operators on their respective left sides (only the "sources" on the respective right sides are different), it follows that the same computational effort is required to solve any of these 2nd-LASS. Thus, since $TF \ll TP$, there are far fewer $(TF)$ large-scale computations needed to solve the 2nd-LASS for obtaining the 2nd-order sensitivities of a response with respect to the components of the "feature" function $\mathbf{f}(\boldsymbol{\alpha}) \triangleq [f_1(\boldsymbol{\alpha}),\dots,f_{TF}(\boldsymbol{\alpha})]^{\dagger}$ within the 2nd-FASAM-N framework, by comparison to the considerably larger number $(TP)$ of large-scale computations needed to solve the 2nd-LASS for obtaining the second-order sensitivities of a response with respect to the components of $\boldsymbol{\alpha} \triangleq (\alpha_1,\dots,\alpha_{TP})^{\dagger}$ within the 2nd-CASAM-N framework.

Performing $TP$ computations using the 2nd-LASS within the 2nd-CASAM-N methodology, each such computation corresponding to one of the $TP$ 1st-order sensitivities computed within the 1st-CASAM-N, will yield a total of $TP \times TP$ 2nd-order response sensitivities with respect to the primary model parameters $\boldsymbol{\alpha} \triangleq (\alpha_1,\dots,\alpha_{TP})^{\dagger}$, thereby computing the mixed 2nd-order sensitivities twice, using distinct 2nd-level adjoint sensitivity functions. This property of the 2nd-CASAM-N provides an intrinsic mechanism for verifying the accuracy of the adjoint sensitivity functions produced by solving the corresponding 2-LASS systems.

The above intrinsic verification mechanism, which ensures that the second-level adjoint sensitivity functions are computed accurately, is also inherent to the 2nd-FASAM-N. Solving the 2nd-LASS corresponding to each of the $TF$ 1st-order sensitivities computed within the 1st-FASAM-N will yield a total of $TF \times TF$ 2nd-order response sensitivities with respect to the components of the feature function $\mathbf{f}(\boldsymbol{\alpha}) \triangleq [f_1(\boldsymbol{\alpha}),\dots,f_{TF}(\boldsymbol{\alpha})]^{\dagger}$, thereby computing the mixed 2nd-order sensitivities twice, using distinct 2nd-level adjoint sensitivity functions. Due to the symmetry of the mixed 2nd-order sensitivities, the two distinct solutions of the respective 2nd-LASS must produce the same numerical result for the 2nd-order sensitivity under consideration. These considerations will be explicitly illustrated in the application to the paradigm particle transport model considered in Section 3.3.

After applying the 1st-FASAM-N and the 2nd-FASAM-N methodologies to obtain the 1st-order and, respectively, 2nd-order sensitivities of a response $R[\mathbf{u}(\mathbf{x});\boldsymbol{\alpha}]$ with respect to the components of the "feature" function $\mathbf{f}(\boldsymbol{\alpha}) \triangleq [f_1(\boldsymbol{\alpha}),\dots,f_{TF}(\boldsymbol{\alpha})]^{\dagger}$, the computation of the 2nd-order sensitivities of the response $R[\mathbf{u}(\mathbf{x});\boldsymbol{\alpha}]$ with respect to the components of the vector $\boldsymbol{\alpha} \triangleq (\alpha_1,\dots,\alpha_{TP})^{\dagger}$ of primary model parameters is obtained at practically no additional computational costs by simply using the "chain rule" to obtain:

$$\frac{\partial^2 R[\mathbf{u}(\mathbf{x});\boldsymbol{\alpha}]}{\partial \alpha_{j_2}\,\partial \alpha_{j_1}} = \frac{\partial}{\partial \alpha_{j_2}} \left\{ \sum_{i=1}^{TF} \frac{\partial R[\mathbf{u}(\mathbf{x});\boldsymbol{\alpha}]}{\partial f_i} \frac{\partial f_i(\boldsymbol{\alpha})}{\partial \alpha_{j_1}} \right\}, \quad j_1 = 1,\dots,TP; \quad j_2 = 1,\dots,j_1. \tag{3.25}$$

## 3.3 APPLYING THE 2ND-FASAM-N TO OBTAIN THE 2ND-ORDER RESPONSE SENSITIVITIES TO FUNCTIONS/ FEATURES OF PARAMETERS FOR THE PARADIGM PARTICLE TRANSPORT MODEL CONSIDERED IN CHAPTER 2

This section illustrates the application of the 2nd-FASAM-N to obtain the exact expressions of the 2nd-order sensitivities of the "kerma" response with respect to the components of the "feature" function $\mathbf{f}(\boldsymbol{\alpha}) \triangleq [g_1(\boldsymbol{\alpha}), g_2(\boldsymbol{\alpha}); h(\boldsymbol{\alpha}); a; b]^\dagger$ for the paradigm particle transport model analyzed in Chapter 2. As has been shown in Section 3.2, the 2nd-FASAM-N obtains the second-order sensitivities by computing "the 1st-order sensitivities of the 1st-order sensitivities." Since the "feature" function $\mathbf{f}(\boldsymbol{\alpha}) \triangleq [g_1(\boldsymbol{\alpha}), g_2(\boldsymbol{\alpha}); h(\boldsymbol{\alpha}); a; b]^\dagger$ comprises $TF = 5$ components, there are five 1st-order sensitivities $R^{(1)}\left[j_1; \varphi; a^{(1)}; \mathbf{f}(\boldsymbol{\alpha})\right] \triangleq \partial R[\mathbf{u}(\mathbf{x}); \mathbf{f}] / \partial f_{j_1}$, $j_1 = 1, \ldots, TF = 5$, which were obtained in Chapter 2. The 2nd-order sensitivities $\partial^2 R[\mathbf{u}(\mathbf{x}); \mathbf{f}] / \partial f_{j_1} \partial f_{j_2}$, $j_1, j_2 = 1, \ldots, TF = 5$ of the model response with respect to the components of $\mathbf{f}(\boldsymbol{\alpha}) \triangleq [g_1(\boldsymbol{\alpha}), g_2(\boldsymbol{\alpha}); h(\boldsymbol{\alpha}); a; b]^\dagger$ will be determined by applying the 2nd-FASAM-N to the 1st-order sensitivities $R^{(1)}\left[j_1; \varphi; a^{(1)}; \mathbf{f}(\boldsymbol{\alpha})\right]$. Thus, for each 1st-order sensitivity, there will correspond a "2nd-level adjoint sensitivity function" that will be used for obtaining the 2nd-order sensitivities that arise from the respective 1st-order sensitivity. In general, a 1st-order sensitivity will depend on both the original forward function and on the corresponding adjoint function; in general, therefore, a "2nd-level adjoint sensitivity function" will be a two-component function, having one component corresponding to the variation in the original forward function and having the 2nd component corresponding to the variation in the 1st-level adjoint sensitivity function. For the illustrative particle transport model considered in Chapter 2, however, an examination of the 1st-order sensitivities obtained in Section 2.4.2 reveals the following characteristics:

1. As indicated in Eq. (2.119), the 1st-order sensitivity $R^{(1)}\left[j_1 = 1; a^{(1)}; \mathbf{f}(\boldsymbol{\alpha})\right] \triangleq \partial R(\varphi; \mathbf{f}) / \partial g_1(\boldsymbol{\alpha})$ depends only on the 1st-level adjoint sensitivity function $a^{(1)}(z, \omega)$. Hence, when applying the 2nd-FASAM-N, the 2nd-order sensitivities that will stem from $R^{(1)}\left[j_1 = 1; a^{(1)}; \mathbf{f}(\boldsymbol{\alpha})\right] \triangleq \partial R(\varphi; \mathbf{f}) / \partial g_1(\boldsymbol{\alpha})$ will be obtained in terms of a "two component second-level adjoint sensitivity function" $\mathbf{A}^{(2)}(2; j_1 = 1; \mathbf{x}) \triangleq \left[a^{(2)}(1; j_1 = 1; z, \omega) \equiv 0, a^{(2)}(2; j_1 = 1; z, \omega)\right]^\dagger$ which will have a single non-zero component, $a^{(2)}(2; j_1 = 1; z, \omega)$, corresponding to $\delta a^{(1)}$.
2. As indicated in Eq. (2.120), the 1st-order sensitivity $R^{(1)}\left[j_1 = 2; \varphi; a^{(1)}; \mathbf{f}(\boldsymbol{\alpha})\right] \triangleq \partial R(\varphi; \mathbf{f}) / \partial g_2(\boldsymbol{\alpha})$ is the only one that depends on both the forward function $\varphi(z, \omega)$ and on the 1st-level adjoint sensitivity function $a^{(1)}(z, \omega)$. Consequently, when applying the 2nd-FASAM-N, the 2nd-order sensitivities

that will stem from $R^{(1)}\left[j_1 = 2;\varphi;a^{(1)};\mathbf{f}(\boldsymbol{\alpha})\right] \triangleq \partial R(\varphi;\mathbf{f})/\partial g_2(\boldsymbol{\alpha})$ will be obtained in terms of a "two component 2nd-level adjoint sensitivity function" $\mathbf{A}^{(2)}(2;j_1 = 2;\mathbf{x}) \triangleq \left[a^{(2)}(1;j_1 = 2;z,\omega),a^{(2)}(2;j_1 = 2;z,\omega)\right]^{\dagger}$ which will have two non-zero components: one component corresponding to $\delta\varphi$, and the 2nd component corresponding to $\delta a^{(1)}$.

3. As indicated in Eq. (2.121), the 1st-order sensitivity $R^{(1)}\left[j_1 = 3;\varphi;\mathbf{f}(\boldsymbol{\alpha})\right] \triangleq \partial R(\varphi;\mathbf{f})/\partial h(\boldsymbol{\alpha})$ depends only on the forward function $\varphi(z,\omega)$ but does not depend on the 1st-level adjoint sensitivity function $a^{(1)}(z,\omega)$. Hence, when applying the 2nd-FASAM-N, the 2nd-order sensitivities that will stem from $R^{(1)}\left[j_1 = 3;\varphi;\mathbf{f}(\boldsymbol{\alpha})\right] \triangleq \partial R(\varphi;\mathbf{f})/\partial h(\boldsymbol{\alpha})$ will be obtained in terms of a "two component second-level adjoint sensitivity function" $\mathbf{A}^{(2)}(2;j_1 = 3;\mathbf{x}) \triangleq \left[a^{(2)}(1;j_1 = 3;z,\omega),a^{(2)}(2;j_1 = 3;z,\omega) \equiv 0\right]^{\dagger}$ which will have a single non-zero component corresponding to $\delta\varphi$.
4. As indicated in Eq. (2.122), the 1st-order sensitivity $R^{(1)}\left[j_1 = 4;a^{(1)};\mathbf{f}(\boldsymbol{\alpha})\right] \triangleq \partial R(\varphi;\mathbf{f})/\partial a$ depends only on the 1st-level adjoint sensitivity function $a^{(1)}(z,\omega)$. Hence, when applying the 2nd-FASAM-N, the 2nd-order sensitivities that will stem from $R^{(1)}\left[j_1 = 4;a^{(1)};\mathbf{f}(\boldsymbol{\alpha})\right] \triangleq \partial R(\varphi;\mathbf{f})/\partial a$ will be obtained in terms of a "two component 2nd-level adjoint sensitivity function" $\mathbf{A}^{(2)}(2;j_1 = 4;\mathbf{x}) \triangleq \left[a^{(2)}(1;j_1 = 4;z,\omega) \equiv 0,a^{(2)}(2;j_1 = 4;z,\omega)\right]^{\dagger}$ which will have a single non-zero component corresponding to $\delta a^{(1)}$. This situation is similar to the determination of the 2nd-order sensitivities stemming from $R^{(1)}\left[j_1 = 1;a^{(1)};\mathbf{f}(\boldsymbol{\alpha})\right] \triangleq \partial R(\varphi;\mathbf{f})/\partial g_1(\boldsymbol{\alpha})$, but the respective 2nd-order adjoint sensitivity functions will differ from each other because, although the respective 2nd-Level Adjoint Sensitivity Systems (2nd-LASS) will have the same left side, the respective 2nd-LASS will have distinct right sides (sources).
5. As indicated in Eq. (2.123), the 1st-order sensitivity $R^{(1)}\left[j_1 = 5;\varphi;\mathbf{f}(\boldsymbol{\alpha})\right] \triangleq \partial R(\varphi;\mathbf{f})/\partial b$ depends only on the forward function $\varphi(z,\omega)$; it does not depend on the 1st-level adjoint sensitivity function $a^{(1)}(z,\omega)$. This situation is similar to the case for $R^{(1)}\left[j_1 = 3;\varphi;\mathbf{f}(\boldsymbol{\alpha})\right] \triangleq \partial R(\varphi;\mathbf{f})/\partial h(\boldsymbol{\alpha})$. Hence, when applying the 2nd-FASAM-N, the 2nd-order sensitivities that will stem from $R^{(1)}\left[j_1 = 5;\varphi;\mathbf{f}(\boldsymbol{\alpha})\right] \triangleq \partial R(\varphi;\mathbf{f})/\partial b$ will be obtained in terms of a "two component 2nd-level adjoint sensitivity function" $\mathbf{A}^{(2)}(2;j_1 = 5;\mathbf{x}) \triangleq \left[a^{(2)}(1;j_1 = 5;z,\omega),a^{(2)}(2;j_1 = 5;z,\omega) \equiv 0\right]^{\dagger}$ which will have a single non-zero component corresponding to $\delta\varphi$. Again, the 2nd-LASS corresponding to $R^{(1)}\left[j_1 = 5;\varphi;\mathbf{f}(\boldsymbol{\alpha})\right] \triangleq \partial R(\varphi;\mathbf{f})/\partial b$ and $R^{(1)}\left[j_1 = 3;\varphi;\mathbf{f}(\boldsymbol{\alpha})\right] \triangleq \partial R(\varphi;\mathbf{f})/\partial h(\boldsymbol{\alpha})$, respectively, will have the same left sides but different right sides (sources).

The above considerations will be detailed in the following five subsections.

### 3.3.1 Applying the 2nd-FASAM-N to Compute the 2nd-Order Response Sensitivities Stemming from the 1st-Order Sensitivity $\partial R(\varphi;\mathbf{f})/\partial g_1(\alpha)$

The 1st-order sensitivity $\partial R(\varphi;\mathbf{f})/\partial g_1(\alpha)$ is defined in Eq. (2.119), which is recast into the following "response-like" form:

$$R^{(1)}\left[1;a^{(1)};\mathbf{f}(\alpha)\right] \triangleq \frac{\partial R(\varphi;\mathbf{f})}{\partial g_1(\alpha)} = \int_0^1 \omega\, d\omega \int_a^b a^{(1)}(z,\omega)\delta(z-a)dz. \tag{3.26}$$

where the 1st-level adjoint function $a^{(1)}(z,\omega)$ has the following expression reproduced below from Eq. (2.116):

$$a^{(1)}(z,\omega) = \left\{\frac{h(\alpha)}{\omega}\left[1-H(z-b)\right]\exp\left[\frac{g_2(\alpha)(z-b)}{\omega}\right]\right\}_{\alpha^0}. \tag{3.27}$$

Recall that $a^{(1)}(z,\omega)$ is the solution of the 1st-LASS, cf. Eqs. (2.114) and (2.115), which are reproduced below, for convenience:

$$\begin{aligned}&\left\{-\omega\frac{d}{dz}a^{(1)}(z,\omega)+g_2(\alpha)a^{(1)}(z,\omega)\right\}_{\alpha^0}\\ &=\left\{h(\alpha)\delta(z-b)\right\}_{\alpha^0}, \quad a^0<z<b^0; \quad 0\le\omega\le 1.\end{aligned} \tag{3.28}$$

$$a^{(1)}(z,\omega)=0, \quad at \quad z=b^0, \quad 0\le\omega\le 1. \tag{3.29}$$

As indicated by the last term in Eq. (3.26), the 1st-order sensitivity $R^{(1)}\left[1;a^{(1)};\mathbf{f}(\alpha)\right] \triangleq \partial R(\varphi;\mathbf{f})/\partial g_1(\alpha)$ does not depend on the original forward function $\varphi(z,\omega)$ (i.e., the particle flux) but only depends on the 1st-level adjoint sensitivity function $a^{(1)}(z,\omega)$. The digit "1" in the list arguments of $R^{(1)}\left[1;a^{(1)};\mathbf{f}(\alpha)\right]$ indicates that $R^{(1)}\left[1;a^{(1)};\mathbf{f}(\alpha)\right]$ is the "1st to be considered" of the five 1st-order sensitivities of the model response with respect to the five components of the "feature" function $\mathbf{f}(\alpha)$, which were obtained in Eqs. (2.119)–(2.123).

According to the principles of the 2nd-CASAM-N presented in Section 3.2, the 2nd-order sensitivities that stem from $\partial R(\varphi;\mathbf{f})/\partial g_1(\alpha)$ are provided by the 1st-order G-differential of $R^{(1)}\left[1;a^{(1)};\mathbf{f}(\alpha)\right] \triangleq \partial R(\varphi;\mathbf{f})/\partial g_1(\alpha)$. Applying the definition of the G-differential to Eq. (3.26) yields the following relation:

$$\begin{aligned}&\left\{\delta R^{(1)}\left[1;a^{(1)};\mathbf{f}(\alpha);\delta a^{(1)};\delta\mathbf{f}\right]\right\}_{\alpha^0}\\ &=\left\{\frac{d}{d\varepsilon}\left[\int_0^1 \omega\, d\omega \int_{a+\varepsilon\delta a}^{b+\varepsilon\delta b}\left(a^{(1)}+\varepsilon\delta a^{(1)}\right)\delta(z-a-\varepsilon\delta a)dz\right]_{\alpha^0}\right\}_{\varepsilon=0}\\ &=\left\{\delta R^{(1)}\left(1;a^{(1)};\mathbf{f};\delta\mathbf{f}\right)_{\alpha^0}\right\}_{dir}+\left\{\delta R^{(1)}\left(1;a^{(1)};\mathbf{f};\delta a^{(1)}\right)_{\alpha^0}\right\}_{ind},\end{aligned} \tag{3.30}$$

where the direct-effect term $\left\{\delta R^{(1)}\left(1;a^{(1)};\mathbf{f};\delta\mathbf{f}\right)_{\alpha^0}\right\}_{dir}$ depends only on variations $\delta\mathbf{f}$ while the indirect-effect term $\left\{\delta R^{(1)}\left(1;a^{(1)};\mathbf{f};\delta a^{(1)}\right)_{\alpha^0}\right\}_{ind}$ depends only on variations $\delta a^{(1)}$; these terms are defined, respectively, as follows:

$$\left\{\delta R^{(1)}\left(1;a^{(1)};\mathbf{f};\delta\mathbf{f}\right)_{\alpha^0}\right\}_{dir} \triangleq \left\{\int_0^1 \omega\, d\omega \left[\frac{da^{(1)}(z,\omega)}{dz}\right]_{z=a}(\delta a)\right\}_{\alpha^0}, \tag{3.31}$$

$$\left\{\delta R^{(1)}\left(1;a^{(1)};\mathbf{f};\delta a^{(1)}\right)_{\alpha^0}\right\}_{ind} \triangleq \left\{\int_0^1 d\omega \int_a^b \omega\, \delta a^{(1)}(z,\omega)\delta(z-a)\,dz\right\}_{\alpha^0}. \tag{3.32}$$

The direct-effect term defined by Eq. (3.31) can be evaluated at this stage by using Eq. (3.27) to obtain:

$$\left[\frac{da^{(1)}(z,\omega)}{dz}\right]_{z=a} = \frac{h(\alpha)g_2(\alpha)}{\omega^2}\exp\left[\frac{g_2(\alpha)(a-b)}{\omega}\right]. \tag{3.33}$$

Inserting the result obtained in Eq. (3.33) into Eq. (3.31) yields:

$$\left\{\delta R^{(1)}\left(1;a^{(1)};\mathbf{f};\delta\mathbf{f}\right)_{\alpha^0}\right\}_{dir} \triangleq (\delta a)\left\{h(\alpha)g_2(\alpha)E_1\left[g_2(\alpha)(a-b)\right]\right\}_{\alpha^0}. \tag{3.34}$$

The indirect-effect term defined in Eq. (3.32) can be evaluated only after having determined the variational function $\delta a^{(1)}(z,\omega)$, which is the solution of the G-differential of the 1st-LASS, which is, in turn, obtained by applying the definition of the (1st-order) G-differential to Eqs. (3.28) and (3.29), namely:

$$\left\{\frac{d}{d\varepsilon}\left[-\omega\frac{d\left(a^{(1)}+\varepsilon\delta a^{(1)}\right)}{dz}+\left(g_2+\varepsilon\delta g_2\right)\left(a^{(1)}+\varepsilon\delta a^{(1)}\right)\right]_{\alpha^0}\right\}_{\varepsilon=0}$$
$$=\left\{\frac{d}{d\varepsilon}\left[(h+\varepsilon\delta h)\delta(z-b-\varepsilon\delta b)\right]_{\alpha^0}\right\}_{\varepsilon=0}, \quad a^0<z<b^0;\quad 0\le\omega\le 1, \tag{3.35}$$

$$\left\{\frac{d}{d\varepsilon}\left[a^{(1)}\left(z=b^0+\varepsilon\delta b;\omega\right)+\varepsilon\delta a^{(1)}\left(z=b^0+\varepsilon\delta b;\omega\right)\right]\right\}_{\varepsilon=0}=0, \tag{3.36}$$

at $z=b^0,\quad 0\le\omega\le 1$.

Carrying out the operations indicated in Eqs. (3.35) and (3.36) yields the following 2nd-Level Forward Variational System (2nd-LVSS) for the function $\delta a^{(1)}(z,\omega)$:

$$\left\{-\omega\frac{d}{dz}\delta a^{(1)}(z,\omega)+g_2(\boldsymbol{\alpha})\delta a^{(1)}(z,\omega)\right\}_{\boldsymbol{\alpha}^0}$$
$$=\left\{-(\delta g_2)a^{(1)}(z,\omega)+(\delta h)\delta(z-b)-(\delta b)h(\boldsymbol{\alpha})\delta'(z-b)\right\}_{\boldsymbol{\alpha}^0}, \tag{3.37}$$
$$a^0<z<b^0;\quad 0\le\omega\le 1,$$

$$\left\{\left[\delta a^{(1)}(z,\omega)\right]_{z=b}\right\}_{\boldsymbol{\alpha}^0}=-\left\{\left[\frac{da^{(1)}(z,\omega)}{dz}\right]_{z=b}\right\}_{\boldsymbol{\alpha}^0}$$
$$=0,\quad \text{at}\quad z=b^0,\quad 0\le\omega\le 1. \tag{3.38}$$

In principle, the 2nd-LVSS could be solved to obtain $\delta a^{(1)}(z,\omega)$ for every possible variation in $\delta g_2$, $\delta h$ and/or $\delta b$, which would be prohibitively expensive and also unnecessary. Instead, solving the 2nd-LVSS is avoided altogether by recasting the expression of the indirect-effect term in terms of a 2nd-level adjoint function, which will be the solution of a 2nd-Level Adjoint Sensitivity System (2nd-LASS) to be constructed by applying the general 2nd-FASAM-N as outlined in Section 3.3.

As has been discussed in the foregoing, the indirect-effect term $\left\{\delta R^{(1)}\left(1;a^{(1)};\mathbf{f};\delta a^{(1)}\right)_{\boldsymbol{\alpha}^0}\right\}_{ind}$, defined in Eq. (3.32), only involves the variational function $\delta a^{(1)}(z,\omega)$ but does not also involve the variational function $\delta\varphi(z,\omega)$. Consequently, the 2nd-level adjoint sensitivity function will comprise a single non-zero component (which will correspond to $\delta a^{(1)}$) rather than two non-zero components –as would have been the case if the 1st-order sensitivity had depended on both the flux $\varphi(z,\omega)$ and the 1st-level adjoint sensitivity function $a^{(1)}(z,\omega)$. Hence, the inner product appropriate for recasting $\left\{\delta R^{(1)}\left(1;a^{(1)};\mathbf{f};\delta a^{(1)}\right)_{\boldsymbol{\alpha}^0}\right\}_{ind}$ in terms of a 2nd-level adjoint sensitivity function is the same as defined in Eq. (2.80), which is recalled below for convenience: for two functions $\chi(z,\omega)\in\mathcal{H}_1$ and $\psi(z,\omega)\in\mathcal{H}_1$ belonging to the Hilbert space $\mathcal{H}_1(\Omega_z),\Omega_z\triangleq\{\omega\in[0,1]\otimes z\in[a,b]\}$, the inner product denoted as $\langle\chi(z,\omega),\psi(z,\omega)\rangle_1$ was defined as follows:

$$\langle\chi(z,\omega),\psi(z,\omega)\rangle_1\triangleq\left\{\int_0^1 d\omega\int_a^b \chi(z,\omega)\psi(z,\omega)dz\right\}_{\boldsymbol{\alpha}^0}. \tag{3.39}$$

Thus, introducing a one-component 2nd-level adjoint sensitivity function denoted as $a^{(2)}(2;1;z,\omega)$, which is yet undefined, and forming the inner product of this function with Eq. (3.37) yields the following relation:

$$\left\{\int_0^1 d\omega \int_a^b a^{(2)}(2;1;z,\omega)\left[-\omega\frac{d}{dz}\delta a^{(1)}(z,\omega)+g_2(\alpha)\delta a^{(1)}(z,\omega)\right]dz\right\}_{\alpha^0}$$

$$=\left\{\int_0^1 d\omega \int_a^b dz a^{(2)}(2;1;z,\omega)\left[-(\delta g_2)a^{(1)}(z,\omega)\right.\right. \tag{3.40}$$

$$\left.\left.+(\delta h)\delta(z-b)-(\delta b)h(\alpha)\delta'(z-b)\right]\right\}_{\alpha^0}.$$

Note that the first argument of $a^{(2)}(2;1;z,\omega)$ indicates that only the "2nd-component" of this "two-component 2nd-level adjoint sensitivity function" is non-zero, while the 2nd argument in $a^{(2)}(2;1;z,\omega)$ refers to the index/label "$j_1 = 1$" indicating that this 2nd-level adjoint sensitivity function will correspond to "the 1st considered" 1st-order sensitivity.

Integrating by parts the left side of Eq. (3.40) over the independent variable $z$ yields the following relation:

$$\left\{\int_0^1 d\omega \int_a^b a^{(2)}(2;1;z,\omega)\left[-\omega\frac{d}{dz}\delta a^{(1)}(z,\omega)+g_2(\alpha)\delta a^{(1)}(z,\omega)\right]dz\right\}_{\alpha^0}$$

$$=\int_0^1 \omega\, d\omega\left\{a^{(2)}(2;1;a,\omega)\delta a^{(1)}(a,\omega)-a^{(2)}(2;1;b,\omega)\delta a^{(1)}(b,\omega)\right\}_{\alpha^0} \tag{3.41}$$

$$+\left\{\int_0^1 d\omega \int_a^b \delta a^{(1)}(z,\omega)\left[\omega\frac{d}{dz}a^{(2)}(2;1;z,\omega)+g_2(\alpha)a^{(2)}(2;1;z,\omega)\right]dz\right\}_{\alpha^0}.$$

The last term on the right side of Eq. (3.41) is now required to represent the indirect-effect term defined in Eq. (3.32) by imposing the following relationship:

$$\begin{aligned}&\left\{\int_0^1 d\omega \int_a^b \delta a^{(1)}(z,\omega)\left[\omega\frac{d}{dz}a^{(2)}(2;1;z,\omega)+g_2(\alpha)a^{(2)}(2;1;z,\omega)\right]dz\right\}_{\alpha^0}\\&=\left\{\int_0^1 d\omega \int_a^b \omega\,\delta a^{(1)}(z,\omega)\delta(z-a)\,dz\right\}_{\alpha^0}.\end{aligned} \tag{3.42}$$

The relation in Eq. (3.42) implies that the following equation holds, in the weak sense, at the nominal parameter values:

$$\left\{\omega\frac{d}{dz}a^{(2)}(2;1;z,\omega)+g_2(\boldsymbol{\alpha})a^{(2)}(2;1;z,\omega)\right\}_{\alpha^0}=\left\{\omega\delta(z-a)\right\}_{\alpha^0}, \tag{3.43}$$

$$a^0<z<b^0;\quad 0\le\omega\le 1.$$

The definition of the function $a^{(2)}(2;1;z,\omega)$ is now completed by requiring that the unknown value of the function $\delta a^{(1)}(a,\omega)$ be eliminated from appearing on the right side of Eq. (3.41) This requirement is met by imposing the following condition to be satisfied by the function $a^{(2)}(2;1;z,\omega)$ on the model's boundary:

$$a^{(2)}(2;1;z,\omega)=0,\quad \text{at}\quad z=a^0,\quad 0\le\omega\le 1. \tag{3.44}$$

Altogether, Eqs. (3.43) and (3.44) constitute the "2nd-Level Adjoint Sensitivity System" (2nd-LASS) for determining the 2nd-level adjoint sensitivity function $a^{(2)}(2;1;z,\omega)$. Since the 2nd-LASS does not depend on the parameter variations, it needs to be solved only once in order to obtain $a^{(2)}(2;1;z,\omega)$. For the paradigm model under consideration, the 1st-LASS given by Eqs. (3.43) and (3.44) can be solved exactly to obtain the following closed-form expression for the 2nd-level adjoint function $a^{(2)}(2;1;z,\omega)$:

$$a^{(2)}(2;1;z,\omega)=H(z-a)\exp\left[\frac{g_2(\boldsymbol{\alpha})(a-z)}{\omega}\right] \tag{3.45}$$

Collecting the results obtained in Eqs. (3.37)–(3.45) leads to the following expression for the indirect-effect term:

$$\begin{aligned}\left\{\delta R^{(1)}\left(1;a^{(1)};\mathbf{f};\delta a^{(1)}\right)_{\alpha^0}\right\}_{ind}&=\left\{\int_0^1 d\omega\int_a^b dza^{(2)}(2;1;z,\omega)\right.\\ &\left.\times\left[-(\delta g_2)a^{(1)}(z,\omega)+(\delta h)\delta(z-b)-(\delta b)h(\boldsymbol{\alpha})\delta'(z-b)\right]\right\}_{\alpha^0}\\ &=\left\{\delta R^{(1)}\left[1;a^{(1)};\mathbf{f};a^{(2)}(2;1;z,\omega)\right]_{\alpha^0}\right\}_{ind}.\end{aligned} \tag{3.46}$$

Note that the dependence of $\left\{\delta R^{(1)}\left[1;a^{(1)};\mathbf{f};a^{(2)}(2;1;z,\omega)\right]_{\alpha^0}\right\}_{ind}$ on $\delta a^{(1)}$ has been replaced in the last equality in Eq. (3.46) by the dependence on $a^{(2)}(2;1;z,\omega)$.

Adding the results obtained in Eqs. (3.46) and (3.34) yields the following relation:

$$\left\{\delta R^{(1)}\left[1;a^{(1)};\mathbf{f}(\alpha);a^{(2)}(2;1;z,\omega);\delta\mathbf{f}\right]\right\}_{\alpha^0}$$
$$=(\delta a)\left\{h(\alpha)g_2(\alpha)E_1\left[g_2(\alpha)(a-b)\right]\right\}_{\alpha^0} \tag{3.47}$$
$$+\left\{\int_0^1 d\omega\int_a^b dza^{(2)}(2;1;z,\omega)\begin{bmatrix}-(\delta g_2)a^{(1)}(z,\omega)+(\delta h)\delta(z-b)\\ -(\delta b)h(\alpha)\delta'(z-b)\end{bmatrix}\right\}_{\alpha^0}.$$

Note that the dependence of $\delta R^{(1)}\left[1;a^{(1)};\mathbf{f}(\alpha);a^{(2)}(2;1;z,\omega);\delta\mathbf{f}\right]$ on the variational function $\delta a^{(1)}$ has been replaced in Eq. (3.47) by the dependence on the 2nd-level adjoint sensitivity function $a^{(2)}(2;1;z,\omega)$. Identifying the expressions that multiply the respective components of $\delta\mathbf{f}(\alpha)\triangleq\left[\delta g_1(\alpha),\delta g_2(\alpha);\delta h(\alpha);\delta a;\delta b\right]^{\dagger}$, and using the expression for the 2nd-level adjoint sensitivity function provided in Eq. (3.45) yields the following results for the 2nd-order sensitivities that stem from the 1st-order sensitivity $R^{(1)}\left[1;\varphi;\mathbf{f}(\alpha)\right]\triangleq\partial R(\varphi;\mathbf{f})/\partial g_1(\alpha)$:

$$\left\{\frac{\partial R^{(1)}\left[1;a^{(1)};\mathbf{f}(\alpha)\right]}{\partial g_1(\alpha)}\right\}_{\alpha^0}=\left\{\frac{\partial^2 R(\varphi;\mathbf{f})}{\partial g_1(\alpha)\partial g_1(\alpha)}\right\}_{\alpha^0}=0; \tag{3.48}$$

$$\left\{\frac{\partial R^{(1)}\left[1;a^{(1)};\mathbf{f}(\alpha)\right]}{\partial g_2(\alpha)}\right\}_{\alpha^0}=\left\{\frac{\partial^2 R(\varphi;\mathbf{f})}{\partial g_2(\alpha)\partial g_1(\alpha)}\right\}_{\alpha^0}$$
$$=-\left\{\int_0^1 d\omega\int_a^b dza^{(2)}(2;1;z,\omega)a^{(1)}(z,\omega)\right\}_{\alpha^0} \tag{3.49}$$
$$=-\left\{h(\alpha)(b-a)E_1\left[(b-a)g_2(\alpha)\right]\right\}_{\alpha^0};$$

$$\left\{\frac{\partial R^{(1)}\left[1;a^{(1)};\mathbf{f}(\alpha)\right]}{\partial h(\alpha)}\right\}_{\alpha^0}=\left\{\frac{\partial^2 R(\varphi;\mathbf{f})}{\partial h(\alpha)\partial g_1(\alpha)}\right\}_{\alpha^0}$$
$$=\left\{\int_0^1 d\omega\int_a^b dza^{(2)}(2;1;z,\omega)\delta(z-b)\right\}_{\alpha^0} \tag{3.50}$$
$$=\left\{E_2\left[(b-a)g_2(\alpha)\right]\right\}_{\alpha^0};$$

$$\left\{\frac{\partial R^{(1)}\left[1;a^{(1)};\mathbf{f}(\boldsymbol{\alpha})\right]}{\partial a}\right\}_{\boldsymbol{\alpha}^0}=\left\{\frac{\partial^2 R(\varphi;\mathbf{f})}{\partial a\,\partial g_1(\boldsymbol{\alpha})}\right\}_{\boldsymbol{\alpha}^0}$$
$$=\left\{\int_0^1 \omega\, d\omega\left[\frac{da^{(1)}(z,\omega)}{dz}\right]_{z=a}\right\}_{\boldsymbol{\alpha}^0} \tag{3.51}$$
$$=\left\{h(\boldsymbol{\alpha})g_2(\boldsymbol{\alpha})E_1\left[(b-a)g_2(\boldsymbol{\alpha})\right]\right\}_{\boldsymbol{\alpha}^0};$$

$$\left\{\frac{\partial R^{(1)}\left[1;a^{(1)};\mathbf{f}(\boldsymbol{\alpha})\right]}{\partial b}\right\}_{\boldsymbol{\alpha}^0}=\left\{\frac{\partial^2 R(\varphi;\mathbf{f})}{\partial b\,\partial g_1(\boldsymbol{\alpha})}\right\}_{\boldsymbol{\alpha}^0}$$
$$=-\left\{h(\boldsymbol{\alpha})\int_0^1 d\omega\int_a^b dz a^{(2)}(2;1;z,\omega)\delta'(z-b)\right\}_{\boldsymbol{\alpha}^0} \tag{3.52}$$
$$=-\left\{h(\boldsymbol{\alpha})g_2(\boldsymbol{\alpha})E_1\left[(b-a)g_2(\boldsymbol{\alpha})\right]\right\}_{\boldsymbol{\alpha}^0}.$$

The validity of the expressions obtained in Eqs. (3.48)–(3.52) can be verified by using the expressions of the 1st-order sensitivities provided in Eqs. (2.119)–(2.123).

### 3.3.2 Applying the 2nd-FASAM-N to Compute the 2nd-Order Response Sensitivities Stemming from the 1st-Order Sensitivity $\partial R(\varphi;\mathbf{f})/\partial g_2(\boldsymbol{\alpha})$

The expression of the 1st-order sensitivity $\partial R(\varphi;\mathbf{f})/\partial g_2(\boldsymbol{\alpha})$, which was determined in Eq. (2.120), is reproduced below for convenience:

$$R^{(1)}\left[2;\varphi;a^{(1)};\mathbf{f}(\boldsymbol{\alpha})\right]\triangleq\frac{\partial R(\varphi;\mathbf{f})}{\partial g_2(\boldsymbol{\alpha})}=-\int_0^1 d\omega\int_a^b a^{(1)}(z,\omega)\varphi(z,\omega)dz. \tag{3.53}$$

The digit "2" in the list of arguments of $R^{(1)}\left[2;\varphi;a^{(1)};\mathbf{f}(\boldsymbol{\alpha})\right]$ indicates that this is the "2nd 1st-order sensitivity to be considered" of the five 1st-order sensitivities that were obtained in Eqs. (2.119)–(2.123). As indicated in Eq. (3.53), the 1st-order sensitivity $R^{(1)}\left[2;\varphi;a^{(1)};\mathbf{f}(\boldsymbol{\alpha})\right]$ depends on both the original forward function $\varphi(z,\omega)$ and also on the 1st-level adjoint sensitivity function $a^{(1)}(z,\omega)$.

The 2nd-order sensitivities that stem from $\partial R(\varphi;\mathbf{f})/\partial g_2(\boldsymbol{\alpha})$ are provided by the 1st-order G-differential of $R^{(1)}\left[2;\varphi;a^{(1)};\mathbf{f}(\boldsymbol{\alpha})\right]\triangleq\partial R(\varphi;\mathbf{f})/\partial g_2(\boldsymbol{\alpha})$. Applying the definition of the G-differential to Eq. (3.53) yields the following relation for variations $v^{(1)}(z,\omega)$ around $\varphi^0(z,\omega)$, variations $\delta a^{(1)}(z,\omega)$ around $\varphi^0(z,\omega)$, and variations $\delta\mathbf{f}$ around $\mathbf{f}^0$:

$$\left\{\delta R^{(1)}\left[2;\varphi;a^{(1)};\mathbf{f};\delta a^{(1)};v^{(1)};\delta\mathbf{f}\right]\right\}_{\alpha^0}$$
$$= -\left\{\frac{d}{d\varepsilon}\left[\int_0^1 d\omega \int_{a+\varepsilon\delta a}^{b+\varepsilon\delta b}\left(a^{(1)}+\varepsilon\delta a^{(1)}\right)\left(\varphi+\varepsilon v^{(1)}\right)dz\right]_{\alpha^0}\right\}_{\varepsilon=0} \tag{3.54}$$
$$= \left\{\delta R^{(1)}\left(2;\varphi;a^{(1)};\mathbf{f};\delta\mathbf{f}\right)_{\alpha^0}\right\}_{dir} + \left\{\delta R^{(1)}\left(2;\varphi;a^{(1)};\mathbf{f};\delta a^{(1)};v^{(1)}\right)_{\alpha^0}\right\}_{ind}.$$

In Eq. (3.54), the direct-effect term $\left\{\delta R^{(1)}\left(2;\varphi;a^{(1)};\mathbf{f};\delta\mathbf{f}\right)_{\alpha^0}\right\}_{dir}$ depends only on variations $\delta\mathbf{f}$ while the indirect-effect term $\left\{\delta R^{(1)}\left(2;\varphi;a^{(1)};\mathbf{f};\delta a^{(1)};v^{(1)}\right)_{\alpha^0}\right\}_{ind}$ depends only on variations $\delta a^{(1)}$ and $v^{(1)}$; these two terms are defined, respectively, as follows:

$$\left\{\delta R^{(1)}\left(2;\varphi;a^{(1)};\mathbf{f};\delta\mathbf{f}\right)_{\alpha^0}\right\}_{dir} \triangleq -\left\{(\delta b)\int_0^1 a^{(1)}(b,\omega)\varphi(b,\omega)d\omega\right\}_{\alpha^0} + \left\{(\delta a)\int_0^1 a^{(1)}(a,\omega)\varphi(a,\omega)d\omega\right\}_{\alpha^0}, \tag{3.55}$$

$$\left\{\delta R^{(1)}\left(2;\varphi;a^{(1)};\mathbf{f};\delta a^{(1)};v^{(1)}\right)_{\alpha^0}\right\}_{ind}$$
$$\triangleq -\left\{\int_0^1 d\omega\int_a^b\left[a^{(1)}(z,\omega)v^{(1)}(z,\omega)+\delta a^{(1)}(z,\omega)\varphi(z,\omega)\right]dz\right\}_{\alpha^0}. \tag{3.56}$$

The expression of the flux $\varphi(z,\omega)$ has been provided in Eq. (2.66) and is reproduced below for convenience.

$$\varphi(z,\omega) = \frac{Q}{\mu_A}\exp\left[\frac{\mu_B}{\omega}(a-z)\right], \quad a \le z \le b, \quad 0 \le \omega \le 1. \tag{3.57}$$

Thus, the direct-effect term defined by Eq. (3.55) can be evaluated at this stage by using Eqs. (3.57) and (3.27) to obtain:

$$\left\{\delta R^{(1)}\left(2;\varphi;a^{(1)};\mathbf{f};\delta\mathbf{f}\right)_{\alpha^0}\right\}_{dir} \triangleq -(\delta b)\left\{\int_0^1 d\omega a^{(1)}(b,\omega)\varphi(b,\omega)\right\}_{\alpha^0}$$
$$+(\delta a)\left\{\int_0^1 d\omega a^{(1)}(a,\omega)\varphi(a,\omega)\right\}_{\alpha^0} \tag{3.58}$$
$$= (\delta a)g_1(\boldsymbol{\alpha})h(\boldsymbol{\alpha})E_1\left[(b-a)g_2(\boldsymbol{\alpha})\right],$$

Note that the 1st term on the right side of Eq. (3.58) vanishes in view of Eq. (3.29).

The indirect-effect term defined in Eq. (3.56) can be evaluated only after having determined the variational functions $\delta a^{(1)}(z,\omega)$ and $v^{(1)}(z,\omega)$. Recall that the variation $\delta a^{(1)}(z,\omega)$ is the solution of Eqs. (3.37) and (3.38), while the variation $v^{(1)}(z,\omega)$ is the solution of the 1st-LVSS defined in Eqs. (2.112) and (2.113), which are reproduced below for convenience.

$$\left\{\omega\frac{dv^{(1)}(z,\omega)}{dz}+g_2(\alpha)v^{(1)}(z,\omega)\right\}_{\alpha^0}=-\left\{(\delta g_2)\varphi(z,\omega)\right\}_{\alpha^0};\quad a^0<z<b^0,\quad 0\le\omega\le 1; \tag{3.59}$$

$$\left\{v^{(1)}(z,\omega)\right\}_{z=a^0}=\left\{\delta g_1(\alpha)-(\delta a)g_1(\alpha)\left[\frac{d\varphi(z,\omega)}{dz}\right]_{z=a}\right\}_{\alpha^0}$$
$$=\left\{\delta g_1(\alpha)\right\}_{\alpha^0}+(\delta a)\left\{\frac{g_1(\alpha)g_2(\alpha)}{\omega}\right\}_{\alpha^0};\quad z=a^0,\quad 0\le\omega\le 1. \tag{3.60}$$

Thus, the 2nd-LVSS for the variational functions $\delta a^{(1)}(z,\omega)$ and $v^{(1)}(z,\omega)$ comprises Eqs. (3.37), (3.38), (3.59), and (3.60). The need for repeatedly solving the 2nd-LVSS to obtain the functions $\delta a^{(1)}(z,\omega)$ and $v^{(1)}(z,\omega)$ for all possible parameter variations is circumvented by applying the 2nd-FASAM-N. Since the indirect-effect term depends on both $v^{(1)}(z,\omega)$ and $\delta a^{(1)}(z,\omega)$, the 2nd-level adjoint function needed for recasting the expression of the indirect-effect term will have two non-zero components, and the inner product will therefore have the form shown in Eq. (3.15). Thus, forming the inner product of a yet undefined 2nd-level function of the form $\mathbf{A}^{(2)}(2;j_1=2;\mathbf{x})\triangleq\left[a^{(2)}(1;j_1=2;z,\omega),a^{(2)}(2;j_1=2;z,\omega)\right]^{\dagger}$ with Eqs. (3.59) and (3.37), respectively, yields the following relation:

$$\left\{\int_0^1 d\omega\int_a^b a^{(2)}(1;2;z,\omega)\left[\omega\frac{dv^{(1)}(z,\omega)}{dz}+g_2(\alpha)v^{(1)}(z,\omega)\right]dz\right\}_{\alpha^0}$$
$$+\left\{\int_0^1 d\omega\int_a^b a^{(2)}(2;2;z,\omega)\left[-\omega\frac{d}{dz}\delta a^{(1)}(z,\omega)+g_2(\alpha)\delta a^{(1)}(z,\omega)\right]dz\right\}_{\alpha^0} \tag{3.61}$$
$$=-\left\{(\delta g_2)\int_0^1 d\omega\int_a^b a^{(2)}(1;2;z,\omega)\varphi(z,\omega)dz\right\}_{\alpha^0}+\left\{\int_0^1 d\omega\int_a^b dz a^{(2)}(2;2;z,\omega)\right.$$
$$\left.\times\left[-(\delta g_2)a^{(1)}(z,\omega)+(\delta h)\delta(z-b)-(\delta b)h(\alpha)\delta'(z-b)\right]\right\}_{\alpha^0}.$$

Integrating by parts the left side of Eq. (3.61) over the independent variable $z$ yields the following relation:

$$\begin{aligned}
&\int_0^1 d\omega \int_a^b a^{(2)}(1;2;z,\omega)\left[\omega\frac{dv^{(1)}(z,\omega)}{dz}+g_2(\alpha)v^{(1)}(z,\omega)\right]dz \\
&+\int_0^1 d\omega \int_a^b a^{(2)}(2;2;z,\omega)\left[-\omega\frac{d}{dz}\delta a^{(1)}(z,\omega)+g_2(\alpha)\delta a^{(1)}(z,\omega)\right]dz \\
&=\int_0^1 \omega\, d\omega\left[a^{(2)}(1;2;b,\omega)v^{(1)}(b,\omega)-a^{(2)}(1;2;a,\omega)v^{(1)}(a,\omega)\right] \\
&+\int_0^1 d\omega \int_a^b v^{(1)}(z,\omega)\left[-\omega\frac{d}{dz}a^{(2)}(1;2;z,\omega)+g_2(\alpha)a^{(2)}(1;2;z,\omega)\right]dz \\
&+\int_0^1 \omega\, d\omega\left[a^{(2)}(2;2;a,\omega)\delta a^{(1)}(a,\omega)-a^{(2)}(2;2;b,\omega)\delta a^{(1)}(b,\omega)\right] \\
&+\int_0^1 d\omega \int_a^b \delta a^{(1)}(z,\omega)\left[\omega\frac{d}{dz}a^{(2)}(2;2;z,\omega)+g_2(\alpha)a^{(2)}(2;2;z,\omega)\right]dz.
\end{aligned} \tag{3.62}$$

The above relation holds at the nominal parameter values, but this fact has not been indicated explicitly in order to simplify the notation.

The 2nd and the last terms on the right side of Eq. (3.62) are now required to represent the indirect-effect term defined in Eq. (3.56). This requirement is achieved by imposing the following relationships, for $a^0 < z < b^0;\ 0 \le \omega \le 1$:

$$\left\{-\omega\frac{d}{dz}a^{(2)}(1;2;z,\omega)+g_2(\alpha)a^{(2)}(1;2;z,\omega)\right\}_{\alpha^0} = -\left\{a^{(1)}(z,\omega)\right\}_{\alpha^0}; \tag{3.63}$$

$$\left\{\omega\frac{d}{dz}a^{(2)}(2;2;z,\omega)+g_2(\alpha)a^{(2)}(2;2;z,\omega)\right\}_{\alpha^0} = -\left\{\varphi(z,\omega)\right\}_{\alpha^0}. \tag{3.64}$$

The definition of the two-component function $\mathbf{A}^{(2)}(2;2;\mathbf{x}) \triangleq \left[a^{(2)}(1;2;z,\omega),a^{(2)}(2;2;z,\omega)\right]^\dagger$ is now completed by requiring that the unknown values of the functions $v^{(1)}(b,\omega)$ and $\delta a^{(1)}(a,\omega)$ be eliminated from appearing on the right side of Eq. (3.62). This requirement is met by imposing the following conditions to be satisfied by the function $\mathbf{A}^{(2)}(2;2;\mathbf{x}) \triangleq \left[a^{(2)}(1;2;z,\omega),a^{(2)}(2;2;z,\omega)\right]^\dagger$ on the model's boundaries:

$$a^{(2)}(1;2;z,\omega)=0, \quad \text{at} \quad z=b^0, \quad 0\le\omega\le 1. \tag{3.65}$$

$$a^{(2)}(2;2;z,\omega)=0, \quad \text{at} \quad z=a^0, \quad 0\le\omega\le 1. \tag{3.66}$$

Altogether, Eqs. (3.63)–(3.66) constitute the "2nd-Level Adjoint Sensitivity System" (2nd-LASS), for determining the 2nd-level adjoint sensitivity function $\mathbf{A}^{(2)}(2;2;\mathbf{x})\triangleq\left[a^{(2)}(1;2;z,\omega),a^{(2)}(2;2;z,\omega)\right]^{\dagger}$. Since the 2nd-LASS does not depend on the parameter variations, it needs to be solved only once in order to obtain $\mathbf{A}^{(2)}(2;2;\mathbf{x})\triangleq\left[a^{(2)}(1;2;z,\omega),a^{(2)}(2;2;z,\omega)\right]^{\dagger}$. For the paradigm model under consideration, the 2nd-LASS given by Eqs. (3.63)–(3.66) can be solved exactly to obtain the following closed-form expression for the two non-zero components of the 2nd-level adjoint function $\mathbf{A}^{(2)}(2;2;z,\omega)\triangleq\left[a^{(2)}(1;2;z,\omega),a^{(2)}(2;2;z,\omega)\right]^{\dagger}$:

$$a^{(2)}(1;2;z,\omega)=\frac{h(\alpha)}{\omega^2}\exp\left[\frac{g_2(\alpha)(z-b)}{\omega}\right]; \tag{3.67}$$

$$a^{(2)}(2;2;z,\omega)=\frac{g_1(\alpha)}{\omega}(a-z)\exp\left[\frac{g_2(\alpha)(a-z)}{\omega}\right]. \tag{3.68}$$

Collecting the results obtained in Eqs. (3.56)–(3.66) while using the boundary conditions provided in Eqs. (3.60) and (3.38) leads to the following expression for the indirect-effect term:

$$\begin{aligned}
\left\{\delta R^{(1)}\left(2;\varphi;a^{(1)};\mathbf{f};\delta a^{(1)};v^{(1)}\right)_{\alpha^0}\right\}_{ind} &= -\left\{(\delta g_2)\int_0^1 d\omega\int_a^b a^{(2)}(1;2;z,\omega)\varphi(z,\omega)\,dz\right\}_{\alpha^0} \\
&+\left\{\int_0^1 d\omega\int_a^b dz a^{(2)}(2;2;z,\omega)\left[-(\delta g_2)a^{(1)}(z,\omega)+(\delta h)\delta(z-b)-(\delta b)h(\alpha)\delta'(z-b)\right]\right\}_{\alpha^0} \\
&+\left\{\int_0^1 \omega\, d\omega a^{(2)}(1;2;a,\omega)\left[\delta g_1+(\delta a)\frac{g_1(\alpha)g_2(\alpha)}{\omega}\right]\right\}_{\alpha^0} \\
&\triangleq\left\{\delta R^{(1)}\left[2;\varphi;a^{(1)};\mathbf{f};\mathbf{A}^{(2)}(2;2;z,\omega)\right]_{\alpha^0}\right\}_{ind}.
\end{aligned} \tag{3.69}$$

Note that the dependence of $\left\{\delta R^{(1)}\left(3;\varphi;\mathbf{f};v^{(1)}\right)_{\alpha^0}\right\}_{ind}$ on the variational functions $v^{(1)}(z,\omega)$ and $\delta a^{(1)}(z,\omega)$ has been replaced in Eq. (3.69) by the dependence on the 2nd-level adjoint sensitivity function $\mathbf{A}^{(2)}(2;2;z,\omega)$.

Adding the results obtained in Eqs. (3.69) and (3.58) yields the following relation:

$$
\begin{aligned}
&\left\{\delta R^{(1)}\left[2;\varphi;a^{(1)};\mathbf{f};\delta a^{(1)};v^{(1)};\delta\mathbf{f}\right]\right\}_{\alpha^0} \\
&=-\left\{(\delta g_2)\int_0^1 d\omega\int_a^b a^{(2)}(1;2;z,\omega)\varphi(z,\omega)\,dz\right\}_{\alpha^0} \\
&+\left\{\int_0^1 d\omega\int_a^b dz a^{(2)}(2;2;z,\omega)\begin{bmatrix}-(\delta g_2)a^{(1)}(z,\omega)+(\delta h)\delta(z-b)\\ -(\delta b)h(\alpha)\delta'(z-b)\end{bmatrix}\right\}_{\alpha^0} \\
&+\left\{\int_0^1 \omega\, d\omega a^{(2)}(1;2;a,\omega)\left[(\delta g_1)+(\delta a)\frac{g_1(\alpha)g_2(\alpha)}{\omega}\right]\right\}_{\alpha^0} \\
&+(\delta a)g_1(\alpha)h(\alpha)E_1\left[(b-a)g_2(\alpha)\right] \\
&\triangleq\left\{\delta R^{(1)}\left[2;\varphi;a^{(1)};\mathbf{f};\mathbf{A}^{(2)}(2;2;z,\omega)\delta\mathbf{f}\right]\right\}_{\alpha^0}.
\end{aligned}
\tag{3.70}
$$

Identifying the expressions that multiply the respective components of $\delta\mathbf{f}(\alpha)\triangleq\left[\delta g_1(\alpha),\delta g_2(\alpha);\delta h(\alpha);\delta a;\delta b\right]^\dagger$ in Eq. (3.70) and using the expressions of the components of the 2nd-level adjoint sensitivity function $\mathbf{A}^{(2)}(2;2;z,\omega)$ yields the following results for the 2nd-order sensitivities that stem from the 1st-order sensitivity $R^{(1)}\left[2;\varphi;a^{(1)};\mathbf{f}\right]\triangleq\partial R(\varphi;\mathbf{f})/\partial g_2(\alpha)$:

$$
\begin{aligned}
\frac{\partial^2 R(\varphi;\mathbf{f})}{\partial g_1(\alpha)\partial g_2(\alpha)}&=\int_0^1 \omega\, d\omega a^{(2)}(1;2;a,\omega)\\
&=(a-b)h(\alpha)E_1\left[(b-a)g_2(\alpha)\right];
\end{aligned}
\tag{3.71}
$$

$$
\begin{aligned}
\frac{\partial^2 R(\varphi;\mathbf{f})}{\partial g_2\,\partial g_2(\alpha)}&=-\int_0^1 d\omega\int_a^b a^{(2)}(1;2;z,\omega)\varphi(z,\omega)\,dz\\
&-\int_0^1 d\omega\int_a^b dz a^{(2)}(2;2;z,\omega)a^{(1)}(z,\omega)\\
&=-g_1(\alpha)h(\alpha)(a-b)(b-a)E_0\left[(b-a)g_2(\alpha)\right]\\
&=\frac{(b-a)g_1(\alpha)h(\alpha)}{g_2(\alpha)}\exp\left[(a-b)g_2(\alpha)\right];
\end{aligned}
\tag{3.72}
$$

$$
\begin{aligned}
\frac{\partial^2 R(\varphi;\mathbf{f})}{\partial h(\boldsymbol{a})\partial g_2(\boldsymbol{a})}&=\int_0^1 d\omega\int_a^b dz a^{(2)}(2;2;z,\omega)\delta(z-b)\\
&=g_1(\boldsymbol{a})(a-b)E_1\left[(b-a)g_2(\boldsymbol{a})\right];
\end{aligned}
\tag{3.73}
$$

$$\frac{\partial^2 R(\varphi;\mathbf{f})}{\partial a\,\partial g_2(\boldsymbol{\alpha})} = g_1(\boldsymbol{\alpha})h(\boldsymbol{\alpha})E_1\left[(b-a)g_2(\boldsymbol{\alpha})\right] + g_1(\boldsymbol{\alpha})g_2(\boldsymbol{\alpha})\int_0^1 d\omega a^{(2)}(1;2;a,\omega)$$

$$= g_1(\boldsymbol{\alpha})h(\boldsymbol{\alpha})E_1\left[(b-a)g_2(\boldsymbol{\alpha})\right]$$

$$+g_1(\boldsymbol{\alpha})h(\boldsymbol{\alpha})(a-b)g_2(\boldsymbol{\alpha})E_0\left[(b-a)g_2(\boldsymbol{\alpha})\right]$$

$$= g_1(\boldsymbol{\alpha})h(\boldsymbol{\alpha})\left\{E_1\left[(b-a)g_2(\boldsymbol{\alpha})\right] - \exp\left[(a-b)g_2(\boldsymbol{\alpha})\right]\right\}; \tag{3.74}$$

$$\frac{\partial^2 R(\varphi;\mathbf{f})}{\partial b\,\partial g_2(\boldsymbol{\alpha})} = -h(\boldsymbol{\alpha})\int_0^1 d\omega \int_a^b dz a^{(2)}(2;2;z,\omega)\delta'(z-b)$$

$$= -g_1(\boldsymbol{\alpha})h(\boldsymbol{\alpha})E_1\left[(b-a)g_2(\boldsymbol{\alpha})\right] \tag{3.75}$$

$$- g_1(\boldsymbol{\alpha})h(\boldsymbol{\alpha})(a-b)g_2(\boldsymbol{\alpha})E_0\left[(b-a)g_2(\boldsymbol{\alpha})\right]$$

$$= g_1(\boldsymbol{\alpha})h(\boldsymbol{\alpha})\left\{\exp\left[(a-b)g_2(\boldsymbol{\alpha})\right] - E_1\left[(b-a)g_2(\boldsymbol{\alpha})\right]\right\}.$$

The validity of the expressions obtained in Eqs. (3.71)–(3.75) can be verified by using the expressions of the 1st-order sensitivities provided in Chapter 2 and effectuating the corresponding differentiations.

### 3.3.3 Applying the 2nd-FASAM-N to Compute the 2nd-Order Response Sensitivities Stemming from the 1st-Order Sensitivity $\partial R(\varphi;\mathbf{f})/\partial h(\boldsymbol{\alpha})$

The expression of the 1st-order sensitivity $\partial R(\varphi;\mathbf{f})/\partial h(\boldsymbol{\alpha})$ has been determined in Eq. (2.121), which is reproduced below for convenience:

$$R^{(1)}\left[3;\varphi;\mathbf{f}(\boldsymbol{\alpha})\right] \triangleq \frac{\partial R(\varphi;\mathbf{f})}{\partial h(\boldsymbol{\alpha})} = \int_0^1 d\omega \int_a^b \varphi(z,\omega)\delta(z-b)dz. \tag{3.76}$$

It is evident that the 1st-order sensitivity $R^{(1)}\left[3;\varphi;\mathbf{f}(\boldsymbol{\alpha})\right] \triangleq \partial R(\varphi;\mathbf{f})/\partial a$ does not depend on the 1st-level adjoint sensitivity function $a^{(1)}(z,\omega)$ but only depends on the original forward function $\varphi(z,\omega)$. The digit "3" in the list of arguments of $R^{(1)}\left[3;\varphi;\mathbf{f}(\boldsymbol{\alpha})\right]$ indicates that this 1st-order sensitivity is the "3rd to be considered" of the five 1st-order sensitivities of the model response with respect to the five components of the "feature" function **f**, which were obtained in Eqs. (2.119) – (2.123).

The 2nd-order sensitivities that stem from $\partial R(\varphi;\mathbf{f})/\partial h(\boldsymbol{\alpha})$ are provided by the 1st-order G-differential of $R^{(1)}\left[3;\varphi;\mathbf{f}(\boldsymbol{\alpha})\right] \triangleq \partial R(\varphi;\mathbf{f})/\partial h$. Applying the definition of the G-differential to Eq. (3.76) yields the following relation:

$$\left\{\delta R^{(1)}\left[3;\varphi;\mathbf{f}(\boldsymbol{\alpha});v^{(1)};\delta\mathbf{f}\right]\right\}_{\boldsymbol{\alpha}^0}$$
$$\triangleq\left\{\frac{d}{d\varepsilon}\left[\int_0^1 d\omega\int_{a+\varepsilon\delta a}^{b+\varepsilon\delta b}\left(\varphi+\varepsilon v^{(1)}\right)\delta\left(z-b-\varepsilon\delta b\right)dz\right]_{\boldsymbol{\alpha}^0}\right\}_{\varepsilon=0} \tag{3.77}$$
$$=\left\{\delta R^{(1)}\left(3;\varphi;\mathbf{f};\delta\mathbf{f}\right)_{\boldsymbol{\alpha}^0}\right\}_{dir}+\left\{\delta R^{(1)}\left(3;\varphi;\mathbf{f};v^{(1)}\right)_{\boldsymbol{\alpha}^0}\right\}_{ind},$$

where the direct-effect term $\left\{\delta R^{(1)}\left(3;\varphi;\mathbf{f};\delta\mathbf{f}\right)_{\boldsymbol{\alpha}^0}\right\}_{dir}$ depends only on variations $\delta\mathbf{f}$ while the indirect-effect term $\left\{\delta R^{(1)}\left(3;\varphi;\mathbf{f};v^{(1)}\right)_{\boldsymbol{\alpha}^0}\right\}_{ind}$ depends only on the variational function $v^{(1)}(z,\omega)$; these terms are defined, respectively, as follows:

$$\left\{\delta R^{(1)}\left(3;\varphi;\mathbf{f};\delta\mathbf{f}\right)_{\boldsymbol{\alpha}^0}\right\}_{dir}\triangleq\left\{(\delta b)\int_0^1 d\omega\left[\frac{d\varphi(z,\omega)}{dz}\right]_{z=b}\right\}_{\boldsymbol{\alpha}^0}, \tag{3.78}$$

$$\left\{\delta R^{(1)}\left(3;\varphi;\mathbf{f};v^{(1)}\right)_{\boldsymbol{\alpha}^0}\right\}_{ind}\triangleq\left\{\int_0^1 d\omega\int_a^b v^{(1)}(z,\omega)\delta(z-b)dz\right\}_{\boldsymbol{\alpha}^0}. \tag{3.79}$$

The direct-effect term defined by Eq. (3.78) can be evaluated at this stage by using Eq. (3.57) to obtain:

$$\left\{\delta R^{(1)}\left(3;\varphi;\mathbf{f};\delta\mathbf{f}\right)_{\boldsymbol{\alpha}^0}\right\}_{dir}\triangleq\left\{(\delta b)\int_0^1 d\omega\left[\frac{d\varphi(z,\omega)}{dz}\right]_{z=b}\right\}_{\boldsymbol{\alpha}^0}$$
$$=-(\delta b)g_1(\boldsymbol{\alpha})g_2(\boldsymbol{\alpha})E_1\left[(b-a)g_2(\boldsymbol{\alpha})\right]. \tag{3.80}$$

The indirect-effect term defined in Eq. (3.79) can be evaluated after having determined the variational function $v^{(1)}(z,\omega)$, which is the solution of Eqs. (3.59) and (3.60). As before, the need for solving the respective variational equations is avoided by constructing an appropriate 2nd-LASS, by applying the principles of the 2nd-FASAM-N. This 2nd-LASS is constructed by forming the inner product of Eq. (3.59) with a yet undefined function $a^{(2)}(1;3;z,\omega)$, to obtain the following relation:

$$\left\{\int_0^1 d\omega\int_a^b a^{(2)}(1;3;z,\omega)\left[\omega\frac{dv^{(1)}(z,\omega)}{dz}+g_2(\boldsymbol{\alpha})v^{(1)}(z,\omega)\right]dz\right\}_{\boldsymbol{\alpha}^0}$$
$$=-\left\{(\delta g_2)\int_0^1 d\omega\int_a^b dz a^{(2)}(1;3;z,\omega)\varphi(z,\omega)\right\}_{\boldsymbol{\alpha}^0}. \tag{3.81}$$

Note that the first argument of $a^{(2)}(1;3;z,\omega)$ indicates that only the "1st-component" of this "two-component 2nd-level adjoint sensitivity function" is non-zero, while the 2nd argument in $a^{(2)}(1;3;z,\omega)$ refers to the index/label "$j_1 = 3$" indicating that this 2nd-level adjoint sensitivity function will correspond to "the 3rd considered" 1st-order sensitivity (namely: $R^{(1)}[3;\varphi;\mathbf{f}(\boldsymbol{\alpha})] \triangleq \partial R(\varphi;\mathbf{f})/\partial h$).

Integrating the left side of Eq. (3.81) by parts over the independent variable $z$ yields the following relation:

$$\left\{\int_0^1 d\omega \int_a^b a^{(2)}(1;3;z,\omega)\left[\omega \frac{dv^{(1)}(z,\omega)}{dz} + g_2(\boldsymbol{\alpha})v^{(1)}(z,\omega)\right]dz\right\}_{\boldsymbol{\alpha}^0}$$

$$= \int_0^1 \omega\, d\omega \left\{a^{(2)}(1;3;b,\omega)v^{(1)}(b,\omega) - a^{(2)}(1;3;a,\omega)v^{(1)}(a,\omega)\right\}_{\boldsymbol{\alpha}^0} \tag{3.82}$$

$$+\left\{\int_0^1 d\omega \int_a^b v^{(1)}(z,\omega)\left[-\omega\frac{d}{dz}a^{(2)}(1;3;z,\omega) + g_2(\boldsymbol{\alpha})a^{(2)}(1;3;z,\omega)\right]dz\right\}_{\boldsymbol{\alpha}^0}.$$

The last term on the right side of Eq. (3.82) is now required to represent the indirect-effect term defined in Eq. (3.79), which is achieved by imposing the following relationship over the domain $a^0 < z < b^0;\ 0 \le \omega \le 1$:

$$\left\{-\omega\frac{d}{dz}a^{(2)}(1;3;z,\omega) + g_2(\boldsymbol{\alpha})a^{(2)}(1;3;z,\omega)\right\}_{\boldsymbol{\alpha}^0} = \left\{\delta(z-b)\right\}_{\boldsymbol{\alpha}^0}. \tag{3.83}$$

The definition of the function $a^{(2)}(1;3;z,\omega)$ is now completed by requiring that the unknown value of the function $v^{(1)}(b,\omega)$ be eliminated from appearing on the right side of Eq. (3.82). This requirement is met by imposing the following condition to be satisfied by the function $a^{(2)}(1;3;z,\omega)$ on the model's outer boundary:

$$a^{(2)}(1;3;z,\omega) = 0, \quad \text{at} \quad z = b^0, \quad 0 \le \omega \le 1. \tag{3.84}$$

Altogether, Eqs. (3.83) and (3.84) constitute the "2nd-Level Adjoint Sensitivity System" (2nd-LASS) for determining the 2nd-level adjoint sensitivity function $a^{(2)}(1;3;z,\omega)$. Since the 2nd-LASS does not depend on the parameter variations, it needs to be solved only once in order to obtain $a^{(2)}(1;3;z,\omega)$. For the paradigm model under consideration, the 1st-LASS given by Eqs. (3.83) and (3.84) can be solved exactly to obtain the following closed-form expression for the 2nd-level adjoint function $a^{(2)}(1;3;z,\omega)$:

$$a^{(2)}(1;3;z,\omega) = \left[1 - H(z-b)\right]\frac{1}{\omega}\exp\left[\frac{g_2(\boldsymbol{\alpha})(z-b)}{\omega}\right]. \tag{3.85}$$

Collecting the results obtained in Eqs. (3.79)–(3.84) leads to the following expression for the indirect-effect term:

$$\left\{\delta R^{(1)}\left(3;\varphi;\mathbf{f};v^{(1)}\right)_{\alpha^0}\right\}_{ind}=-\left\{(\delta g_2)\int_0^1 d\omega\int_a^b a^{(2)}(1;3;z,\omega)\varphi(z,\omega)dz\right\}_{\alpha^0}$$
$$+\left\{\int_0^1 \omega\, d\omega a^{(2)}(1;3;a,\omega)\left[\delta g_1+(\delta a)\frac{g_1(\alpha)g_2(\alpha)}{\omega}\right]\right\}_{\alpha^0}$$
$$\triangleq\left\{\delta R^{(1)}\left(3;\varphi;\mathbf{f};a^{(2)}(1;3;z,\omega)\right)_{\alpha^0}\right\}_{ind}. \tag{3.86}$$

Note that the dependence of $\left\{\delta R^{(1)}\left(3;\varphi;\mathbf{f};v^{(1)}\right)_{\alpha^0}\right\}_{ind}$ on the variational function $v^{(1)}(z,\omega)$ has been replaced in Eq. (3.86) by the dependence on the 2nd-level adjoint sensitivity function $a^{(2)}(1;3;z,\omega)$.

Adding the results obtained in Eqs. (3.86) and (3.80) yields the following relation:

$$\left\{\delta R^{(1)}\left[3;\varphi;\mathbf{f}(\alpha);v^{(1)};\delta\mathbf{f}\right]\right\}_{\alpha^0}=-\left\{(\delta g_2)\int_0^1 d\omega\int_a^b a^{(2)}(1;3;z,\omega)\varphi(z,\omega)dz\right\}_{\alpha^0}$$
$$+\left\{\int_0^1 \omega\, d\omega a^{(2)}(1;3;a,\omega)\left[\delta g_1+(\delta a)\frac{g_1(\alpha)g_2(\alpha)}{\omega}\right]\right\}_{\alpha^0}$$
$$+\left\{(\delta b)\int_0^1 d\omega\left[\frac{d\varphi(z,\omega)}{dz}\right]_{z=b}\right\}_{\alpha^0}\triangleq\left\{\delta R^{(1)}\left[3;\varphi;\mathbf{f};a^{(2)}(1;3;z,\omega);\delta\mathbf{f}\right]\right\}_{\alpha^0}. \tag{3.87}$$

Note that the dependence of $\left\{\delta R^{(1)}\left[3;\varphi;\mathbf{f}(\alpha);v^{(1)};\delta\mathbf{f}\right]\right\}_{\alpha^0}$ on the variational function $v^{(1)}(z,\omega)$ has been replaced in Eq. (3.87) by the dependence on $a^{(2)}(1;3;z,\omega)$.

Identifying the expressions that multiply the respective components of $\delta\mathbf{f}(\alpha)\triangleq\left[\delta g_1(\alpha),\delta g_2(\alpha);\delta h(\alpha);\delta a;\delta b\right]^{\dagger}$ and using the expression for the 2nd-level adjoint sensitivity function provided in Eq. (3.85) yields the following results for the 2nd-order sensitivities that stem from the 1st-order sensitivity $R^{(1)}\left[3;\varphi;\mathbf{f}(\alpha)\right]\triangleq\partial R(\varphi;\mathbf{f})/\partial h$:

$$\frac{\partial^2 R(\varphi;\mathbf{f})}{\partial g_1(\alpha)\partial h(\alpha)}=\int_0^1 \omega\, d\omega a^{(2)}(1;3;a,\omega)=E_2\left[(b-a)g_2(\alpha)\right], \tag{3.88}$$

$$\frac{\partial^2 R(\varphi;\mathbf{f})}{\partial g_2(\boldsymbol{\alpha})\partial h(\boldsymbol{\alpha})} = -\int_0^1 d\omega \int_a^b a^{(2)}(1;3;z,\omega)\varphi(z,\omega)dz$$
$$= -g_1(\boldsymbol{\alpha})(b-a)E_1\left[(b-a)g_2(\boldsymbol{\alpha})\right]; \tag{3.89}$$

$$\frac{\partial^2 R(\varphi;\mathbf{f})}{\partial h(\boldsymbol{\alpha})\partial h(\boldsymbol{\alpha})} = 0; \tag{3.90}$$

$$\frac{\partial^2 R(\varphi;\mathbf{f})}{\partial a\,\partial h(\boldsymbol{\alpha})} = g_1(\boldsymbol{\alpha})g_2(\boldsymbol{\alpha})\int_0^1 d\omega a^{(2)}(1;3;a,\omega)$$
$$= g_1(\boldsymbol{\alpha})g_2(\boldsymbol{\alpha})E_1\left[(b-a)g_2(\boldsymbol{\alpha})\right]; \tag{3.91}$$

$$\frac{\partial^2 R(\varphi;\mathbf{f})}{\partial b\,\partial h(\boldsymbol{\alpha})} = \int_0^1 d\omega \left[\frac{d\varphi(z,\omega)}{dz}\right]_{z=b} = -g_1(\boldsymbol{\alpha})g_2(\boldsymbol{\alpha})E_1\left[(b-a)g_2(\boldsymbol{\alpha})\right]. \tag{3.92}$$

The validity of the expressions obtained in Eqs. (3.88)–(3.92) can be verified by using the expressions of the 1st-order sensitivities provided in Eqs. (2.119)–(2.123).

### 3.3.4 Applying the 2nd-FASAM-N to Compute the 2nd-Order Response Sensitivities Stemming from the 1st-Order Sensitivity $\partial R(\varphi;\mathbf{f})/\partial a$

The 1st-order sensitivity $\partial R(\varphi;\mathbf{f})/\partial a$ is defined in Eq. (2.122), which is reproduced below for convenience:

$$R^{(1)}\left[4;a^{(1)};\mathbf{f}(\boldsymbol{\alpha})\right] \triangleq \frac{\partial R(\varphi;\mathbf{f})}{\partial a}$$
$$= g_1(\boldsymbol{\alpha})g_2(\boldsymbol{\alpha})\int_0^1 d\omega \int_a^b a^{(1)}(z,\omega)\delta(z-a)dz. \tag{3.93}$$

It is evident that the 1st-order sensitivity $R^{(1)}\left[4;a^{(1)};\mathbf{f}(\boldsymbol{\alpha})\right] \triangleq \partial R(\varphi;\mathbf{f})/\partial a$ does not depend on the original forward function $\varphi(z,\omega)$ but only depends on the 1st-level adjoint sensitivity function $a^{(1)}(z,\omega)$. The digit "4" in the list of arguments of $R^{(1)}\left[4;a^{(1)};\mathbf{f}(\boldsymbol{\alpha})\right]$ indicates that this 1st-order sensitivity is the "4th to be considered" of the five 1st-order sensitivities of the model response with respect to the five components of the "feature" function $\mathbf{f}(\boldsymbol{\alpha})$, which were obtained in Eqs. (2.119)–(2.123).

The 2nd-order sensitivities that stem from $\partial R(\varphi;\mathbf{f})/\partial a$ are provided by the 1st-order G-differential of $R^{(1)}\left[4;a^{(1)};\mathbf{f}(\boldsymbol{\alpha})\right] \triangleq \partial R(\varphi;\mathbf{f})/\partial a$. Applying the definition of the G-differential to Eq. (3.93) yields the following relation:

$$\left\{\delta R^{(1)}\left[4;a^{(1)};\mathbf{f}(\boldsymbol{\alpha});\delta a^{(1)};\delta\mathbf{f}\right]\right\}_{\boldsymbol{\alpha}^0}$$

$$\triangleq\left\{\frac{d}{d\varepsilon}\left[(g_1+\varepsilon\delta g_1)(g_2+\varepsilon\delta g_2)\int_0^1 d\omega\int_{a+\varepsilon\delta a}^{b+\varepsilon\delta b}\left(a^{(1)}+\varepsilon\delta a^{(1)}\right)\delta(z-a-\varepsilon\delta a)dz\right]_{\boldsymbol{\alpha}^0}\right\}_{\varepsilon=0}$$

$$=\left\{\delta R^{(1)}\left(4;a^{(1)};\mathbf{f};\delta\mathbf{f}\right)_{\boldsymbol{\alpha}^0}\right\}_{dir}+\left\{\delta R^{(1)}\left(4;a^{(1)};\mathbf{f};\delta a^{(1)}\right)_{\boldsymbol{\alpha}^0}\right\}_{ind},$$

(3.94)

where the direct-effect term $\left\{\delta R^{(1)}\left(4;a^{(1)};\mathbf{f};\delta\mathbf{f}\right)_{\boldsymbol{\alpha}^0}\right\}_{dir}$ depends only on variations $\delta\mathbf{f}$ while the indirect-effect term $\left\{\delta R^{(1)}\left(4;a^{(1)};\mathbf{f};\delta a^{(1)}\right)_{\boldsymbol{\alpha}^0}\right\}_{ind}$ depends only on variations $\delta a^{(1)}$; these terms are defined, respectively, as follows:

$$\left\{\delta R^{(1)}\left(4;a^{(1)};\mathbf{f};\delta\mathbf{f}\right)_{\boldsymbol{\alpha}^0}\right\}_{dir}\triangleq\left\{(\delta g_1)g_2\int_0^1 d\omega\int_a^b a^{(1)}(z,\omega)\delta(z-a)dz\right\}_{\boldsymbol{\alpha}^0}$$
$$+\left\{(\delta g_2)g_1\int_0^1 d\omega\int_a^b a^{(1)}(z,\omega)\delta(z-a)dz\right\}_{\boldsymbol{\alpha}^0} \tag{3.95}$$
$$+\left\{(\delta a)g_1(\boldsymbol{\alpha})g_2(\boldsymbol{\alpha})\int_0^1 d\omega\left[\frac{da^{(1)}(z,\omega)}{dz}\right]_{z=a}\right\}_{\boldsymbol{\alpha}^0},$$

$$\left\{\delta R^{(1)}\left(4;a^{(1)};\mathbf{f};\delta a^{(1)}\right)_{\boldsymbol{\alpha}^0}\right\}_{ind}$$
$$\triangleq\left\{g_1(\boldsymbol{\alpha})g_2(\boldsymbol{\alpha})\int_0^1 d\omega\int_a^b \delta a^{(1)}(z,\omega)\delta(z-a)dz\right\}_{\boldsymbol{\alpha}^0}. \tag{3.96}$$

The direct-effect term defined by Eq. (3.95) can be evaluated at this stage by using Eqs. (3.27) and (3.33) to obtain:

$$\left\{\delta R^{(1)}\left(4;a^{(1)};\mathbf{f};\delta\mathbf{f}\right)_{\boldsymbol{\alpha}^0}\right\}_{dir}\triangleq(\delta a)\left\{\frac{g_1(\boldsymbol{\alpha})g_2(\boldsymbol{\alpha})h(\boldsymbol{\alpha})}{(b-a)}\exp[(a-b)g_2(\boldsymbol{\alpha})]\right\}_{\boldsymbol{\alpha}^0}$$
$$+\left\{\left[(\delta g_1)g_2(\boldsymbol{\alpha})+(\delta g_2)g_1(\boldsymbol{\alpha})\right]h(\boldsymbol{\alpha})E_1\left[(b-a)g_2(\boldsymbol{\alpha})\right]\right\}_{\boldsymbol{\alpha}^0}.$$

(3.97)

The indirect-effect term defined in Eq. (3.96) can be evaluated after having determined the variational function $\delta a^{(1)}(z,\omega)$, which is the solution of Eqs. (3.37) and (3.38), but the need for solving variational equations is avoided by constructing an appropriate 2nd-LASS, by applying the principles of the 2nd-FASAM-N. This 2nd-LASS is constructed by forming the inner product of Eq. (3.37) with a yet undefined function $a^{(2)}(2; j_1 = 4; z,\omega)$, to obtain the following relation:

$$\left\{\int_0^1 d\omega \int_a^b a^{(2)}(2;4;z,\omega)\left[-\omega\frac{d}{dz}\delta a^{(1)}(z,\omega)+g_2(\alpha)\delta a^{(1)}(z,\omega)\right]dz\right\}_{\alpha^0}$$
$$=\left\{\int_0^1 d\omega \int_a^b dz a^{(2)}(2;4;z,\omega)\begin{bmatrix}-(\delta g_2)a^{(1)}(z,\omega)+(\delta h)\delta(z-b)\\-(\delta b)h(\alpha)\delta'(z-b)\end{bmatrix}\right\}_{\alpha^0}. \quad (3.98)$$

Note that the first argument of $a^{(2)}(2;1;z,\omega)$ indicates that only the "2nd-component" of this "two-component 2nd-level adjoint sensitivity function" is non-zero, while the 2nd argument in $a^{(2)}(2;4;z,\omega)$ refers to the index/label $j_1 = 4$ indicating that this 2nd-level adjoint sensitivity function will correspond to "the 4th considered" 1st-order sensitivity.

Integrating by parts the left side of Eq. (3.98) over the independent variable $z$ yields the following relation:

$$\left\{\int_0^1 d\omega \int_a^b a^{(2)}(2;4;z,\omega)\left[-\omega\frac{d}{dz}\delta a^{(1)}(z,\omega)+g_2(\alpha)\delta a^{(1)}(z,\omega)\right]dz\right\}_{\alpha^0}$$
$$=\int_0^1 \omega\, d\omega\left\{a^{(2)}(2;4;a,\omega)\delta a^{(1)}(a,\omega)-a^{(2)}(2;1;b,\omega)\delta a^{(1)}(b,\omega)\right\}_{\alpha^0} \quad (3.99)$$
$$+\left\{\int_0^1 d\omega \int_a^b \delta a^{(1)}(z,\omega)\left[\omega\frac{d}{dz}a^{(2)}(2;4;z,\omega)+g_2(\alpha)a^{(2)}(2;4;z,\omega)\right]dz\right\}_{\alpha^0}.$$

The last term on the right side of Eq. (3.99) is now required to represent the indirect-effect term defined in Eq. (3.96) by imposing the following relationship:

$$\left\{\int_0^1 d\omega \int_a^b \delta a^{(1)}(z,\omega)\left[\omega\frac{d}{dz}a^{(2)}(2;1;z,\omega)+g_2(\alpha)a^{(2)}(2;1;z,\omega)\right]dz\right\}_{\alpha^0}$$
$$=\left\{g_1(\alpha)g_2(\alpha)\int_0^1 d\omega\int_a^b \delta a^{(1)}(z,\omega)\delta(z-a)\,dz\right\}_{\alpha^0}. \quad (3.100)$$

The relation in Eq. (3.100) implies that the following equation holds, in the weak sense, at the nominal parameter values for $a^0 < z < b^0; 0 \le \omega \le 1$:

$$\left\{\omega \frac{d}{dz} a^{(2)}(2;4;z,\omega) + g_2(\alpha) a^{(2)}(2;4;z,\omega)\right\}_{\alpha^0}$$
$$= \{g_1(\alpha) g_2(\alpha) \delta(z-a)\}_{\alpha^0}. \tag{3.101}$$

The definition of the function $a^{(2)}(2;4;z,\omega)$ is now completed by requiring that the unknown value of the function $\delta a^{(1)}(a,\omega)$ be eliminated from appearing on the right side of Eq. (3.99). This requirement is met by imposing the following condition to be satisfied by the function $a^{(2)}(2;4;z,\omega)$ on the model's boundary:

$$a^{(2)}(2;4;z,\omega) = 0, \quad \text{at} \quad z = a^0, \quad 0 \le \omega \le 1. \tag{3.102}$$

Altogether, Eqs. (3.101) and (3.102) constitute the "2nd-Level Adjoint Sensitivity System" (2nd-LASS) for determining the 2nd-level adjoint sensitivity function $a^{(2)}(2;4;z,\omega)$. Since the 2nd-LASS does not depend on the parameter variations, it needs to be solved only once in order to obtain $a^{(2)}(2;4;z,\omega)$. For the paradigm model under consideration, the 1st-LASS given by Eqs. (3.101) and (3.102) can be solved exactly to obtain the following closed-form expression for the 2nd-level adjoint function $a^{(2)}(2;4;z,\omega)$:

$$a^{(2)}(2;4;z,\omega) = H(z-a) g_1(\alpha) g_2(\alpha) \frac{1}{\omega} \exp\left[\frac{g_2(\alpha)(a-z)}{\omega}\right]. \tag{3.103}$$

Collecting the results obtained in Eqs. (3.98)–(3.103) leads to the following expression for the indirect-effect term:

$$\left\{\delta R^{(1)}\left(4;a^{(1)};\mathbf{f};\delta a^{(1)}\right)_{\alpha^0}\right\}_{ind} = \left\{\int_0^1 d\omega \int_a^b dz a^{(2)}(2;4;z,\omega)\right.$$
$$\left.\times\left[-(\delta g_2) a^{(1)}(z,\omega) + (\delta h)\delta(z-b) - (\delta b) h(\alpha)\delta'(z-b)\right]\right\}_{\alpha^0} \tag{3.104}$$
$$= \left\{\delta R^{(1)}\left[4;a^{(1)};\mathbf{f};a^{(2)}(2;4;z,\omega)\right]_{\alpha^0}\right\}_{ind}.$$

Note that the dependence of $\left\{\delta R^{(1)}\left[4;a^{(1)};\mathbf{f};a^{(2)}(2;4;z,\omega)\right]_{\alpha^0}\right\}_{ind}$ on the variational function $\delta a^{(1)}$ has been replaced in Eq. (3.104) by the dependence on the 2nd-level adjoint sensitivity function $a^{(2)}(2;1;z,\omega)$.

Adding the results obtained in Eqs. (3.104) and (3.97) yields the following relation:

$$\left\{\delta R^{(1)}\left[4;a^{(1)};\mathbf{f}(\alpha);a^{(2)}(2;4;z,\omega);\delta\mathbf{f}\right]\right\}_{\alpha^0}$$
$$=(\delta a)\left\{\frac{g_1(\alpha)g_2(\alpha)h(\alpha)}{(b-a)}\exp\left[(a-b)g_2(\alpha)\right]\right\}_{\alpha^0}$$
$$+\left\{\left[(\delta g_1)g_2(\alpha)+(\delta g_2)g_1(\alpha)\right]h(\alpha)E_1\left[(b-a)g_2(\alpha)\right]\right\}_{\alpha^0} \tag{3.105}$$
$$+\left\{\int_0^1 d\omega\int_a^b dz a^{(2)}(2;4;z,\omega)\begin{bmatrix}-(\delta g_2)a^{(1)}(z,\omega)+(\delta h)\delta(z-b)\\-(\delta b)h(\alpha)\delta'(z-b)\end{bmatrix}\right\}_{\alpha^0}.$$

Identifying the expressions that multiply the respective components of $\delta\mathbf{f}(\alpha)\triangleq\left[\delta g_1(\alpha),\delta g_2(\alpha);\delta h(\alpha);\delta a;\delta b\right]^{\dagger}$ and using the expression for the 2nd-level adjoint sensitivity function provided in Eq. (3.103) yields the following results for the 2nd-order sensitivities that stem from the 1st-order sensitivity $\partial R(\varphi;\mathbf{f})/\partial a$:

$$\frac{\partial^2 R(\varphi;\mathbf{f})}{\partial g_1(\alpha)\partial a}=g_2(\alpha)\int_0^1 d\omega\int_a^b a^{(1)}(z,\omega)\delta(z-a)dz$$
$$=g_2(\alpha)h(\alpha)E_1\left[(b-a)g_2(\alpha)\right], \tag{3.106}$$

$$\frac{\partial^2 R(\varphi;\mathbf{f})}{\partial g_2(\alpha)\partial a}=g_1(\alpha)\int_0^1 d\omega\int_a^b a^{(1)}(z,\omega)\delta(z-a)dz-\int_0^1 d\omega\int_a^b dz a^{(1)}(z,\omega)a^{(2)}(2;4;z,\omega)$$
$$=g_1(\alpha)h(\alpha)E_1\left[(b-a)g_2(\alpha)\right]$$
$$-g_1(\alpha)h(\alpha)g_2(\alpha)(b-a)E_0\left[(b-a)g_2(\alpha)\right]$$
$$=g_1(\alpha)h(\alpha)\left\{E_1\left[(b-a)g_2(\alpha)\right]-\exp\left[(a-b)g_2(\alpha)\right]\right\}. \tag{3.107}$$

$$\frac{\partial^2 R(\varphi;\mathbf{f})}{\partial h(\alpha)\partial a}=\int_0^1 d\omega\int_a^b dz a^{(2)}(2;4;z,\omega)\delta(z-b)$$
$$=g_1(\alpha)g_2(\alpha)E_1\left[(b-a)g_2(\alpha)\right]; \tag{3.108}$$

$$\frac{\partial^2 R(\varphi;\mathbf{f})}{\partial a\,\partial a}=g_1(\alpha)g_2(\alpha)\int_0^1 d\omega\left[\frac{da^{(1)}(z,\omega)}{dz}\right]_{z=a}$$
$$=g_1(\alpha)\left[g_2(\alpha)\right]^2 h(\alpha)E_0\left[(b-a)g_2(\alpha)\right] \tag{3.109}$$
$$=\frac{g_1(\alpha)g_2(\alpha)h(\alpha)}{(b-a)}\exp\left[(a-b)g_2(\alpha)\right];$$

$$\frac{\partial^2 R(\varphi;\mathbf{f})}{\partial b\,\partial a} = -h(\alpha)\int_0^1 d\omega \int_a^b dz a^{(2)}(2;4;z,\omega)\delta'(z-b)$$

$$= -g_1(\alpha)[g_2(\alpha)]^2 h(\alpha) E_0[(b-a)g_2(\alpha)] \tag{3.110}$$

$$= -\frac{g_1(\alpha)g_2(\alpha)h(\alpha)}{(b-a)}\exp[(a-b)g_2(\alpha)].$$

The validity of the expressions obtained in Eqs. (3.106)–(3.110) can be verified by using the expressions of the 1st-order sensitivities provided in Eqs. (2.119)–(2.123).

### 3.3.5 Applying the 2nd-FASAM-N to Compute the 2nd-Order Response Sensitivities Stemming from the 1st-Order Sensitivities $\partial R(\varphi;\mathbf{f})/\partial b$

The expression of the 1st-order sensitivity $\partial R(\varphi;\mathbf{f})/\partial b$ was obtained in Eq. (2.123), which is reproduced below for convenience:

$$R^{(1)}[5;\varphi;\mathbf{f}(\alpha)] \triangleq \left\{\frac{\partial R(\varphi;\mathbf{f})}{\partial b}\right\}_{\alpha^0}$$

$$= -\left\{h(\alpha)\int_0^1 d\omega \int_a^b \varphi(z,\omega)\delta'(z-b)dz\right\}_{\alpha^0}. \tag{3.111}$$

As Eq. (3.111) indicates, the 1st-order sensitivity $R^{(1)}[5;\varphi;\mathbf{f}(\alpha)] \triangleq \partial R(\varphi;\mathbf{f})/\partial b$ does not depend on the first-level adjoint sensitivity function $a^{(1)}(z,\omega)$ but only depends on the original forward function $\varphi(z,\omega)$. The digit "5" in the list of arguments of $R^{(1)}[5;\varphi;\mathbf{f}(\alpha)]$ indicates that this 1st-order sensitivity is the "5th to be considered" – of the five 1st-order sensitivities of the model response with respect to the five components of the "feature" function $\mathbf{f}$ – which were obtained in Eqs. (2.119)–(2.123).

The 2nd-order sensitivities that stem from $\partial R(\varphi;\mathbf{f})/\partial b$ are provided by the 1st-order G-differential of $R^{(1)}[5;\varphi;\mathbf{f}(\alpha)] \triangleq \partial R(\varphi;\mathbf{f})/\partial b$. Applying the definition of the G-differential to Eq. (3.111) yields the following relation:

$$\left\{\delta R^{(1)}\left[5;\varphi;\mathbf{f}(\alpha);v^{(1)};\delta\mathbf{f}\right]\right\}_{\alpha^0}$$

$$\triangleq -\left\{\frac{d}{d\varepsilon}\left[(h+\varepsilon\delta h)\int_0^1 d\omega \int_{a+\varepsilon\delta a}^{b+\varepsilon\delta b}\left(\varphi+\varepsilon v^{(1)}\right)\delta'(z-b-\varepsilon\delta b)dz\right]_{\alpha^0}\right\}_{\varepsilon=0} \tag{3.112}$$

$$= \left\{\delta R^{(1)}(5;\varphi;\mathbf{f};\delta\mathbf{f})_{\alpha^0}\right\}_{dir} + \left\{\delta R^{(1)}\left(5;\varphi;\mathbf{f};v^{(1)}\right)_{\alpha^0}\right\}_{ind},$$

where the direct-effect term $\left\{\delta R^{(1)}(5;\varphi;\mathbf{f};\delta\mathbf{f})_{\alpha^0}\right\}_{dir}$ depends only on variations $\delta\mathbf{f}$ while the indirect-effect term $\left\{\delta R^{(1)}\left(5;\varphi;\mathbf{f};v^{(1)}\right)_{\alpha^0}\right\}_{ind}$ depends only on variations $v^{(1)}(z,\omega)$; these terms are defined, respectively, as follows:

$$\left\{\delta R^{(1)}(5;\varphi;\mathbf{f};\delta\mathbf{f})_{\alpha^0}\right\}_{dir} \triangleq -\left\{(\delta h)\int_0^1 d\omega\int_a^b \varphi(z,\omega)\delta'(z-b)dz\right\}_{\alpha^0} + (\delta b)h(\alpha)\int_0^1 d\omega\int_a^b \varphi(z,\omega)\delta''(z-b)dz; \tag{3.113}$$

$$\left\{\delta R^{(1)}\left(5;\varphi;\mathbf{f};v^{(1)}\right)_{\alpha^0}\right\}_{ind} \triangleq -\left\{h(\alpha)\int_0^1 d\omega\int_a^b v^{(1)}(z,\omega)\delta'(z-b)dz\right\}_{\alpha^0}. \tag{3.114}$$

The direct-effect term defined by Eq. (3.113) can be evaluated at this stage by using Eq. (3.57) to obtain:

$$\begin{aligned}\left\{\delta R^{(1)}(5;\varphi;\mathbf{f};\delta\mathbf{f})_{\alpha^0}\right\}_{dir} &= \left\{\int_0^1 d\omega\left[\frac{d\varphi(z,\omega)}{dz}\right]_{z=b}\right\}_{\alpha^0} \\ &+ \left\{(\delta b)h(\alpha)\int_0^1 d\omega\left[\frac{d^2\varphi(z,\omega)}{dz^2}\right]_{z=b}\right\}_{\alpha^0} \\ &= -\left\{(\delta h)g_1(\alpha)g_2(\alpha)E_1\left[(b-a)g_2(\alpha)\right]\right\}_{\alpha^0} \\ &+ \left\{(\delta b)g_1(\alpha)h(\alpha)\left[g_2(\alpha)\right]^2 E_0\left[(b-a)g_2(\alpha)\right]\right\}_{\alpha^0}.\end{aligned} \tag{3.115}$$

The indirect-effect term defined in Eq. (3.114) can be evaluated after having determined the variational function $v^{(1)}(z,\omega)$, which is the solution of Eqs. (3.59) and (3.60), but the need for solving the respective variational equations is avoided by constructing an appropriate 2nd-LASS, by applying the principles of the 2nd-FASAM-N. This 2nd-LASS is constructed by forming the inner product of Eq. (3.59) with a yet undefined function $a^{(2)}(1;5;z,\omega)$, to obtain the following relation:

$$\begin{aligned}&\left\{\int_0^1 d\omega\int_a^b a^{(2)}(1;5;z,\omega)\left[\omega\frac{dv^{(1)}(z,\omega)}{dz} + g_2(\alpha)v^{(1)}(z,\omega)\right]dz\right\}_{\alpha^0} \\ &= -\left\{(\delta g_2)\int_0^1 d\omega\int_a^b dz a^{(2)}(1;5;z,\omega)\varphi(z,\omega)\right\}_{\alpha^0}.\end{aligned} \tag{3.116}$$

Note that the 1st argument of $a^{(2)}(1;5;z,\omega)$ indicates that only the "1st-component" of this "two-component 2nd-level adjoint sensitivity function" is non-zero, while the 2nd argument in $a^{(2)}(1;5;z,\omega)$ refers to the index/label $j_1 = 5$ indicating that this 2nd-level adjoint sensitivity function will correspond to "the 5th considered" 1st-order sensitivity (namely: $R^{(1)}[5;\varphi;\mathbf{f}(\alpha)] \triangleq \partial R(\varphi;\mathbf{f})/\partial h$).

Integrating the left side of Eq. (3.116) by parts over the independent variable $z$ yields the following relation:

$$\left\{\int_0^1 d\omega \int_a^b a^{(2)}(1;5;z,\omega)\left[\omega\frac{dv^{(1)}(z,\omega)}{dz} + g_2(\alpha)v^{(1)}(z,\omega)\right]dz\right\}_{\alpha^0}$$

$$= \int_0^1 \omega\, d\omega \left\{a^{(2)}(1;5;b,\omega)v^{(1)}(b,\omega) - a^{(2)}(1;5;a,\omega)v^{(1)}(a,\omega)\right\}_{\alpha^0} \tag{3.117}$$

$$+\left\{\int_0^1 d\omega \int_a^b v^{(1)}(z,\omega)\left[-\omega\frac{d}{dz}a^{(2)}(1;5;z,\omega) + g_2(\alpha)a^{(2)}(1;5;z,\omega)\right]dz\right\}_{\alpha^0}.$$

The last term on the right side of Eq. (3.117) is now required to represent the indirect-effect term defined in Eq. (3.114), which is achieved by imposing the following relationship over the domain $a^0 < z < b^0;\ 0 \le \omega \le 1$:

$$\left\{-\omega\frac{d}{dz}a^{(2)}(1;5;z,\omega) + g_2(\alpha)a^{(2)}(1;5;z,\omega)\right\}_{\alpha^0} = -\left\{h(\alpha)\delta'(z-b)\right\}_{\alpha^0}. \tag{3.118}$$

The definition of the function $a^{(2)}(1;5;z,\omega)$ is now completed by requiring that the unknown value of the function $v^{(1)}(b,\omega)$ be eliminated from appearing on the right side of Eq. (3.117). This requirement is fulfilled by imposing the following condition to be satisfied by the function $a^{(2)}(1;5;z,\omega)$ on the model's outer boundary:

$$a^{(2)}(1;5;z,\omega) = 0, \quad \text{at} \quad z = b^0, \quad 0 \le \omega \le 1. \tag{3.119}$$

Altogether, Eqs. (3.118) and (3.119) constitute the "2nd-Level Adjoint Sensitivity System" (2nd-LASS) for determining the 2nd-level adjoint sensitivity function $a^{(2)}(1;5;z,\omega)$. Since the 2nd-LASS does not depend on the parameter variations, it needs to be solved only once in order to obtain $a^{(2)}(1;5;z,\omega)$. For the paradigm model under consideration, the 1st-LASS given by Eqs. (3.118) and (3.119) can be solved exactly to obtain the following closed-form expression for the 2nd-level adjoint function $a^{(2)}(1;5;z,\omega)$:

$$a^{(2)}(1;3;z,\omega) = \frac{h(\alpha)}{\omega}\delta(z-b)$$

$$+\left[H(z-b)-1\right]\frac{h(\alpha)g_2(\alpha)}{\omega^2}\exp\left[\frac{g_2(\alpha)(z-b)}{\omega}\right]. \tag{3.120}$$

Collecting the results obtained in Eqs. (3.114)–(3.119) leads to the following expression for the indirect-effect term:

$$\left\{\delta R^{(1)}\left(5;\varphi;\mathbf{f};v^{(1)}\right)_{\alpha^0}\right\}_{ind} = -\left\{(\delta g_2)\int_0^1 d\omega\int_a^b a^{(2)}(1;5;z,\omega)\varphi(z,\omega)dz\right\}_{\alpha^0}$$
$$+\left\{\int_0^1 \omega\, d\omega a^{(2)}(1;5;a,\omega)\left[(\delta g_1)+(\delta a)\frac{g_1(\alpha)g_2(\alpha)}{\omega}\right]\right\}_{\alpha^0}$$
$$\triangleq \left\{\delta R^{(1)}\left(5;\varphi;\mathbf{f};a^{(2)}(1;5;z,\omega)\right)_{\alpha^0}\right\}_{ind}. \tag{3.121}$$

Note that the dependence of $\left\{\delta R^{(1)}\left(5;\varphi;\mathbf{f};v^{(1)}\right)_{\alpha^0}\right\}_{ind}$ on the variational function $v^{(1)}(z,\omega)$ has been replaced in Eq. (3.121) by the dependence on the second-level adjoint sensitivity function $a^{(2)}(1;5;z,\omega)$.

Adding the results obtained in Eqs. (3.121) and (3.115) yields the following relation:

$$\left\{\delta R^{(1)}\left[5;\varphi;\mathbf{f}(\alpha);v^{(1)};\delta\mathbf{f}\right]\right\}_{\alpha^0} = -\left\{(\delta g_2)\int_0^1 d\omega\int_a^b a^{(2)}(1;5;z,\omega)\varphi(z,\omega)dz\right\}_{\alpha^0}$$
$$+\left\{\int_0^1 \omega\, d\omega a^{(2)}(1;5;a,\omega)\left[(\delta g_1)+(\delta a)\frac{g_1(\alpha)g_2(\alpha)}{\omega}\right]\right\}_{\alpha^0}$$
$$+\left\{(\delta h)\int_0^1 d\omega\left[\frac{d\varphi(z,\omega)}{dz}\right]_{z=b}\right\}_{\alpha^0}+\left\{(\delta b)h(\alpha)\int_0^1 d\omega\left[\frac{d^2\varphi(z,\omega)}{dz^2}\right]_{z=b}\right\}_{\alpha^0}$$
$$\triangleq \left\{\delta R^{(1)}\left[5;\varphi;\mathbf{f};a^{(2)}(1;3;z,\omega);\delta\mathbf{f}\right]\right\}_{\alpha^0}. \tag{3.122}$$

Identifying the expressions that multiply the respective components of $\delta\mathbf{f}(\alpha)\triangleq\left[\delta g_1(\alpha),\delta g_2(\alpha);\delta h(\alpha);\delta a;\delta b\right]^\dagger$ and using the expression for the 2nd-level adjoint sensitivity function provided in Eq. (3.120) yields the following results for the 2nd-order sensitivities that stem from the 1st-order sensitivity $R^{(1)}\left[5;\varphi;\mathbf{f}(\alpha)\right]\triangleq\partial R(\varphi;\mathbf{f})/\partial b$:

$$\frac{\partial^2 R(\varphi;\mathbf{f})}{\partial g_1(\alpha)\partial b} = \int_0^1 a^{(2)}(1;5;a,\omega)\omega\, d\omega = -g_2(\alpha)h(\alpha)E_1\left[(b-a)g_2(\alpha)\right]; \tag{3.123}$$

$$\frac{\partial^2 R(\varphi;\mathbf{f})}{\partial g_2(\alpha)\partial b} = -\int_0^1 d\omega \int_a^b a^{(2)}(1;5;z,\omega)\varphi(z,\omega)\,dz$$
$$= -g_1(\alpha)h(\alpha)E_1\left[(b-a)g_2(\alpha)\right] \tag{3.124}$$
$$+ g_1(\alpha)g_2(\alpha)h(\alpha)(b-a)E_0\left[(b-a)g_2(\alpha)\right]$$
$$= g_1(\alpha)h(\alpha)\left\{-E_1\left[(b-a)g_2(\alpha)\right] + \exp\left[(a-b)g_2(\alpha)\right]\right\};$$

$$\frac{\partial^2 R(\varphi;\mathbf{f})}{\partial h(\alpha)\partial b} = \int_0^1 d\omega \left[\frac{d\varphi(z,\omega)}{dz}\right]_{z=b} = -g_1(\alpha)g_2(\alpha)E_1\left[(b-a)g_2(\alpha)\right]; \tag{3.125}$$

$$\frac{\partial^2 R(\varphi;\mathbf{f})}{\partial a\,\partial b} = g_1(\alpha)g_2(\alpha)\int_0^1 a^{(2)}(1;5;a,\omega)\,d\omega$$
$$= -g_1(\alpha)h(\alpha)\left[g_2(\alpha)\right]^2 E_0\left[(b-a)g_2(\alpha)\right] \tag{3.126}$$
$$= -\frac{g_1(\alpha)g_2(\alpha)h(\alpha)}{(b-a)}\exp\left[(b-a)g_2(\alpha)\right];$$

$$\frac{\partial^2 R(\varphi;\mathbf{f})}{\partial b\,\partial b} = h(\alpha)\int_0^1 d\omega\left[\frac{d^2\varphi(z,\omega)}{dz^2}\right]_{z=b}$$
$$= g_1(\alpha)h(\alpha)\left[g_2(\alpha)\right]^2 E_0\left[(b-a)g_2(\alpha)\right] \tag{3.127}$$
$$= \frac{g_1(\alpha)g_2(\alpha)h(\alpha)}{(b-a)}\exp\left\{(a-b)g_2(\alpha)\right\}.$$

The validity of the expressions obtained in Eqs. (3.123)–(3.127) can be verified by using the expressions of the 1st-order sensitivities provided in Eqs. (2.119)–(2.123).

### 3.3.6 Comparing Computational Efficiencies: the 2nd-FASAM-N versus the 2nd-CASAM-N

As has been shown in the foregoing Sections 3.3.1–3.3.5, each of the five 1st-order sensitivities of the model response with respect to the components of the feature function $\mathbf{f}(\alpha) \triangleq \left[g_1(\alpha), g_2(\alpha); h(\alpha); a; b\right]^{\dagger}$ has given rise to five 2nd-order sensitivities. The ten mixed 2nd-order sensitivities have been computed twice, each time using two distinct 2nd-level adjoint sensitivity functions, as follows:

$$\frac{\partial^2 R(\varphi;\mathbf{f})}{\partial g_2(\boldsymbol{\alpha})\partial g_1(\boldsymbol{\alpha})} = -\int_0^1 d\omega \int_a^b dz a^{(2)}(2;1;z,\omega)a^{(1)}(z,\omega)$$

$$= \frac{\partial^2 R(\varphi;\mathbf{f})}{\partial g_1(\boldsymbol{\alpha})\partial g_2(\boldsymbol{\alpha})}$$

$$= \int_0^1 \omega\, d\omega a^{(2)}(1;2;a,\omega) \tag{3.128}$$

$$= (a-b)h(\boldsymbol{\alpha})E_1\left[(b-a)g_2(\boldsymbol{\alpha})\right];$$

$$\frac{\partial^2 R(\varphi;\mathbf{f})}{\partial h(\boldsymbol{\alpha})\partial g_1(\boldsymbol{\alpha})} = \int_0^1 d\omega \int_a^b dz a^{(2)}(2;1;z,\omega)\delta(z-b) = \frac{\partial^2 R(\varphi;\mathbf{f})}{\partial g_1(\boldsymbol{\alpha})\partial h(\boldsymbol{\alpha})}$$
$$= \int_0^1 \omega\, d\omega a^{(2)}(1;3;a,\omega) = E_2\left[(b-a)g_2(\boldsymbol{\alpha})\right]; \tag{3.129}$$

$$\frac{\partial^2 R(\varphi;\mathbf{f})}{\partial a\, \partial g_1(\boldsymbol{\alpha})} = \int_0^1 \omega\, d\omega \left[\frac{da^{(1)}(z,\omega)}{dz}\right]_{z=a} = \frac{\partial^2 R(\varphi;\mathbf{f})}{\partial g_1(\boldsymbol{\alpha})\partial a}$$
$$= g_2(\boldsymbol{\alpha})\int_0^1 d\omega \int_a a^{(1)}(z,\omega)\delta(z-a)dz = g_2(\boldsymbol{\alpha})h(\boldsymbol{\alpha})E_1\left[(b-a)g_2(\boldsymbol{\alpha})\right]; \tag{3.130}$$

$$\frac{\partial^2 R(\varphi;\mathbf{f})}{\partial b\, \partial g_1(\boldsymbol{\alpha})} = -h(\boldsymbol{\alpha})\int_0^1 d\omega \int_a^b dz a^{(2)}(2;1;z,\omega)\delta'(z-b) = \frac{\partial^2 R(\varphi;\mathbf{f})}{\partial g_1(\boldsymbol{\alpha})\partial b}$$
$$= \int_0^1 a^{(2)}(1;5;a,\omega)\omega\, d\omega = -g_2(\boldsymbol{\alpha})h(\boldsymbol{\alpha})E_1\left[(b-a)g_2(\boldsymbol{\alpha})\right]; \tag{3.131}$$

$$\frac{\partial^2 R(\varphi;\mathbf{f})}{\partial h(\boldsymbol{\alpha})\partial g_2(\boldsymbol{\alpha})} = \int_0^1 d\omega \int_a^b dz a^{(2)}(2;2;z,\omega)\delta(z-b) = \frac{\partial^2 R(\varphi;\mathbf{f})}{\partial g_2(\boldsymbol{\alpha})\partial h(\boldsymbol{\alpha})}$$

$$= -\int_0^1 d\omega \int_a^b a^{(2)}(1;3;z,\omega)\varphi(z,\omega)dz \tag{3.132}$$

$$= g_1(\boldsymbol{\alpha})(a-b)E_1\left[(b-a)g_2(\boldsymbol{\alpha})\right];$$

$$\frac{\partial^2 R(\varphi;\mathbf{f})}{\partial a\,\partial g_2(\boldsymbol{\alpha})} = g_1(\boldsymbol{\alpha})h(\boldsymbol{\alpha})E_1\left[(b-a)g_2(\boldsymbol{\alpha})\right] + g_1(\boldsymbol{\alpha})g_2(\boldsymbol{\alpha})\int_0^1 d\omega a^{(2)}(1;2;a,\omega)$$

$$= \frac{\partial^2 R(\varphi;\mathbf{f})}{\partial g_2(\boldsymbol{\alpha})\partial a} = g_1(\boldsymbol{\alpha})\int_0^1 d\omega \int_a^b a^{(1)}(z,\omega)\delta(z-a)dz - \int_0^1 d\omega \int_a^b dz a^{(1)}(z,\omega)a^{(2)}(2;4;z,\omega)$$

$$= g_1(\boldsymbol{\alpha})h(\boldsymbol{\alpha})\left\{E_1\left[(b-a)g_2(\boldsymbol{\alpha})\right] - \exp\left[(a-b)g_2(\boldsymbol{\alpha})\right]\right\}; \tag{3.133}$$

$$\frac{\partial^2 R(\varphi;\mathbf{f})}{\partial b\,\partial g_2(\boldsymbol{\alpha})} = -h(\boldsymbol{\alpha})\int_0^1 d\omega \int_a^b dz a^{(2)}(2;2;z,\omega)\delta'(z-b)$$

$$= \frac{\partial^2 R(\varphi;\mathbf{f})}{\partial g_2(\boldsymbol{\alpha})\partial b} = -\int_0^1 d\omega \int_a^b a^{(2)}(1;5;z,\omega)\varphi(z,\omega)dz \tag{3.134}$$

$$= g_1(\boldsymbol{\alpha})h(\boldsymbol{\alpha})\left\{\exp\left[(a-b)g_2(\boldsymbol{\alpha})\right] - E_1\left[(b-a)g_2(\boldsymbol{\alpha})\right]\right\};$$

$$\frac{\partial^2 R(\varphi;\mathbf{f})}{\partial a\,\partial h(\boldsymbol{\alpha})} = g_1(\boldsymbol{\alpha})g_2(\boldsymbol{\alpha})\int_0^1 d\omega a^{(2)}(1;3;a,\omega) = \frac{\partial^2 R(\varphi;\mathbf{f})}{\partial h(\boldsymbol{\alpha})\partial a}$$

$$= \int_0^1 d\omega \int_a^b dz a^{(2)}(2;4;z,\omega)\delta(z-b) = g_1(\boldsymbol{\alpha})g_2(\boldsymbol{\alpha})E_1\left[(b-a)g_2(\boldsymbol{\alpha})\right]; \tag{3.135}$$

$$\frac{\partial^2 R(\varphi;\mathbf{f})}{\partial b\,\partial h(\boldsymbol{\alpha})} = \int_0^1 d\omega \left[\frac{d\varphi(z,\omega)}{dz}\right]_{z=b} = \frac{\partial^2 R(\varphi;\mathbf{f})}{\partial h(\boldsymbol{\alpha})\partial b}$$

$$= \int_0^1 d\omega \left[\frac{d\varphi(z,\omega)}{dz}\right]_{z=b} = -g_1(\boldsymbol{\alpha})g_2(\boldsymbol{\alpha})E_1\left[(b-a)g_2(\boldsymbol{\alpha})\right]; \tag{3.136}$$

$$\frac{\partial^2 R(\varphi;\mathbf{f})}{\partial b\,\partial a} = -h(\alpha)\int_0^1 d\omega \int_a^b dz a^{(2)}(2;4;z,\omega)\delta'(z-b) = \frac{\partial^2 R(\varphi;\mathbf{f})}{\partial a\,\partial b}$$

$$= g_1(\alpha)g_2(\alpha)\int_0^1 a^{(2)}(1;5;a,\omega)d\omega \tag{3.137}$$

$$= -\frac{g_1(\alpha)g_2(\alpha)h(\alpha)}{(b-a)}\exp\left[(a-b)g_2(\alpha)\right].$$

As the relations shown in Eqs. (3.128)–(3.137) indicate, the 2nd-FASAM-N provides an intrinsic verification, based on the symmetries inherent to the mixed 2nd-order sensitivities, of having correctly solved the original forward system, the 1st-LASS, and the five 2nd-LASS for the five 2nd-level adjoint sensitivity functions, all of which are involved in the expressions of the mixed 2nd-order sensitivities.

The determination of the 2nd-order response sensitivities with respect to the primary model parameters $\boldsymbol{\alpha} \triangleq (\alpha_1,\ldots,\alpha_{TP})^\dagger$, where $TP = 2MA + 2MB + 2MD + MQ + 2$, is accomplished by using the chain rule provided in Eq. (3.25) together with the expressions of the five 1st-order response sensitivities with respect to the components of the feature function $\mathbf{f}(\alpha) \triangleq [g_1(\alpha), g_2(\alpha); h(\alpha); a; b]^\dagger$, as provided in Eqs. (2.119)–(2.123), and the 2nd-order response sensitivities with respect to the components of the feature function provided in Sections 3.3.1–3.3.5.

Alternatively, it would have been possible to apply the 2nd-CASAM-N (Cacuci 2016, 2023a) to determine the 2nd-order sensitivities (of the response) directly with respect to the primary parameters $\boldsymbol{\alpha} \triangleq (\alpha_1,\ldots,\alpha_{TP})^\dagger$, by using the $TP = 2MA + 2MB + 2MD + MQ + 2$ 1st-order sensitivities with respect to the primary parameters. Evidently, following this procedure would have required $TP = 2MA + 2MB + 2MD + MQ + 2$ large-scale computations, needed for solving the TP 2nd-LASS systems, each of which would have corresponded to one of the $TP$ 1st-order sensitivities. It is evident that two of these computations would stem from the 1st-order sensitivities to the boundary parameters $a$ and $b$, which are also responsible for two of the five 2nd-LASS computations involved in the 2nd-FASAM-N; recall that the "feature" function $\mathbf{f}(\alpha) \triangleq [g_1(\alpha), g_2(\alpha); h(\alpha); a; b]^\dagger$ comprises just three bona-fide "features," namely $g_1(\alpha)$, $g_2(\alpha)$ and $h(\alpha)$.

Even for the simple paradigm shielding model considered in this work, it is evident that the 2nd-FASAM-N is computationally significantly more advantageous to use than the 2nd-CASAM-N: the 2nd-FASAM-N requires five large-scale computations (plus insignificant additional analytic computations for applying the chain rule) for obtaining all of the 2nd-order sensitivities with respect to the primary model parameters, while the 2nd-CASAM-N would require $TP > 70$ large-scale computations to obtain these 2nd-order sensitivities. By comparison, even the simplest-minded four-point finite-difference scheme would require $4 \times TP \times (TP+1)/2 = O(10^4)$ large-scale computations and would at best provide approximate values (within 2nd-order errors of the finite-difference step-size) for the respective 2nd-order sensitivities.

## 3.4 CHAPTER SUMMARY

This chapter has presented the "2nd-Order Function/Feature Adjoint Sensitivity Analysis Methodology for Nonlinear Systems" (2nd-FASAM-N), which enables the most efficient computation of exactly obtained expressions of the 2nd-order sensitivities of a model response with respect to a multi-component *function* ("feature") of primary model parameters. To compute all of these second-order sensitivities, the 2nd-FASAM-N requires only as many large-scale (adjoint) computations as there are components of the "features function." In contradistinction, the 2nd-CASAM-N (Cacuci 2016, 2023a) requires as many large-scale (adjoint) computations as there are primary parameters. Both the 2nd-FASAM-N and the 2nd-CASAM-N methodologies compute the respective mixed 2nd-order sensitivities twice, involving distinct 2nd-level adjoint sensitivity functions, a fact that is essential for the mutual verification of the computational accuracy of the respective adjoint functions. Since the number of "features" in a computational model is invariably smaller than the number of primary model parameters, the 2nd-FASAM-N is always more advantageous to use, computationally, than the 2nd-CASAM-N. All of these considerations were demonstrated in this chapter by using a simple particle transport/shielding model, which admits closed-form analytic expressions for all of the 1st- and 2nd-order sensitivities. Both the 2nd-FASAM-N and the 2nd-CASAM-N are vastly more efficient to use than conventional finite-difference or statistical methods, which are not only exorbitantly expensive computationally but also cannot produce exact expressions for any sensitivity because of their very theoretical foundation.

# 4 The Mathematical Framework of the *n*th-Order Feature/Function Adjoint Sensitivity Analysis Methodology for Nonlinear Systems (*n*th-FASAM-N)

## 4.1 INTRODUCTION

The "1st-Order Function/Feature Adjoint Sensitivity Analysis Methodology for Nonlinear Systems" (1st-FASAM-N) has been presented in Chapter 2, while the "2nd-Order Function/Feature Adjoint Sensitivity Analysis Methodology for Nonlinear Systems" (2nd-FASAM-N) has been presented in Chapter 3. The construction of the 1st-FASAM-N is based on the same principles as those underlying the construction of the 1st-CASAM-N while the construction of the 2nd-FASAM-N is based on the same principles as those underlying the construction of the 2nd-CASAM-N. These facts indicate that the principles which enabled the construction of the *n*th-CASAM-N methodology (Cacuci 2022a, 2023a) could also be used to generalize the 2nd-FASAM-N so as to enable the most efficient computation of the exact expressions of arbitrarily high ("*n*th") order sensitivities of model responses with respect to features/functions of model parameters. This is indeed the case, as will be shown in Section 4.2, which presents the construction of the general mathematical framework underlying the *n*th-FASAM-N. This construction uses "mathematical induction" mirroring the construction of the *n*th-CASAM-N methodology (Cacuci 2022a, 2023a). Section 4.3 proves that the general framework surmised in Section 4.2 for the *n*th-FASAM-N is valid for the lowest value of $n$, i.e., for $n = 1$, by showing that the mathematical framework underlying the *n*th-FASAM-N reduces to the 1st-FASAM-N methodology presented in Chapter 2. Continuing the proof by mathematical induction, Section 4.4 proves that assuming the framework underlying the *n*th-FASAM-N to be valid for an arbitrarily high-order, $n$, the framework also

DOI: 10.1201/9781003478171-4

remains valid for $(n+1)$, namely for the framework of the $(n+1)$th-FASAM-N methodology. Section 4.5 summarizes the significance of the $n$th-CASAM-N and prepares the groundwork for a paradigm illustrative application to the Nordheim-Fuchs reactor dynamics/safety model (Hetrick 1993; Stacey 2001) to be presented in Chapter 5.

## 4.2 THE *N*TH-ORDER FEATURE/FUNCTION ADJOINT SENSITIVITY ANALYSIS METHODOLOGY FOR NONLINEAR SYSTEMS (*N*TH-FASAM-N)

Mirroring the establishment by Cacuci (2022a, 2023a) of the mathematical framework of the "$n$th-Order Comprehensive Adjoint Sensitivity Analysis Methodology for Nonlinear Systems" ($n$th-CASAM-N), the mathematical framework underlying the $n$th-FASAM-N methodology will be established in this section by using the "proof by mathematical induction" comprising the usual steps, as follows:

1. Surmise the general pattern underlying the $n$th-FASAM-N methodology for an arbitrarily high-order denoted as "$n$". It is expected that the general pattern underlying the $n$th-FASAM-N can be surmised based on the expected similarities with the general pattern underlying the $n$th-CASAM-N (Cacuci 2022a, 2023a).
2. Prove that the general pattern underlying the $n$th-FASAM-N is valid for the lowest values of $n$, i.e., for $n = 1$, which must reduce to the mathematical framework underlying the 1st-FASAM-N that was presented in Chapter 2.
3. Assuming that the pattern underlying the $n$th-FASAM-N is valid for an arbitrarily high-order, $n$, prove that this pattern is also valid for $(n+1)$, namely for the framework of the $(n+1)$th-FASAM-N methodology.

Comparing the mathematical framework of the 1st-FASAM-N presented in Chapter 2 to the framework of the 1st-CASAM-N (Cacuci 2016, 2022a) suggests that the component "features" $f_i(\boldsymbol{\alpha}), i = 1,\ldots,TF$, of the vector-valued "feature function" $\mathbf{f}(\boldsymbol{\alpha}) \triangleq [f_1(\boldsymbol{\alpha}),\ldots,f_{TF}(\boldsymbol{\alpha})]^\dagger$, where $TF$ denotes the total number of the components, play within the 1st-FASAM-N the same role as played by the components $\alpha_j$, $j = 1,\ldots,TP$, of the "vector of primary model parameters" $\boldsymbol{\alpha} \triangleq (\alpha_1,\ldots,\alpha_{TP})^\dagger$ within the framework of the 1st-CASAM-N. It is important to underscore at the outset that the total number of model parameters is by definition much larger than the total number of components of the feature function $\mathbf{f}(\boldsymbol{\alpha})$, i.e., $TP \gg TF$; $TF$ coincides with $TP$ if the mathematical/computational model under consideration possesses no "features."

An examination of the pattern underlying the $n$th-CASAM-N methodology (Cacuci 2022a, 2023a) indicates that the $n$th-order sensitivity of the model's response $R[\mathbf{u}(\mathbf{x});\mathbf{f}(\boldsymbol{\alpha})]$ with respect to the components $f_{j_1},\ldots,f_{j_n}$ of the "feature" function $\mathbf{f}(\boldsymbol{\alpha}) \triangleq [f_{j_1}(\boldsymbol{\alpha}),\ldots,f_{j_n}(\boldsymbol{\alpha})]^\dagger$, for each $j_1,\ldots,j_n = 1,\ldots,TF$, is expected to have the functional form $R^{(n)}[j_n;\ldots;j_1;\mathbf{U}^{(n)};\mathbf{A}^{(n)};\mathbf{f}(\boldsymbol{\alpha})] \triangleq \partial^n R[\mathbf{u}(\mathbf{x});\mathbf{f}(\boldsymbol{\alpha})]/\partial f_{j_n}\ldots\partial f_{j_1}$. It can also be surmised that the $n$th-order sensitivity $R^{(n)}[j_n;\ldots;j_1;\mathbf{U}^{(n)};\mathbf{A}^{(n)};\mathbf{f}(\boldsymbol{\alpha})]$ stems from the total first-order G-differential of each of the $(n-1)$th-order sensitivities, each of which is expected to have the following expression for $j_1,\ldots,j_{n-1}=1,\ldots,TF$:

$$R^{(n-1)}\left[j_{n-1};\ldots;j_1;\mathbf{U}^{(n-1)};\mathbf{A}^{(n-1)};\mathbf{f}(\boldsymbol{\alpha})\right]\equiv\partial^{n-1}R\left[\mathbf{u}(\mathbf{x});\mathbf{f}(\boldsymbol{\alpha})\right]/\partial\alpha_{j_{n-1}}\ldots\partial\alpha_{j_1}$$
$$\triangleq\int_{\lambda_1(\boldsymbol{\alpha})}^{\omega_1(\boldsymbol{\alpha})}\cdots\int_{\lambda_{TI}(\boldsymbol{\alpha})}^{\omega_{TI}(\boldsymbol{\alpha})}S^{(n-1)}\left[j_{n-1};\ldots;j_1;\mathbf{U}^{(n-1)};\mathbf{A}^{(n-1)};\mathbf{f}(\boldsymbol{\alpha})\right]dx_1\ldots dx_{TI}. \tag{4.1}$$

Analogous with the pattern underlying the $n$th-CASAM-N methodology (Cacuci 2022a, 2023a), it is surmised that the total first-order G-differential of the $(n-1)$th-order sensitivity $R^{(n-1)}\left[j_{n-1};\ldots;j_1;\mathbf{U}^{(n-1)};\mathbf{A}^{(n-1)};\mathbf{f}(\boldsymbol{\alpha})\right]$ with respect to the feature-function $\mathbf{f}(\boldsymbol{\alpha})$ is expected to have the following expression:

$$\left\{\delta R^{(n-1)}\left[j_{n-1};\ldots;j_1;\mathbf{U}^{(n-1)};\mathbf{A}^{(n-1)};\mathbf{f}\right]\right\}_{\boldsymbol{\alpha}^0}\equiv\sum_{j_n=1}^{TF}\left\{\frac{\partial^n R\left[\mathbf{u}(\mathbf{x});\mathbf{f}(\boldsymbol{\alpha})\right]}{\partial f_{j_n}\ldots\partial f_{j_1}}\right\}_{\boldsymbol{\alpha}^0}\delta f_{j_n}$$
$$\triangleq\left\{\frac{d}{d\varepsilon}\delta R^{(n-1)}\left[j_{n-1};\ldots;j_1;\mathbf{U}^{(n-1)}+\varepsilon\mathbf{V}^{(n-1)};\mathbf{A}^{(n-1)}+\varepsilon\delta\mathbf{A}^{(n-1)};\mathbf{f}+\varepsilon\delta\mathbf{f}\right]\right\}_{\varepsilon=0}$$
$$=\sum_{j_n=1}^{TF}\left\{\frac{\partial R^{(n-1)}\left[j_{n-1};\ldots;j_1;\mathbf{U}^{(n-1)};\mathbf{A}^{(n-1)};\mathbf{f}\right]}{\partial f_{j_n}}\right\}_{\boldsymbol{\alpha}^0}\delta f_{j_n}$$
$$+\left\{\delta R^{(n-1)}\left[j_{n-1};\ldots;j_1;\mathbf{U}^{(n-1)};\mathbf{A}^{(n-1)};\mathbf{f};\mathbf{V}^{(n-1)};\delta\mathbf{A}^{(n-1)}\right]\right\}_{ind}, \tag{4.2}$$

where the quantity $\left\{\delta R^{(n-1)}\left[j_{n-1};\ldots;j_1;\mathbf{U}^{(n-1)};\mathbf{A}^{(n-1)};\mathbf{f};\mathbf{V}^{(n-1)};\delta\mathbf{A}^{(n-1)}\right]\right\}_{ind}$ denotes the so-called "indirect-effect term" which is defined as follows:

$$\left\{\delta R^{(n-1)}\left[j_{n-1};\ldots;j_1;\mathbf{U}^{(n-1)};\mathbf{A}^{(n-1)};\mathbf{f};\mathbf{V}^{(n-1)};\delta\mathbf{A}^{(n-1)}\right]\right\}_{ind}$$
$$\triangleq\int_{\lambda_1(\boldsymbol{\alpha})}^{\omega_1(\boldsymbol{\alpha})}\cdots\int_{\lambda_{TI}(\boldsymbol{\alpha})}^{\omega_{TI}(\boldsymbol{\alpha})}\left\{\begin{array}{l}\dfrac{\partial S^{(n-1)}}{\partial\mathbf{U}^{(n-1)}(\mathbf{x})}\mathbf{V}^{(n-1)}\left(2^{n-2};\mathbf{x}\right)\\+\dfrac{\partial S^{(n-1)}}{\partial\mathbf{A}^{(n-1)}(\mathbf{x})}\delta\mathbf{A}^{(n-1)}\left(2^{n-2};\mathbf{x}\right)\end{array}\right\}_{\boldsymbol{\alpha}^0}dx_1\ldots dx_{TI}. \tag{4.3}$$

The vectors $\mathbf{V}^{(n-1)}\left(2^{n-2};j_{n-3};\ldots;j_1;\mathbf{x}\right)\triangleq\delta\mathbf{U}^{(n-1)}\left(2^{n-2};j_{n-3};\ldots;j_1;\mathbf{x}\right)$ and $\delta\mathbf{A}^{(n-1)}\left(2^{n-2};j_{n-2};\ldots;j_1;\mathbf{x}\right)$ are the solution of the following $n$th-Level Variational Sensitivity System ($n$th-LVSS), which is obtained by concatenating the $(n-1)$th-LVSS together with the G-differentiated $(n-1)$th-Level Adjoint Sensitivity System, for $j_1,\ldots,j_{n-1}=1,\ldots,TF$:

$$\left\{\mathbf{VM}^{(n)}\left[2^{n-1}\times 2^{n-1};\mathbf{U}^{(n)}\left(2^{n-1};j_{n-2};\ldots;j_1;\mathbf{x}\right);\mathbf{f}\right]\mathbf{V}^{(n)}\left(2^{n-1};j_{n-2};\ldots;j_1;\mathbf{x}\right)\right\}_{\alpha^0}$$
$$=\left\{\mathbf{Q}_V^{(n)}\left[2^{n-1};\mathbf{U}^{(n)}\left(2^{n-1};j_{n-2};\ldots;j_1;\mathbf{x}\right);\mathbf{f};\delta\mathbf{f}\right]\right\}_{\alpha^0},\quad \mathbf{x}\in\Omega_x, \tag{4.4}$$

$$\left\{\mathbf{B}_V^{(n)}\left[2^{n-1};\mathbf{U}^{(n)}\left(2^{n-1};\mathbf{x}\right);\mathbf{V}^{(n)}\left(2^{n-1};\mathbf{x}\right);\mathbf{f};\delta\mathbf{f}\right]\right\}_{\alpha^0}=\mathbf{0}\left[2^{n-1}\right];\quad \mathbf{x}\in\partial\Omega_x\left(\alpha^0\right), \tag{4.5}$$

The various matrices and vectors appearing in Eqs. (4.4) and (4.5) are defined as follows:

1. The variational matrix $\mathbf{VM}^{(n)}\left[2^{n-1}\times 2^{n-1};\mathbf{U}^{(n)}\left(2^{n-1};j_{n-2};\ldots;j_1;\mathbf{x}\right);\mathbf{f}\right]$ comprises $\left(2^{n-1}\times 2^{n-1}\right)$ block-matrices, each comprising $TD^2$ components/elements, defined as follows:

$$\mathbf{VM}^{(n)}\left[2^{n-1}\times 2^{n-1};\mathbf{x};\mathbf{f}\right]\triangleq\begin{pmatrix}\mathbf{VM}^{(n-1)}\left[2^{n-2}\times 2^{n-2};\mathbf{x}\right] & \mathbf{0}\left[2^{n-2}\times 2^{n-2}\right]\\ \mathbf{VM}_{21}^{(n)}\left(2^{n-2}\times 2^{n-2};\mathbf{x}\right) & \mathbf{VM}_{22}^{(n)}\left(2^{n-2}\times 2^{n-2};\mathbf{x}\right)\end{pmatrix}. \tag{4.6}$$

2. The matrix $\mathbf{VM}^{(n)}\left[2^{n-1}\times 2^{n-1};\mathbf{U}^{(n)}\left(2^{n-1};j_{n-2};\ldots;j_1;\mathbf{x}\right);\mathbf{f}\right]$ comprises a total of $\left(2^{n-1}\times 2^{n-1}\right)TD^2$ components/elements, as indicated by its first argument. The submatrix $\mathbf{VM}^{(n-1)}\left[2^{n-2}\times 2^{n-2};\mathbf{x}\right]$ is defined recursively, while the remaining submatrices are defined as follows:

$$\mathbf{VM}_{21}^{(n)}\left(2^{n-2}\times 2^{n-2};j_{n-2};\ldots;j_1;\mathbf{x}\right)$$
$$\triangleq-\frac{\partial\mathbf{Q}_A^{(n-1)}\left[2^{n-2};j_{n-2};\ldots;j_1;\mathbf{U}^{(n-2)}\left(2^{n-2};j_{n-3};\ldots;j_1;\mathbf{x}\right);\mathbf{f}\right]}{\partial\mathbf{U}^{(n-1)}\left(2^{n-2};j_{n-3};\ldots;j_1;\mathbf{x}\right)} \tag{4.7}$$
$$+\frac{\partial\left\{\mathbf{AM}^{(n-1)}\left[2^{n-2}\times 2^{n-2};\mathbf{U}^{(n-1)}\left(2^{n-2};\mathbf{x}\right);\mathbf{f}\right]\mathbf{A}^{(n-1)}\left(2^{n-2};j_{n-2};\ldots;j_1;\mathbf{x}\right)\right\}}{\partial\mathbf{U}^{(n-1)}\left(2^{n-2};j_{n-3};\ldots;j_1;\mathbf{x}\right)};$$

$$\mathbf{VM}_{22}^{(n)}\left(2^{n-2}\times 2^{n-2};\mathbf{x}\right)\triangleq\mathbf{AM}^{(n-1)}\left[2^{n-2}\times 2^{n-2};\mathbf{U}^{(n-1)}\left(2^{n-2};j_{n-3};\ldots;j_1;\mathbf{x}\right);\mathbf{f}\right] \tag{4.8}$$

3. The vector $\mathbf{V}^{(n)}\left(2^{n-1};j_{n-2};\ldots;j_1;\mathbf{x}\right)$ is defined recursively, below, and comprises $2^{n-1}$ blocks of $TD$-dimensional vectors [thus comprising a total of $\left(2^{n-1}\times TD\right)$ components/elements]; this fact is indicated by the first argument of each of these quantities.

$$\mathbf{V}^{(n)}\left(2^{n-1};j_{n-2};\dots;j_1;\mathbf{x}\right)\triangleq\delta\mathbf{U}^{(n)}\left(2^{n-1};j_{n-2};\dots;j_1;\mathbf{x}\right)$$
$$=\begin{pmatrix}\mathbf{V}^{(n-1)}\left(2^{n-2};j_{n-3};\dots;j_1;\mathbf{x}\right)\\ \delta\mathbf{A}^{(n-1)}\left(2^{n-2};j_{n-2};\dots;j_1;\mathbf{x}\right)\end{pmatrix}$$
$$=\left[\mathbf{v}^{(1)}(\mathbf{x}),\delta\mathbf{a}^{(1)}(\mathbf{x}),\delta\mathbf{a}^{(2)}(1;j_1;\mathbf{x}),\delta\mathbf{a}^{(2)}(2;j_1;\mathbf{x}),\right.\tag{4.9}$$
$$\dots,\delta\mathbf{a}^{(n-1)}\left(1;j_{n-2};\dots;j_1;\mathbf{x}\right),\dots,$$
$$\left.\delta\mathbf{a}^{(n-1)}\left(2^{n-2};j_{n-2};\dots;j_1;\mathbf{x}\right)\right]^{\dagger};$$

4. The vector $\mathbf{U}^{(n)}\left(2^{n-1};j_{n-2};\dots;j_1;\mathbf{x}\right)$ has the same structure as the vector $\mathbf{V}^{(n)}\left(2^{n-1};j_{n-2};\dots;j_1;\mathbf{x}\right)$, comprising $2^{n-1}$ blocks of $TD$-dimensional vectors, as defined below.

$$\mathbf{U}^{(n)}\left(2^{n-1};j_{n-2};\dots;j_1;\mathbf{x}\right)\triangleq\begin{pmatrix}\mathbf{U}^{(n-1)}\left(2^{n-2};j_{n-3};\dots;j_1;\mathbf{x}\right)\\ \mathbf{A}^{(n-1)}\left(2^{n-2};j_{n-2};\dots;j_1;\mathbf{x}\right)\end{pmatrix};\tag{4.10}$$

5. The vector $\mathbf{Q}_V^{(n)}\left[2^{n-1};\mathbf{U}^{(n)}\left(2^{n-1};j_{n-2};\dots;j_1;\mathbf{x}\right);\mathbf{f};\delta\mathbf{f}\right]$ also comprises $2^{n-1}$ blocks of $TD$-dimensional vectors, as defined recursively below:

$$\mathbf{Q}_V^{(n)}\left[2^{n-1};j_1;\mathbf{U}^{(n)}\left(2^{n-1};j_{n-2};\dots;j_1;\mathbf{x}\right);\mathbf{f};\delta\mathbf{f}\right]$$
$$\triangleq\begin{pmatrix}\mathbf{Q}_V^{(n-2)}\left[2^{n-2};\mathbf{U}^{(n-1)}\left(2^{n-2};\mathbf{x}\right);\mathbf{f};\delta\mathbf{f}\right]\\ \mathbf{Q}_2^{(n)}\left[2^{n-2};\mathbf{U}^{(n)}\left(2^{n-1};\mathbf{x}\right);\mathbf{f};\delta\mathbf{f}\right]\end{pmatrix}\tag{4.11}$$
$$\triangleq\left\{\mathbf{q}_V^{(n)}\left[1;\mathbf{U}^{(n)}\left(2^{n-1};j_{n-2};\dots;j_1;\mathbf{x}\right);\mathbf{f};\delta\mathbf{f}\right],\dots\right.$$
$$\left.\dots,\mathbf{q}_V^{(n)}\left[2^{n-1};\mathbf{U}^{(n)}\left(2^{n-1};j_{n-2};\dots;j_1;\mathbf{x}\right);\mathbf{f};\delta\mathbf{f}\right]\right\}^{\dagger};$$

$$\mathbf{q}_V^{(n)}\left[i;\mathbf{U}^{(n)}(\mathbf{x});\mathbf{f};\delta\mathbf{f}\right]\equiv\sum_{j_n=1}^{TP}\mathbf{s}_V^{(n)}\left[i;j_{n-2};\dots;j_1;\mathbf{U}^{(n)}(\mathbf{x});\mathbf{f}\right]\delta f_{j_n};\quad i=1,\dots;2^{n-1};\tag{4.12}$$

$$\mathbf{Q}_2^{(n)}\left[2^{n-2};\mathbf{U}^{(n)}\left(2^{n-1};j_{n-2};\ldots;j_1;\mathbf{x}\right);\mathbf{f};\delta\mathbf{f}\right]$$
$$\triangleq \frac{\partial \mathbf{Q}_A^{(n-1)}\left[2^{n-2};j_1;\mathbf{U}^{(n-1)}(\mathbf{x});\mathbf{f}\right]}{\partial \mathbf{f}}\partial\mathbf{f} \tag{4.13}$$
$$-\frac{\partial\left\{\mathbf{AM}^{(n-1)}\left[2^{n-2}\times 2^{n-2};\mathbf{U}^{(n-1)}\left(2^{n-2};\mathbf{x}\right);\mathbf{f}\right]\mathbf{A}^{(n-1)}\left(2^{n-2};\mathbf{x}\right)\right\}}{\partial\mathbf{f}}\partial\mathbf{f};$$

6. The vector block-vector $\mathbf{0}\left[2^{n-1}\right]$ comprises $2^{n-1}$ components, each component being a $TD$-dimensional vector having identically zero components.
7. The boundary terms are represented by the block-vector $\mathbf{B}_V^{(n)}\left[2^{n-1};\mathbf{U}^{(n)}\left(2^{n-1};j_{n-2};\ldots;j_1;\mathbf{x}\right);\mathbf{V}^{(n)}\left(2^{n-1};j_{n-2};\ldots;j_1;\mathbf{x}\right);\mathbf{f};\delta\mathbf{f}\right]$, which is defined recursively as shown below.

$$\mathbf{B}_V^{(n)}\left[2^{n-1};\mathbf{U}^{(n)}\left(2^{n-1};j_{n-2};\ldots;j_1;\mathbf{x}\right);\mathbf{V}^{(n)}\left(2^{n-1};j_{n-2};\ldots;j_1;\mathbf{x}\right);\mathbf{f};\delta\mathbf{f}\right]$$
$$\triangleq\begin{pmatrix}\mathbf{B}_V^{(n-1)}\left[2^{n-2};\mathbf{U}^{(n-1)}\left(2^{n-2};j_{n-3};\ldots;j_1;\mathbf{x}\right);\mathbf{V}^{(n-1)}\left(2^{n-2};j_{n-3};\ldots;j_1;\mathbf{x}\right);\mathbf{f};\delta\mathbf{f}\right]\\ \delta\mathbf{B}_A^{(n-1)}\left[2^{n-2};\mathbf{U}^{(n)}\left(2^{n-1};j_{n-2};\ldots;j_1;\mathbf{x}\right);\mathbf{V}^{(n)}\left(2^{n-1};j_{n-2};\ldots;j_1;\mathbf{x}\right);\mathbf{f};\delta\mathbf{f}\right]\end{pmatrix}. \tag{4.14}$$

The need for solving repeatedly the $n$th-Level Variational Sensitivity System ($n$th-LVSS) defined by Eqs. (4.4) and (4.5) is circumvented by deriving an alternative expression for the indirect-effect term defined in Eq. (4.3), in which the function $\mathbf{V}^{(n)}\left(2^{n-1};j_{n-2};\ldots;j_1;\mathbf{x}\right)$ is replaced by a $n$th-level adjoint function that is independent of parameter variations and is denoted as $\mathbf{A}^{(n)}\left(2^{n-1};j_{n-1};\ldots;j_1;\mathbf{x}\right)\triangleq\left[\mathbf{a}^{(n)}\left(1;j_{n-1};\ldots;j_1;\mathbf{x}\right),\ldots,\mathbf{a}^{(n)}\left(2^{n-1};j_{n-1};\ldots;j_1;\mathbf{x}\right)\right]^{\dagger}\in\mathcal{H}_n\left(\Omega_x\right)$. The elements of the Hilbert space $\mathcal{H}_n\left(\Omega_x\right)$ are surmised to be block-vectors of the form $\mathbf{\Psi}^{(n)}\left(2^{n-1};j_{n-1};\ldots;j_1;\mathbf{x}\right)\triangleq\left[\boldsymbol{\psi}^{(n)}\left(1;j_{n-1};\ldots;j_1;\mathbf{x}\right),\ldots,\boldsymbol{\psi}^{(n)}\left(2^{n-1};j_{n-1};\ldots;j_1;\mathbf{x}\right)\right]^{\dagger}$, comprising as elements a total of $2^{n-1}$ $TD$-dimensional vectors of the form $\boldsymbol{\psi}^{(n)}(i;\mathbf{x})\triangleq\left[\psi_1^{(n)}(i;\mathbf{x}),\ldots,\psi_{TD}^{(n)}(i;\mathbf{x})\right]^{\dagger}\in\mathcal{H}_1\left(\Omega_x\right)$, $i=1,\ldots,2^n$. The inner product of two vectors $\mathbf{\Psi}^{(n)}\left(2^{n-1};j_{n-1};\ldots;j_1;\mathbf{x}\right)$ and $\mathbf{\Phi}^{(n)}\left(2^{n-1};j_{n-1};\ldots;j_1;\mathbf{x}\right)$ in the Hilbert space $\mathcal{H}_n\left(\Omega_x\right)$ is denoted as $\left\langle\mathbf{\Psi}^{(n)}\left(2^{n-1};\mathbf{x}\right),\mathbf{\Phi}^{(n)}\left(2^{n-1};\mathbf{x}\right)\right\rangle_n$ and defined as follows:

$$\left\langle\mathbf{\Psi}^{(n)}\left(2^{n-1};\mathbf{x}\right),\mathbf{\Phi}^{(n)}\left(2^{n-1};\mathbf{x}\right)\right\rangle_n\triangleq\sum_{i=1}^{2^{n-1}}\left\langle\boldsymbol{\psi}^{(n)}(i;\mathbf{x}),\boldsymbol{\varphi}^{(n)}(i;\mathbf{x})\right\rangle_1. \tag{4.15}$$

The $n$th-Level Adjoint Sensitivity System ($n$th-LASS) for the $n$th-level adjoint function $\mathbf{A}^{(n)}\left(2^{n-1}; j_{n-1};\ldots; j_1;\mathbf{x}\right)$ is obtained by using the inner product defined in Eq. (4.15), as follows:

$$\left\{\left\langle \mathbf{A}^{(n)}\left(2^{n-1};\mathbf{x}\right), \mathbf{VM}^{(n)}\left(2^{n-1};\mathbf{x}\right)\right\rangle_2\right\}_{\alpha^0} = \left\{\left[P^{(n)}\left(\mathbf{U}^{(n)};\mathbf{A}^{(n)};\mathbf{V}^{(n)};\alpha\right)\right]_{\partial\Omega_x}\right\}_{\alpha^0}$$
$$+\left\{\left\langle \mathbf{V}^{(n)}\left(2^{n-1};\mathbf{x}\right), \mathbf{AM}^{(n)}\left[2^{n-1}\times 2^{n-1};\mathbf{U}^{(n)}\left(2^{n-1};\mathbf{x}\right);\alpha\right]\mathbf{A}^{(n)}\left(2^{n-1};\mathbf{x}\right)\right\rangle_n\right\}_{\alpha^0}, \tag{4.16}$$

where the quantity $\left\{\left[P^{(n)}\left(\mathbf{U}^{(n)};\mathbf{A}^{(n)};\mathbf{V}^{(n)};\alpha\right)\right]_{\partial\Omega_x}\right\}_{\alpha^0}$ denotes the corresponding bilinear concomitant on the domain's boundary, evaluated at the nominal values for the parameters and respective state functions.

In terms of the $n$th-level adjoint function $\mathbf{A}^{(n)}\left(2^{n-1}; j_{n-1};\ldots; j_1;\mathbf{x}\right)$, the indirect-effect term defined by Eq. (4.3) will have the following expression:

$$\left\{\delta R^{(n-1)}\left[j_{n-1};\ldots;j_1;\mathbf{U}^{(n-1)};\mathbf{A}^{(n-1)};\alpha;\mathbf{V}^{(n-1)};\delta\mathbf{A}^{(n-1)}\right]\right\}_{ind} =$$
$$+\left\{\left\langle \mathbf{A}^{(n)}\left(2^{n-1};j_{n-1};\ldots;j_1;\mathbf{x}\right), \mathbf{Q}_V^{(n)}\left[2^{n-1};\mathbf{U}^{(n)}\left(2^{n-1};j_{n-2};\ldots;j_1;\mathbf{x}\right);\alpha;\delta\alpha\right]\right\rangle_n\right\}_{\alpha^0} \tag{4.17}$$
$$-\left\{\left[\hat{P}^{(n)}\left(\mathbf{U}^{(n)};\mathbf{A}^{(n)};\delta\alpha\right)\right]_{\partial\Omega_x}\right\}_{\alpha^0},$$

where the vector $\mathbf{A}^{(n)}\left(2^{n-1}; j_{n-1};\ldots; j_1;\mathbf{x}\right)$ is the solution of the following $n$th-Level Adjoint Sensitivity System ($n$th-LASS):

$$\mathbf{AM}^{(n)}\left[2^{n-1}\times 2^{n-1};\mathbf{U}^{(n)}\left(2^{n-1};\mathbf{x}\right);\alpha\right]\mathbf{A}^{(n)}\left(2^{n-1};j_{n-1};\ldots;j_1;\mathbf{x}\right)$$
$$= \mathbf{Q}_A^{(n)}\left[2^{n-1};j_{n-1};\ldots;j_1;\mathbf{U}^{(n)}\left(2^{n-1};j_{n-2};\ldots;j_1;\mathbf{x}\right);\alpha\right], \tag{4.18}$$

subject to boundary conditions represented in operator form as follows:

$$\left\{\mathbf{B}_A^{(n)}\left[2^{n-1};\mathbf{U}^{(n)}\left(2^{n-1};j_{n-2};\ldots;j_1;\mathbf{x}\right);\mathbf{A}^{(n)}\left(2^{n-1};j_{n-1};\ldots;j_1;\mathbf{x}\right);\alpha\right]\right\}_{\alpha^0} = \mathbf{0}\left[2^{n-1}\right],$$
$$\mathbf{x}\in\partial\Omega_x\left(\alpha^0\right);\quad j_1 = 1,\ldots,TP;\quad j_2 = 1,\ldots,j_1;\quad \ldots;\quad j_{n-1} = 1,\ldots,j_{n-2}. \tag{4.19}$$

In Eq. (4.17), the quantity $\left\{\left[\hat{P}^{(n)}\left(\mathbf{U}^{(n)};\mathbf{A}^{(n)};\delta\mathbf{f}\right)\right]_{\partial\Omega_x}\right\}_{\alpha^0}$ denotes residual boundary terms which may have not vanished automatically after using the boundary conditions provided in Eqs. (4.5) and (4.19) to eliminate from Eq. (4.17) all unknown values of the $n$th-level variational function $\mathbf{V}^{(n)}\left(2^{n-1}; j_{n-2};\ldots; j_1;\mathbf{x}\right)$.

The quantities which appear in the definition of the $n$th-LASS represented by Eqs. (4.18) and (4.19) are defined as follows:

$$\mathbf{AM}^{(n)}\left[2^{n-1}\times 2^{n-1};\mathbf{U}^{(n)}\left(2^{n-1};j_{n-2};\ldots;j_1;\mathbf{x}\right);\mathbf{f}\right]\triangleq\left[\mathbf{VM}^{(n)}\left(2^{n-1}\times 2^{n-1};\mathbf{U}^{(n)};\mathbf{f}\right)\right]^*$$

$$=\begin{pmatrix}\left\{\left[\mathbf{VM}^{(n-1)}\left(2^{n-2}\times 2^{n-2}\right)\right]^*\right\}^\dagger & \left\{\left[\mathbf{VM}_{21}^{(n)}\left(2^{n-2}\times 2^{n-2}\right)\right]^*\right\}^\dagger \\ \mathbf{0}\left[2^{n-2}\times 2^{n-2}\right] & \left\{\left[\mathbf{VM}_{22}^{(n)}\left(2^{n-2}\times 2^{n-2}\right)\right]^*\right\}^\dagger\end{pmatrix}, \tag{4.20}$$

$$\mathbf{Q}_A^{(n)}\left[2^{n-1};j_{n-1};\ldots;j_1;\mathbf{U}^{(n)}\left(2^{n-1};\mathbf{x}\right);\mathbf{f}\right]\triangleq\left\{\mathbf{q}_A^{(n)}\left[1;j_{n-1};\ldots;j_1;\mathbf{U}^{(n)}\left(2^{n-1};\mathbf{x}\right);\mathbf{f}\right],\ldots\right.$$

$$\left.\ldots,\mathbf{q}_A^{(n)}\left[2^{n-1};j_{n-1};\ldots;j_1;\mathbf{U}^{(n)}\left(2^{n-1};\mathbf{x}\right);\mathbf{f}\right]\right\}^\dagger, \tag{4.21}$$

$$j_1=1,\ldots,TF;\quad\ldots;\quad j_{n-1}=1,\ldots,j_{n-2};$$

$$\mathbf{q}_A^{(n)}\left[1;j_{n-1};\ldots;j_1;\mathbf{U}^{(n)}(\mathbf{x});\mathbf{f}\right]\triangleq\partial S^{(n-1)}\left[j_{n-1};\ldots;j_1;\mathbf{U}^{(n)};\mathbf{f}\right]/\partial\mathbf{u}(\mathbf{x}); \tag{4.22}$$

$$\mathbf{q}_A^{(n)}\left[2;j_{n-1};\ldots;j_1;\mathbf{U}^{(n)}(\mathbf{x});\mathbf{f}\right]\triangleq\partial S^{(n-1)}\left[j_{n-1};\ldots;j_1;\mathbf{U}^{(n)};\mathbf{f}\right]/\partial\mathbf{a}^{(1)}(\mathbf{x}); \tag{4.23}$$

$$\text{For}\quad n\geq 3:\quad \mathbf{q}_A^{(n)}\left[2^k+i;j_{n-1};\ldots;j_1;\mathbf{U}^{(n)}(\mathbf{x});\mathbf{f}\right]$$

$$\triangleq\frac{\partial S^{(n-1)}\left[j_{n-1};\ldots;j_1;\mathbf{U}^{(n)};\mathbf{f}\right]}{\partial\mathbf{a}^{(k+1)}\left(i;j_k;\ldots;j_1;\mathbf{x}\right)};\quad k=1,\ldots,n-2;\quad i=1,\ldots,2^k. \tag{4.24}$$

The final expression of the total differential expressed by Eq. (4.17) is obtained by inserting in Eq. (4.2) the expression for the indirect-effect term obtained in Eq. (4.17), which yields the following expression for the $n$th-order sensitivities $R^{(n)}\left[j_n;\ldots;j_1;\mathbf{U}^{(n)};\mathbf{A}^{(n)};\mathbf{f}\right]\equiv\partial^n R\left[\mathbf{u}(\mathbf{x});\mathbf{f}\right]/\partial f_{j_n}\ldots\partial f_{j_1}$ of the response $R\left[\mathbf{u}(\mathbf{x});\alpha\right]$ with respect to the parameters $\alpha_{j_1},\ldots,\alpha_{j_n}$, for $j_1=1,\ldots,TF;\ldots;j_{n-1}=1,\ldots,j_{n-2}$:

$$R^{(n)}\left[j_n;\ldots;j_1;\mathbf{U}^{(n)};\mathbf{A}^{(n)};\mathbf{f}\right]\equiv\partial^n R\left[\mathbf{u}(\mathbf{x});\mathbf{f}\right]/\partial f_{j_n}\ldots\partial f_{j_1}$$

$$=\frac{\partial R^{(n-1)}\left[j_{n-1};\ldots;j_1;\mathbf{U}^{(n-1)}(\mathbf{x});\mathbf{A}^{(n-1)}(\mathbf{x});\mathbf{f}\right]}{\partial f_{j_n}}-\frac{\left[\partial\hat{P}^{(n)}\left(\mathbf{U}^{(n)};\mathbf{A}^{(n)};\delta\mathbf{f}\right)\right]_{\partial\Omega_x}}{\partial f_{j_n}}$$

$$+\sum_{i=1}^{2^{n-1}}\left\langle\mathbf{a}^{(n)}\left(i;j_{n-1};\ldots;j_1;\mathbf{x}\right),\mathbf{s}_V^{(n)}\left[i;j_n;\ldots;j_1;\mathbf{U}^{(n)}(\mathbf{x});\mathbf{f}\right]\right\rangle_1 \tag{4.25}$$

$$\triangleq\int_{\lambda_1(\alpha)}^{\omega_1(\alpha)}\ldots\int_{\lambda_{TI}(\alpha)}^{\omega_{TI}(\alpha)}S^{(n)}\left[j_n;\ldots;j_1;\mathbf{U}^{(n)}\left(2^{n-1};\mathbf{x}\right);\mathbf{A}^{(n)}\left(2^{n-1};\mathbf{x}\right);\mathbf{f}\right]dx_1\ldots dx_{TI}.$$

## 4.3 THE PARTICULAR FORM OF THE *NTH*-FASAM-N FRAMEWORK FOR *n* = 1

Setting $n = 1$ into the mathematical framework of the *n*th-FASAM-N conjectured in Section 4.2, yields the following expressions for the particular case $n = 1$:

1. Expression of the model response:

$$R^{(0)}\left[\mathbf{U}^{(0)};\mathbf{f}\right] \equiv R\left[\mathbf{u}(\mathbf{x});\mathbf{f}(\alpha)\right] \triangleq \int_{\lambda_1(\alpha)}^{\omega_1(\alpha)} \dots \int_{\lambda_{TI}(\alpha)}^{\omega_{TI}(\alpha)} S\left[\mathbf{u}(\mathbf{x});\mathbf{g}(\alpha);\mathbf{h}(\alpha);\mathbf{x}\right] dx_1 \dots dx_{TI}. \tag{4.26}$$

2. Expression of the 1st-order response sensitivities to the components of the feature function of model parameters, for $j_1 = 1,\dots,TF$:

$$R^{(1)}\left[j_1;\mathbf{U}^{(1)}\left(2^0;\mathbf{x}\right);\mathbf{A}^{(1)}\left(2^0;\mathbf{x}\right);\mathbf{f}(\alpha)\right] = \frac{\partial R\left[\mathbf{u}(\mathbf{x});\mathbf{f}(\alpha)\right]}{\partial f_{j_1}}$$
$$= -\frac{\left[\partial \hat{P}^{(1)}\left(\mathbf{U}^{(1)};\mathbf{A}^{(1)};\delta\mathbf{f}\right)\right]_{\partial\Omega_x}}{\partial f_{j_1}} + \left\langle \mathbf{a}^{(1)}(\mathbf{x}), \mathbf{s}_V^{(1)}\left[\mathbf{U}^{(1)}(\mathbf{x});\alpha\right]\right\rangle_1. \tag{4.27}$$

The various vectors which appear in Eq. (4.27) take on the following particular forms for $n = 1$:

$$\mathbf{U}^{(1)}\left(2^0;\mathbf{x}\right) \equiv \mathbf{u}(\mathbf{x}); \quad \mathbf{A}^{(1)}\left(2^0;\mathbf{x}\right) \triangleq \mathbf{a}^{(1)}(\mathbf{x}); \quad \mathbf{s}_V^{(1)}\left(j_1;\mathbf{u};\alpha\right) \triangleq \frac{\partial\left[\mathbf{Q}(\alpha) - \mathbf{N}(\mathbf{u};\alpha)\right]}{\partial f_{j_1}}. \tag{4.28}$$

The 1st-level adjoint sensitivity function $\mathbf{A}^{(1)}\left(2^0;\mathbf{x}\right) \equiv \mathbf{a}^{(1)}(\mathbf{x})$ is the solution of the following 1st-LASS obtained by setting $n = 1$ in Eqs. (4.18) and (4.19):

$$\left\{\mathbf{AM}^{(1)}\left[2^0 \times 2^0;\mathbf{U}^{(1)}\left(2^0;\mathbf{x}\right);\mathbf{f}\right]\mathbf{A}^{(1)}\left(2^0;\mathbf{x}\right)\right\}_{\alpha^0}$$
$$= \left\{\mathbf{Q}_A^{(1)}\left[2^0;\mathbf{U}^{(1)}\left(2^0;\mathbf{x}\right);\alpha;\delta\mathbf{f}\right]\right\}_{\alpha^0}, \quad \mathbf{x} \in \Omega_x, \tag{4.29}$$

$$\left\{\mathbf{B}_A^{(1)}\left[2^0;\mathbf{U}^{(1)}\left(2^0;\mathbf{x}\right);\mathbf{V}^{(1)}\left(2^0;\mathbf{x}\right);\alpha;\delta\mathbf{f}\right]\right\}_{\alpha^0} = \mathbf{0}, \quad \mathbf{x} \in \partial\Omega_x\left(\alpha^0\right). \tag{4.30}$$

where:

$$\mathbf{AM}^{(1)}\left[2^0 \times 2^0;\mathbf{U}^{(1)}\left(2^0;\mathbf{x}\right);\mathbf{f}\right] \equiv \left[\mathbf{N}^{(1)}(\mathbf{u};\mathbf{f})\right]^*; \tag{4.31}$$

$$\mathbf{Q}_A^{(1)}\left[2^0;\mathbf{U}^{(1)}\left(2^0;\mathbf{x}\right);\alpha;\delta\mathbf{f}\right] \equiv \mathbf{q}_A^{(1)}\left[\mathbf{u}(\mathbf{x});\mathbf{f}\right] \triangleq \left\{\partial S(\mathbf{u};\mathbf{f})/\partial\mathbf{u}\right\}_{\alpha^0}; \tag{4.32}$$

$$\mathbf{B}_V^{(1)}\left[2^0;\mathbf{U}^{(1)}\left(2^0;\mathbf{x}\right);\mathbf{V}^{(1)}\left(2^0;\mathbf{x}\right);\mathbf{f};\delta\mathbf{f}\right]\equiv\mathbf{b}_A^{(1)}\left(\mathbf{u};\mathbf{a}^{(1)};\mathbf{f}\right). \tag{4.33}$$

Comparing the expressions obtained above in Eqs. (4.26)–(4.33) with the expressions obtained in Chapter 2 reveals that the corresponding expressions are identical to each other, which proves the correctness of the conjectured general expressions obtained within the $n$th-FASAM-N methodology for the particular case $n = 1$.

## 4.4 MATHEMATICAL FRAMEWORK OF THE (*n* + 1) TH-FASAM-N METHODOLOGY

It will be assumed that, for each $j_1,\ldots,j_n = 1,\ldots,TF$, the 1st-order total G-differential of the $n$th-order sensitivities $R^{(n)}\left[j_n;\ldots;j_1;\mathbf{U}^{(n)};\mathbf{A}^{(n)};\mathbf{f}\right]$ will exist and will be linear in the variations $\mathbf{V}^{(n)}\left(2^{n-1};j_{n-2};\ldots;j_1;\mathbf{x}\right)\triangleq\delta\mathbf{U}^{(n-1)}\left(2^{n-1};j_{n-2};\ldots;j_1;\mathbf{x}\right)$ and $\delta\mathbf{A}^{(n)}\left(2^{n-1};j_{n-1};\ldots;j_1;\mathbf{x}\right)$ in a neighborhood around the nominal values of the "feature functions," the parameters, and the respective state functions. By definition, the 1st-order total G-differential of $R^{(n)}\left[j_n;\ldots;j_1;\mathbf{U}^{(n)};\mathbf{A}^{(n)};\mathbf{f}\right]$ is given by the following expression:

$$\begin{aligned}\left\{\delta R^{(n)}\left[j_n;\ldots;j_1;\mathbf{U}^{(n)};\mathbf{A}^{(n)};\mathbf{f}\right]\right\}_{\alpha^0} &\triangleq \frac{d}{d\varepsilon}\left\{R^{(n)}\left[j_n;\ldots;j_1;\mathbf{U}^{(n)}+\varepsilon\mathbf{V}^{(n)};\right.\right.\\ &\left.\left.\mathbf{A}^{(n)}+\varepsilon\delta\mathbf{A}^{(n)};\mathbf{f}+\varepsilon\delta\mathbf{f}\right]\right\}_{\varepsilon=0} \triangleq \sum_{j_{n+1}=1}^{TP}\left\{\frac{\partial R^{(n)}\left[\ldots;\mathbf{U}^{(n)};\mathbf{A}^{(n)};\mathbf{f}\right]}{\partial f_{j_{n+1}}}\right\}_{\alpha^0}\delta f_{j_{n+1}}\\ &+\left\{\delta R^{(n)}\left[j_n;\ldots;j_1;\mathbf{U}^{(n)};\mathbf{A}^{(n)};\mathbf{f};\mathbf{V}^{(n)}(\mathbf{x});\delta\mathbf{A}^{(n)}(\mathbf{x})\right]\right\}_{ind},\end{aligned} \tag{4.34}$$

where the quantity $\left\{\delta R^{(n)}\left[j_n;\ldots;j_1;\mathbf{U}^{(n)};\mathbf{A}^{(n)};\mathbf{f};\mathbf{V}^{(n)};\delta\mathbf{A}^{(n)}\right]\right\}_{ind}$ denotes the "indirect-effect term" which is defined as follows:

$$\begin{aligned}&\left\{\delta R^{(n)}\left[j_n;\ldots;j_1;\mathbf{U}^{(n)};\mathbf{A}^{(n)};\mathbf{f};\mathbf{V}^{(n)};\delta\mathbf{A}^{(n)}\right]\right\}_{ind}\\ &\triangleq\int_{\lambda_1(\alpha)}^{\omega_1(\alpha)}\cdots\int_{\lambda_{TI}(\alpha)}^{\omega_{TI}(\alpha)}\left\{\frac{\partial S^{(n)}}{\partial\mathbf{U}^{(n)}(\mathbf{x})}\mathbf{V}^{(n)}\left(2^{n-1};\mathbf{x}\right)+\frac{\partial S^{(n)}}{\partial\mathbf{A}^{(n)}(\mathbf{x})}\delta\mathbf{A}^{(n)}\left(2^{n-1};\mathbf{x}\right)\right\}_{\alpha^0}dx_1\ldots dx_{TI}.\end{aligned} \tag{4.35}$$

The vectors $\mathbf{V}^{(n)}\left(2^{n-1};j_{n-2};\ldots;j_1;\mathbf{x}\right)\triangleq\delta\mathbf{U}^{(n)}\left(2^{n-1};j_{n-2};\ldots;j_1;\mathbf{x}\right)$ and $\delta\mathbf{A}^{(n)}\left(2^{n-1};j_{n-1};\ldots;j_1;\mathbf{x}\right)$, which are needed in order to evaluate the indirect-effect term $\left\{\delta R^{(n)}\left[j_n;\ldots;j_1;\mathbf{U}^{(n)};\mathbf{A}^{(n)};\mathbf{f};\mathbf{V}^{(n)};\delta\mathbf{A}^{(n)}\right]\right\}_{ind}$, are the solutions of the following $(n + 1)$

th-Level Variational Sensitivity System, which is obtained by concatenating the $n$th-LVSS defined by Eqs. (4.4) and (4.5) together with the G-differentiated $n$th-LASS, for $j_1,\ldots,j_n = 1,\ldots,TP$:

$$\left\{\mathbf{VM}^{(n+1)}\left[2^n \times 2^n;\mathbf{U}^{(n+1)}\left(2^n;j_{n-2};\ldots;j_1;\mathbf{x}\right);\mathbf{f}\right]\mathbf{V}^{(n+1)}\left(2^n;j_{n-2};\ldots;j_1;\mathbf{x}\right)\right\}_{\alpha^0}$$
$$=\left\{\mathbf{Q}_V^{(n+1)}\left[2^n;\mathbf{U}^{(n)}\left(2^n;j_{n-2};\ldots;j_1;\mathbf{x}\right);\mathbf{f};\delta\mathbf{f}\right]\right\}_{\alpha^0}, \quad \mathbf{x}\in\Omega_x, \tag{4.36}$$

$$\left\{\mathbf{B}_V^{(n+1)}\left[2^n;\mathbf{U}^{(n+1)}\left(2^n;\mathbf{x}\right);\mathbf{V}^{(n+1)}\left(2^n;\mathbf{x}\right);\mathbf{f};\delta\mathbf{f}\right]\right\}_{\alpha^0} = \mathbf{0}\left[2^n\right]; \quad \mathbf{x}\in\partial\Omega_x\left(\alpha^0\right), \tag{4.37}$$

where:

$$\mathbf{U}^{(n+1)}\left(2^n;j_{n-1};\ldots;j_1;\mathbf{x}\right) \triangleq \begin{pmatrix} \mathbf{U}^{(n)}\left(2^{n-1};j_{n-2};\ldots;j_1;\mathbf{x}\right) \\ \mathbf{A}^{(n)}\left(2^{n-1};j_{n-1};\ldots;j_1;\mathbf{x}\right) \end{pmatrix}; \tag{4.38}$$

$$\mathbf{V}^{(n+1)}\left(2^n;j_{n-1};\ldots;j_1;\mathbf{x}\right) \triangleq \delta\mathbf{U}^{(n+1)}\left(2^n;j_{n-1};\ldots;j_1;\mathbf{x}\right) = \begin{pmatrix} \mathbf{V}^{(n)}\left(2^{n-1};j_{n-2};\ldots;j_1;\mathbf{x}\right) \\ \delta\mathbf{A}^{(n)}\left(2^{n-1};j_{n-1};\ldots;j_1;\mathbf{x}\right) \end{pmatrix}$$
$$=\left[\mathbf{v}^{(1)}(\mathbf{x}),\delta\mathbf{a}^{(1)}(\mathbf{x}),\delta\mathbf{a}^{(2)}\left(1;j_1;\mathbf{x}\right),\delta\mathbf{a}^{(2)}\left(2;j_1;\mathbf{x}\right),\ldots\right.$$
$$\left.\ldots,\delta\mathbf{a}^{(n)}\left(1;j_{n-1};\ldots;j_1;\mathbf{x}\right),\ldots,\delta\mathbf{a}^{(n)}\left(2^{n-1};j_{n-1};\ldots;j_1;\mathbf{x}\right)\right]^{\dagger}; \tag{4.39}$$

$$\mathbf{VM}^{(n+1)}\left[2^n\times 2^n;\mathbf{U}^{(n+1)}\left(2^n;j_{n-1};\ldots;j_1;\mathbf{x}\right);\mathbf{f}\right]$$
$$\triangleq \begin{pmatrix} \mathbf{VM}^{(n)}\left[2^{n-1}\times 2^{n-1};\mathbf{x};\mathbf{f}\right] & \mathbf{0}\left[2^{n-1}\times 2^{n-1}\right] \\ \mathbf{VM}_{21}^{(n+1)}\left(2^{n-1}\times 2^{n-1};\mathbf{x};\mathbf{f}\right) & \mathbf{VM}_{22}^{(n+1)}\left(2^{n-1}\times 2^{n-1};\mathbf{x};\mathbf{f}\right) \end{pmatrix}; \tag{4.40}$$

$$\mathbf{VM}_{21}^{(n+1)}\left(2^{n-1}\times 2^{n-1};j_{n-1};\ldots;j_1;\mathbf{x};\mathbf{f}\right)$$
$$\triangleq -\frac{\partial\mathbf{Q}_A^{(n)}\left[2^{n-1};j_{n-1};\ldots;j_1;\mathbf{U}^{(n)}\left(2^{n-1};j_{n-2};\ldots;j_1;\mathbf{x}\right);\mathbf{f}\right]}{\partial\mathbf{U}^{(n)}\left(2^{n-1};j_{n-2};\ldots;j_1;\mathbf{x}\right)} \tag{4.41}$$
$$+\frac{\partial\left\{\mathbf{AM}^{(n)}\left[2^{n-1}\times 2^{n-1};\mathbf{U}^{(n)}\left(2^{n-1};\mathbf{x}\right);\mathbf{f}\right]\mathbf{A}^{(n)}\left(2^{n-1};j_{n-1};\ldots;j_1;\mathbf{x}\right)\right\}}{\partial\mathbf{U}^{(n)}\left(2^{n-1};j_{n-2};\ldots;j_1;\mathbf{x}\right)};$$

$$\mathbf{VM}_{22}^{(n+1)}\left(2^{n-1}\times 2^{n-1};\mathbf{x};\mathbf{f}\right)\triangleq \mathbf{AM}^{(n)}\left[2^{n-1}\times 2^{n-1};\mathbf{U}^{(n)}\left(2^{n-1};j_{n-2};\ldots;j_1;\mathbf{x}\right);\mathbf{f}\right]; \quad (4.42)$$

$$\begin{aligned}&\mathbf{Q}_V^{(n+1)}\left[2^n;j_1;\mathbf{U}^{(n+1)}\left(2^n;j_{n-1};\ldots;j_1;\mathbf{x}\right);\mathbf{f};\delta\mathbf{f}\right]\\ &\triangleq\begin{pmatrix}\mathbf{Q}_V^{(n)}\left[2^{n-1};\mathbf{U}^{(n)}\left(2^{n-1};\mathbf{x}\right);\mathbf{f};\delta\mathbf{f}\right]\\ \mathbf{Q}_2^{(n+1)}\left[2^{n-1};\mathbf{U}^{(n+1)}\left(2^n;\mathbf{x}\right);\mathbf{f};\delta\mathbf{f}\right]\end{pmatrix}\\ &\triangleq\left\{\mathbf{q}_V^{(n+1)}\left[1;\mathbf{U}^{(n+1)}\left(2^n;j_{n-1};\ldots;j_1;\mathbf{x}\right);\mathbf{f};\delta\mathbf{f}\right],\right.\\ &\left.\ldots,\mathbf{q}_V^{(n+1)}\left[2^n;\mathbf{U}^{(n+1)}\left(2^n;j_{n-1};\ldots;j_1;\mathbf{x}\right);\mathbf{f};\delta\mathbf{f}\right]\right\}^\dagger;\end{aligned} \quad (4.43)$$

$$\mathbf{q}_V^{(n)}\left[i;\mathbf{U}^{(n)}(\mathbf{x});\alpha;\delta\mathbf{f}\right]\equiv\sum_{j_n=1}^{TP}\mathbf{s}_V^{(n)}\left[i;j_{n-2};\ldots;j_1;\mathbf{U}^{(n)}(\mathbf{x});\mathbf{f}\right]\delta f_{j_n};\quad i=1,\ldots;2^{n-1}; \quad (4.44)$$

$$\begin{aligned}&\mathbf{Q}_2^{(n+1)}\left[2^{n-1};\mathbf{U}^{(n+1)}\left(2^n;j_{n-1};\ldots;j_1;\mathbf{x}\right);\mathbf{f};\delta\mathbf{f}\right]\\ &\triangleq\frac{\partial\mathbf{Q}_A^{(n)}\left[2^{n-1};j_{n-1};\ldots;j_1;\mathbf{U}^{(n)}\left(2^{n-1};j_{n-2};\ldots;j_1;\mathbf{x}\right);\mathbf{f}\right]}{\partial\mathbf{f}}\partial\mathbf{f}\\ &-\frac{\partial\left\{\mathbf{AM}^{(n)}\left[2^{n-1}\times 2^{n-1};\mathbf{U}^{(n)}\left(2^{n-1};\mathbf{x}\right);\mathbf{f}\right]\mathbf{A}^{(n)}\left(2^{n-1};j_{n-1};\ldots;j_1;\mathbf{x}\right)\right\}}{\partial\mathbf{f}}\partial\mathbf{f};\end{aligned} \quad (4.45)$$

$$\begin{aligned}&\mathbf{B}_V^{(n+1)}\left[2^n;\mathbf{U}^{(n+1)}\left(2^n;j_{n-1};j_1;\mathbf{x}\right);\mathbf{V}^{(n+1)}\left(2^n;j_{n-1};\ldots;j_1;\mathbf{x}\right);\mathbf{f};\delta\mathbf{f}\right]\\ &\triangleq\begin{pmatrix}\mathbf{B}_V^{(n)}\left[2^{n-1};\mathbf{U}^{(n)}\left(2^{n-1};j_{n-2};\ldots;j_1;\mathbf{x}\right);\mathbf{V}^{(n)}\left(2^{n-1};j_{n-2};\ldots;j_1;\mathbf{x}\right);\mathbf{f};\delta\mathbf{f}\right]\\ \delta\mathbf{B}_A^{(n)}\left[2^{n-1};\mathbf{U}^{(n)}\left(2^{n-1};j_{n-2};\ldots;j_1;\mathbf{x}\right);\mathbf{A}^{(n)}\left(2^{n-1};j_{n-1};\ldots;j_1;\mathbf{x}\right);\mathbf{f}\right]\end{pmatrix}.\end{aligned} \quad (4.46)$$

Solving the $(n+1)$th-LVSS would require $O\left(TP^{n+1}\right)$ large-scale computations, which is unrealistic for large-scale systems comprising many parameters. The $(n+1)$th-FASAM-N circumvents the need for solving the $(n+1)$th-LVSS by deriving an alternative expression for the indirect-effect term defined in Eq. (4.35), in which the function $\mathbf{V}^{(n+1)}\left(2^n;j_{n-1};\ldots;j_1;\mathbf{x}\right)$ is replaced by a $(n+1)$th-level adjoint function, denoted as $\mathbf{A}^{(n+1)}\left(2^n;j_n;\ldots;j_1;\mathbf{x}\right)\triangleq\left[\mathbf{a}^{(n+1)}\left(1;j_n;\ldots;j_1;\mathbf{x}\right),\ldots,\mathbf{a}^{(n+1)}\left(2^n;j_n;\ldots;j_1;\mathbf{x}\right)\right]^\dagger\in\mathcal{H}_{n+1}(\Omega_x)$, which is independent of parameter variations. The elements of the Hilbert space $\mathcal{H}_{n+1}(\Omega_x)$ are block-vectors of the form $\boldsymbol{\Psi}^{(n+1)}\left(2^n;j_n;\ldots;j_1;\mathbf{x}\right)\triangleq\left[\boldsymbol{\psi}^{(n+1)}\left(1;j_n;\ldots;j_1;\mathbf{x}\right),\ldots,\boldsymbol{\psi}^{(n+1)}\left(2^n;j_n;\ldots;j_1;\mathbf{x}\right)\right]^\dagger$, comprising a total of $2^n$ *TD*-dimensional block-vectors of the form $\boldsymbol{\psi}^{(n+1)}(i;\mathbf{x}) \triangleq \left[\psi_1^{(n+1)}(i;\mathbf{x}),\ldots,\psi_{TD}^{(n+1)}(i;\mathbf{x})\right]^\dagger \in \mathcal{H}_1(\Omega_x)$, $i=1,\ldots,2^{n+1}$. The inner product of two vectors $\boldsymbol{\Psi}^{(n+1)}\left(2^n;j_n;\ldots;j_1;\mathbf{x}\right)$ and $\boldsymbol{\Phi}^{(n+1)}\left(2^n;j_n;\ldots;j_1;\mathbf{x}\right)$ in the Hilbert space $\mathcal{H}_{n+1}(\Omega_x)$ is denoted as $\left\langle \boldsymbol{\Psi}^{(n+1)}\left(2^n;\mathbf{x}\right),\boldsymbol{\Phi}^{(n+1)}\left(2^n;\mathbf{x}\right)\right\rangle_{n+1}$ and defined as follows:

$$\left\langle \boldsymbol{\Psi}^{(n+1)}\left(2^n;\mathbf{x}\right),\boldsymbol{\Phi}^{(n+1)}\left(2^n;\mathbf{x}\right)\right\rangle_{n+1} \triangleq \sum_{i=1}^{2^n}\left\langle \boldsymbol{\psi}^{(n+1)}(i;\mathbf{x}),\boldsymbol{\varphi}^{(n+1)}(i;\mathbf{x})\right\rangle_1. \tag{4.47}$$

The $(n+1)$th-Level Adjoint Sensitivity System – abbreviated as $(n+1)$th-LASS – for the $(n+1)$th-level adjoint function $\mathbf{A}^{(n+1)}\left(2^n;j_n;\ldots;j_1;\mathbf{x}\right)$ is obtained by using the inner product defined in Eq. (4.47), as follows:

$$\begin{aligned}
&\left\{\left\langle \mathbf{A}^{(n+1)}\left(2^n;\mathbf{x}\right),\mathbf{VM}^{(n+1)}\left(2^n;\mathbf{x}\right)\right\rangle_2\right\}_{\alpha^0} \\
&=\left\{\left[P^{(n+1)}\left(\mathbf{U}^{(n+1)};\mathbf{A}^{(n+1)};\mathbf{V}^{(n+1)};\mathbf{f}\right)\right]_{\partial\Omega_x}\right\}_{\alpha^0} \\
&+\left\{\left\langle \mathbf{V}^{(n+1)}\left(2^n;\mathbf{x}\right),\mathbf{AM}^{(n+1)}\left[2^n\times 2^n;\mathbf{U}^{(n+1)}\left(2^n;\mathbf{x}\right);\mathbf{f}\right]\mathbf{A}^{(n+1)}\left(2^n;\mathbf{x}\right)\right\rangle_n\right\}_{\alpha^0},
\end{aligned} \tag{4.48}$$

where the quantity $\left\{\left[P^{(n+1)}\left(\mathbf{U}^{(n+1)};\mathbf{A}^{(n+1)};\mathbf{V}^{(n+1)};\mathbf{f}\right)\right]_{\partial\Omega_x}\right\}_{\alpha^0}$ denotes the corresponding bilinear concomitant on the domain's boundary, evaluated at the nominal values for the parameters and respective state functions.

In terms of the $(n+1)$th-level adjoint function $\mathbf{A}^{(n+1)}\left(2^n;j_n;\ldots;j_1;\mathbf{x}\right)$, the indirect-effect term defined by Eq. (4.35) will have the following expression:

$$\begin{aligned}
&\left\{\delta R^{(n)}\left[j_n;\ldots;j_1;\mathbf{U}^{(n)};\mathbf{A}^{(n)};\boldsymbol{\alpha};\mathbf{V}^{(n)};\delta\mathbf{A}^{(n)}\right]\right\}_{ind} \\
&=-\left\{\left[\hat{P}^{(n+1)}\left(\mathbf{U}^{(n+1)};\mathbf{A}^{(n+1)};\delta\mathbf{f}\right)\right]_{\partial\Omega_x}\right\}_{\alpha^0} \\
&+\left\{\left\langle \mathbf{A}^{(n+1)}\left(2^n;j_n;\ldots;j_1;\mathbf{x}\right),\mathbf{Q}_V^{(n+1)}\left[2^n;\mathbf{U}^{(n+1)}\left(2^n;j_{n-1};\ldots;j_1;\mathbf{x}\right);\mathbf{f};\delta\mathbf{f}\right]\right\rangle_n\right\}_{\alpha^0},
\end{aligned} \tag{4.49}$$

where the block-vector $\mathbf{A}^{(n+1)}\left(2^n;j_n;\ldots;j_1;\mathbf{x}\right)$ is the solution of the following $(n+1)$th-LASS:

$$\begin{aligned}
&\mathbf{AM}^{(n+1)}\left[2^n\times 2^n;\mathbf{U}^{(n+1)}\left(2^n;\mathbf{x}\right);\mathbf{f}\right]\mathbf{A}^{(n+1)}\left(2^n;j_n;\ldots;j_1;\mathbf{x}\right) \\
&=\mathbf{Q}_A^{(n+1)}\left[2^n;j_n;\ldots;j_1;\mathbf{U}^{(n+1)}\left(2^n;j_{n-1};\ldots;j_1;\mathbf{x}\right);\mathbf{f}\right],
\end{aligned} \tag{4.50}$$

subject to boundary conditions represented in operator form as follows:

$$\left\{\mathbf{B}_A^{(n+1)}\left[2^n;\mathbf{U}^{(n)}\left(2^n;j_{n-1};\ldots;j_1;\mathbf{x}\right);\mathbf{A}^{(n)}\left(2^n;j_n;\ldots;j_1;\mathbf{x}\right);\mathbf{f}\right]\right\}_{\alpha^0}=\mathbf{0}\left[2^n\right],$$
$$\mathbf{x}\in\partial\Omega_x\left(\alpha^0\right);\quad j_1=1,\ldots,TF;\quad j_2=1,\ldots,j_1;\quad \ldots;\quad j_n=1,\ldots,j_{n-1}. \tag{4.51}$$

In Eq. (4.49), the quantity $\left\{\left[\hat{P}^{(n+1)}\left(\mathbf{U}^{(n+1)};\mathbf{A}^{(n+1)};\delta\mathbf{f}\right)\right]_{\partial\Omega_x}\right\}_{\alpha^0}$ denotes residual boundary terms which may have not vanished automatically after having used the boundary conditions provided in Eqs. (4.37) and (4.51) to eliminate from Eq. (4.48) all unknown values of the $(n+1)$th-level variational function $\mathbf{V}^{(n+1)}\left(2^n;j_{n-1};\ldots;j_1;\mathbf{x}\right)$.

The quantities which appear in the definition of the $(n+1)$th-LASS represented by Eqs. (4.50) and (4.51) are defined as follows:

$$\mathbf{AM}^{(n+1)}\left[2^n\times 2^n;\mathbf{U}^{(n+1)}\left(2^n;j_{n-1};\ldots;j_1;\mathbf{x}\right);\mathbf{f}\right]\triangleq\left[\mathbf{VM}^{(n+1)}\left(2^n\times 2^n;\mathbf{U}^{(n+1)};\mathbf{f}\right)\right]^*$$
$$=\begin{pmatrix}\left\{\left[\mathbf{VM}^{(n)}\left(2^{n-1}\times 2^{n-1}\right)\right]^*\right\}^\dagger & \left\{\left[\mathbf{VM}_{21}^{(n+1)}\left(2^{n-1}\times 2^{n-1}\right)\right]^*\right\}^\dagger \\ \mathbf{0}\left[2^{n-1}\times 2^{n-1}\right] & \left\{\left[\mathbf{VM}_{22}^{(n+1)}\left(2^{n-1}\times 2^{n-1}\right)\right]^*\right\}^\dagger\end{pmatrix}, \tag{4.52}$$

$$\mathbf{Q}_A^{(n+1)}\left[2^n;j_n;\ldots;j_1;\mathbf{U}^{(n+1)}\left(2^n;\mathbf{x}\right);\mathbf{f}\right]\triangleq\left\{\mathbf{q}_A^{(n+1)}\left[1;j_n;\ldots;j_1;\mathbf{U}^{(n+1)}\left(2^n;\mathbf{x}\right);\mathbf{f}\right],\ldots\right.$$
$$\left.\ldots,\mathbf{q}_A^{(n+1)}\left[2^n;j_n;\ldots;j_1;\mathbf{U}^{(n+1)}\left(2^n;\mathbf{x}\right);\mathbf{f}\right]\right\}^\dagger,\quad j_1=1,\ldots,TF;\quad\ldots;\quad j_n=1,\ldots,j_{n-1}; \tag{4.53}$$

$$\mathbf{q}_A^{(n+1)}\left[1;j_n;\ldots;j_1;\mathbf{U}^{(n+1)}(\mathbf{x});\mathbf{f}\right]\triangleq\partial S^{(n)}\left[j_n;\ldots;j_1;\mathbf{U}^{(n+1)};\mathbf{f}\right]/\partial\mathbf{u}(\mathbf{x}); \tag{4.54}$$

$$\mathbf{q}_A^{(n+1)}\left[2;j_n;\ldots;j_1;\mathbf{U}^{(n+1)}(\mathbf{x});\alpha\right]\triangleq\partial S^{(n)}\left[j_n;\ldots;j_1;\mathbf{U}^{(n+1)};\alpha\right]/\partial\mathbf{a}^{(1)}(\mathbf{x}); \tag{4.55}$$

$$\text{For}\quad n\geq 3:\quad \mathbf{q}_A^{(n+1)}\left[2^k+i;j_n;\ldots;j_1;\mathbf{U}^{(n+1)}(\mathbf{x});\mathbf{f}\right]$$
$$\triangleq\frac{\partial S^{(n)}\left[j_n;\ldots;j_1;\mathbf{U}^{(n+1)};\mathbf{f}\right]}{\partial\mathbf{a}^{(k+1)}\left(i;j_k;\ldots;j_1;\mathbf{x}\right)};\quad k=1,\ldots,n-1;\quad i=1,\ldots,2^k. \tag{4.56}$$

The final expression of the total differential expressed by Eq. (4.17) is obtained by inserting in Eq. (4.34) the expression for the indirect-effect term obtained in Eq. (4.49), which yields the following expression for the $n$th-order sensitivities $R^{(n+1)}\left[j_{n+1};\ldots;j_1;\mathbf{U}^{(n+1)};\mathbf{A}^{(n+1)};\mathbf{f}\right]\equiv\partial^{n+1}R\left[\mathbf{u}(\mathbf{x});\mathbf{f}(\alpha)\right]/\partial f_{j_{n+1}}\ldots\partial f_{j_1}$ of the response $R\left[\mathbf{u}(\mathbf{x});\mathbf{f}(\alpha)\right]$ with respect to the components $f_{j_1},\ldots,f_{j_{n+1}}$ of the "features function" $\mathbf{f}(\alpha)$, for $j_1=1,\ldots,TF;\ldots;j_n=1,\ldots,j_{n-1}$:

$$R^{(n+1)}\left[j_{n+1};...;j_1;\mathbf{U}^{(n+1)};\mathbf{A}^{(n+1)};\mathbf{f}(\boldsymbol{\alpha})\right] \equiv \partial^{n+1}R\left[\mathbf{u}(\mathbf{x});\boldsymbol{\alpha}\right]/\partial f_{j_{n+1}}...\partial f_{j_1}$$

$$= \frac{\partial R^{(n)}\left[j_n;...;j_1;\mathbf{U}^{(n)}(\mathbf{x});\mathbf{A}^{(n)}(\mathbf{x});\mathbf{f}(\boldsymbol{\alpha})\right]}{\partial f_{j_{n+1}}} - \frac{\left[\partial \hat{P}^{(n+1)}\left(\mathbf{U}^{(n+1)};\mathbf{A}^{(n+1)};\delta\mathbf{f}\right)\right]_{\partial\Omega_x}}{\partial f_{j_{n+1}}}$$

$$+\sum_{i=1}^{2^n}\left\langle \mathbf{a}^{(n+1)}\left(i;j_n;...;j_1;\mathbf{x}\right),\mathbf{s}_V^{(n+1)}\left[i;j_{n+1};...;j_1;\mathbf{U}^{(n+1)}(\mathbf{x});\mathbf{f}\right]\right\rangle_1 . \tag{4.57}$$

$$\triangleq \int_{\lambda_1(\boldsymbol{\alpha})}^{\omega_1(\boldsymbol{\alpha})} ... \int_{\lambda_{TI}(\boldsymbol{\alpha})}^{\omega_{TI}(\boldsymbol{\alpha})} S^{(n+1)}\left[j_{n+1};...;j_1;\mathbf{U}^{(n+1)}\left(2^n;\mathbf{x}\right);\mathbf{A}^{(n+1)}\left(2^n;\mathbf{x}\right);\mathbf{f}\right]dx_1...dx_{TI}.$$

The expression obtained in Eq. (4.57) is identical with the expression that would be obtained by replacing the index $n$ with $(n+1)$ in the expression obtained in Eq. (4.25), thus completing the proof, by mathematical induction, of the validity/correctness of the conjectured general expressions underlying the $n$th-FASAM-N.

## 4.5 CHAPTER SUMMARY

This work has presented the "$n$th-Order Feature Adjoint Sensitivity Analysis Methodology for Nonlinear Systems" (abbreviated as "$n$th-FASAM-N"), which is the most efficient methodology for computing exact expressions of sensitivities of model responses with respect to features of model parameters and, subsequently, with respect to the model parameters themselves. This efficiency stems from the reduction of the number of adjoint computations – which are considered to be "large-scale" computations – by comparison to the high-order adjoint sensitivity analysis methodology $n$th-CASAM-N (the "$n$th-Order Comprehensive Adjoint Sensitivity Analysis Methodology for Nonlinear Systems"), as follows:

1. Comparing the mathematical framework of the $n$th-FASAM-N methodology presented in this chapter to the framework of the $n$th-CASAM-N methodology (Cacuci 2022a, 2023a) indicates that the components $f_i(\boldsymbol{\alpha}), i = 1,...,TF$ of the "feature function" $\mathbf{f}(\boldsymbol{\alpha}) \triangleq \left[f_1(\boldsymbol{\alpha}),...,f_{TF}(\boldsymbol{\alpha})\right]^\dagger$ play within the $n$th-FASAM-N the same role as played by the components $\alpha_j, j = 1,...,TP$ of the "vector of primary model parameters" $\boldsymbol{\alpha} \triangleq (\alpha_1,...,\alpha_{TP})^\dagger$ within the framework of the $n$th-CASAM-N (Cacuci 2022a, 2023a). It is paramount to underscore, at the outset, that the total number of model parameters is always larger (usually by a wide margin) than the total number of components of the feature function $\mathbf{f}(\boldsymbol{\alpha})$, i.e., $TP \gg TF$.
2. The 1st-FASAM-N and the 1st-CASAM-N methodologies require a single large-scale "adjoint" computations for solving the 1st-LASS (1st-Level Adjoint Sensitivity System), so they are equally efficient for computing the exact expressions of the first-order sensitivities of a model response to the model's uncertain parameters, boundaries, and internal interfaces.

3. For computing the exact expressions of the 2nd-order response sensitivities with respect to the primary model parameters, the 2nd-FASAM-N methodology requires as many large-scale "adjoint" computations as there are "feature functions of parameters" $f_i(\alpha)$, $i = 1,\ldots,TF$ (where $TF$ denotes the total number of feature functions) for solving the left side of the 2nd-LASS with $TF$ distinct sources on its right side. By comparison, the 2nd-CASAM-N methodology requires $TP$ (where $TP$ denotes the total number of model parameters) large-scale computations for solving the same left side of the 2nd-LASS but with $TP$ distinct sources. Since $TF \ll TP$, the 2nd-FASAM-N methodology is considerably more efficient than the 2nd-CASAM-N methodology for computing the exact expressions of the 2nd-order sensitivities of a model response to the model's uncertain parameters, boundaries, and internal interfaces.
4. For computing the exact expressions of the 3rd-order response sensitivities with respect to the primary model's parameters, it can be deduced from the general theory presented in this Chapter that, for $n = 3$, the 3rd-FASAM-N requires at most $TF(TF+1)/2$ large-scale "adjoint" computations while the 3rd-CASAM-N methodology (Cacuci 2022a, 2023a) requires at most $TP(TP+1)/2$. The same computational-count of "large-scale computations" carries over when computing the higher-order sensitivities, i.e., the formula for calculating the "number of large-scale adjoint computations" is formally the same for both the *n*th-FASAM-N and the *n*th-CASAM-N methodologies, but the "variable" in the formula for determining the number of adjoint computations for the *n*th-FASAM-N methodology is $TF$ (i.e., the total number of feature functions), while the counterpart for the formula for determining the number of adjoint computations for the *n*th-CASAM-N is methodology is $TP$ (i.e., the total number of model parameters). Since $TF \ll TP$, it follows that the higher the order of computed sensitivities, the more efficient the *n*th-FASAM-N methodology becomes by comparison to the *n*th-CASAM-N methodology.
5. When a model has no "feature" functions of parameters, but only comprises primary parameters, the *n*th-FASAM-N methodology becomes identical to the *n*th-CASAM-N methodology. All of the characteristics of the *n*th-FASAM-N methodology will be illustrated in Chapter 5 by applying this methodology to the Nordheim-Fuchs reactor dynamics/safety model (Hetrick 1993; Stacey 2001).
6. Both the *n*th-FASAM-N and the *n*th-CASAM-N methodologies are formulated in linearly increasing higher-dimensional Hilbert spaces – as opposed to exponentially increasing parameter-dimensional spaces – thus overcoming the curse of dimensionality in sensitivity analysis of nonlinear systems. Both the *n*th-FASAM-N and the *n*th-CASAM-N methodologies are incomparably more efficient and more accurate than any other methods (statistical, finite differences, etc.) for computing exact expressions of response sensitivities (of any order) with respect to the model's uncertain parameters, boundaries, and internal interfaces.

The question of "when to stop computing progressively higher-order sensitivities?" has been addressed in Chapter 1 in conjunction with the question of convergence of the Taylor-series expansion of the response in terms of the uncertain model parameters. This Taylor-series expansion is the fundamental premise for obtaining meaningful (convergent) expressions provided by the "propagation of errors" methodology, originally proposed heuristically by Tukey (1957) for the cumulants of the model response distribution in the phase-space of model parameters. The convergence of the Taylor-series expansion of the response in terms of the uncertain model parameters, which depends on both the response sensitivities to parameters and the uncertainties associated with the parameter distribution, must be ensured. This can be done by ensuring that the combination of parameter uncertainties and response sensitivities is sufficiently small to fall inside the radius of convergence of this Taylor-series expansion.

# 5 Illustrative Application of the *n*th-FASAM-N Methodology to the Nordheim-Fuchs Reactor Safety Model

## 5.1 INTRODUCTION

As shown in Chapter 4, the "*n*th-Order Function/Feature Adjoint Sensitivity Analysis Methodology for Nonlinear Systems" (*n*th-FASAM-N) enables the most efficient computation of exactly determined expressions of arbitrarily high-order sensitivities of generic nonlinear system responses with respect to functions (features) of model parameters, which enables, in turn, the most efficient computation of the corresponding high-order sensitivities with respect to primary model parameters, uncertain boundaries and internal interfaces in the model's phase-space. It was shown in Chapter 4 that the *n*th-FASAM-N methodology requires the fewest possible number of large-scale computations of any method for determining exactly the respective high-order sensitivities. The unparalleled efficiency of the *n*th-FASAM-N methodology stems from its formulation in the phase-space of "feature-functions," which is always smaller than the phase-space of model parameters. The application of the *n*th-FASAM-N is illustrated in this chapter by considering a well-known paradigm model that describes a short-time self-limiting power excursion in a nuclear reactor system having a negative temperature coefficient in which a large amount of reactivity is suddenly inserted, either intentionally or by accident. In the textbook by Lamarsh (1966), this model is called the "Fuchs model," while in the textbook of Hetrick (1993), this model is called the "Nordheim-Fuchs model." This nonlinear paradigm model is sufficiently complex to realistically model self-limiting power excursions for short times while admitting closed-form exact expressions for the time-dependent neutron flux, temperature distribution, and energy released during the transient power burst.

This work is structured as follows: Section 5.2 presents the balance equations that underlie the Nordheim-Fuchs phenomenological model describing a prompt-critical reactor transient. Section 5.3 illustrates the computation of 1st-order sensitivities of the Nordheim-Fuchs model's response (which is chosen to be the total energy


DOI: 10.1201/9781003478171-5

released during the modeled power-burst) with respect to the parameters underlying this model by applying the well-known 1st-CASAM-N methodology (in Section 5.3.1) and subsequently comparing (in Section 5.3.2) this methodology with the 1st-FASAM-N methodology. Both methodologies require just one large-scale adjoint computation to obtain the exact analytical expressions of the 1st-order sensitivities of the model's response with respect to the model's parameters.

Section 5.4 illustrates the computation of the 2nd-order response sensitivities with respect to the Nordheim-Fuchs model's parameters by showing that applying the 2nd-FASAM-N (in Section 5.4.1) requires just two large-scale computations, while applying the 2nd-CASAM-N (in Section 5.4.2) requires seven large-scale computations, to obtain all of the seven distinct 2nd-order sensitivities.

Section 5.5 illustrates the computation of the 3rd-order response sensitivities with respect to the Nordheim-Fuchs model's parameters by applying the 3rd-FASAM-N versus applying the 3rd-CASAM-N. It is shown (in Section 5.5.1) that the 3rd-FASAM-N requires just two large-scale computations to obtain all of the 84 distinct 3rd-order sensitivities. By comparison, applying the 3rd-CASAM-N (in Section 5.5.2) requires at least 22 large-scale computations for determining the 84 distinct 3rd-order sensitivities. Evidently, the 3rd-FASAM methodology is significantly more efficient for computing the 2nd-and higher-order sensitivities than the 3rd-CASAM-N methodology. Both the 3rd-FASAM-N and the 3rd-CASAM-N methodologies yield exact values for the expressions of the 3rd-order sensitivities.

The concluding discussion presented in Section 5.6 commences by noting that when no feature functions of parameters can be identified, the mathematical frameworks of the *n*th-FASAM-N and the *n*th-CASAM-N methodologies coincide. When feature-functions of parameters can be identified within the model, the *n*th-FASAM-N methodology requires the least number of large-scale computations of any practical methodology for computing exact expressions of 2nd- and higher-order sensitivities. In general, the number of large-scale computations required when applying the *n*th-FASAM-N is proportional to the number of feature-functions underlying the model being analyzed, while the number of large-scale computations required when applying the *n*th-CASAM-N is proportional to the number of parameters underlying the respective model. Since the number of feature-functions is necessarily smaller than the number of model parameters, it follows that the *n*th-FASAM-N is computationally more efficient than the *n*th-CASAM-N. Both the *n*th-FASAM-N and the *n*th-CASAM-N are vastly more efficient computationally than finite-difference schemes, particularly for computing sensitivities of order higher than 1st for large-scale models with many parameters. Furthermore, the finite difference-schemes are approximate, while the *n*th-FASAM-N and the *n*th-CASAM-N accurately compute exact expressions of the arbitrarily high-order sensitivities of model responses with respect to the model's parameters. Altogether, the *n*th-FASAM-N and the *n*th-CASAM-N methodologies remain the most practical methodologies for computing response sensitivities comprehensively and accurately, overcoming the curse of dimensionality in sensitivity analysis of nonlinear systems.

## 5.2 THE NORDHEIM-FUCHS PHENOMENOLOGICAL REACTOR DYNAMICS/SAFETY MODEL

The Nordheim-Fuchs phenomenological model (Lamarsh 1966; Hetrick 1993; Stacey 2001) describes a short-time self-limiting power transient in a nuclear reactor system having a negative temperature coefficient, in which a large amount of reactivity is suddenly inserted, either intentionally or by accident. The response of such a reactor system can be estimated by considering that the reactivity insertion is sufficiently large and the time-span of the transient phenomena under consideration is of the order of the lifetime of prompt-neutrons, which is sufficiently small to neglect the effects of delayed neutrons. For such short times, the local spatial variations of the neutron distribution in the reactor are negligible, and the heat generated during the transient remains within the reactor.

Using the notation of Lamarsh (1966), the Nordheim-Fuchs paradigm model describing the aforementioned self-limiting power transient comprises the following balance equations:

1. The time-dependent neutron balance (point kinetics) equation for the neutron flux $\varphi(t)$:

$$\frac{d\varphi(t)}{dt}=\frac{k(t)-1}{l_p}\varphi(t), \quad t>0, \tag{5.1}$$

$$\varphi(0)=\varphi_0, \quad t=0, \tag{5.2}$$

where $l_p$ denotes the prompt-neutron lifetime, $k(t)$ denotes the reactor's multiplication factor, and $\varphi_0$ denotes the initial (i.e., extant) flux prior to initiating the transient at time $t=0$.

2. The energy conservation equation:

$$c_p\left[T(t)-T_0\right]=E(t), \tag{5.3}$$

where $E(t)$ denotes the total energy released (per cm$^3$) at time $t$ in the reactor since the onset of reactivity change; $c_p$ denotes the specific heat (per cm$^3$) of the reactor.

3. The energy production equation:

$$E(t)=\gamma\Sigma_f\int_0^t\varphi(x)dx, \tag{5.4}$$

where $\gamma$ denotes the recoverable energy per fission; $\Sigma_f \triangleq \sigma_f N_f$ denotes the reactor's effective macroscopic fission cross-section, where $\sigma_f$ denotes the reactor's equivalent microscopic fission cross-section while $N_f$ denotes the reactor's equivalent atomic number density.

4. The reactivity-temperature feedback equation: $k(t) = k_0 - \alpha_T k_0 \left[T(t) - T_0\right]$, where $k_0 \triangleq k(0) \geq 1$ denotes the changed multiplication factor following the reactivity insertion at $t = 0$, $\alpha_T$ denotes the magnitude of the negative temperature coefficient, $T(t)$ denotes the reactor's temperature, and $T_0$ denotes the reactor's initial temperature at time $t = 0$. For illustrating the application of the 1st-FASAM-N methodology, it suffices to consider the special case of a "prompt critical transient," when the reactor becomes prompt critical after the reactivity insertion, i.e., when $k_0 = 1$, so that the reactivity-temperature feedback equation takes on the following particular form:

$$k(t) = 1 - \alpha_T \left[T(t) - T_0\right]. \tag{5.5}$$

The Nordheim-Fuchs model, comprising Eqs. (5.1)–(5.5), is representative of the types of equations that underly large-scale computational models, including a combination of algebraic, differential, and integral equations. Typically, such equations are discretized into systems of nonlinear algebraic and/or differential equations, which are solved numerically by using standard solvers for such systems of equations. Equations (5.1)–(5.5) can be transformed into the following system of nonlinear differential equations:

$$\frac{d\varphi(t)}{dt} = -\frac{\alpha_T}{l_p c_p} E(t)\varphi(t), \quad t > 0. \quad \varphi(0) = \varphi_0, \quad t = 0, \tag{5.6}$$

$$\frac{dE(t)}{dt} = \gamma\sigma_f N_f \varphi(t), \quad E(0) = 0, \tag{5.7}$$

$$\frac{dT(t)}{dt} = \frac{\gamma\sigma_f N_f}{c_p}\varphi(t); \quad T(0) = T_0. \tag{5.8}$$

The Nordheim-Fuchs model described by Eqs. (5.6)–(5.8) can be solved analytically to obtain closed-form expressions for the state functions $\varphi(t)$, $E(t)$, and $T(t)$. Thus, eliminating the function $\varphi(t)$ from Eqs. (5.6) and (5.7) yields a nonlinear equation which can be integrated directly to obtain the following relation:

$$\varphi(t) = -\frac{\alpha_T}{2 l_p c_p \gamma\sigma_f N_f} E^2(t) + \varphi_0. \tag{5.9}$$

Using Eq. (5.9) in Eq. (5.7) yields the following nonlinear equation for the released energy $E(t)$:

$$\frac{dE(t)}{dt} = -\frac{\alpha_T}{2 l_p c_p} E^2(t) + \varphi_0 \gamma\sigma_f N_f, \tag{5.10}$$

$$E(0) = 0. \tag{5.11}$$

The most important quantity of interest (i.e., "model response") for the Nordheim-Fuchs model is the total energy per cm$^3$, $E(\tau)$, released at a user-chosen "final time" instance denoted as $t = \tau$, after the initiation at $t = 0$ of the prompt-critical power transient. This response can be defined mathematically in several equivalent ways, the simplest of which is as follows:

$$E(\tau) = \int_0^\tau E(t)\delta(t-\tau)\,dt, \tag{5.12}$$

where $\delta(t-\tau)$ denotes the Dirac-delta functional.

The response $E(\tau)$ defined in Eq. (5.12) is an implicit function of seven uncertain parameters, which are considered to be the components of a "vector of model parameters" denoted as $\boldsymbol{\alpha}$ and defined as follows:

$$\boldsymbol{\alpha} \triangleq (\alpha_1,\ldots,\alpha_7)^\dagger \triangleq (\alpha_T, l_p, c_p, \varphi_0, \gamma, \sigma_f, N_f)^\dagger. \tag{5.13}$$

As usual, all vectors in this chapter are considered to be column vectors, and the dagger symbol ($\dagger$) will be used to denote "transposition." The model parameters are considered to be uncertain (i.e., imprecisely known), but having known nominal values which will be denoted using a superscript "zero," as follows:

$$\boldsymbol{\alpha}^0 \triangleq (\alpha_1^0,\ldots,\alpha_7^0)^\dagger \triangleq (\alpha_T^0, l_p^0, c_p^0, \varphi_0^0, \gamma^0, \sigma_f^0, N_f^0)^\dagger. \tag{5.14}$$

For subsequent reference, the closed-form solution of Eq. (5.10) has the following expression:

$$E(t) = K(\boldsymbol{\alpha})\tanh\left[t\theta(\boldsymbol{\alpha})\right], \tag{5.15}$$

where:

$$K(\boldsymbol{\alpha}) \triangleq \left[\frac{2\varphi_0\gamma\sigma_f N_f l_p c_p}{\alpha_T}\right]^{1/2}; \quad \theta(\boldsymbol{\alpha}) \triangleq \left[\frac{\alpha_T\varphi_0\gamma\sigma_f N_f}{2 l_p c_p}\right]^{1/2}. \tag{5.16}$$

The closed-form expression of $\varphi(t)$ is obtained by replacing Eq. (5.15) into Eq. (5.9) to obtain:

$$\varphi(t) = \varphi_0\left\{1 - \tanh^2\left[t\theta(\boldsymbol{\alpha})\right]\right\} = \frac{\varphi_0}{\cosh^2\left[t\theta(\boldsymbol{\alpha})\right]}. \tag{5.17}$$

The closed-form expression of $T(t)$ is obtained by replacing Eq. (5.15) into Eq. (5.3) to obtain:

$$T(t) = T_0 + \frac{K(\boldsymbol{\alpha})}{c_p}\tanh\left[t\theta(\boldsymbol{\alpha})\right]. \tag{5.18}$$

The remainder of the material in this chapter will use the model response $E(\tau)$ to illustrate the advantages of applying the *n*th-FASAM-N versus the *n*th-CASAM-N for the higher-order sensitivity analysis of this response to the underlying model parameters.

## 5.3 COMPUTATION OF 1ST-ORDER SENSITIVITIES OF THE MODEL RESPONSE TO MODEL PARAMETERS: APPLICATION OF THE 1ST-CASAM-N VERSUS THE 1ST-FASAM-N

This section presents the computation of the 1st-order sensitivities of the selected model response by following two alternative pathways. Thus, Section 5.3.1 presents the application of the conventional 1st-CASAM, while Section 5.3.2 presents the application of the 1st-FASAM. The advantages of applying the 1st-FASAM versus the 1st-CASAM are summarized in Section 5.3.3.

### 5.3.1 Application of the 1st-CASAM-N to Obtain the 1st-Order Sensitivities of the Response *E*(*τ*) Directly with Respect to the Model Parameters

The 1st-order sensitivities of $E(\tau)$ with respect to variations in the model parameters are obtained by determining the 1st-order Gateaux (G)-differential $\delta E(\tau)$ of $E(\tau)$ for known variations $\delta E(t) \triangleq E(t) - E^0(t)$ and $\delta\alpha \triangleq \alpha - \alpha^0$ around the nominal values $(E^0;\alpha^0)$. Considering that the final observation time is perfectly well known, the 1st-order Gateaux (G)-differential $\delta E(\tau)$ is obtained, by definition, as follows:

$$\delta E(\tau) = \frac{d}{d\varepsilon}\left\{\int_0^\tau \left[E^0 + \varepsilon\delta E(t)\right]\delta(t-\tau)\right\}_{\varepsilon=0} dt = \int_0^\tau \delta E(t)\delta(t-\tau)dt. \tag{5.19}$$

The variational function $\delta E(t)$ is the solution of the 1st-order G-differentials of Eqs. (5.10) and (5.11), which are obtained, by definition, as follows:

$$\frac{d}{d\varepsilon}\left\{\frac{d\left[E(t)+\varepsilon\delta E(t)\right]}{dt}\right\}_{\varepsilon=0} = \frac{d}{d\varepsilon}\left\{-\frac{(\alpha_T+\varepsilon\delta\alpha_T)}{2(l_p+\varepsilon\delta l_p)(c_p+\varepsilon\delta c_p)}\left[E(t)+\varepsilon\delta E(t)\right]^2\right.$$
$$\left.+(\varphi_0+\varepsilon\delta\varphi_0)(\gamma+\varepsilon\delta\gamma)(\sigma_f+\varepsilon\delta\sigma_f)(N_f+\varepsilon\delta N_f)\right\}_{\varepsilon=0}, \tag{5.20}$$

$$\frac{d}{d\varepsilon}\left\{\left[E(t)+\varepsilon\delta E(t)\right]_{t=0}\right\}_{\varepsilon=0} = 0. \tag{5.21}$$

Performing the operations indicated in Eqs. (5.20) and (5.21) yields the following differential equation, which constitutes the 1st-Level Variational Sensitivity System (1st-LVSS) for the 1st-level variational function $\delta E(t)$:

$$\left\{\left[\frac{d}{dt}+\frac{\alpha_T}{l_p c_p}E(t)\right]\delta E(t)\right\}_{\alpha^0}$$
$$=\left\{-\frac{\delta\alpha_T}{2l_p c_p}+\frac{\alpha_T}{2l_p (c_p)^2}\delta c_p+\frac{\alpha_T}{2(l_p)^2 c_p}\delta l_p\right\}_{\alpha^0}E^2(t) \tag{5.22}$$
$$+\left\{\varphi_0\sigma_f N_f(\delta\gamma)+\varphi_0\gamma N_f(\delta\sigma_f)+\varphi_0\gamma\sigma_f(\delta N_f)+\gamma\sigma_f N_f(\delta\varphi_0)\right\}_{\alpha^0},\quad t>0,$$

$$\delta E(0)=0,\quad t=0. \tag{5.23}$$

In Eq. (5.22), the parameter variations are known, and the notation $\{\ \}_{\alpha^0}$ indicates that the quantity enclosed within the braces is to be evaluated at the nominal parameter values $\boldsymbol{\alpha}^0$. For every parameter variation, the 1st-LVSS would need to be solved anew. This need for repeatedly solving the 1st-LVSS can be avoided by applying the 1st-CASAM-N methodology to construct the corresponding 1st-Level Adjoint Sensitivity System (1st-LASS). Thus, the Hilbert space appropriate for the construction of the 1st-LASS corresponding to the above 1st-LVSS is endowed with the following inner product, denoted as $\langle u(t),v(t)\rangle$, between two square integrable functions $u(t)$ and $v(t)$:

$$\langle u(t),v(t)\rangle \triangleq \int_0^\tau u(t)v(t)dt. \tag{5.24}$$

Forming the inner product of Eq. (5.22) with a yet undefined function $a^{(1)}(t)$ yields the following relation:

$$\left\{\int_0^\tau a^{(1)}(t)\left[\frac{d}{dt}+\frac{\alpha_T}{l_p c_p}E(t)\right]\delta E(t)dt\right\}_{\alpha^0}$$
$$=\left\{\varphi_0\sigma_f N_f(\delta\gamma)+\varphi_0\gamma N_f(\delta\sigma_f)+\varphi_0\gamma\sigma_f(\delta N_f)+\gamma\sigma_f N_f(\delta\varphi_0)\right\}_{\alpha^0}\int_0^\tau a^{(1)}(t)dt$$
$$+\left\{-\frac{\delta\alpha_T}{2l_p c_p}+\frac{\alpha_T}{2l_p (c_p)^2}\delta c_p+\frac{\alpha_T}{2(l_p)^2 c_p}\delta l_p\right\}_{\alpha^0}\int_0^\tau a^{(1)}(t)E^2(t)dt. \tag{5.25}$$

Integrating by parts the left side of Eq. (5.25) yields the following relation:

$$\left\{\int_0^\tau a^{(1)}(t)\left[\frac{d}{dt}+\frac{\alpha_T}{l_p c_p}E(t)\right]\delta E(t)dt\right\}_{\alpha^0}$$
$$=a^{(1)}(\tau)\delta E(\tau)-a^{(1)}(0)\delta E(0) \tag{5.26}$$
$$+\left\{\int_0^\tau \delta E(t)\left[-\frac{da^{(1)}(t)}{dt}+\frac{\alpha_T}{l_p c_p}E(t)a^{(1)}(t)\right]dt\right\}_{\alpha^0}.$$

Identifying the left side of Eq. (5.26) with the G-differential $\delta E(\tau)$ of the response $E(\tau)$ and eliminating the unknown value $\delta E(\tau)$ from the right side of Eq. (5.26) by setting $a^{(1)}(\tau)=0$ yields the following 1st-Level Adjoint Sensitivity System (1st-LASS) for the 1st-level adjoint sensitivity function $a^{(1)}(t)$:

$$\left\{\left[-\frac{d}{dt}+\frac{\alpha_T}{l_p c_p}E(t)\right]a^{(1)}(t)\right\}_{\alpha^0}=\delta(t-\tau),\quad t>0, \tag{5.27}$$

$$a^{(1)}(\tau)=0,\quad t=\tau. \tag{5.28}$$

For further reference, solving the above 1st-LASS yields the following closed-form expression for the 1st-level adjoint sensitivity function $a^{(1)}(t)$:

$$a^{(1)}(t)=H(\tau-t)\left\{\frac{\cosh[t\theta(\boldsymbol{\alpha})]}{\cosh[\tau\theta(\boldsymbol{\alpha})]}\right\}^2, \tag{5.29}$$

where $H(t-t_f)$ denotes the Heaviside functional.

Using the relations provided by the 1st-LVSS and the 1st-LASS in Eq. (5.25), and recalling Eq. (5.19), yields the following alternate expression of $\delta E(\tau)$ in terms of the 1st-level adjoint sensitivity function $a^{(1)}(t)$:

$$\begin{aligned}\delta E(\tau)=&\left\{-\frac{\delta\alpha_T}{2l_p c_p}+\frac{\alpha_T}{2l_p(c_p)^2}\delta c_p+\frac{\alpha_T}{2(l_p)^2 c_p}\delta l_p\right\}_{\alpha^0}\int_0^\tau a^{(1)}(t)E^2(t)\,dt\\&+\left\{\begin{matrix}\varphi_0\sigma_f N_f(\delta\gamma)+\varphi_0\gamma N_f(\delta\sigma_f)\\+\varphi_0\gamma\sigma_f(\delta N_f)+\gamma\sigma_f N_f(\delta\varphi_0)\end{matrix}\right\}_{\alpha^0}\int_0^\tau a^{(1)}(t)\,dt.\end{aligned} \tag{5.30}$$

It follows from Eq. (5.30) that the 1st-order sensitivities of the response $E(\tau)$ with respect to the model parameters have the following expressions in terms of the 1st-level adjoint sensitivity function $a^{(1)}(t)$:

$$\begin{aligned}\frac{\partial E(\tau)}{\partial\alpha_T}&=-\frac{1}{2l_p c_p}\int_0^\tau a^{(1)}(t)E^2(t)\,dt=-\frac{K^2(\boldsymbol{\alpha})}{2l_p c_p}\left\{\frac{\tanh[\tau\theta(\boldsymbol{\alpha})]}{2\theta(\boldsymbol{\alpha})}-\frac{\tau}{2\cosh^2[\tau\theta(\boldsymbol{\alpha})]}\right\}\\&=\frac{\tau\varphi_0\gamma\sigma_f N_f}{(2\alpha_T)\cosh^2[\tau\theta(\boldsymbol{\alpha})]}-\left[\frac{2\varphi_0\gamma\sigma_f N_f l_p c_p}{\alpha_T}\right]^{1/2}\frac{\tanh[\tau\theta(\boldsymbol{\alpha})]}{2\alpha_T};\end{aligned} \tag{5.31}$$

$$\begin{aligned}\frac{\partial E(\tau)}{\partial l_p}&=\frac{\alpha_T}{2(l_p)^2 c_p}\int_0^\tau a^{(1)}(t)E^2(t)\,dt\\&=\left[\frac{\varphi_0\gamma\sigma_f N_f c_p}{2\alpha_T l_p}\right]^{1/2}\tanh[\tau\theta(\boldsymbol{\alpha})]-\frac{\tau\varphi_0\gamma\sigma_f N_f}{(2l_p)\cosh^2[\tau\theta(\boldsymbol{\alpha})]};\end{aligned} \tag{5.32}$$

$$\frac{\partial E(\tau)}{\partial c_p} = \frac{\alpha_T}{2l_p\left(c_p\right)^2}\int_0^\tau a^{(1)}(t)E^2(t)\,dt$$
$$= \left[\frac{\varphi_0\gamma\sigma_f N_f l_p}{2\alpha_T c_p}\right]^{1/2}\tanh\left[\tau\theta(\boldsymbol{\alpha})\right] - \frac{\tau\varphi_0\gamma\sigma_f N_f}{\left(2c_p\right)\cosh^2\left[\tau\theta(\boldsymbol{\alpha})\right]}; \tag{5.33}$$

$$\frac{\partial E(\tau)}{\partial \varphi_0} = \gamma\sigma_f N_f\int_0^\tau a^{(1)}(t)\,dt = \gamma\sigma_f N_f\left\{\frac{\tanh\left[\tau\theta(\boldsymbol{\alpha})\right]}{2\theta(\boldsymbol{\alpha})} + \frac{\tau}{2\cosh^2\left[\tau\theta(\boldsymbol{\alpha})\right]}\right\}; \tag{5.34}$$

$$\frac{\partial E(\tau)}{\partial \gamma} = \varphi_0\sigma_f N_f\int_0^\tau a^{(1)}(t)\,dt = \varphi_0\sigma_f N_f\left\{\frac{\tanh\left[\tau\theta(\boldsymbol{\alpha})\right]}{2\theta(\boldsymbol{\alpha})} + \frac{\tau}{2\cosh^2\left[\tau\theta(\boldsymbol{\alpha})\right]}\right\}; \tag{5.35}$$

$$\frac{\partial E(\tau)}{\partial \sigma_f} = \varphi_0\gamma N_f\int_0^\tau a^{(1)}(t)\,dt = N_f\varphi_0\gamma\left\{\frac{\tanh\left[\tau\theta(\boldsymbol{\alpha})\right]}{2\theta(\boldsymbol{\alpha})} + \frac{\tau}{2\cosh^2\left[\tau\theta(\boldsymbol{\alpha})\right]}\right\}; \tag{5.36}$$

$$\frac{\partial E(\tau)}{\partial N_f} = \varphi_0\gamma\sigma_f\int_0^\tau a^{(1)}(t)\,dt = \sigma_f\varphi_0\gamma\left\{\frac{\tanh\left[\tau\theta(\boldsymbol{\alpha})\right]}{2\theta(\boldsymbol{\alpha})} + \frac{\tau}{2\cosh^2\left[\tau\theta(\boldsymbol{\alpha})\right]}\right\}. \tag{5.37}$$

The following formulas have been used to obtain the expressions in Eqs. (5.31)–(5.34): $\int \sinh^2(ax)\,dx = \sinh(2ax)/4a - x/2$ and $\int \cosh^2(ax)\,dx = \sinh(2ax)/4a + x/2$.

### 5.3.2 Application of the 1st-FASAM-N to Obtain the 1st-Order Sensitivities of the Response $E(\tau)$ with Respect to the Features and Model Parameters

The form of Eq. (5.10) indicates that the "features" (i.e., functions) of model parameters characterizing this balance equation can be chosen as follows:

$$f_1(\boldsymbol{\alpha}) \triangleq \frac{\alpha_T}{2l_p c_p}; \quad f_2(\boldsymbol{\alpha}) \triangleq \varphi_0\gamma\sigma_f N_f; \quad \mathbf{f}(\boldsymbol{\alpha}) \triangleq \left[f_1(\boldsymbol{\alpha}), f_2(\boldsymbol{\alpha})\right]^\dagger. \tag{5.38}$$

Consequently, Eq. (5.10) can alternatively be written in terms of the "feature function" $\mathbf{f}(\boldsymbol{\alpha}) \triangleq \left[f_1(\boldsymbol{\alpha}), f_2(\boldsymbol{\alpha})\right]^\dagger$ as follows:

$$\frac{dE(t)}{dt} = -f_1(\boldsymbol{\alpha})E^2(t) + f_2(\boldsymbol{\alpha}), \quad E(0) = 0. \tag{5.39}$$

In terms of the feature function $\mathbf{f}(\boldsymbol{\alpha}) \triangleq \left[f_1(\boldsymbol{\alpha}), f_2(\boldsymbol{\alpha})\right]^\dagger$, the solution of Eq. (5.39) has the following form:

$$E(t) = \left[\frac{f_2(\boldsymbol{\alpha})}{f_1(\boldsymbol{\alpha})}\right]^{1/2}\tanh\left[t g(\boldsymbol{\alpha})\right]; \quad g(\boldsymbol{\alpha}) \triangleq \sqrt{f_1(\boldsymbol{\alpha}) f_2(\boldsymbol{\alpha})}. \tag{5.40}$$

Taking the G-differential of Eq. (5.39) yields the following 1st-Level Variational Sensitivity System (1st-LVSS) for the variational function $\delta E(t)$:

$$\frac{d}{d\varepsilon}\left\{\frac{d\left[E^0(t)+\varepsilon\delta E(t)\right]}{dt}\right\}_{\varepsilon=0} + \frac{d}{d\varepsilon}\left\{\left[f_1^0+\varepsilon(\delta f_1)\right]\left[E^0(t)+\varepsilon\delta E(t)\right]^2-\left[f_2^0+\varepsilon(\delta f_2)\right]\right\}_{\varepsilon=0}=0, \tag{5.41}$$

$$\frac{d}{d\varepsilon}\left\{\left[E^0(t)+\varepsilon\delta E(t)\right]_{t=0}\right\}_{\varepsilon=0}=0. \tag{5.42}$$

Performing the operations indicated in Eqs. (5.41) and (5.42) yields the following expressions for the 1st-LVSS:

$$\left\{\left[\frac{d}{dt}+2f_1E(t)\right]\delta E(t)\right\}_{\mathbf{f}^0}=\left\{-\delta f_1E^2(t)+\delta f_2\right\}_{\mathbf{f}^0},\quad t>0, \tag{5.43}$$

$$\delta E(0)=0,\quad t=0. \tag{5.44}$$

In Eq. (5.43), the notation $\{\ \}_{\mathbf{f}^0}$ indicates that the quantity enclosed within the braces is to be evaluated at the nominal values $E^0(t)$ of the dependent variable and $\mathbf{f}^0 \triangleq \left(f_1^0, f_2^0\right)^\dagger$, $f_1^0 \triangleq f_1\left(\boldsymbol{\alpha}^0\right)$, $f_2^0 \triangleq f_2\left(\boldsymbol{\alpha}^0\right)$, of the components of the feature function $\mathbf{f}(\boldsymbol{\alpha})$. The 1st-LVSS would need to be solved anew for all variations $\delta f_1, \delta f_2$. This need for repeatedly solving the 1st-LVSS can be avoided by constructing the corresponding 1st-Level Adjoint Sensitivity System (1st-LASS). Note that the left side of Eq. (5.43) is the same as the left side of Eq. (5.22). Therefore, the Hilbert space appropriate for the construction of the 1st-LASS corresponding to Eq. (5.43) is the same as for the application of the 1st-CASAM-N, being endowed with the inner product defined in Eq. (5.24). It is therefore also expected that the left side of the 1st-LASS to be constructed for Eq. (5.43) will be the same as the left side of Eq. (5.27). Thus, forming the inner product of Eq. (5.43) with a yet undefined function $a^{(1)}(t)$ yields the following relation:

$$\left\{\int_0^\tau a^{(1)}(t)\left[\frac{d}{dt}+2f_1E(t)\right]\delta E(t)\,dt\right\}_{\mathbf{f}^0} = \left\{-(\delta f_1)\int_0^\tau a^{(1)}(t)E^2(t)\,dt+(\delta f_2)\int_0^\tau a^{(1)}(t)\,dt\right\}_{\mathbf{f}^0}. \tag{5.45}$$

Integrating by parts the left side of Eq. (5.45) yields the following relation:

$$\left\{\int_0^\tau a^{(1)}(t)\left[\frac{d}{dt}+2f_1E(t)\right]\delta E(t)\,dt\right\}_{\boldsymbol{\alpha}^0} = a^{(1)}(\tau)\delta E(\tau) - a^{(1)}(0)\delta E(0)+\left\{\int_0^\tau \delta E(t)\left[-\frac{da^{(1)}(t)}{dt}+2f_1E(t)a^{(1)}(t)\right]dt\right\}_{\boldsymbol{\alpha}^0}. \tag{5.46}$$

Identifying the left side of Eq. (5.46) with the G-differential $\delta E(\tau)$ of the response $E(\tau)$ obtained in Eq. (5.19), and eliminating the unknown value $\delta E(\tau)$ from the right side of Eq. (5.46) by setting $a^{(1)}(\tau)=0$ yields the following 1st-Level Adjoint Sensitivity System (1st-LASS) for the 1st-level adjoint sensitivity function $a^{(1)}(t)$:

$$\left\{\left[-\frac{d}{dt}+2f_1E(t)\right]a^{(1)}(t)\right\}_{\mathbf{f}^0}=\delta(t-\tau), \quad t>0, \tag{5.47}$$

$$a^{(1)}(\tau)=0, \quad t=\tau. \tag{5.48}$$

Note that the above 1st-LASS, comprising Eqs. (5.47) and (5.48), is the same as the 1st-LASS obtained for determining the sensitivities of the response with respect to the model parameters, i.e., Eqs. (5.27) and (5.28). This outcome is expected since the 1st-LASS is independent of any variations in the model parameters and hence, variations in the feature functions. Therefore, it is justified to use the same symbol, $a^{(1)}(t)$, for the 1st-level adjoint sensitivity function, which is the solution of either of these 1st-LASS. In terms of the feature function $\mathbf{f}(\alpha)$, the 1st-level adjoint sensitivity function $a^{(1)}(t)$ has the following closed-form expression:

$$a^{(1)}(t)=H(\tau-t)\left\{\frac{\cosh\left[t(f_1f_2)^{1/2}\right]}{\cosh\left[\tau(f_1f_2)^{1/2}\right]}\right\}^2, \tag{5.49}$$

where $H(t-\tau)$ denotes the Heaviside functional.

Using Eqs. (5.46)–(5.48) in Eq. (5.45) yields the expression below for the 1st-order total G-differential $\delta E(\tau)$ of the response $E(\tau)$ in terms of the 1st-level adjoint function $a^{(1)}(t)$:

$$\delta E(\tau)=\left\{-(\delta f_1)\int_0^\tau a^{(1)}(t)E^2(t)dt+(\delta f_2)\int_0^\tau a^{(1)}(t)dt\right\}_{\alpha^0}. \tag{5.50}$$

It follows from Eqs. (5.50), (5.49), and (5.40) that the two sensitivities of the response $E(\tau)$ with respect to the two components of the feature function $\mathbf{f}\triangleq(f_1,f_2)^\dagger$ have the following expressions:

$$\frac{\partial E(\tau)}{\partial f_1}=-\int_0^\tau a^{(1)}(t)E^2(t)dt=\frac{1}{2}\left[\frac{f_2(\alpha)}{f_1(\alpha)}\right]^{1/2}\left\{\frac{\tau}{\cosh^2[\tau g(\alpha)]}-\frac{\tanh[\tau g(\alpha)]}{g(\alpha)}\right\}; \tag{5.51}$$

$$\frac{\partial E(\tau)}{\partial f_2}=\int_0^\tau a^{(1)}(t)dt=\frac{1}{2g(\alpha)}\tanh[\tau g(\alpha)]+\frac{\tau}{2\cosh^2[\tau g(\alpha)]}. \tag{5.52}$$

The above expressions are to be evaluated at the nominal parameter values, but the notation $\{\ \}_{\alpha^0}$ has been omitted, for simplicity. The expressions obtained in Eqs. (5.51) and (5.52) can be verified by differentiating the expression provided in Eq. (5.40), evaluated at a user-chosen time $t=\tau$ within the interval $0<\tau<\infty$.

The sensitivities of the response $E(\tau)$ with respect to the model parameters are obtained by using the general relationship:

$$\frac{\partial E(\tau;f_1;f_2)}{\partial \alpha_i}=\frac{\partial E(\tau)}{\partial f_1}\frac{\partial f_1(\boldsymbol{\alpha})}{\partial \alpha_i}+\frac{\partial E(\tau)}{\partial f_2}\frac{\partial f_2(\boldsymbol{\alpha})}{\partial \alpha_i};\quad i=1,\dots,7. \tag{5.53}$$

Using Eqs. (5.51) and (5.52) while recalling the definitions of the feature functions $f_1(\boldsymbol{\alpha})$ and $f_2(\boldsymbol{\alpha})$ defined in Eq. (5.38) yields the following explicit formulas for the particular cases of Eq. (5.53):

$$\frac{\partial E(\tau)}{\partial \alpha_T}=\frac{\partial E(\tau)}{\partial f_1}\frac{\partial f_1}{\partial \alpha_T}+\frac{\partial E(\tau)}{\partial f_2}\frac{\partial f_2}{\partial \alpha_T}=\frac{1}{2l_p c_p}\frac{\partial E(\tau)}{\partial f_1}; \tag{5.54}$$

$$\frac{\partial E(\tau)}{\partial l_p}=\frac{\partial E(\tau)}{\partial f_1}\frac{\partial f_1}{\partial l_p}+\frac{\partial E(\tau)}{\partial f_2}\frac{\partial f_2}{\partial l_p}=-\frac{\alpha_T}{2\left(l_p\right)^2 c_p}\frac{\partial E(\tau)}{\partial f_1}; \tag{5.55}$$

$$\frac{\partial E(\tau)}{\partial c_p}=\frac{\partial E(\tau)}{\partial f_1}\frac{\partial f_1}{\partial c_p}+\frac{\partial E(\tau)}{\partial f_2}\frac{\partial f_2}{\partial c_p}=-\frac{\alpha_T}{2\left(c_p\right)^2 l_p}\frac{\partial E(\tau)}{\partial f_1}; \tag{5.56}$$

$$\frac{\partial E(\tau)}{\partial \varphi_0}=\frac{\partial E(\tau)}{\partial f_1}\frac{\partial f_2}{\partial \varphi_0}+\frac{\partial E(\tau)}{\partial f_2}\frac{\partial f_2}{\partial \varphi_0}=\gamma\sigma_f N_f\frac{\partial E(\tau)}{\partial f_2}; \tag{5.57}$$

$$\frac{\partial E(\tau)}{\partial \gamma}=\frac{\partial E(\tau)}{\partial f_1}\frac{\partial f_1}{\partial \gamma}+\frac{\partial E(\tau)}{\partial f_2}\frac{\partial f_2}{\partial \gamma}=\varphi_0\sigma_f N_f\frac{\partial E(\tau)}{\partial f_2}; \tag{5.58}$$

$$\frac{\partial E(\tau)}{\partial \sigma_f}=\frac{\partial E(\tau)}{\partial f_1}\frac{\partial f_1}{\partial \sigma_f}+\frac{\partial E(\tau)}{\partial f_2}\frac{\partial f_2}{\partial \sigma_f}=\varphi_0\gamma N_f\frac{\partial E(\tau)}{\partial f_2}; \tag{5.59}$$

$$\frac{\partial E(\tau)}{\partial N_f}=\frac{\partial E(\tau)}{\partial f_1}\frac{\partial f_1}{\partial N_f}+\frac{\partial E(\tau)}{\partial f_2}\frac{\partial f_2}{\partial N_f}=\varphi_0\gamma\sigma_f\frac{\partial E(\tau)}{\partial f_2}. \tag{5.60}$$

### 5.3.3 Comparative Discussion: Applying the 1st-CASAM-N versus the 1st-FASAM-N for Computing the 1st-Order Response Sensitivities to Model Parameters

Both the 1st-CASAM-N and the 1st-FASAM-N require the solution of the same 1st-LASS. Hence, the application of the 1st-CASAM-N necessitates a single large-scale computation (for solving the 1st-LASS) to obtain all of the seven 1st-order sensitivities for the instant energy response to the model parameters. The application of the 1st-FASAM-N also necessitates a single large-scale computation (for solving the same 1st-LASS as for the 1st-CASAM-N) to obtain the two 1st-order sensitivities of the model's response to the model's chosen feature functions. This equivalence between the application of the 1st-FASAM-N and the 1st-CASAM-N is expected, since the 1st-LASS is independent of parameter variations or, equivalently, of variations in the feature functions. After the 1st-level adjoint function has been computed, the computation of the sensitivities to the model parameters using the 1st-FASAM-N additionally requires two quadratures to compute $\partial E(\tau)/\partial f_1$ and $\partial E(\tau)/\partial f_2$ using Eqs. (5.51) and (5.52), respectively, followed by simple differentiations of the feature functions with respect to the component parameters, as shown in Eqs. (5.54)–(5.60). Alternatively, computing the response sensitivities to the same model parameters using the 1st-CASAM-N additionally requires seven integrations (quadratures), as shown in Eqs. (5.31)–(5.37). Neither these differentiations nor these quadratures require "large-scale" computations, so the differences in the computational resources needed to apply the 1st-FASAM-N versus applying the 1st-CASAM-N are minimal, with a slight advantage towards the 1st-FASAM-N, since the respective differentiations are computationally somewhat less demanding than the integrations/quadratures required by the application of the 1st-CASAM-N.

## 5.4 COMPUTATION OF THE 2ND-ORDER RESPONSE SENSITIVITIES WITH RESPECT TO MODEL PARAMETERS: APPLYING THE 2ND-FASAM-N VERSUS THE 2ND-CASAM-N

The fundamental principle underlying both the 2nd-FASAM-N and the 2nd-CASAM-N methodologies is to determine the 2nd-order sensitivities by employing their definition of being the "1st-order sensitivities of the 1st-order sensitivities." Thus, each 1st-order sensitivity is treated as a "model response," and the G-differential of each of these "model responses" subsequently provides the partial 2nd-order sensitivities that stem from the respective 1st-order sensitivity. As will be highlighted in Section 5.4.1, below, the computation of the 49 2nd-order sensitivities (of which 28 are distinct) of the response $E(\tau)$ with respect to the seven model parameters will require just two large-scale computations (namely, solving the corresponding two 2nd-Level Adjoint Sensitivity Systems) when using the 2nd-FASAM-N methodology, since there are only two sensitivities of the response $E(\tau)$ with respect to the "feature functions" $f_1(\alpha)$ and $f_2(\alpha)$. In contradistinction, as will be highlighted in Section 5.4.2, applying the 2nd-CASAM-N methodology requires seven large-scale computations for solving the seven 2nd-Level Adjoint Sensitivity Systems (one 2nd-LASS for each of the seven 1st-order sensitivities with respect to the primary model parameters) for obtaining all 2nd-order sensitivities of the response $E(\tau)$ with respect to the model parameters.

### 5.4.1 Computation of 2nd-Order Sensitivities Using the 2nd-FASAM-N

The determination of the 2nd-order sensitivities that stem from the 1st-order sensitivity $\partial E(\tau)/\partial f_1$ of the response $E(\tau)$ with respect to the "feature function" $f_1(\alpha)$ is presented in Section 5.4.1.1, while the 2nd-order sensitivities that arise from the 1st-order sensitivity $\partial E(\tau)/\partial f_2$ of the response $E(\tau)$ with respect to the "feature function" $f_2(\alpha)$ will be presented in Section 5.4.1.2.

#### 5.4.1.1 Computation of 2nd-Order Sensitivities Stemming from the 1st-Order Sensitivity $\partial E(\tau)/\partial f_1$

The 2nd-order sensitivities that stem from the 1st-order sensitivity $\partial E(\tau)/\partial f_1$ defined in Eq. (5.51) will be obtained by determining the 1st-order G-differential $\delta\{\partial E(\tau)/\partial f_1\}$ of $\partial E(\tau)/\partial f_1$. By definition, the 1st-order G-differential of $\partial E(\tau)/\partial f_1$ is obtained as follows:

$$\delta\left\{\frac{\partial E(\tau;f_1,f_2)}{\partial f_1}\right\}=-\left\{\frac{d}{d\varepsilon}\left[\int_0^\tau\left(a^{(1)}+\varepsilon\delta a^{(1)}\right)\left[E(t)+\varepsilon\delta E(t)\right]^2dt\right]\right\}_{\varepsilon=0}$$

$$=-\int_0^\tau\left[2a^{(1)}(t)E(t)\delta E(t)+\delta a^{(1)}(t)E^2(t)\right]dt \tag{5.61}$$

$$=\frac{\partial^2E(\tau;f_1,f_2)}{\partial f_1\partial f_1}\delta f_1+\frac{\partial^2E(\tau;f_1,f_2)}{\partial f_2\partial f_1}\delta f_2.$$

The relation obtained in Eq. (5.61) holds at the nominal values of the parameters/feature functions, and nominal values of the functions $a^{(1)}(t)$ and $E(t)$, but the superscript "zero" has been omitted, for simplicity. The variational function $\delta a^{(1)}(t)$ is the solution of the system of equations obtained by G-differentiating the 1st-LASS defined in Eqs. (5.47) and (5.48). Performing the G-differentiation of this 1st-LASS yields the following equations:

$$\left\{\left[-\frac{d}{dt}+2f_1E(t)\right]\delta a^{(1)}(t)+2f_1a^{(1)}(t)\left[\delta E(t)\right]\right\}_{\mathbf{f}^0}$$
$$=-2\left\{(\delta f_1)a^{(1)}(t)E(t)\right\}_{\mathbf{f}^0},\quad 0<t<\tau, \tag{5.62}$$

$$\delta a^{(1)}(\tau)=0,\quad t=\tau. \tag{5.63}$$

Concatenating Eqs. (5.62) and (5.63) with the 1st-LVSS for $\delta E(t)$ defined in Eqs. (5.43) and (5.44) yields the following 2nd-Level Variational Sensitivity System (2nd-LVSS) for the 2nd-level variational function $\mathbf{V}^{(2)}(2;t)\triangleq\left[v^{(2)}(1;t),v^{(2)}(2;t)\right]^\dagger\triangleq\left[\delta E(t),\delta a^{(1)}(t)\right]^\dagger$:

$$\left\{\mathbf{VM}^{(2)}[2\times2;\mathbf{f}]\mathbf{V}^{(2)}(2;t)\right\}_{\mathbf{f}^0}=\left\{\mathbf{Q}_V^{(2)}[2;\mathbf{f};\delta\mathbf{f}]\right\}_{\mathbf{f}^0},\quad 0<t<\tau, \tag{5.64}$$

$$\left\{\mathbf{B}_V^{(2)}\left[2;\mathbf{V}^{(2)}(2;t);\mathbf{f};\delta\mathbf{f}\right]\right\}_{\mathbf{f}^0}=\mathbf{0}[2],\quad \mathbf{0}[2]\triangleq[0,0]^\dagger, \tag{5.65}$$

where:

$$\mathbf{VM}^{(2)}[2\times2;\mathbf{f}]\triangleq\begin{pmatrix}\frac{d}{dt}+2f_1E(t) & 0\\ 2f_1a^{(1)}(t) & -\frac{d}{dt}+2f_1E(t)\end{pmatrix}; \tag{5.66}$$

$$\mathbf{Q}_V^{(2)}[2;\mathbf{f};\delta\mathbf{f}] \triangleq \begin{bmatrix} -(\delta f_1)E^2(t)+(\delta f_2) \\ -2(\delta f_1)a^{(1)}(t)E(t) \end{bmatrix}; \tag{5.67}$$

$$\mathbf{B}_V^{(2)}\left[2;\mathbf{V}^{(2)}(2;t);\mathbf{f};\delta\mathbf{f}\right] \triangleq \begin{pmatrix} \delta E(0) \\ \delta a^{(1)}(t_f) \end{pmatrix} = \begin{pmatrix} 0 \\ 0 \end{pmatrix}. \tag{5.68}$$

The need for solving the 2nd-LVSS is circumvented by deriving an alternative expression for the 1st-order G-differential $\delta\{\partial E(\tau)/\partial f_1\}$ defined in Eq. (5.61), in which the variational function $\mathbf{V}^{(2)}(2;t) \triangleq \left[v^{(2)}(1;t), v^{(2)}(2;t)\right]^\dagger \triangleq \left[\delta E(t), \delta a^{(1)}(t)\right]^\dagger$ is replaced by a 2nd-level adjoint function, which will be denoted as $\mathbf{A}^{(2)}(2;1;t) \triangleq \left[a^{(2)}(1;1;t), a^{(2)}(2;1;t)\right]^\dagger \in \mathcal{H}_2$. The notation for $\mathbf{A}^{(2)}(2;1;t) \triangleq \left[a^{(2)}(1;1;t), a^{(2)}(2;1;t)\right]^\dagger \in \mathcal{H}_2$ has the following significance: (i) the bold letter "A" indicates a vector-valued "adjoint" function; (ii) the superscript "(2)" indicates "2nd-level;" (iii) the 1st argument, denoted as "2," in $\mathbf{A}^{(2)}(2;1;t)$ indicates that this vector has two components, denoted as $a^{(2)}(i;1;t)$, $i = 1,2$, each of which is a scalar-valued function of time; (iv) the 2nd argument of $\mathbf{A}^{(2)}(2;1;t)$, denoted as "1," indicates that this 2nd-level adjoint function corresponds to the 1st-order sensitivity $\partial E(\tau)/\partial f_1$ of the response with respect to the "1st feature function," $f_1(\boldsymbol{\alpha})$. In Eq. (5.64) and in the remainder of this chapter, matrices will be denoted by using two bold capital letters.

The 2nd-level adjoint function $\mathbf{A}^{(2)}(2;1;t)$ will be the solution of a 2nd-Level Adjoint Sensitivity System (2nd-LASS) to be constructed by applying the 2nd-FASAM-N methodology. This 2nd-LASS is constructed in a Hilbert space, denoted as $\mathcal{H}_2$, which comprises as elements block-vectors of the same form as $\mathbf{V}^{(2)}(2;t)$, and is endowed with the following inner product, denoted as $\left\langle \boldsymbol{\Psi}^{(2)}(2;t), \boldsymbol{\Phi}^{(2)}(2;t)\right\rangle_2$, of two vectors $\boldsymbol{\Psi}^{(2)}(2;t) \triangleq \left[\psi^{(2)}(1;t), \psi^{(2)}(2;t)\right]^\dagger \in \mathcal{H}_2$ and $\boldsymbol{\Phi}^{(2)}(t) \triangleq \left[\varphi^{(2)}(1;t), \varphi^{(2)}(2;t)\right]^\dagger \in \mathcal{H}_2$:

$$\left\langle \boldsymbol{\Psi}^{(2)}(2;t), \boldsymbol{\Phi}^{(2)}(2;t)\right\rangle_2 \triangleq \sum_{i=1}^{2} \int_0^\tau \psi^{(2)}(i;t)\varphi^{(2)}(i;t)\,dt. \tag{5.69}$$

The inner product defined in Eq. (5.69) is now used to construct the 2nd-Level Adjoint Sensitivity System (2nd-LASS) for the 2nd-level adjoint function $\mathbf{A}^{(2)}(2;t) \triangleq \left[a^{(2)}(1;t), a^{(2)}(2;t)\right]^\dagger \in \mathcal{H}_2$, as follows:

1. Using the inner product defined in Eq. (5.69), form the inner product of $\mathbf{A}^{(2)}(2;1;t) \triangleq \left[a^{(2)}(1;1;t), a^{(2)}(2;1;t)\right]^\dagger \in \mathcal{H}_2$ with Eq. (5.64), and subsequently integrate by parts the left side of the resulting equation to obtain the following relation:

$$\begin{aligned} &\left\{\left\langle \mathbf{A}^{(2)}(2;1;t), \mathbf{VM}^{(2)}[2\times 2;\mathbf{f}]\mathbf{V}^{(2)}(2;t)\right\rangle_2\right\}_{\mathbf{f}^0} = \left\{\left\langle \mathbf{A}^{(2)}(2;1;t), \mathbf{Q}_V^{(2)}(2;\mathbf{f};\delta\mathbf{f})\right\rangle_2\right\}_{\mathbf{f}^0} \\ &= \left\{a^{(2)}(1;1;t)\delta E(t) - a^{(2)}(2;1;t)\delta a^{(1)}(t)\right\}_{t=0}^{t=\tau} \\ &+ \left\{\left\langle \mathbf{V}^{(2)}(2;t), \mathbf{AM}^{(2)}[2\times 2;\mathbf{f}]\mathbf{A}^{(2)}(2;1;t)\right\rangle_2\right\}_{\mathbf{f}^0}. \end{aligned} \tag{5.70}$$

where the operator $\mathbf{AM}^{(2)}[2\times 2;\mathbf{f}]$ represents the formal adjoint of the operator $\mathbf{VM}^{(2)}(2\times 2;\mathbf{f})$, i.e., $\mathbf{AM}^{(2)}[2\times 2;\mathbf{f}]\triangleq\left[\mathbf{VM}^{(2)}(2\times 2;\mathbf{f})\right]^*$, and is defined as follows:

$$\mathbf{AM}^{(2)}[2\times 2;\mathbf{f}]\triangleq\left[\mathbf{VM}^{(2)}(2\times 2;\mathbf{f})\right]^*\triangleq\begin{pmatrix} -\dfrac{d}{dt}+2f_1(\alpha)E(t) & 2f_1(\alpha)a^{(1)}(t) \\ 0 & \dfrac{d}{dt}+2f_1(\alpha)E(t) \end{pmatrix}. \tag{5.71}$$

2. Eliminate the boundary terms on the right side of the 2nd equality in Eq. (5.70) and require the last term on the right side of the 2nd equality in Eq. (5.70) to represent the right side of Eq. (5.61) by imposing the following relations:

$$\left\{\mathbf{AM}^{(2)}[2\times 2;\mathbf{f}]\mathbf{A}^{(2)}(2;1;t)\right\}_{\mathbf{f}^0}=\left\{\begin{pmatrix} -2a^{(1)}(t)E(t) \\ -E^2(t) \end{pmatrix}\right\}_{\mathbf{f}^0}, \quad 0<t<\tau, \tag{5.72}$$

$$\left\{\mathbf{B}_A^{(2)}\left[2;\mathbf{A}^{(2)}(2;1;t);\alpha\right]\right\}_{\mathbf{f}^0}\triangleq\begin{pmatrix} a^{(2)}(1;1;\tau) \\ a^{(2)}(2;1;0) \end{pmatrix}_{\mathbf{f}^0}=\begin{pmatrix} 0 \\ 0 \end{pmatrix}. \tag{5.73}$$

The relations represented by Eqs. (5.72) and (5.73) constitute the 2nd-LASS for the 2nd-level adjoint function $\mathbf{A}^{(2)}(2;1;t)$. Notably, the 2nd-LASS is independent of variations in the feature functions (and/or parameter variations), so it needs to be solved just once to obtain the 2nd-level adjoint function $\mathbf{A}^{(2)}(2;1;t)$. Furthermore, the 2nd-LASS is an upper-triangular system, so the equations need not be solved simultaneously, but can be solved sequentially, initially for the component $a^{(2)}(2;1;t)$ and subsequently for the component $a^{(2)}(1;1;t)$.

Solving Eqs. (5.72) and (5.73) yields the following closed-form expressions for the components of the 2nd-level adjoint sensitivity function $\mathbf{A}^{(2)}(2;1;t)$:

$$\begin{aligned} a^{(2)}(1;1;t)&=\frac{\int_t^\tau\left[2a^{(1)}(x)E(x)-2f_1a^{(2)}(2;1;x)\right]\cosh^2[xg(\alpha)]dx}{\cosh^2[tg(\alpha)]} \\ &=\cosh^{-2}[tg(\alpha)]\left\{\frac{2}{f_1(\alpha)}\left[1-\frac{\cosh[tg(\alpha)]}{\cosh[\tau g(\alpha)]}\right]\right. \\ &\quad\left.-\frac{2}{3f_1(\alpha)}\left[\frac{1}{2}\cosh^2[\tau g(\alpha)]-\frac{1}{2}\cosh^2[tg(\alpha)]-\ln\frac{\cosh[\tau g(\alpha)]}{\cosh[tg(\alpha)]}\right]\right\}; \end{aligned} \tag{5.74}$$

$$a^{(2)}(2;1;t) = \cosh^2[tg(\alpha)]\int_0^t E^2(x)\cosh^{-2}[xg(\alpha)]dx$$
$$= \frac{1}{3g(\alpha)}\frac{f_2(\alpha)}{f_1(\alpha)}\tanh[tg(\alpha)]\sinh^2[tg(\alpha)]. \tag{5.75}$$

3. Use the relations provided by the 2nd-LVSS and the 2nd-LASS in Eq. (5.70) to obtain the following expression for the variation $\delta\{\partial E(\tau)/\partial f_1\}$ in terms of the 2nd-level adjoint function $\mathbf{A}^{(2)}(2;1;t)$:

$$\delta\left\{\frac{\partial E(\tau)}{\partial f_1}\right\} = \frac{\partial^2 E(\tau;f_1,f_2)}{\partial f_1 \partial f_1}\delta f_1 + \frac{\partial^2 E(\tau;f_1,f_2)}{\partial f_2 \partial f_1}\delta f_2$$
$$= \left\{\left\langle \mathbf{A}^{(2)}(2;1;t), \mathbf{Q}_V^{(2)}(2;\mathbf{f};\delta\mathbf{f})\right\rangle_2\right\}_{\mathbf{f}^0} \tag{5.76}$$
$$= \int_0^\tau dt \begin{Bmatrix} a^{(2)}(1;1;t)\left[-(\delta f_1)E^2(t) + (\delta f_2)\right] \\ -a^{(2)}(2;1;t)2(\delta f_1)a^{(1)}(t)E(t) \end{Bmatrix}.$$

It follows from Eq. (5.76) that:

$$\frac{\partial^2 E(\tau)}{\partial f_1 \partial f_1} = -\int_0^\tau \left[a^{(2)}(1;1;t)E(t) + 2a^{(2)}(2;1;t)a^{(1)}(t)\right]E(t)dt; \tag{5.77}$$

$$\frac{\partial^2 E(\tau)}{\partial f_2 \partial f_1} = \int_0^\tau a^{(2)}(1;1;t)dt. \tag{5.78}$$

#### 5.4.1.2 Computation of 2nd-Order Sensitivities Stemming from the 1st-Order Sensitivity $\partial E(\tau)/\partial f_2$

The 2nd-order sensitivities which stem from the 1st-order sensitivity $\partial E(\tau)/\partial f_2$ defined in Eq. (5.52) will be obtained from the 1st-order G-differential $\delta\{\partial E(\tau)/\partial f_2\}$ of $\partial E(\tau)/\partial f_2$. By definition, the 1st-order G-differential $\delta\{\partial E(\tau)/\partial f_2\}$ is obtained as follows:

$$\delta\left\{\frac{\partial E(\tau;f_1,f_2)}{\partial f_2}\right\} = \left\{\frac{d}{d\varepsilon}\left[\int_0^\tau \left(a^{(1)} + \varepsilon\delta a^{(1)}\right)dt\right]\right\}_{\varepsilon=0}$$
$$= \int_0^\tau \delta a^{(1)}(t)dt = \frac{\partial^2 E(\tau;f_1,f_2)}{\partial f_1 \partial f_2}\delta f_1 + \frac{\partial^2 E(\tau;f_1,f_2)}{\partial f_2 \partial f_2}\delta f_2. \tag{5.79}$$

The variational function $\delta a^{(1)}(t)$ is the solution of Eqs. (5.62) and (5.63). Notably, the right side of Eq. (5.79) depends only on the variational function $\delta a^{(1)}(t)$, but does not depend directly on the variational function $\delta E(t)$. Nevertheless, since the variational function $\delta a^{(1)}(t)$ is related to the variational function $\delta E(t)$ through Eqs. (5.62) and (5.63), the 2nd-level adjoint function that will be constructed in order to eliminate the appearance of $\delta a^{(1)}(t)$ on the right side of Eq. (5.79) will be the solution of a 2nd-LASS which will correspond to the 2nd-LVSS defined by Eqs. (5.64) and (5.65). The construction of the 2nd-LASS that will be used to eliminate the appearance of the variational function $\delta a^{(1)}(t)$ from Eq. (5.79) follows the same steps as in Section 5.4.1.1. The 2nd-level adjoint function that will be defined for this purpose will be denoted as $\mathbf{A}^{(2)}(2;2;t) \triangleq \left[a^{(2)}(1;2;t), a^{(2)}(2;2;t)\right]^{\dagger} \in \mathcal{H}_2$, where the notation has the following significance: (i) the bold letter "A" indicates a vector-valued "adjoint" function within the 2nd-FASAM-N formalism; (ii) the superscript "(2)" indicates "2nd-level;" (iii) the 1st argument, i.e., "2," in $\mathbf{A}^{(2)}(2;2;t)$ indicates that this vector has two components, denoted as $a^{(2)}(i;2;t)$, $i = 1,2$, each of which is a scalar-valued function of time; (iv) the 2nd-argument of $\mathbf{A}^{(2)}(2;2;t)$, denoted as "2," indicates that this 2nd-level adjoint function corresponds to the 1st-order sensitivity $\partial E(\tau)/\partial f_2$ of the response with respect to the "2nd-feature function," i.e., $f_2(\alpha)$.

The inner product defined in Eq. (5.69) is now used to construct the 2nd-Level Adjoint Sensitivity System (2nd-LASS) for the 2nd-level adjoint function $\mathbf{A}^{(2)}(2;2;t)$ by following the same sequence of steps as used in Section 5.4.1.1, but using the expression provided in Eq. (5.79) to determine the right side ("source") for the 2nd-LASS. This procedure leads to the following 2nd-LASS for the 2nd-level adjoint function $\mathbf{A}^{(2)}(2;2;t)$:

$$\left\{\mathbf{AM}^{(2)}[2\times 2;\mathbf{f}]\mathbf{A}^{(2)}(2;2;t)\right\}_{\mathbf{f}^0} = \left\{\begin{pmatrix} 0 \\ 1 \end{pmatrix}\right\}_{\mathbf{f}^0}, \quad 0 < t < \tau, \tag{5.80}$$

$$\left\{\mathbf{B}_A^{(2)}\left[2;\mathbf{A}^{(2)}(2;2;t);\alpha\right]\right\}_{\mathbf{f}^0} \triangleq \left\{\begin{pmatrix} a^{(2)}(1;2;\tau) \\ a^{(2)}(2;2;0) \end{pmatrix}\right\}_{\mathbf{f}^0} = \begin{pmatrix} 0 \\ 0 \end{pmatrix}. \tag{5.81}$$

Solving Eqs. (5.80) and (5.81) yields the following closed-form expressions for the components of the 2nd-level adjoint sensitivity function $\mathbf{A}^{(2)}(2;2;t)$:

$$a^{(2)}(1;2;t) = \frac{1}{2f_2(\alpha)}\left\{-\cosh^2\left[tg(\alpha)\right] + \frac{\cosh^4\left[\tau g(\alpha)\right]}{\cosh^2\left[tg(\alpha)\right]}\right\}; \tag{5.82}$$

$$a^{(2)}(2;2;t) = -\frac{1}{2g(\alpha)}\sinh\left[2tg(\alpha)\right]. \tag{5.83}$$

The expressions in Eqs. (5.82) and (5.83) are to be evaluated at the nominal values of the feature functions (and, implicitly, at the nominal parameter values) but the notation $\{\ \}_{\mathbf{f}^0}$ has been omitted for simplicity.

Using the relations provided by the 2nd-LVSS – comprising Eqs. (5.64) and (5.65) – and the 2nd-LASS – comprising Eqs. (5.80) and (5.81) – leads to the following expression for the variation $\delta\left\{\partial E(\tau)/\partial f_2\right\}$ in terms of the 2nd-level adjoint function $\mathbf{A}^{(2)}(2;2;t)$:

$$\delta\left\{\frac{\partial E(\tau)}{\partial f_2}\right\}=\frac{\partial^2 E(\tau;f_1,f_2)}{\partial f_1\partial f_2}\delta f_1+\frac{\partial^2 E(\tau;f_1,f_2)}{\partial f_2\partial f_2}\delta f_2$$
$$=\left\{\left\langle \mathbf{A}^{(2)}(2;2;t),\mathbf{Q}_V^{(2)}(2;\mathbf{f};\delta\mathbf{f})\right\rangle_2\right\}_{\mathbf{f}^0} \tag{5.84}$$
$$=\int_0^\tau dt\begin{Bmatrix} a^{(2)}(1;2;t)\left[-(\delta f_1)E^2(t)+(\delta f_2)\right] \\ -a^{(2)}(2;2;t)2(\delta f_1)a^{(1)}(t)E(t) \end{Bmatrix}.$$

It follows from Eq. (5.84) that:

$$\frac{\partial^2 E(\tau)}{\partial f_1\partial f_2}=-\int_0^\tau\left[a^{(2)}(1;2;t)E(t)+2a^{(2)}(2;2;t)a^{(1)}(t)\right]E(t)dt; \tag{5.85}$$

$$\frac{\partial^2 E(\tau)}{\partial f_2\partial f_2}=\int_0^\tau a^{(2)}(1;2;t)dt. \tag{5.86}$$

It is important to note that the mixed 2nd-order partial derivative $\partial^2 E(\tau)/\partial f_1\partial f_2$ can be obtained by using either Eq. (5.85) or Eq. (5.78). The equivalence between the two respective expressions provides a stringent verification of the accuracy of solving the two 2nd-LASS, one for $\mathbf{A}^{(2)}(2;1;t)$ comprising Eqs. (5.72) and (5.73), and the other 2nd-LASS comprising Eqs. (5.80) and (5.81) for $\mathbf{A}^{(2)}(2;2;t)$.

The 2nd-order sensitivities of the response $E(\tau)$ with respect to the primary model parameters are obtained by using the parameter-dependencies of the functions $f_1(\boldsymbol{\alpha})$ and $f_2(\boldsymbol{\alpha})$, cf. Eq. (5.38), in conjunction with the expressions obtained in Eqs. (5.77), (5.78), (5.85), and (5.86) by using the following general formula, which is obtained by taking the total differential of the expression provided in Eq. (5.53):

$$\frac{\partial^2 E(\tau;f_1;f_2)}{\partial\alpha_j\partial\alpha_i}=\left[\frac{\partial^2 E(\tau)}{\partial f_1\partial f_1}\frac{\partial f_1(\boldsymbol{\alpha})}{\partial\alpha_j}+\frac{\partial^2 E(\tau)}{\partial f_2\partial f_1}\frac{\partial f_2(\boldsymbol{\alpha})}{\partial\alpha_j}\right]\frac{\partial f_1(\boldsymbol{\alpha})}{\partial\alpha_i}+\frac{\partial E(\tau)}{\partial f_1}\frac{\partial^2 f_1(\boldsymbol{\alpha})}{\partial\alpha_j\partial\alpha_i}$$
$$+\left[\frac{\partial^2 E(\tau)}{\partial f_1\partial f_2}\frac{\partial f_1(\boldsymbol{\alpha})}{\partial\alpha_j}+\frac{\partial^2 E(\tau)}{\partial f_2\partial f_2}\frac{\partial f_2(\boldsymbol{\alpha})}{\partial\alpha_j}\right]\frac{\partial f_2(\boldsymbol{\alpha})}{\partial\alpha_i}+\frac{\partial E(\tau)}{\partial f_2}\frac{\partial^2 f_2(\boldsymbol{\alpha})}{\partial\alpha_j\partial\alpha_i};$$
$$i,j=1,\ldots,7. \tag{5.87}$$

For example, the 2nd-order sensitivities of the response $E(\tau)$ with respect to the parameter $\alpha_T$ are obtained as follows:

$$\frac{\partial^2 E(\tau)}{\partial\alpha_T\partial\alpha_T}=\frac{\partial}{\partial\alpha_T}\left(\frac{\partial E(\tau)}{\partial f_1}\frac{\partial f_1}{\partial\alpha_T}\right)=\frac{\partial^2 E(\tau)}{\partial f_1\partial f_1}\left(\frac{\partial f_1}{\partial\alpha_T}\right)^2=\left(\frac{1}{2l_p c_p}\right)^2\frac{\partial^2 E(\tau)}{\partial f_1\partial f_1}; \tag{5.88}$$

$$\frac{\partial^2 E(\tau)}{\partial c_p \partial \alpha_T} = \frac{\partial}{\partial c_p}\left(\frac{\partial E(\tau)}{\partial f_1}\frac{\partial f_1}{\partial \alpha_T}\right) = \frac{\partial^2 E(\tau)}{\partial f_1 \partial f_1}\frac{\partial f_1}{\partial c_p}\frac{\partial f_1}{\partial \alpha_T} + \frac{\partial E(\tau)}{\partial f_1}\frac{\partial^2 f_1}{\partial c_p \partial \alpha_T}$$
$$= -\frac{\alpha_T}{4(l_p)^2(c_p)^3}\frac{\partial^2 E(\tau)}{\partial f_1 \partial f_1} - \frac{1}{2l_p(c_p)^2}\frac{\partial E(\tau)}{\partial f_1}. \tag{5.89}$$

$$\frac{\partial^2 E(\tau)}{\partial l_p \partial \alpha_T} = \frac{\partial}{\partial l_p}\left(\frac{\partial E(\tau)}{\partial f_1}\frac{\partial f_1}{\partial \alpha_T}\right) = \frac{\partial^2 E(\tau)}{\partial f_1 \partial f_1}\frac{\partial f_1}{\partial l_p}\frac{\partial f_1}{\partial \alpha_T} + \frac{\partial E(\tau)}{\partial f_1}\frac{\partial^2 f_1}{\partial l_p \partial \alpha_T}$$
$$= -\frac{\alpha_T}{4(c_p)^2(l_p)^3}\frac{\partial^2 E(\tau)}{\partial f_1 \partial f_1} - \frac{1}{2c_p(l_p)^2}\frac{\partial E(\tau)}{\partial f_1}. \tag{5.90}$$

$$\frac{\partial^2 E(\tau)}{\partial \varphi_0 \partial \alpha_T} = \frac{\partial}{\partial \varphi_0}\left(\frac{\partial E(\tau)}{\partial f_1}\frac{\partial f_1}{\partial \alpha_T}\right) = \frac{1}{2l_p c_p}\frac{\partial^2 E(\tau)}{\partial f_1 \partial f_2}\frac{\partial f_2}{\partial \varphi_0} = \frac{\gamma \sigma_f N_f}{2l_p c_p}\frac{\partial^2 E(\tau)}{\partial f_1 \partial f_2}; \tag{5.91}$$

$$\frac{\partial^2 E(t_f)}{\partial \gamma \partial \alpha_T} = \frac{\partial}{\partial \gamma}\left(\frac{\partial E(t_f)}{\partial f_1}\frac{\partial f_1}{\partial \alpha_T}\right) = \frac{1}{2l_p c_p}\frac{\partial^2 E(t_f)}{\partial f_1 \partial f_2}\frac{\partial f_2}{\partial \gamma} = \frac{\varphi_0 \sigma_f N_f}{2l_p c_p}\frac{\partial^2 E(t_f)}{\partial f_1 \partial f_2}; \tag{5.92}$$

$$\frac{\partial^2 E(\tau)}{\partial \sigma_f \partial \alpha_T} = \frac{\partial}{\partial \sigma_f}\left(\frac{\partial E(\tau)}{\partial f_1}\frac{\partial f_1}{\partial \alpha_T}\right) = \frac{1}{2l_p c_p}\frac{\partial^2 E(\tau)}{\partial f_1 \partial f_2}\frac{\partial f_2}{\partial \sigma_f} = \frac{\gamma \varphi_0 N_f}{2l_p c_p}\frac{\partial^2 E(\tau)}{\partial f_1 \partial f_2}; \tag{5.93}$$

$$\frac{\partial^2 E(\tau)}{\partial N_f \partial \alpha_T} = \frac{\partial}{\partial N_f}\left(\frac{\partial E(\tau)}{\partial f_1}\frac{\partial f_1}{\partial \alpha_T}\right) = \frac{1}{2l_p c_p}\frac{\partial^2 E(\tau)}{\partial f_1 \partial f_2}\frac{\partial f_2}{\partial N_f} = \frac{\gamma \varphi_0 \sigma_f}{2l_p c_p}\frac{\partial^2 E(\tau)}{\partial f_1 \partial f_2}. \tag{5.94}$$

Because of symmetry, the mixed 2nd-order sensitivities $\partial^2 E(\tau)/\partial f_1 \partial f_2$ or $\partial^2 E(\tau)/\partial f_2 \partial f_1$ can be obtained by using distinct but equivalent expressions in terms of the 2nd-level adjoint functions $\mathbf{A}^{(2)}(2;1;t)$ and $\mathbf{A}^{(2)}(2;2;t)$, since the expressions obtained in Eqs. (5.78) and (5.85) represent the same quantity, because $\partial^2 E(\tau)/\partial f_1 \partial f_2 = \partial^2 E(\tau)/\partial f_2 \partial f_1$ by definition. Notably, only two "large-scale computations" are necessary for solving the two distinct 2nd-LASS for obtaining the 2nd-level adjoint functions $\mathbf{A}^{(2)}(2;1;t)$ and $\mathbf{A}^{(2)}(2;2;t)$ involved in the computation of the three distinct 2nd-order response sensitivities, i.e., $\partial^2 E(\tau)/\partial f_1 \partial f_1$, $\partial^2 E(\tau)/\partial f_1 \partial f_2$ and $\partial^2 E(\tau)/\partial f_2 \partial f_2$, with respect to the two feature functions $f_1$ and $f_2$. The subsequent use of Eq. (5.87) to obtain the 49 2nd-order sensitivities $\partial^2 E(\tau)/\partial \alpha_i \partial \alpha_j$ of the response with respect to the primary model parameters involves only inexpensive differentiations that are performed exactly, analytically, since the exact dependence of the feature functions on the model parameters is explicitly known. For verification purposes, all of the mixed 2nd-order sensitivities $\partial^2 E(\tau)/\partial \alpha_i \partial \alpha_j$, $i,j = 1,\ldots,7$, with respect to the model parameters can be computed twice, using distinct expressions in terms of the 2nd-level adjoint functions $\mathbf{A}^{(2)}(2;1;t)$ and $\mathbf{A}^{(2)}(2;2;t)$.

### 5.4.2 Computation of 2nd-Order Sensitivities Using the 2nd-CASAM-N

The conventional 2nd-CASAM-N methodology applies the same fundamental principle (namely, that the 2nd-order sensitivities are the "1st-order sensitivities of the 1st-order sensitivities") as the 2nd-FASAM. However, this principle is applied within the 2nd-CASAM-N methodology directly to the 1st-order sensitivities with respect to the primary model parameters, as opposed to applying this principle to the 1st-order response sensitivities with respect to the feature functions, as implemented within the application of the 2nd-FASAM-N methodology. This subsection illustrates the application of the conventional 2nd-CASAM-N methodology to obtain the 2nd-order response sensitivities from the seven 1st-order sensitivities with respect to the underlying model parameters (as obtained in Section 5.3.1). It will be shown in this subsection that the application of the 2nd-CASAM-N methodology will require solving *seven* distinct 2nd-Level Adjoint Sensitivity Systems (2nd-LASS), each system comprising a distinct source term which corresponds to one of the seven distinct 1st-order sensitivities, in order to obtain all of the 2nd-order sensitivities. In contradistinction, the application of the *2nd-FASAM-N* requires solving only *two* 2nd-LASS, as has been shown in Section 5.3.

#### 5.4.2.1 Computation of 2nd-Order Sensitivities Stemming from the 1st-Order Sensitivity $\partial E(\tau)/\partial\alpha_T$

The 2nd-order sensitivities which stem from the 1st-order sensitivity $\partial E(\tau)/\partial\alpha_T$ are the components of the 1st-order G-differential of Eq. (5.31). By definition, the 1st-order G-differential of Eq. (5.31) is obtained as follows:

$$\delta\left\{\partial E(\tau)/\partial\alpha_T\right\}_{\alpha^0} \triangleq \left\{\delta\left[\partial E(\tau)/\partial\alpha_T\right]\right\}_{dir} + \left\{\delta\left[\partial E(\tau)/\partial\alpha_T\right]\right\}_{ind}$$

$$\triangleq -\left\{\frac{d}{d\varepsilon}\left[\frac{1}{2\left(l_p+\varepsilon\delta l_p\right)\left(c_p+\varepsilon\delta c_p\right)}\int_0^\tau \left(a^{(1)}+\varepsilon\delta a^{(1)}\right)\left[E(t)+\varepsilon\delta E(t)\right]^2 dt\right]_{\alpha^0}\right\}_{\varepsilon=0}, \tag{5.95}$$

where the "direct-effect" term $\left\{\delta\left[\partial E(\tau)/\partial\alpha_T\right]\right\}_{dir}$ can be determined immediately and is defined as follows:

$$\left\{\delta\left[\partial E(\tau)/\partial\alpha_T\right]\right\}_{dir} \triangleq \left\{\left[\frac{\delta l_p}{2\left(l_p\right)^2 c_p}+\frac{\delta c_p}{2l_p\left(c_p\right)^2}\right]\int_0^\tau a^{(1)}(t)E^2(t)\,dt\right\}_{\alpha^0}, \tag{5.96}$$

and where the "indirect-effect" term $\left\{\delta\left[\partial E(\tau)/\partial\alpha_T\right]\right\}_{ind}$ is defined as follows:

$$\left\{\delta\left[\partial E(\tau)/\partial\alpha_T\right]\right\}_{ind} \triangleq -\left\{\frac{1}{2l_p c_p}\int_0^\tau \left[\delta a^{(1)}(t)E^2(t)+2a^{(1)}(t)E(t)\delta E(t)\right]dt\right\}_{\alpha^0}. \tag{5.97}$$

The variational function $\delta E(t)$ in Eq. (5.97) is the solution of the 1st-LVSS provided in Eqs. (5.22) and (5.23). The variational function $\delta a^{(1)}(t)$ in Eq. (5.97) is the solution of Eqs. (5.62) and (5.63), but written in terms of the primary model parameters and variations thereof, as follows:

$$\left\{\left[-\frac{d}{dt}+\frac{\alpha_T}{l_p c_p}E(t)\right]\delta a^{(1)}(t)+\frac{\alpha_T}{l_p c_p}[\delta E(t)]\right\}_{\alpha^0}$$
$$=\left\{\left[\frac{\delta\alpha_T}{l_p c_p}-\frac{\alpha_T}{l_p (c_p)^2}\delta c_p-\frac{\alpha_T}{(l_p)^2 c_p}\delta l_p\right]a^{(1)}(t)E(t)\right\}_{\alpha^0}, \quad 0<t<\tau, \tag{5.98}$$

$$\delta a^{(1)}(\tau)=0, \quad t=\tau. \tag{5.99}$$

Concatenating Eqs. (5.98) and (5.99) with the 1st-LVSS for $\delta E(t)$ represented by Eqs. (5.22) and (5.23) yields the following 2nd-LVSS for the 2nd-level variational function $\mathbf{V}^{(2)}(2;t)\triangleq\left[v^{(2)}(1;t),v^{(2)}(2;t)\right]^{\dagger}\triangleq\left[\delta E(t),\delta a^{(1)}(t)\right]^{\dagger}$:

$$\left\{\mathbf{VM}^{(2)}[2\times 2;\boldsymbol{\alpha}]\mathbf{V}^{(2)}(2;t)\right\}_{\alpha^0}=\left\{\mathbf{S}_V^{(2)}[2;\boldsymbol{\alpha};\delta\boldsymbol{\alpha}]\right\}_{\alpha^0}, \quad 0<t<\tau, \tag{5.100}$$

$$\left\{\mathbf{B}_V^{(2)}\left[2;\mathbf{V}^{(2)}(2;t);\boldsymbol{\alpha};\delta\boldsymbol{\alpha}\right]\right\}_{\alpha^0}=\mathbf{0}[2], \quad \mathbf{0}[2]\triangleq[0,0]^{\dagger}, \tag{5.101}$$

where

$$\mathbf{VM}^{(2)}[2\times 2;\boldsymbol{\alpha}]\triangleq\begin{pmatrix}\frac{d}{dt}+\frac{\alpha_T}{l_p c_p}E(t) & 0\\ \frac{\alpha_T}{l_p c_p} & -\frac{d}{dt}+\frac{\alpha_T}{l_p c_{p\,1}}E(t)\end{pmatrix}; \tag{5.102}$$

$$\mathbf{S}_V^{(2)}[2;\boldsymbol{\alpha};\delta\boldsymbol{\alpha}]\triangleq\begin{bmatrix}s_V^{(2)}(1;\boldsymbol{\alpha};\delta\boldsymbol{\alpha})\\ s_V^{(2)}(2;\boldsymbol{\alpha};\delta\boldsymbol{\alpha})\end{bmatrix}; \tag{5.103}$$

$$s_V^{(2)}(1;\boldsymbol{\alpha};\delta\boldsymbol{\alpha})\triangleq\left\{-\frac{\delta\alpha_T}{2l_p c_p}+\frac{\alpha_T}{2(l_p)^2 c_p}\delta l_p+\frac{\alpha_T}{2l_p (c_p)^2}\delta c_p\right\}_{\alpha^0}E^2(t) \tag{5.104}$$
$$+\left\{\gamma\sigma_f N_f(\delta\varphi_0)+\varphi_0\sigma_f N_f(\delta\gamma)+\varphi_0\gamma N_f(\delta\sigma_f)+\varphi_0\gamma\sigma_f(\delta N_f)\right\}_{\alpha^0};$$

$$s_V^{(2)}(2;\boldsymbol{\alpha};\delta\boldsymbol{\alpha}) \triangleq \left\{ -\frac{\delta\alpha_T}{l_p c_p} + \frac{\alpha_T}{(l_p)^2 c_p}\delta l_p + \frac{\alpha_T}{l_p (c_p)^2}\delta c_p \right\}_{\boldsymbol{\alpha}^0} a^{(1)}(t)E(t); \quad (5.105)$$

$$\mathbf{B}_V^{(2)}\left[2;\mathbf{V}^{(2)}(2;t);\boldsymbol{\alpha};\delta\boldsymbol{\alpha}\right] \triangleq \begin{pmatrix} \delta E(0) \\ \delta a^{(1)}(\tau) \end{pmatrix}. \quad (5.106)$$

Except for the distinct notation, the 2nd-LVSS defined by Eqs. (5.100) and (5.101) is identical with the 2nd-LVSS defined by Eqs. (5.64) and (5.65), which is the reason for having used for both systems of equations the same notation for the respective 2nd-level variational function, namely, $\mathbf{V}^{(2)}(2;t) \triangleq \left[v^{(2)}(1;t), v^{(2)}(2;t)\right]^{\dagger} \triangleq \left[\delta E(t), \delta a^{(1)}(t)\right]^{\dagger}$, which is the solution of this 2nd-LVSS. However, the dependence on the components of the vector of parameters, $\boldsymbol{\alpha} \triangleq \left(\gamma, \sigma_f, N_f, \varphi_0, l_p, \alpha_T, c_p\right)^{\dagger}$, is emphasized in Eqs. (5.100) and (5.101), because this explicit dependence is necessary to distinguish the developments of the 2nd-Level Adjoint Sensitivity Systems to follow, which will be distinct from each other depending on the specific expression of each of the seven 1st-order sensitivities of the response with respect to the primary model parameters.

As discussed in Section 5.4.1, the computationally expensive path of solving the 2nd-LVSS repeatedly for every possible parameter variation will be avoided by replacing the variational function $\mathbf{V}^{(2)}(2;t)$ in the expression of the "indirect-effect" term defined in Eq. (5.97) by a corresponding 2nd-level adjoint function, which will be denoted as $\mathbf{C}^{(2)}(2;1;t) \triangleq \left[c^{(2)}(1;1;t), c^{(2)}(2;1;t)\right]^{\dagger} \in \mathcal{H}_2$. This vector-valued function will be the solution of a 2nd-Level Adjoint Sensitivity System (2nd-LASS) to be constructed by applying the 2nd-CASAM-N. The notation used for $\mathbf{C}^{(2)}(2;1;t) \triangleq \left[c^{(2)}(1;1;t), c^{(2)}(2;1;t)\right]^{\dagger} \in \mathcal{H}_2$ has the following significance: (i) the bold letter "C" indicates a vector-valued "adjoint" function within the 2nd-CASAM-N formalism; (ii) the superscript "(2)" indicates "2nd-level;" (iii) the 1st argument, namely "2," in $\mathbf{C}^{(2)}(2;1;t)$ indicates that this vector has two components, denoted as $c^{(2)}(i;2;t)$, $i = 1,2$, each of which is a scalar-valued function of time; (iv) the 2nd argument of $\mathbf{C}^{(2)}(2;1;t)$ is denoted as "1" and indicates that this 2nd-level adjoint function corresponds to the 1st-order sensitivity $\partial E(\tau)/\partial \alpha_T$ of the response with respect to the *1st* component of the vector of model parameters $\boldsymbol{\alpha} \triangleq (\alpha_1, \ldots, \alpha_7)^{\dagger} \triangleq \left(\alpha_T, l_p, c_p, \varphi_0, \gamma, \sigma_f, N_f\right)^{\dagger}$, namely, $\alpha_T$.

The 2nd-LASS for the function $\mathbf{C}^{(2)}(2;1;t) \triangleq \left[c^{(2)}(1;1;t), c^{(2)}(2;1;t)\right]^{\dagger} \in \mathcal{H}_2$ is constructed in the same Hilbert space which was denoted as $\mathcal{H}_2$ in the previous subsection, and which is endowed with the inner product defined in Eq. (5.69). This inner product is used to construct the 2nd-Level Adjoint Sensitivity System (2nd-LASS) for the 2nd-level adjoint function $\mathbf{C}^{(2)}(2;1;t) \triangleq \left[c^{(2)}(1;1;t), c^{(2)}(2;1;t)\right]^{\dagger} \in \mathcal{H}_2$, as follows:

1. Using Eq. (5.69), form the inner product of $\mathbf{C}^{(2)}(2;1;t)$ with Eq. (5.100) to obtain the following relation which has the same form as shown in Eq. (5.70), namely:

$$\begin{aligned}&\left\{\left\langle \mathbf{C}^{(2)}(2;1;t),\mathbf{VM}^{(2)}[2\times 2;\alpha]\mathbf{V}^{(2)}(2;t)\right\rangle_2\right\}_{\alpha^0}\\&=\left\{\left\langle \mathbf{C}^{(2)}(2;1;t),\mathbf{S}_V^{(2)}(2;\alpha;\delta\alpha)\right\rangle_2\right\}_{\alpha^0}\\&=\left\{c^{(2)}(1;1;t)\delta E(t)-c^{(2)}(2;1;t)\delta a^{(1)}(t)\right\}_{t=0}^{t=\tau}\\&+\left\{\left\langle \mathbf{V}^{(2)}(2;t),\mathbf{AM}^{(2)}[2\times 2;\alpha]\mathbf{C}^{(2)}(2;1;t)\right\rangle_2\right\}_{\alpha^0},\end{aligned}\tag{5.107}$$

where the adjoint operator $\mathbf{AM}^{(2)}[2\times 2;\alpha]$ is the same as defined in Eq. (5.71).

2. Eliminate the boundary terms on the right side of Eq. (5.107) and require the 2nd term on the right side of the 2nd equality in Eq. (5.107) to represent the "indirect-effect" term defined in Eq. (5.97) by imposing the following relations:

$$\left\{\mathbf{AM}^{(2)}[2\times 2;\alpha]\mathbf{C}^{(2)}(2;1;t)\right\}_{\alpha^0}=\begin{pmatrix}-\dfrac{a^{(1)}(t)E(t)}{l_p c_p}\\ -\dfrac{E^2(t)}{2l_p c_p}\end{pmatrix},\quad 0<t<\tau,\tag{5.108}$$

$$\left\{\mathbf{B}_A^{(2)}\left[2;\mathbf{C}^{(2)}(2;1;t);\alpha\right]\right\}_{\alpha^0}\triangleq\begin{pmatrix}c^{(2)}(1;1;\tau)\\ c^{(2)}(2;1;0)\end{pmatrix}_{\alpha^0}=\begin{pmatrix}0\\0\end{pmatrix}.\tag{5.109}$$

The relations represented by Eqs. (5.108) and (5.109) constitute the 2nd-LASS for the 2nd-level adjoint function $\mathbf{C}^{(2)}(2;1;t)\triangleq\left[c^{(2)}(1;1;t),c^{(2)}(2;1;t)\right]^{\dagger}\in\mathcal{H}_2$.

Inserting the equations underlying the 2nd-LVSS, i.e., Eqs. (5.100) and (5.101), together with those underlying the 2nd-LASS, i.e., Eqs. (5.108) and (5.109), into Eq. (5.107) and recalling Eq. (5.97) yields the following expression for the "indirect-effect" term as a function of $\mathbf{C}^{(2)}(2;1;t)$:

$$\left\{\delta\left[\partial E(\tau)/\partial\alpha_T\right]\right\}_{ind}=\left\{\left\langle \mathbf{C}^{(2)}(2;1;t),\mathbf{S}_V^{(2)}(2;\alpha;\delta\alpha)\right\rangle_2\right\}_{\alpha^0}.\tag{5.110}$$

Inserting the result for the indirect-effect term obtained in Eq. (5.110) together with the expression for the direct-effect term shown in Eq. (5.96) into Eq. (5.95) yields the following expression for the 1st-order G-differential $\delta\left\{\partial E(\tau)/\partial\alpha_T\right\}_{\alpha^0}$:

$$\begin{aligned}\delta\left\{\partial E(\tau)/\partial\alpha_T\right\}_{\alpha^0}&=\left\{\left[\frac{\delta l_p}{2(l_p)^2 c_p}+\frac{\delta c_p}{2l_p(c_p)^2}\right]\int_0^{\tau}a^{(1)}(t)E^2(t)\,dt\right\}_{\alpha^0}\\&+\left\{\left\langle \mathbf{C}^{(2)}(2;1;t),\mathbf{S}_V^{(2)}(2;\alpha;\delta\alpha)\right\rangle_2\right\}_{\alpha^0}=\left\{\sum_{i=1}^{7}\left[\partial^2 E(\tau)/\partial\alpha_i\,\partial\alpha_T\right]\delta\alpha_i\right\}_{\alpha^0}.\end{aligned}\tag{5.111}$$

Inserting into Eq. (5.111) the expressions provided in Eq. (5.104) and (5.105) for the respective components of the source $\mathbf{S}_V^{(2)}(2;\alpha;\delta\alpha)$ and collecting the terms that multiply the respective parameter variations yields the following expressions for the 2nd-order partial sensitivities that stem from $\delta\{\partial E(\tau)/\partial\alpha_T\}_{\alpha^0}$:

$$\frac{\partial^2 E(\tau)}{\partial\alpha_T\,\partial\alpha_T} = -\frac{1}{2l_p c_p}\int_0^\tau \left[c^{(2)}(1;1;t)E^2(t)+2c^{(2)}(2;1;t)a^{(1)}(t)E(t)\right]dt; \quad (5.112)$$

$$\frac{\partial^2 E(\tau)}{\partial l_p\,\partial\alpha_T} = \frac{1}{2(l_p)^2 c_p}\int_0^\tau a^{(1)}(t)E^2(t)\,dt$$
$$+\frac{\alpha_T}{2(l_p)^2 c_p}\int_0^\tau \left[c^{(2)}(1;1;t)E^2(t)+2c^{(2)}(2;1;t)a^{(1)}(t)E(t)\right]dt; \quad (5.113)$$

$$\frac{\partial^2 E(\tau)}{\partial c_p\,\partial\alpha_T} = \frac{1}{2(c_p)^2 l_p}\int_0^\tau a^{(1)}(t)E^2(t)\,dt$$
$$+\frac{\alpha_T}{2(c_p)^2 l_p}\int_0^\tau \left[c^{(2)}(1;1;t)E^2(t)+2c^{(2)}(2;1;t)a^{(1)}(t)E(t)\right]dt; \quad (5.114)$$

$$\frac{\partial^2 E(\tau)}{\partial\varphi_0\,\partial\alpha_T} = \gamma\sigma_f N_f\int_0^\tau c^{(2)}(1;1;t)\,dt; \quad (5.115)$$

$$\frac{\partial^2 E(\tau)}{\partial\gamma\,\partial\alpha_T} = \varphi_0\sigma_f N_f\int_0^\tau c^{(2)}(1;1;t)\,dt; \quad (5.116)$$

$$\frac{\partial^2 E(\tau)}{\partial\sigma_f\,\partial\alpha_T} = \varphi_0\gamma N_f\int_0^\tau c^{(2)}(1;1;t)\,dt; \quad (5.117)$$

$$\frac{\partial^2 E(\tau)}{\partial N_f\,\partial\alpha_T} = \varphi_0\gamma\sigma_f\int_0^\tau c^{(2)}(1;1;t)\,dt. \quad (5.118)$$

#### 5.4.2.2 Computation of 2nd-Order Sensitivities Stemming from the 1st-Order Sensitivity $\partial E(\tau)/\partial l_p$

The 2nd-order sensitivities which stem from the 1st-order sensitivity $\partial E(\tau)/\partial l_p$ are the components of the 1st-order G-differential of Eq. (5.32). By definition, the 1st-order G-differential of Eq. (5.32) is obtained as follows:

$$\delta\{\partial E(\tau)/\partial l_p\}_{\alpha^0} \triangleq \{\delta[\partial E(\tau)/\partial l_p]\}_{dir} + \{\delta[\partial E(\tau)/\partial l_p]\}_{ind}$$
$$\triangleq \left\{\frac{d}{d\varepsilon}\left[\frac{\alpha_T+\varepsilon\delta\alpha_T}{2(l_p+\varepsilon\delta l_p)^2(c_p+\varepsilon\delta c_p)}\int_0^\tau \left(a^{(1)}+\varepsilon\delta a^{(1)}\right)\left[E(t)+\varepsilon\delta E(t)\right]^2 dt\right]_{\alpha^0}\right\}_{\varepsilon=0}, \quad (5.119)$$

where the "direct-effect" term $\left\{\delta\left[\partial E(\tau)/\partial l_p\right]\right\}_{dir}$ can be determined immediately and is defined as follows:

$$\left\{\delta\left[\partial E(\tau)/\partial l_p\right]\right\}_{dir} \triangleq \left\{\left[\frac{\delta\alpha_T}{2(l_p)^2 c_p} - \frac{\alpha_T \delta l_p}{(l_p)^3 c_p} - \frac{\alpha_T \delta c_p}{2(l_p c_p)^2}\right]\int_0^\tau a^{(1)}(t)E^2(t)\,dt\right\}_{\alpha^0}, \tag{5.120}$$

and where the "indirect-effect" term $\left\{\delta\left[\partial E(\tau)/\partial l_p\right]\right\}_{ind}$ is defined as follows:

$$\left\{\delta\left[\partial E(\tau)/\partial l_p\right]\right\}_{ind} \triangleq \left\{\frac{\alpha_T}{2(l_p)^2 c_p}\int_0^\tau \left[\delta a^{(1)}(t)E^2(t) + 2a^{(1)}(t)E(t)\delta E(t)\right]dt\right\}_{\alpha^0}. \tag{5.121}$$

Just as in Section 5.4.2.1, the 2nd-level variational function $\mathbf{V}^{(2)}(2;t) \triangleq \left[\delta E(t), \delta a^{(1)}(t)\right]^\dagger$, which is needed to evaluate the "indirect-effect" term $\left\{\delta\left[\partial E(\tau)/\partial l_p\right]\right\}_{ind}$, is the solution of the 2nd-LVSS defined by Eqs. (5.100) and (5.101). The computationally expensive path of solving the 2nd-LVSS repeatedly for every possible parameter variation is avoided by replacing the dependence of the "indirect-effect" term defined in Eq. (5.121) on the variational function $\mathbf{V}^{(2)}(2;t)$ by a dependence on a corresponding 2nd-level adjoint function, which will be denoted as $\mathbf{C}^{(2)}(2;2;t) \triangleq \left[c^{(2)}(1;2;t), c^{(2)}(2;2;t)\right]^\dagger \in \mathcal{H}_2$, where the notation has the following significance: (i) the bold letter "C" indicates a vector-valued "adjoint" function within the 2nd-CASAM-N formalism; (ii) the superscript "(2)" indicates "2nd-level;" (iii) the 1st argument, namely, "2," in $\mathbf{C}^{(2)}(2;2;t)$ indicates that this vector has two components, denoted as $c^{(2)}(i;2;t)$, $i = 1,2$, each of which is a scalar-valued function of time; (iv) the 2nd argument of $\mathbf{C}^{(2)}(2;2;t)$ is denoted as "2" and indicates that this 2nd-level adjoint function corresponds to the 1st-order sensitivity $\partial E(\tau)/\partial l_p$ of the response with respect to the *2nd* component of the vector of model parameters $\boldsymbol{\alpha} \triangleq (\alpha_1,\ldots,\alpha_7)^\dagger \triangleq \left(\alpha_T, l_p, c_p, \varphi_0, \gamma, \sigma_f, N_f\right)^\dagger$, namely, $l_p$.

The 2nd-LASS for the function $\mathbf{C}^{(2)}(2;2;t) \triangleq \left[c^{(2)}(1;2;t), c^{(2)}(2;2;t)\right]^\dagger \in \mathcal{H}_2$ is constructed by following the same procedure as in Section 5.4.1.1, except for the source term (i.e., right side) of the 2nd-LASS; this source-term now corresponds to the "indirect-effect" term $\left\{\delta\left[\partial E(\tau)/\partial l_p\right]\right\}_{ind}$ defined in Eq. (5.121). This procedure leads to the following 2nd-LASS for $\mathbf{C}^{(2)}(2;2;t)$:

$$\left\{\mathbf{AM}^{(2)}[2\times 2;\boldsymbol{\alpha}]\mathbf{C}^{(2)}(2;2;t)\right\}_{\alpha^0} = \begin{pmatrix} \dfrac{\alpha_T}{(l_p)^2 c_p} a^{(1)}(t)E(t) \\ \dfrac{\alpha_T}{2(l_p)^2 c_p} E^2(t) \end{pmatrix}, \quad 0 < t < \tau, \tag{5.122}$$

$$\left\{\mathbf{B}_A^{(2)}\left[2;\mathbf{C}^{(2)}(2;2;t);\boldsymbol{\alpha}\right]\right\}_{\alpha^0} \triangleq \left(\begin{array}{c} c^{(2)}(1;2;\tau) \\ c^{(2)}(2;2;0) \end{array}\right)_{\alpha^0} = \left(\begin{array}{c} 0 \\ 0 \end{array}\right). \tag{5.123}$$

Furthermore, the 2nd-order partial sensitivities that are obtained in terms of the components of $\mathbf{C}^{(2)}(2;2;t)$ have expressions that are formally similar to those obtained in Section 5.4.1.1, except for the contributions stemming from the direct effect term defined in Eq. (5.120). Omitting these repetitive derivations, the final expressions for the 2nd-order partial sensitivities that stem from $\partial E(\tau)/\partial l_p$ are as follows:

$$\begin{aligned} \frac{\partial^2 E(\tau)}{\partial \alpha_T \partial l_p} &= \frac{1}{2(l_p)^2 c_p}\int_0^\tau a^{(1)}(t)E^2(t)\,dt \\ &- \frac{1}{2l_p c_p}\int_0^\tau \left[c^{(2)}(1;2;t)E^2(t) + 2c^{(2)}(2;2;t)a^{(1)}(t)E(t)\right]dt; \end{aligned} \tag{5.124}$$

$$\begin{aligned} \frac{\partial^2 E(\tau)}{\partial l_p \partial l_p} &= -\frac{\alpha_T}{(l_p)^3 c_p}\int_0^\tau a^{(1)}(t)E^2(t)\,dt \\ &+ \frac{\alpha_T}{2(l_p)^2 c_p}\int_0^\tau \left[c^{(2)}(1;2;t)E^2(t) + 2c^{(2)}(2;2;t)a^{(1)}(t)E(t)\right]dt; \end{aligned} \tag{5.125}$$

$$\begin{aligned} \frac{\partial^2 E(\tau)}{\partial c_p \partial l_p} &= -\frac{\alpha_T}{2(l_p c_p)^2}\int_0^\tau a^{(1)}(t)E^2(t)\,dt \\ &+ \frac{\alpha_T}{2(c_p)^2 l_p}\int_0^\tau \left[c^{(2)}(1;2;t)E^2(t) + 2c^{(2)}(2;2;t)a^{(1)}(t)E(t)\right]dt; \end{aligned} \tag{5.126}$$

$$\frac{\partial^2 E(\tau)}{\partial \varphi_0 \partial l_p} = \gamma\sigma_f N_f \int_0^\tau c^{(2)}(1;2;t)\,dt; \tag{5.127}$$

$$\frac{\partial^2 E(\tau)}{\partial \gamma \partial l_p} = \varphi_0\sigma_f N_f \int_0^\tau c^{(2)}(1;2;t)\,dt; \tag{5.128}$$

$$\frac{\partial^2 E(\tau)}{\partial \sigma_f \partial l_p} = \varphi_0\gamma N_f \int_0^\tau c^{(2)}(1;2;t)\,dt; \tag{5.129}$$

$$\frac{\partial^2 E(\tau)}{\partial N_f \partial l_p} = \varphi_0\gamma\sigma_f \int_0^\tau c^{(2)}(1;2;t)\,dt. \tag{5.130}$$

#### 5.4.2.3 Computation of 2nd-Order Sensitivities Stemming from the 1st-Order Sensitivity $\partial E(\tau)/\partial c_p$

The 2nd-order sensitivities which stem from the 1st-order sensitivity $\partial E(\tau)/\partial c_p$ are the components of the 1st-order G-differential of Eq. (5.33), which has by definition the following expression:

$$\delta\left\{\partial E(\tau)/\partial c_p\right\}_{\alpha^0} \triangleq \left\{\delta\left[\partial E(\tau)/\partial c_p\right]\right\}_{dir} + \left\{\delta\left[\partial E(\tau)/\partial c_p\right]\right\}_{ind}$$

$$\triangleq \left\{\frac{d}{d\varepsilon}\left[\frac{\alpha_T + \varepsilon\delta\alpha_T}{2\left(l_p+\varepsilon\delta l_p\right)\left(c_p+\varepsilon\delta c_p\right)^2}\int_0^\tau \left(a^{(1)}+\varepsilon\delta a^{(1)}\right)\left[E(t)+\varepsilon\delta E(t)\right]^2 dt\right]_{\alpha^0}\right\}_{\varepsilon=0}, \tag{5.131}$$

where the "direct-effect" term $\left\{\delta\left[\partial E(\tau)/\partial c_p\right]\right\}_{dir}$ can be determined immediately and is defined as follows:

$$\left\{\delta\left[\partial E(\tau)/\partial c_p\right]\right\}_{dir} \triangleq \left\{\left[\frac{\delta\alpha_T}{2l_p\left(c_p\right)^2} - \frac{\alpha_T\delta l_p}{2\left(l_p c_p\right)^2} - \frac{\alpha_T\delta c_p}{l_p\left(c_p\right)^3}\right]\int_0^\tau a^{(1)}(t)E^2(t)\,dt\right\}_{\alpha^0}, \tag{5.132}$$

and where the "indirect-effect" term $\left\{\delta\left[\partial E(\tau)/\partial c_p\right]\right\}_{ind}$ is defined as follows:

$$\left\{\delta\left[\partial E(\tau)/\partial c_p\right]\right\}_{ind} \triangleq \left\{\frac{\alpha_T}{2l_p\left(c_p\right)^2}\int_0^\tau\left[\delta a^{(1)}(t)E^2(t)+2a^{(1)}(t)E(t)\delta E(t)\right]dt\right\}_{\alpha^0}. \tag{5.133}$$

Just as in the previous subsections of Section 5.4.2, the 2nd-level variational function $\mathbf{V}^{(2)}(2;t) \triangleq \left[\delta E(t), \delta a^{(1)}(t)\right]^\dagger$, which is needed to evaluate the "indirect-effect" term $\left\{\delta\left[\partial E(\tau)/\partial c_p\right]\right\}_{ind}$, is the solution of the 2nd-LVSS defined by Eqs. (5.100) and (5.101). The computationally expensive path of solving the 2nd-LVSS repeatedly for every possible parameter variation is avoided by replacing the dependence of the "indirect-effect" term defined in Eq. (5.133) on the variational function $\mathbf{V}^{(2)}(2;t)$ by a dependence on a corresponding 2nd-level adjoint function, which will be denoted as $\mathbf{C}^{(2)}(2;3;t) \triangleq \left[c^{(2)}(1;3;t), c^{(2)}(2;3;t)\right]^\dagger \in \mathcal{H}_2$, where the notation is as in the previous subsections, except that the 2nd argument of $\mathbf{C}^{(2)}(2;3;t)$ is denoted as "3" to indicate that this 2nd-level adjoint function corresponds to the 1st-order sensitivity $\partial E(\tau)/\partial c_p$ of the response with respect to the *3rd* component of the vector of model parameters $\boldsymbol{\alpha} \triangleq (\alpha_1,\ldots,\alpha_7)^\dagger \triangleq \left(\alpha_T, l_p, c_p, \varphi_0, \gamma, \sigma_f, N_f\right)^\dagger$, namely, $c_p$.

The 2nd-LASS for the function $\mathbf{C}^{(2)}(2;3;t) \triangleq \left[c^{(2)}(1;3;t), c^{(2)}(2;3;t)\right]^\dagger \in \mathcal{H}_2$ is constructed by following the same procedure as in the previous subsections, except for the source term (i.e., right side) of the 2nd-LASS which corresponds to the "indirect-effect" term $\left\{\delta\left[\partial E(\tau)/\partial c_p\right]\right\}_{ind}$ defined in Eq. (5.133). This procedure leads to the following 2nd-LASS for $\mathbf{C}^{(2)}(2;3;t)$:

$$\left\{\mathbf{AM}^{(2)}[2\times 2;\boldsymbol{\alpha}]\mathbf{C}^{(2)}(2;3;t)\right\}_{\alpha^0} = \begin{pmatrix} \dfrac{\alpha_T}{l_p\left(c_p\right)^2}a^{(1)}(t)E(t) \\[2ex] \dfrac{\alpha_T}{2l_p\left(c_p\right)^2}E^2(t) \end{pmatrix}, \quad 0<t<\tau, \tag{5.134}$$

$$\left\{\mathbf{B}_A^{(2)}\left[2;\mathbf{C}^{(2)}(2;3;t);\boldsymbol{\alpha}\right]\right\}_{\alpha^0} \triangleq \begin{pmatrix} c^{(2)}(1;3;\tau) \\ c^{(2)}(2;3;0) \end{pmatrix}_{\alpha^0} = \begin{pmatrix} 0 \\ 0 \end{pmatrix}. \tag{5.135}$$

The 2nd-order partial sensitivities are obtained in terms of the components of $\mathbf{C}^{(2)}(2;3;t)$ by applying the same procedure as in Sections 5.4.2.1 and 5.4.2.2, and will have expressions that are formally similar to those obtained in the previous subsections except for the contributions stemming from the direct effect term defined in Eq. (5.132). The final expressions for the 2nd-order partial sensitivities that stem from $\partial E(\tau)/\partial c_p$ are as follows:

$$\begin{aligned}\frac{\partial^2 E(\tau)}{\partial \alpha_T\,\partial c_p} &= \frac{1}{2l_p(c_p)^2}\int_0^\tau a^{(1)}(t)E^2(t)\,dt \\ &-\frac{1}{2l_p c_p}\int_0^\tau \left[c^{(2)}(1;3;t)E^2(t)+2c^{(2)}(2;3;t)a^{(1)}(t)E(t)\right]dt;\end{aligned} \tag{5.136}$$

$$\begin{aligned}\frac{\partial^2 E(\tau)}{\partial l_p\,\partial c_p} &= -\frac{\alpha_T}{2(l_p c_p)^2}\int_0^\tau a^{(1)}(t)E^2(t)\,dt \\ &+\frac{\alpha_T}{2(l_p)^2 c_p}\int_0^\tau \left[c^{(2)}(1;3;t)E^2(t)+2c^{(2)}(2;3;t)a^{(1)}(t)E(t)\right]dt;\end{aligned} \tag{5.137}$$

$$\begin{aligned}\frac{\partial^2 E(\tau)}{\partial c_p\,\partial c_p} &= -\frac{\alpha_T}{l_p(c_p)^3}\int_0^\tau a^{(1)}(t)E^2(t)\,dt \\ &+\frac{\alpha_T}{2(c_p)^2 l_p}\int_0^\tau \left[c^{(2)}(1;3;t)E^2(t)+2c^{(2)}(2;3;t)a^{(1)}(t)E(t)\right]dt;\end{aligned} \tag{5.138}$$

$$\frac{\partial^2 E(\tau)}{\partial \varphi_0\,\partial c_p} = \gamma\sigma_f N_f \int_0^\tau c^{(2)}(1;3;t)\,dt; \tag{5.139}$$

$$\frac{\partial^2 E(\tau)}{\partial \gamma\,\partial c_p} = \varphi_0\sigma_f N_f \int_0^\tau c^{(2)}(1;3;t)\,dt; \tag{5.140}$$

$$\frac{\partial^2 E(\tau)}{\partial \sigma_f\,\partial c_p} = \varphi_0\gamma N_f \int_0^\tau c^{(2)}(1;3;t)\,dt; \tag{5.141}$$

$$\frac{\partial^2 E(\tau)}{\partial N_f\,\partial c_p} = \varphi_0\gamma\sigma_f \int_0^\tau c^{(2)}(1;3;t)\,dt. \tag{5.142}$$

### 5.4.2.4 Computation of 2nd-Order Sensitivities Stemming from the 1st-Order Sensitivity $\partial E(\tau)/\partial \varphi_0$

The 2nd-order sensitivities that stem from the 1st-order sensitivity $\partial E(\tau)/\partial \varphi_0$ are the components of the 1st-order G-differential of Eq. (5.34), which has by definition the following expression:

$$\delta\{\partial E(\tau)/\partial\varphi_0\}_{\alpha^0} \triangleq \{\delta[\partial E(\tau)/\partial\varphi_0]\}_{dir} + \{\delta[\partial E(\tau)/\partial\varphi_0]\}_{ind}$$
$$\triangleq \left\{\frac{d}{d\varepsilon}\left[(\gamma+\varepsilon\delta\gamma)(\sigma_f+\varepsilon\delta\sigma_f)(N_f+\varepsilon\delta N_f)\int_0^\tau\left(a^{(1)}+\varepsilon\delta a^{(1)}\right)dt\right]_{\alpha^0}\right\}_{\varepsilon=0}, \tag{5.143}$$

where the "direct-effect" term $\{\delta[\partial E(\tau)/\partial\varphi_0]\}_{dir}$ can be determined immediately and is defined as follows:

$$\{\delta[\partial E(\tau)/\partial\varphi_0]\}_{dir} \triangleq \left\{\left[(\delta\gamma)\sigma_f N_f + (\delta\sigma_f)\gamma N_f + (\delta N_f)\gamma\sigma_f\right]\int_0^\tau a^{(1)}(t)dt\right\}_{\alpha^0}, \tag{5.144}$$

and where the "indirect-effect" term $\{\delta[\partial E(\tau)/\partial\varphi_0]\}_{ind}$ is defined as follows:

$$\{\delta[\partial E(\tau)/\partial\varphi_0]\}_{ind} \triangleq \left\{\gamma\sigma_f N_f\int_0^\tau \delta a^{(1)}(t)dt\right\}_{\alpha^0}. \tag{5.145}$$

Just as in the previous subsections of Section 5.4.2, the 2nd-level variational function $\mathbf{V}^{(2)}(2;t) \triangleq \left[\delta E(t), \delta a^{(1)}(t)\right]^\dagger$, which is needed to evaluate the "indirect-effect" term $\{\delta[\partial E(\tau)/\partial\varphi_0]\}_{ind}$, is the solution of the 2nd-LVSS defined by Eqs. (5.100) and (5.101). The computationally expensive path of solving the 2nd-LVSS repeatedly for every possible parameter variation is avoided by replacing the dependence of the "indirect-effect" term defined in Eq. (5.145) on the variational function $\mathbf{V}^{(2)}(2;t)$ by a dependence on a corresponding 2nd-level adjoint function, which will be denoted as $\mathbf{C}^{(2)}(2;4;t) \triangleq \left[c^{(2)}(1;4;t), c^{(2)}(2;4;t)\right]^\dagger \in \mathcal{H}_2$, where the notation is as in the previous subsections, except that the 2nd argument of $\mathbf{C}^{(2)}(2;4;t)$ is denoted as "4" to indicate that this 2nd-level adjoint function corresponds to the 1st-order sensitivity $\partial E(\tau)/\partial\varphi_0$ of the response with respect to the *4th* component of the vector of model parameters $\boldsymbol{\alpha} \triangleq (\alpha_1,\ldots,\alpha_7)^\dagger \triangleq \left(\alpha_T, l_p, c_p, \varphi_0, \gamma, \sigma_f, N_f\right)^\dagger$, namely, $\varphi_0$.

The 2nd-LASS for the function $\mathbf{C}^{(2)}(2;4;t) \triangleq \left[c^{(2)}(1;4;t), c^{(2)}(2;4;t)\right]^\dagger \in \mathcal{H}_2$ is constructed by following the same procedure as in the previous subsections, except for the source term (i.e., right side) of the 2nd-LASS, which corresponds to the "indirect-effect" term $\{\delta[\partial E(\tau)/\partial\varphi_0]\}_{ind}$ defined in Eq. (5.145). This procedure leads to the following 2nd-LASS for $\mathbf{C}^{(2)}(2;4;t)$:

$$\left\{\mathbf{AM}^{(2)}[2\times 2;\boldsymbol{\alpha}]\mathbf{C}^{(2)}(2;4;t)\right\}_{\alpha^0} = \begin{pmatrix} 0 \\ \gamma\sigma_f N_f \end{pmatrix}, \quad 0 < t < \tau, \tag{5.146}$$

$$\left\{\mathbf{B}_A^{(2)}\left[2;\mathbf{C}^{(2)}(2;4;t);\boldsymbol{\alpha}\right]\right\}_{\alpha^0} \triangleq \begin{pmatrix} c^{(2)}(1;4;\tau) \\ c^{(2)}(2;4;0) \end{pmatrix}_{\alpha^0} = \begin{pmatrix} 0 \\ 0 \end{pmatrix}. \tag{5.147}$$

The 2nd-order partial sensitivities are obtained in terms of the components of $\mathbf{C}^{(2)}(2;4;t)$ by applying the same procedure as in Sections 5.4.2.1 and 5.4.2.2, and will have expressions that are formally similar to those obtained in the previous subsections, except for the contributions stemming from the direct-effect term defined in Eq. (5.144). The final expressions for the 2nd-order partial sensitivities that stem from $\partial E(\tau)/\partial\varphi_0$ are as follows:

$$\frac{\partial^2 E(\tau)}{\partial\alpha_T\,\partial\varphi_0} = -\frac{1}{2l_p c_p}\int_0^\tau \left[c^{(2)}(1;4;t)E^2(t) + 2c^{(2)}(2;4;t)a^{(1)}(t)E(t)\right]dt; \tag{5.148}$$

$$\frac{\partial^2 E(\tau)}{\partial l_p\,\partial\varphi_0} = \frac{\alpha_T}{2\left(l_p\right)^2 c_p}\int_0^\tau \left[c^{(2)}(1;4;t)E^2(t) + 2c^{(2)}(2;4;t)a^{(1)}(t)E(t)\right]dt; \tag{5.149}$$

$$\frac{\partial^2 E(\tau)}{\partial c_p\,\partial\varphi_0} = \frac{\alpha_T}{2\left(c_p\right)^2 l_p}\int_0^\tau \left[c^{(2)}(1;4;t)E^2(t) + 2c^{(2)}(2;4;t)a^{(1)}(t)E(t)\right]dt; \tag{5.150}$$

$$\frac{\partial^2 E(\tau)}{\partial\varphi_0\,\partial\varphi_0} = \gamma\sigma_f N_f\int_0^\tau c^{(2)}(1;4;t)\,dt; \tag{5.151}$$

$$\frac{\partial^2 E(\tau)}{\partial\gamma\,\partial\varphi_0} = \sigma_f N_f\int_0^\tau a^{(1)}(t)dt + \varphi_0\sigma_f N_f\int_0^\tau c^{(2)}(1;4;t)\,dt; \tag{5.152}$$

$$\frac{\partial^2 E(\tau)}{\partial\sigma_f\,\partial\varphi_0} = \gamma N_f\int_0^\tau a^{(1)}(t)dt + \varphi_0\gamma N_f\int_0^\tau c^{(2)}(1;4;t)\,dt; \tag{5.153}$$

$$\frac{\partial^2 E(\tau)}{\partial N_f\,\partial\varphi_0} = \gamma\sigma_f\int_0^\tau a^{(1)}(t)dt + \varphi_0\gamma\sigma_f\int_0^\tau c^{(2)}(1;4;t)\,dt. \tag{5.154}$$

#### 5.4.2.5 Computation of 2nd-Order Sensitivities Stemming from the 1st-Order Sensitivity $\partial E(\tau)/\partial\gamma$

The 2nd-order sensitivities that stem from the 1st-order sensitivity $\partial E(\tau)/\partial\gamma$ are the components of the 1st-order G-differential of Eq. (5.35), which has by definition the following expression:

$$\delta\{\partial E(\tau)/\partial\gamma\}_{\alpha^0} \triangleq \{\delta[\partial E(\tau)/\partial\gamma]\}_{dir} + \{\delta[\partial E(\tau)/\partial\gamma]\}_{ind}$$

$$\triangleq \left\{\frac{d}{d\varepsilon}\left[(\varphi_0+\varepsilon\delta\varphi_0)(\sigma_f+\varepsilon\delta\sigma_f)(N_f+\varepsilon\delta N_f)\int_0^\tau\left(a^{(1)}+\varepsilon\delta a^{(1)}\right)dt\right]_{\alpha^0}\right\}_{\varepsilon=0}, \tag{5.155}$$

where the "direct-effect" term $\{\delta[\partial E(\tau)/\partial\gamma]\}_{dir}$ can be determined immediately and is defined as follows:

$$\{\delta[\partial E(\tau)/\partial\gamma]\}_{dir} \triangleq \left\{\left[(\delta\varphi_0)\sigma_f N_f + (\delta\sigma_f)\varphi_0 N_f + (\delta N_f)\varphi_0\sigma_f\right]\int_0^\tau a^{(1)}(t)dt\right\}_{\alpha^0}, \tag{5.156}$$

and where the "indirect-effect" term $\{\delta[\partial E(\tau)/\partial\gamma]\}_{ind}$ is defined as follows:

$$\{\delta[\partial E(\tau)/\partial\gamma]\}_{ind} \triangleq \left\{\varphi_0\sigma_f N_f \int_0^\tau \delta a^{(1)}(t)dt\right\}_{\alpha^0}. \tag{5.157}$$

Just as in the previous subsections of Section 5.4.2, the 2nd-level variational function $\mathbf{V}^{(2)}(2;t) \triangleq \left[\delta E(t), \delta a^{(1)}(t)\right]^\dagger$, which is needed to evaluate the "indirect-effect" term $\{\delta[\partial E(\tau)/\partial\gamma]\}_{ind}$, is the solution of the 2nd-LVSS defined by Eqs. (5.100) and (5.101). The computationally expensive path of solving the 2nd-LVSS repeatedly for every possible parameter variation is avoided by replacing the dependence of the "indirect-effect" term defined in Eq. (5.157) on the variational function $\mathbf{V}^{(2)}(2;t)$ by a dependence on a corresponding 2nd-level adjoint function, which will be denoted as $\mathbf{C}^{(2)}(2;5;t) \triangleq \left[c^{(2)}(1;5;t), c^{(2)}(2;5;t)\right]^\dagger \in \mathcal{H}_2$, where the notation is as in the previous subsections, except that the second argument of $\mathbf{C}^{(2)}(2;5;t)$ is denoted as "5" to indicate that this 2nd-level adjoint function corresponds to the 1st-order sensitivity $\partial E(\tau)/\partial\gamma$ of the response with respect to the *5th* component of the vector of model parameters $\boldsymbol{\alpha} \triangleq (\alpha_1,\ldots,\alpha_7)^\dagger \triangleq (\alpha_T, l_p, c_p, \varphi_0, \gamma, \sigma_f, N_f)^\dagger$, namely, $\gamma$.

The 2nd-LASS for the function $\mathbf{C}^{(2)}(2;5;t) \triangleq \left[c^{(2)}(1;5;t), c^{(2)}(2;5;t)\right]^\dagger \in \mathcal{H}_2$ is constructed by following the same procedure as in the previous subsections, except for the source term (i.e., right side) of the 2nd-LASS, which now corresponds to the "indirect-effect" term $\{\delta[\partial E(\tau)/\partial\gamma]\}_{ind}$ defined in Eq. (5.157). This procedure leads to the following 2nd-LASS for $\mathbf{C}^{(2)}(2;5;t)$:

$$\left\{\mathbf{AM}^{(2)}[2\times 2;\boldsymbol{\alpha}]\mathbf{C}^{(2)}(2;5;t)\right\}_{\alpha^0} = \begin{pmatrix} 0 \\ \varphi_0\sigma_f N_f \end{pmatrix}, \quad 0 < t < \tau, \tag{5.158}$$

$$\left\{\mathbf{B}_A^{(2)}\left[2;\mathbf{C}^{(2)}(2;5;t);\boldsymbol{\alpha}\right]\right\}_{\alpha^0} \triangleq \begin{pmatrix} c^{(2)}(1;5;\tau) \\ c^{(2)}(2;5;0) \end{pmatrix}_{\alpha^0} = \begin{pmatrix} 0 \\ 0 \end{pmatrix}. \tag{5.159}$$

The 2nd-order partial sensitivities are obtained in terms of the components of $\mathbf{C}^{(2)}(2;5;t)$ by applying the same procedure as in Sections 4.2.1–4.2.4, and will have expressions that are formally similar to those obtained in the previous subsections, except for the contributions stemming from the direct-effect term defined in Eq. (5.156). The final expressions for the 2nd-order partial sensitivities that stem from $\partial E(\tau)/\partial\gamma$ are as follows:

$$\frac{\partial^2 E(\tau)}{\partial \alpha_T \, \partial \gamma} = -\frac{1}{2 l_p c_p} \int_0^\tau \left[ c^{(2)}(1;5;t) E^2(t) + 2c^{(2)}(2;5;t) a^{(1)}(t) E(t) \right] dt; \quad (5.160)$$

$$\frac{\partial^2 E(\tau)}{\partial l_p \, \partial \gamma} = \frac{\alpha_T}{2 (l_p)^2 c_p} \int_0^\tau \left[ c^{(2)}(1;5;t) E^2(t) + 2c^{(2)}(2;5;t) a^{(1)}(t) E(t) \right] dt; \quad (5.161)$$

$$\frac{\partial^2 E(\tau)}{\partial c_p \, \partial \gamma} = \frac{\alpha_T}{2 (c_p)^2 l_p} \int_0^\tau \left[ c^{(2)}(1;5;t) E^2(t) + 2c^{(2)}(2;5;t) a^{(1)}(t) E(t) \right] dt; \quad (5.162)$$

$$\frac{\partial^2 E(\tau)}{\partial \varphi_0 \, \partial \gamma} = \sigma_f N_f \int_0^\tau a^{(1)}(t) dt + \gamma \sigma_f N_f \int_0^\tau c^{(2)}(1;5;t) dt; \quad (5.163)$$

$$\frac{\partial^2 E(\tau)}{\partial \gamma \, \partial \gamma} = \varphi_0 \sigma_f N_f \int_0^\tau c^{(2)}(1;5;t) dt; \quad (5.164)$$

$$\frac{\partial^2 E(\tau)}{\partial \sigma_f \, \partial \gamma} = \varphi_0 N_f \int_0^\tau a^{(1)}(t) dt + \varphi_0 \gamma N_f \int_0^\tau c^{(2)}(1;5;t) dt; \quad (5.165)$$

$$\frac{\partial^2 E(\tau)}{\partial N_f \, \partial \gamma} = \varphi_0 \sigma_f \int_0^\tau a^{(1)}(t) dt + \varphi_0 \gamma \sigma_f \int_0^\tau c^{(2)}(1;5;t) dt. \quad (5.166)$$

#### 5.4.2.6 Computation of 2nd-Order Sensitivities Stemming from the 1st-Order Sensitivity $\partial E(\tau)/\partial \sigma_f$

The 2nd-order sensitivities that stem from the 1st-order sensitivity $\partial E(\tau)/\partial \sigma_f$ are the components of the 1st-order G-differential of Eq. (5.36), which has by definition the following expression:

$$\begin{aligned} \delta\left\{\partial E(\tau)/\partial \sigma_f\right\}_{\alpha^0} &\triangleq \left\{\delta\left[\partial E(\tau)/\partial \sigma_f\right]\right\}_{dir} + \left\{\delta\left[\partial E(\tau)/\partial \sigma_f\right]\right\}_{ind} \\ &\triangleq \left\{ \frac{d}{d\varepsilon} \left[ (\varphi_0 + \varepsilon \delta \varphi_0)(\gamma + \varepsilon \delta \gamma)(N_f + \varepsilon \delta N_f) \int_0^\tau \left( a^{(1)} + \varepsilon \delta a^{(1)} \right) dt \right]_{\alpha^0} \right\}_{\varepsilon=0}, \end{aligned} \quad (5.167)$$

where the "direct-effect" term $\left\{\delta\left[\partial E(\tau)/\partial \sigma_f\right]\right\}_{dir}$ can be determined immediately and is defined as follows:

$$\left\{\delta\left[\partial E(\tau)/\partial \sigma_f\right]\right\}_{dir} \triangleq \left\{ \left[ (\delta \varphi_0) \gamma N_f + (\delta \gamma) \varphi_0 N_f + (\delta N_f) \varphi_0 \gamma \right] \int_0^\tau a^{(1)}(t) dt \right\}_{\alpha^0}, \quad (5.168)$$

and where the "indirect-effect" term $\left\{\delta\left[\partial E(\tau)/\partial \sigma_f\right]\right\}_{ind}$ is defined as follows:

$$\left\{\delta\left[\partial E(\tau)/\partial\sigma_f\right]\right\}_{ind} \triangleq \left\{\varphi_0\gamma N_f\int_0^{\tau}\delta a^{(1)}(t)dt\right\}_{\alpha^0}. \tag{5.169}$$

Just as in the previous subsections of Section 5.4.2, the dependence of the "indirect-effect" term defined in Eq. (5.169) on the variational function $\mathbf{V}^{(2)}(2;t)$ is replaced by a dependence on a corresponding 2nd-level adjoint function, which will be denoted as $\mathbf{C}^{(2)}(2;6;t)\triangleq\left[c^{(2)}(1;6;t),c^{(2)}(2;6;t)\right]^{\dagger}\in\mathcal{H}_2$, where the notation is as in the previous subsections, except that the 2nd argument of $\mathbf{C}^{(2)}(2;6;t)$ is denoted as "6" to indicate that this 2nd-level adjoint function corresponds to the 1st-order sensitivity $\partial E(\tau)/\partial\sigma_f$ of the response with respect to the *6th* component of the vector of model parameters $\boldsymbol{\alpha}\triangleq(\alpha_1,\ldots,\alpha_7)^{\dagger}\triangleq\left(\alpha_T,l_p,c_p,\varphi_0,\gamma,\sigma_f,N_f\right)^{\dagger}$, namely, $\sigma_f$.

The 2nd-LASS for the function $\mathbf{C}^{(2)}(2;6;t)\triangleq\left[c^{(2)}(1;6;t),c^{(2)}(2;6;t)\right]^{\dagger}\in\mathcal{H}_2$ is constructed by following the same procedure as in the previous subsections, except for the source term (i.e., right side) of the 2nd-LASS, which corresponds to the "indirect-effect" term $\left\{\delta\,\partial E(\tau)/\partial\sigma_f\right\}_{ind}$ defined in Eq. (5.169). This procedure leads to the following 2nd-LASS for $\mathbf{C}^{(2)}(2;6;t)$:

$$\left\{\mathbf{AM}^{(2)}[2\times 2;\boldsymbol{\alpha}]\mathbf{C}^{(2)}(2;6;t)\right\}_{\alpha^0} = \begin{pmatrix} 0 \\ \varphi_0\gamma N_f \end{pmatrix}, \quad 0<t<\tau, \tag{5.170}$$

$$\left\{\mathbf{B}_A^{(2)}\left[2;\mathbf{C}^{(2)}(2;6;t);\boldsymbol{\alpha}\right]\right\}_{\alpha^0} \triangleq \begin{pmatrix} c^{(2)}(1;6;\tau) \\ c^{(2)}(2;6;0) \end{pmatrix}_{\alpha^0} = \begin{pmatrix} 0 \\ 0 \end{pmatrix}. \tag{5.171}$$

The 2nd-order partial sensitivities are obtained in terms of the components of $\mathbf{C}^{(2)}(2;6;t)$ by applying the same procedure as in subsections 5.4.2.1–5.4.2.5, and will have expressions that are formally similar to those obtained in the previous subsections except for the contributions stemming from the direct-effect term defined in Eq. (5.168). The final expressions for the 2nd-order partial sensitivities that stem from $\partial E(\tau)/\partial\sigma_f$ are as follows:

$$\frac{\partial^2 E(\tau)}{\partial\alpha_T\,\partial\sigma_f} = -\frac{1}{2l_pc_p}\int_0^{\tau}\left[c^{(2)}(1;6;t)E^2(t)+2c^{(2)}(2;6;t)a^{(1)}(t)E(t)\right]dt; \tag{5.172}$$

$$\frac{\partial^2 E(\tau)}{\partial l_p\,\partial\sigma_f} = \frac{\alpha_T}{2(l_p)^2c_p}\int_0^{\tau}\left[c^{(2)}(1;6;t)E^2(t)+2c^{(2)}(2;6;t)a^{(1)}(t)E(t)\right]dt; \tag{5.173}$$

$$\frac{\partial^2 E(\tau)}{\partial c_p\,\partial\sigma_f} = \frac{\alpha_T}{2(c_p)^2l_p}\int_0^{\tau}\left[c^{(2)}(1;6;t)E^2(t)+2c^{(2)}(2;6;t)a^{(1)}(t)E(t)\right]dt; \tag{5.174}$$

$$\frac{\partial^2 E(\tau)}{\partial\varphi_0\,\partial\sigma_f} = \gamma N_f\int_0^{\tau}a^{(1)}(t)dt+\gamma\sigma_fN_f\int_0^{\tau}c^{(2)}(1;6;t)dt; \tag{5.175}$$

$$\frac{\partial^2 E(\tau)}{\partial \gamma \, \partial \sigma_f} = \varphi_0 N_f \int_0^\tau a^{(1)}(t)dt + \varphi_0 \sigma_f N_f \int_0^\tau c^{(2)}(1;6;t)dt; \tag{5.176}$$

$$\frac{\partial^2 E(\tau)}{\partial \sigma_f \, \partial \sigma_f} = \varphi_0 \gamma N_f \int_0^\tau c^{(2)}(1;6;t)dt; \tag{5.177}$$

$$\frac{\partial^2 E(\tau)}{\partial N_f \, \partial \sigma_f} = \varphi_0 \gamma \int_0^\tau a^{(1)}(t)dt + \varphi_0 \gamma \sigma_f \int_0^\tau c^{(2)}(1;6;t)dt. \tag{5.178}$$

#### 5.4.2.7 Computation of 2nd-Order Sensitivities Stemming from the 1st-Order Sensitivity $\partial E(\tau)/\partial N_f$

The 2nd-order sensitivities that stem from the 1st-order sensitivity $\partial E(\tau)/\partial N_f$ are the components of the 1st-order G-differential of Eq. (5.37), which has by definition the following expression:

$$\begin{aligned} \delta\{\partial E(\tau)/\partial N_f\}_{\alpha^0} &\triangleq \{\delta[\partial E(\tau)/\partial N_f]\}_{dir} + \{\delta[\partial E(\tau)/\partial N_f]\}_{ind} \\ &\triangleq \left\{ \frac{d}{d\varepsilon} \left[ (\varphi_0 + \varepsilon\delta\varphi_0)(\gamma + \varepsilon\delta\gamma)(\sigma_f + \varepsilon\delta\sigma_f) \int_0^\tau \left(a^{(1)} + \varepsilon\delta a^{(1)}\right) dt \right]_{\alpha^0} \right\}_{\varepsilon=0}, \end{aligned} \tag{5.179}$$

where the "direct-effect" term $\{\delta[\partial E(\tau)/\partial N_f]\}_{dir}$ can be determined immediately and is defined as follows:

$$\{\delta[\partial E(\tau)/\partial N_f]\}_{dir} \triangleq \left\{ \left[ (\delta\varphi_0)\gamma\sigma_f + (\delta\gamma)\varphi_0\sigma_f + (\delta\sigma_f)\varphi_0\gamma \right] \int_0^\tau a^{(1)}(t)dt \right\}_{\alpha^0}, \tag{5.180}$$

and where the "indirect-effect" term $\{\delta[\partial E(\tau)/\partial N_f]\}_{ind}$ is defined as follows:

$$\{\delta \, \partial E(\tau)/\partial N_f\}_{ind} \triangleq \left\{ \varphi_0 \gamma \sigma_f \int_0^\tau \delta a^{(1)}(t) \, dt \right\}_{\alpha^0}. \tag{5.181}$$

Just as in the previous subsections of Section 5.4.2, the dependence of the "indirect-effect" term defined in Eq. (5.181) on the variational function $\mathbf{V}^{(2)}(2;t)$ is replaced by a dependence on a corresponding 2nd-level adjoint function, which will be denoted as $\mathbf{C}^{(2)}(2;7;t) \triangleq \left[c^{(2)}(1;7;t), c^{(2)}(2;7;t)\right]^\dagger \in \mathcal{H}_2$, where the notation is as in the previous subsections, except that the 2nd argument of $\mathbf{C}^{(2)}(2;7;t)$ is denoted as "7" to indicate that this 2nd-level adjoint function corresponds to the 1st-order sensitivity $\partial E(\tau)/\partial N_f$ of the response with respect to the *7th* component of the vector of model parameters $\boldsymbol{\alpha} \triangleq (\alpha_1, \ldots, \alpha_7)^\dagger \triangleq (\alpha_T, l_p, c_p, \varphi_0, \gamma, \sigma_f, N_f)^\dagger$, namely, $N_f$.

The 2nd-LASS for the function $\mathbf{C}^{(2)}(2;7;t) \triangleq \left[c^{(2)}(1;7;t), c^{(2)}(2;7;t)\right]^\dagger \in \mathcal{H}_2$ is constructed by following the same procedure as in the previous subsections, except for the source term (i.e., right side) of the 2nd-LASS, which now corresponds to the "indirect-effect" term $\left\{\delta\left[\partial E(\tau)/\partial N_f\right]\right\}_{ind}$ defined in Eq. (5.181). This procedure leads to the following 2nd-LASS for $\mathbf{C}^{(2)}(2;7;t)$:

$$\left\{\mathbf{AM}^{(2)}[2\times 2;\alpha]\mathbf{C}^{(2)}(2;7;t)\right\}_{\alpha^0}=\begin{pmatrix}0\\ \varphi_0\gamma\sigma_f\end{pmatrix},\quad 0<t<\tau, \tag{5.182}$$

$$\left\{\mathbf{B}_A^{(2)}\left[2;\mathbf{C}^{(2)}(2;7;t);\alpha\right]\right\}_{\alpha^0}\triangleq\begin{pmatrix}c^{(2)}(1;7;\tau)\\ c^{(2)}(2;7;0)\end{pmatrix}_{\alpha^0}=\begin{pmatrix}0\\0\end{pmatrix}. \tag{5.183}$$

The 2nd-order partial sensitivities are obtained in terms of the components of $\mathbf{C}^{(2)}(2;6;t)$ by applying the same procedure as in subsections 5.4.2.1–5.4.2.5, and will have expressions that are formally similar to those obtained in the previous subsections, except for the contributions stemming from the direct-effect term defined in Eq. (5.168). The final expressions for the 2nd-order partial sensitivities that stem from $\partial E(\tau)/\partial\sigma_f$ are as follows:

$$\frac{\partial^2 E(\tau)}{\partial\alpha_T\,\partial N_f}=-\frac{1}{2l_p c_p}\int_0^\tau\left[c^{(2)}(1;7;t)E^2(t)+2c^{(2)}(2;7;t)a^{(1)}(t)E(t)\right]dt; \tag{5.184}$$

$$\frac{\partial^2 E(\tau)}{\partial l_p\,\partial N_f}=\frac{\alpha_T}{2\left(l_p\right)^2 c_p}\int_0^\tau\left[c^{(2)}(1;7;t)E^2(t)+2c^{(2)}(2;7;t)a^{(1)}(t)E(t)\right]dt; \tag{5.185}$$

$$\frac{\partial^2 E(\tau)}{\partial c_p\,\partial N_f}=\frac{\alpha_T}{2\left(c_p\right)^2 l_p}\int_0^\tau\left[c^{(2)}(1;7;t)E^2(t)+2c^{(2)}(2;7;t)a^{(1)}(t)E(t)\right]dt; \tag{5.186}$$

$$\frac{\partial^2 E(\tau)}{\partial\varphi_0\,\partial N_f}=\gamma\sigma_f\int_0^\tau a^{(1)}(t)dt+\gamma\sigma_f N_f\int_0^\tau c^{(2)}(1;7;t)dt; \tag{5.187}$$

$$\frac{\partial^2 E(\tau)}{\partial\gamma\,\partial N_f}=\varphi_0\sigma_f\int_0^\tau a^{(1)}(t)dt+\varphi_0\sigma_f N_f\int_0^\tau c^{(2)}(1;7;t)dt; \tag{5.188}$$

$$\frac{\partial^2 E(\tau)}{\partial\sigma_f\,\partial N_f}=\varphi_0\gamma\int_0^\tau a^{(1)}(t)dt+\varphi_0\gamma N_f\int_0^\tau c^{(2)}(1;7;t)dt; \tag{5.189}$$

$$\frac{\partial^2 E(\tau)}{\partial N_f\,\partial N_f}=\varphi_0\gamma\sigma_f\int_0^\tau c^{(2)}(1;7;t)dt. \tag{5.190}$$

### 5.4.3 Computational Advantages of Using the 2nd-FASAM-N versus the 2nd-CASAM-N

"Large-scale" computations are those needed to solve systems of equations (algebraic, differential, integral) such as those underlying the original model and the adjoint sensitivity systems of various levels (1st-LASS, 2nd-LASS, etc.). By comparison, the computational efforts involved in evaluating integrals by means of quadrature formulas are "small-scale" computations. If no "feature" functions of parameters can be identified in the model, then the formalisms of 1st-CASAM-N and the 1st-FASAM-N become identical to one another, requiring a single large-scale computation, which is needed for solving the 1st-LASS to obtain the 1st-level adjoint sensitivity function. Subsequently, the 1st-order sensitivities of the model's response with respect to the underlying model parameters are computed inexpensively using quadrature formulas to evaluate numerically the respective integrals involving the 1st-level adjoint sensitivity function. On the other hand, if "feature" functions of parameters can be identified in the model, then the 1st-FASAM-N methodology is marginally more efficient than the 1st-CASAM-N: both methodologies require a single large-scale computation, but the 1st-FASAM-N requires fewer numerical quadratures (only as many as there are feature-functions) than the 1st-CASAM-N (which requires as many quadratures as there are parameters), because the sensitivities with respect to the parameters are obtained from the sensitivities with respect to the feature-functions by analytical differentiations. In both cases, the 1st-LASS to be solved involves the same operators to be inverted (on the left side of the 1st-LASS); only the source terms on the right side of the 1st-LASS within the 1st-FASAM-N differ from the source terms on the right side of the 1st-LASS within the 1st-CASAM-N.

The conventional (e.g., "statistical" or "finite-difference") methods are impractical for computing response sensitivities higher than 1st-order. Both the 2nd-FASAM-N and the 2nd-CASAM-N methodologies are constructed by using the fundamental definition that "the 2nd-order differential is the 1st-order differential of the 1st-order differential." Thus, each of the 1st-order sensitivities becomes the "model response" for the application of either the 2nd-FASAM-N or the 2nd-CASAM-N. Consequently, there would be as many large-scale computations for solving the 2nd-LASS as there are 1st-order sensitivities. Consequently, since the number of feature functions is always smaller than the number of model parameters, there would be fewer large-scale computations (to solve the 2nd-LASS) to be performed within the 2nd-FASAM-N methodology than there would be within the 2nd-CASAM-N methodology. In particular, for the Nordheim-Fuchs model analyzed in this work, it was shown that the 2nd-FASAM-N methodology requires $TF = 2$ large-scale computations (as there are two feature-functions), whereas the 2nd-CASAM-N methodology requires $TP = 7$ large-scale computations (i.e., as many computations as there are primary model parameters) for obtaining all of the $TP^2 = 49$ 2nd-order sensitivities to the model parameters; $TP(TP+1)/2 = 28$ (out of the 49) of these 2nd-order sensitivities are distinct from each other. It is also important to note that the mixed 2nd-order sensitivities are computed twice, using distinct adjoint functions, within either the 2nd-FASAM-N or the 2nd-CASAM-N methodology. This characteristic of the 2nd-FASAM-N and 2nd-CASAM-N methodologies provides an intrinsic mechanism for verifying the accuracy of the respective 1st- and 2nd-level adjoint functions. Furthermore, the user can select which of the alternative – but equivalent – expressions of the 2nd-order mixed sensitivity under consideration is computationally more advantageous to use.

## 5.5 COMPUTATION OF THE 3RD-ORDER RESPONSE SENSITIVITIES WITH RESPECT TO MODEL PARAMETERS: APPLYING THE 3RD-FASAM-N VERSUS THE 3RD-CASAM-N

The fundamental principle underlying both the 3rd-FASAM-N and the 3rd-CASAM-N methodologies is to determine the 3rd-order sensitivities by employing their definition of being the "1st-order sensitivities of the 2nd-order sensitivities." Thus, each 2nd-order sensitivity is treated as a "model response," and the G-differential of each of these "model responses" subsequently provides the partial 3rd-order sensitivities that stem from the respective 3rd-order sensitivity. As will be highlighted in Section 5.5.1, the computation of the 343 3rd-order sensitivities (of which 84 are distinct) of the response $E(\tau)$ with respect to the seven model parameters will require just three large-scale computations for solving the corresponding three 3rd-Level Adjoint Sensitivity Systems (3rd-LASS) that correspond to the three distinct sensitivities of the response $E(\tau)$ with respect to the "feature functions" $f_1(\alpha)$ and $f_2(\alpha)$. In contradistinction, as will be highlighted in Section 5.5.2, applying the 3rd-CASAM-N methodology requires 28 large-scale computations for solving the 28 3rd-Level Adjoint Sensitivity Systems (3rd-LASS) that correspond to the distinct 2nd-order sensitivities of the response $E(\tau)$ with respect to the model parameters.

### 5.5.1 COMPUTATION OF 3RD-ORDER SENSITIVITIES USING THE 3RD-FASAM-N

As will be shown in Section 5.5.1.1, the 2nd-order sensitivity $\partial^2 E(\tau)/\partial f_1 \partial f_1$ will give rise to two 3rd-order sensitivities, $\partial^3 E(\tau)/\partial f_1 \partial f_1 \partial f_1$ and $\partial^3 E(\tau)/\partial f_2 \partial f_1 \partial f_1$, which will be determined by solving a corresponding 3rd-Level Adjoint Sensitivity System (3rd-LASS). For bookkeeping purposes, the quantity $\partial^2 E(\tau)/\partial f_1 \partial f_1$ will the labeled "the *1st* 2nd-order sensitivity" and the solution of the 3rd-LASS that corresponds to it will be labeled "the *1st* 3rd-level adjoint function."

Similarly, the 2nd-order sensitivity $\partial^2 E(\tau)/\partial f_2 \partial f_2$ will give rise to two 3rd-order sensitivities, $\partial^3 E(\tau)/\partial f_1 \partial f_2 \partial f_2$ and $\partial^3 E(\tau)/\partial f_2 \partial f_2 \partial f_2$, which will be determined in Section 5.5.1.2 by solving a corresponding 3rd-Level Adjoint Sensitivity System (3rd-LASS). For bookkeeping purposes, the quantity $\partial^2 E(\tau)/\partial f_2 \partial f_2$ will the labeled "the *2nd* 2nd-order sensitivity" and the solution of the 3rd-LASS that corresponds to it will be labeled "the *2nd* 3rd-level adjoint function."

Finally, the 2nd-order sensitivity $\partial^2 E(\tau)/\partial f_2 \partial f_1$ or, equivalently, the 2nd-order sensitivity $\partial^2 E(\tau)/\partial f_1 \partial f_2$, will give rise to two 3rd-order sensitivities, $\partial^3 E(\tau)/\partial f_1 \partial f_1 \partial f_2$ and $\partial^3 E(\tau)/\partial f_2 \partial f_1 \partial f_2$, which provide alternative but equivalent expressions for the respective mixed 3rd-order sensitivities obtained as mentioned above. These alternative computations will be discussed in Section 5.5.1.3.

#### 5.5.1.1 Computation of 3rd-Order Sensitivities Stemming from the 2nd-Order Sensitivity $\partial^2 E(\tau)/\partial f_1 \partial f_1$

The 3rd-order sensitivities, which stem from $\partial^2 E(\tau)/\partial f_1 \partial f_1$ are obtained by applying the definition of the 1st-order G-differential to Eq. (5.77), which yields the following expression:

$$\delta\left\{\frac{\partial^2 E(\tau)}{\partial f_1 \partial f_1}\right\} \triangleq \frac{\partial^3 E(\tau)}{\partial f_1 \partial f_1 \partial f_1}\delta f_1 + \frac{\partial^3 E(\tau)}{\partial f_2 \partial f_1 \partial f_1}\delta f_2$$

$$\triangleq -\left\{\frac{d}{d\varepsilon}\left\{\int_0^\tau \left[a^{(2)}(1;1;t)+\varepsilon\delta a^{(2)}(1;1;t)\right]\left[E(t)+\varepsilon\delta E(t)\right]^2 dt\right\}_{\mathbf{f}^0}\right\}_{\varepsilon=0}$$

$$-2\left\{\frac{d}{d\varepsilon}\left\{\int_0^\tau \left[a^{(2)}(2;1;t)+\varepsilon\delta a^{(2)}(2;1;t)\right]\left(a^{(1)}+\varepsilon\delta a^{(1)}\right)\left(E+\varepsilon\delta E\right)dt\right\}_{\mathbf{f}^0}\right\}_{\varepsilon=0}$$

$$= -\left\{\int_0^\tau \delta a^{(2)}(1;1;t)E^2(t)dt\right\}_{\mathbf{f}^0}$$

$$-2\left\{\int_0^\tau \left[a^{(2)}(1;1;t)E(t)+a^{(2)}(2;1;t)a^{(1)}(t)\right]\delta E(t)dt\right\}_{\mathbf{f}^0}$$

$$-2\left\{\int_0^\tau \delta a^{(2)}(2;1;t)a^{(1)}(t)E(t)dt\right\}_{\mathbf{f}^0} - 2\left\{\int_0^\tau \delta a^{(1)}(t)a^{(2)}(2;1;t)E(t)dt\right\}_{\mathbf{f}^0}. \tag{5.191}$$

The expression obtained in Eq. (5.191) comprises no direct-effect term, and it can be evaluated after having determined the vector-valued variational functions $\mathbf{V}^{(2)}(2;t) \triangleq \left[\delta E(t), \delta a^{(1)}(t)\right]^\dagger$ and $\delta\mathbf{A}^{(2)}(2;1;t) \triangleq \left[\delta a^{(2)}(1;1;t), \delta a^{(2)}(2;1;t)\right]^\dagger$. The vector-valued variational function $\mathbf{V}^{(2)}(2;t) \triangleq \left[\delta E(t), \delta a^{(1)}(t)\right]^\dagger$ is the solution of the 2nd-Level Variational Sensitivity System (2nd-LVSS) defined by Eqs. (5.64) and (5.65). On the other hand, the vector-valued variational function $\delta\mathbf{A}^{(2)}(2;1;t) \triangleq \left[\delta a^{(2)}(1;1;t), \delta a^{(2)}(2;1;t)\right]^\dagger$ is the solution of the G-differentiated 2nd-LASS defined by Eqs. (5.72) and (5.73), evaluated at the nominal values of the feature functions and state functions (i.e., dependent variables). Using the superscript "zero" to denote the respective nominal values of the various quantities and applying the definition of the G-differential to Eqs. (5.72) and (5.73) yields the following system of equations:

$$\left\{\frac{d}{d\varepsilon}\left[-\frac{d}{dt}+2\left(f_1^0+\varepsilon\delta f_1\right)\left(E^0+\varepsilon\delta E\right)\right]\left[a^{(2,0)}(1;1;t)+\varepsilon\delta a^{(2)}(1;1;t)\right]\right\}_{\varepsilon=0}$$

$$+2\left\{\frac{d}{d\varepsilon}\left(f_1^0+\varepsilon\delta f_1\right)\left(a^{(1,0)}+\varepsilon\delta a^{(1)}\right)\left[a^{(2,0)}(2;1;t)+\varepsilon\delta a^{(2)}(2;1;t)\right]\right\}_{\varepsilon=0} \tag{5.192}$$

$$=-2\left\{\frac{d}{d\varepsilon}\left[\left(a^{(1,0)}+\varepsilon\delta a^{(1)}\right)\left(E^0+\varepsilon\delta E\right)\right]\right\}_{\varepsilon=0},$$

$$\left\{\frac{d}{d\varepsilon}\left[\frac{d}{dt}+2\left(f_1^0+\varepsilon\delta f_1\right)\left(E^0+\varepsilon\delta E\right)\right]\left[a^{(2,0)}(2;1;t)+\varepsilon\delta a^{(2)}(2;1;t)\right]\right\}_{\varepsilon=0}$$

$$=-\left\{\left(E^0+\varepsilon\delta E\right)^2\right\}_{\varepsilon=0}, \tag{5.193}$$

$$\left\{\delta\mathbf{B}_A^{(2)}\left[2;\delta\mathbf{A}^{(2)}(2;1;t);\mathbf{f}\right]\right\}_{\mathbf{f}^0}\triangleq\left\{\begin{pmatrix}\delta a^{(2)}(1;1;\tau)\\ \delta a^{(2)}(2;1;0)\end{pmatrix}\right\}_{\mathbf{f}^0}=\begin{pmatrix}0\\0\end{pmatrix}. \tag{5.194}$$

Carrying out the operations with respect to $\varepsilon$ indicated in Eqs. (5.192)–(5.194) yields the following system of equations:

$$\left[2f_1^0a^{(2,0)}(1;1;t)+2a^{(1,0)}(t)\right]\delta E(t)+\left[2f_1^0a^{(2,0)}(2;1;t)+2E^0\right]\delta a^{(1)}(t)$$

$$+\left[-\frac{d}{dt}+2f_1^0E^0(t)\right]\delta a^{(2)}(1;1;t)+2f_1^0a^{(1,0)}(t)\delta a^{(2)}(2;1;t) \tag{5.195}$$

$$=-2\left(\delta f_1\right)\left[E^0(t)a^{(2,0)}(1;1;t)+a^{(1,0)}(t)a^{(2,0)}(2;1;t)\right],$$

$$\left[f_1^0a^{(2,0)}(2;1;t)+2E^0(t)\right]\delta E(t)+\left[\frac{d}{dt}+2f_1^0E^0(t)\right]\delta a^{(2)}(2;1;t)$$

$$=-2E^0(t)a^{(2,0)}(2;1;t)\left(\delta f_1\right). \tag{5.196}$$

Concatenating Eqs. (5.195) and (5.196) with the equations underlying the 2nd-LVSS provided in Eqs. (5.64) and (5.65) yields the following 3rd-Level Variational Sensitivity System (3rd-LVSS) for the 3rd-level vector-valued variational function $\mathbf{V}^{(3)}(4;1;t)\triangleq\left[\mathbf{V}^{(2)}(2;t),\delta\mathbf{A}^{(2)}(2;1;t)\right]^\dagger$:

$$\left\{\mathbf{VM}^{(3)}[4\times4;1;\mathbf{f}]\mathbf{V}^{(3)}(4;1;t)\right\}_{\mathbf{f}^0}=\left\{\mathbf{Q}_V^{(3)}[4;1;\mathbf{f};\delta\mathbf{f}]\right\}_{\mathbf{f}^0},\quad 0<t<\tau, \tag{5.197}$$

together with the following boundary conditions:

$$\left[\delta E(0),\delta a^{(1)}(\tau),\delta a^{(2)}(1;1;\tau),\delta a^{(2)}(2;1;0)\right]^\dagger=[0,0,0,0]^\dagger. \tag{5.198}$$

The quantities which appear in Eq. (5.197) are defined as follows:

$$\mathbf{VM}^{(3)}[4\times 4;1;\mathbf{f}] \triangleq \begin{pmatrix} \mathbf{VM}^{(2)}[2\times 2;\mathbf{f}] & \mathbf{0}[2\times 2] \\ \mathbf{VM}_{21}^{(3)}[2\times 2;1;\mathbf{f}] & \mathbf{VM}_{22}^{(3)}[2\times 2;1;\mathbf{f}] \end{pmatrix}; \tag{5.199}$$

$$\mathbf{VM}_{21}^{(3)}[2\times 2;1;\mathbf{f}] \triangleq \begin{pmatrix} 2f_1^0 a^{(2,0)}(1;1;t)+2a^{(1,0)}(t) & 2f_1^0 a^{(2,0)}(2;1;t)+2E^0 \\ 2f_1(\boldsymbol{\alpha})a^{(2)}(2;1;t)+2E(t) & 0 \end{pmatrix};$$
$$\mathbf{0}[2\times 2] \triangleq \begin{pmatrix} 0 & 0 \\ 0 & 0 \end{pmatrix}; \tag{5.200}$$

$$\mathbf{VM}_{22}^{(3)}[2\times 2;1;\mathbf{f}] \triangleq \mathbf{AM}^{(2)}[2\times 2;\mathbf{f}] \triangleq \begin{pmatrix} -\frac{d}{dt}+2f_1(\boldsymbol{\alpha})E(t) & 2f_1(\boldsymbol{\alpha})a^{(1)}(t) \\ 0 & \frac{d}{dt}+2f_1(\boldsymbol{\alpha})E(t) \end{pmatrix}; \tag{5.201}$$

$$\mathbf{Q}_V^{(3)}[4;1;\mathbf{f};\delta\mathbf{f}] \triangleq \begin{pmatrix} \mathbf{Q}_V^{(2)}[2;\mathbf{f};\delta\mathbf{f}] \\ \mathbf{Q}_2^{(3)}[2;1;\mathbf{f};\delta\mathbf{f}] \end{pmatrix};$$
$$\mathbf{Q}_2^{(3)}[2;1;\mathbf{f};\delta\mathbf{f}] \triangleq \begin{pmatrix} -2(\delta f_1)\left[E(t)a^{(2)}(1;1;t)+a^{(1)}(t)a^{(2)}(2;1;t)\right] \\ -2(\delta f_1)E(t)a^{(2)}(2;1;t) \end{pmatrix}. \tag{5.202}$$

The notation used for the above quantities is as follows: (i) the argument "$4\times 4$" of the block-matrix $\mathbf{VM}^{(3)}[4\times 4;1;\mathbf{f}]$ indicates the dimensions of this matrix; (ii) the 2nd argument, namely "1," of the block-matrix $\mathbf{VM}^{(3)}[4\times 4;1;\mathbf{f}]$ indicates that this matrix corresponds to the quantity $\partial^2 E(\tau)/\partial f_1 \partial f_1$, which was labeled (by convention) the "*1st* 2nd-order sensitivity;" (iii) the 1st argument (i.e., "4") of the 3rd-level variational vector $\mathbf{V}^{(3)}(4;1;t)$ indicates that this vector has four components; (iv) the 2nd argument of $\mathbf{V}^{(3)}(4;1;t)$, namely, "1," indicates that this vector corresponds to $\partial^2 E(\tau)/\partial f_1 \partial f_1$, which was labeled (by convention) the "*1st* 2nd-order sensitivity." The notation for the arguments of the vector-function $\mathbf{Q}_V^{(3)}[4;1;\mathbf{f};\delta\mathbf{f}]$ is similar to the notation used for the block-vector $\mathbf{V}^{(3)}(4;1;t)$.

The need for repeatedly solving the 3rd-LVSS represented by Eqs. (5.197) and (5.198) for all parameter variations of interest is circumvented by applying the

3rd-FASAM-N to eliminate the appearance of the variational function $\mathbf{V}^{(3)}(4;1;t)$ in Eq. (5.191) by expressing this indirect-effect term in terms of a 3rd-level adjoint sensitivity function, which will have the same number of components as $\mathbf{V}^{(3)}(4;1;t)$ but would be independent of parameter variations. This 3rd-level adjoint function will be denoted as $\mathbf{A}^{(3)}(4;1;t) \triangleq \left[a^{(3)}(1;1;t), a^{(3)}(2;1;t), a^{(3)}(3;1;t), a^{(3)}(4;1;t)\right]^{\dagger}$, where: (i) the 1st argument of $\mathbf{A}^{(3)}(4;1;t)$, namely, "4," indicates that this vector-valued function has four components; (ii) the 2nd argument of $\mathbf{A}^{(3)}(4;1;t)$, namely, "1," indicates that this vector-valued function corresponds to $\partial^2 E(\tau)/\partial f_1 \partial f_1$, which was labeled (by convention) the "*1st* 2nd-order sensitivity."

The 3rd-level adjoint function $\mathbf{A}^{(3)}(4;1;t)$ will be obtained as the solution of a 3rd-Level Adjoint Sensitivity System (3rd-LASS), which will be constructed in a Hilbert space denoted as $\mathcal{H}_3$, and which comprises as elements block-vectors of the same form as $\mathbf{V}^{(3)}(4;1;t)$. The inner product, denoted as $\left\langle \mathbf{\Psi}^{(3)}(4;t), \mathbf{\Phi}^{(3)}(4;t)\right\rangle_3$, of two generic vectors $\mathbf{\Psi}^{(3)}(4;t) \triangleq \left[\psi^{(3)}(1;t),\ldots,\psi^{(3)}(4;t)\right]^{\dagger} \in \mathcal{H}_3$ and $\mathbf{\Phi}^{(3)}(4;t) \triangleq \left[\varphi^{(3)}(1;t),\ldots,\varphi^{(3)}(4;t)\right]^{\dagger} \in \mathcal{H}_3$ in the Hilbert space $\mathcal{H}_3$ is defined as follows:

$$\left\langle \mathbf{\Psi}^{(3)}(4;t), \mathbf{\Phi}^{(3)}(4;t)\right\rangle_3 \triangleq \sum_{i=1}^{4} \int_0^{\tau} \psi^{(3)}(i;t)\varphi^{(3)}(i;t)\,dt. \tag{5.203}$$

The 3rd-Level Adjoint Sensitivity System (3rd-LASS) for the 3rd-level adjoint function $\mathbf{A}^{(3)}(4;1;t) \triangleq \left[a^{(3)}(1;1;t), a^{(3)}(2;1;t), a^{(3)}(3;1;t), a^{(3)}(4;1;t)\right]^{\dagger} \in \mathcal{H}_3$, is constructed as follows:

1. Using Eq. (5.203), form the inner product of the vector $\mathbf{A}^{(3)}(4;1;t)$ with Eq. (5.197) to obtain the following relation:

$$\begin{aligned}
&\left\{\left\langle \mathbf{A}^{(3)}(4;1;t), \mathbf{VM}^{(3)}[4\times 4;1]\mathbf{V}^{(3)}(4;1;t)\right\rangle_3\right\}_{\mathbf{f}^0} = \left\{a^{(3)}(1;1;t)\delta E(t)\right.\\
&\left. -a^{(3)}(2;1;t)\delta a^{(1)}(t) - a^{(3)}(3;1;t)\delta a^{(2)}(1;1;t) + a^{(3)}(4;1;t)\delta a^{(2)}(2;1;t)\right\}_{t=0}^{t=\tau}\\
&+\left\{\left\langle \mathbf{V}^{(3)}(4;1;t), \mathbf{AM}^{(3)}[4\times 4;1]\mathbf{A}^{(3)}(4;1;t)\right\rangle_3\right\}_{\mathbf{f}^0}\\
&=\left\{\left\langle \mathbf{A}^{(3)}(4;1;t), \mathbf{Q}_V^{(3)}[4;1;\mathbf{f};\delta\mathbf{f}]\right\rangle_3\right\}_{\mathbf{f}^0}.
\end{aligned} \tag{5.204}$$

2. Eliminate the boundary terms on the right side of Eq. (5.204) and require the 2nd term on the right side of the 1st equality in Eq. (5.204) to represent the G-differential defined in Eq. (5.191) by imposing the following relations:

$$\left\{\mathbf{AM}^{(3)}[4\times 4;1]\mathbf{A}^{(3)}(4;1;t)\right\}_{\mathbf{f}^0} = \left\{\mathbf{Q}_A^{(3)}[4;1;\mathbf{f}]\right\}_{\mathbf{f}^0}, \quad 0<t<\tau, \tag{5.205}$$

$$\left\{\mathbf{B}_A^{(3)}\left[4;\mathbf{A}^{(3)}(4;1;t);\mathbf{f}\right]\right\}_{\mathbf{f}^0} \triangleq \begin{pmatrix} a^{(3)}(1;1;\tau) \\ a^{(3)}(2;1;0) \\ a^{(3)}(3;1;0) \\ a^{(3)}(4;1;\tau) \end{pmatrix} = \begin{pmatrix} 0 \\ 0 \\ 0 \\ 0 \end{pmatrix}; \tag{5.206}$$

where the block-matrix $\mathbf{AM}^{(3)}[4\times 4;1] \triangleq \left[\mathbf{VM}^{(3)}[4\times 4;1]\right]^*$ is the formal adjoint of the block-matrix $\mathbf{VM}^{(3)}[4\times 4;1]$, obtained by transposing the adjoints of the elements of $\mathbf{VM}^{(3)}[4\times 4;1]$, namely:

$$\mathbf{AM}^{(3)}[4\times 4;1] \triangleq \left[\mathbf{VM}^{(3)}[4\times 4;1]\right]^* = \begin{pmatrix} \left\{\left[\mathbf{VM}^{(2)}\right]^*\right\}^\dagger & \left\{\left[\mathbf{VM}_{21}^{(3)}\right]^*\right\}^\dagger \\ \mathbf{0}[2\times 2] & \left\{\left[\mathbf{VM}_{22}^{(3)}\right]^*\right\}^\dagger \end{pmatrix}; \tag{5.207}$$

and where the following definitions were used:

$$\mathbf{Q}_A^{(3)}[4;1;\mathbf{f}] \triangleq \left[\mathbf{q}_A^{(3)}(1;1;\mathbf{f}),\ldots,\mathbf{q}_A^{(3)}(4;1;\mathbf{f})\right]^\dagger; \tag{5.208}$$

$$\mathbf{q}_A^{(3)}(1;1;\mathbf{f}) \triangleq -2\left[a^{(2)}(1;1;t)E(t) + a^{(2)}(2;1;t)a^{(1)}(t)\right]; \tag{5.209}$$

$$\mathbf{q}_A^{(3)}(2;1;\mathbf{f}) \triangleq -2a^{(2)}(2;1;t)E(t); \tag{5.210}$$

$$\mathbf{q}_A^{(3)}(3;1;\mathbf{f}) \triangleq -E^2(t); \tag{5.211}$$

$$\mathbf{q}_A^{(3)}(4;1;\mathbf{f}) \triangleq -2a^{(1)}(t)E(t). \tag{5.212}$$

Replacing Eqs. (5.197), (5.198), and (5.191) into Eq. (5.204) reduces the latter to the following expression:

$$\frac{\partial^3 E(\tau)}{\partial f_1 \partial f_1 \partial f_1}\delta f_1 + \frac{\partial^3 E(\tau)}{\partial f_2 \partial f_1 \partial f_1}\delta f_2 = \left\{ \left\langle \mathbf{A}^{(3)}(4;1;t), \mathbf{Q}_V^{(3)}[4;1;\mathbf{f};\delta\mathbf{f}] \right\rangle_3 \right\}_{\mathbf{f}^0}$$
$$= \left\{ \int_0^\tau a^{(3)}(1;1;t)\left[ -(\delta f_1) E^2(t) + (\delta f_2) \right] dt \right\}_{\mathbf{f}^0}$$
$$+ \left\{ \int_0^\tau a^{(3)}(2;1;t)\left[ -2(\delta f_1) a^{(1)}(t) E(t) \right] dt \right\}_{\mathbf{f}^0} \tag{5.213}$$
$$-2(\delta f_1)\left\{ \int_0^\tau a^{(3)}(3;1;t)\left[ E(t) a^{(2)}(1;1;t) + a^{(1)}(t) a^{(2)}(2;1;t) \right] dt \right\}_{\mathbf{f}^0}$$
$$-2(\delta f_1)\left\{ \int_0^\tau a^{(3)}(4;1;t) E(t) a^{(2)}(2;1;t) dt \right\}_{\mathbf{f}^0}.$$

It follows from Eq. (5.213) that:

$$\frac{\partial^3 E(\tau)}{\partial f_1 \partial f_1 \partial f_1} = -\left\{ \int_0^\tau a^{(3)}(1;1;t) E^2(t) dt \right\}_{\mathbf{f}^0} - 2\left\{ \int_0^\tau a^{(3)}(2;1;t) a^{(1)}(t) E(t) dt \right\}_{\mathbf{f}^0}$$
$$-2\left\{ \int_0^\tau a^{(3)}(3;1;t)\left[ E(t) a^{(2)}(1;1;t) + a^{(1)}(t) a^{(2)}(2;1;t) \right] dt \right\}_{\mathbf{f}^0}$$
$$-2\left\{ \int_0^\tau a^{(3)}(4;1;t) E(t) a^{(2)}(2;1;t) dt \right\}_{\mathbf{f}^0}. \tag{5.214}$$

$$\frac{\partial^3 E(\tau)}{\partial f_2 \partial f_1 \partial f_1} = \left\{ \int_0^\tau a^{(3)}(1;1;t) dt \right\}_{\mathbf{f}^0}. \tag{5.215}$$

#### 5.5.1.2 Computation of 3rd-Order Sensitivities Stemming from the 2nd-Order Sensitivity $\partial^2 E(\tau)/\partial f_2 \partial f_2$

The 3rd-order sensitivities that stem from $\partial^2 E(\tau)/\partial f_2 \partial f_2$ are obtained by applying the definition of the 1st-order G-differential to Eq. (5.86), which yields the following expression:

$$\delta\left\{ \frac{\partial^2 E(\tau)}{\partial f_2 \partial f_2} \right\} \triangleq \frac{\partial^3 E(\tau)}{\partial f_1 \partial f_2 \partial f_2}\delta f_1 + \frac{\partial^3 E(\tau)}{\partial f_2 \partial f_2 \partial f_2}\delta f_2$$
$$\triangleq \left\{ \frac{d}{d\varepsilon}\left\{ \int_0^\tau \left[ a^{(2)}(1;2;t) + \varepsilon \delta a^{(2)}(1;2;t) \right] dt \right\}_{\mathbf{f}^0} \right\}_{\varepsilon=0} \tag{5.216}$$
$$= \left\{ \int_0^\tau \delta a^{(2)}(1;2;t) dt \right\}_{\mathbf{f}^0}.$$

Evidently, the 2nd-order sensitivity $\partial^2 E(\tau)/\partial f_2 \partial f_2$ gives rise to two 3rd-order sensitivities, $\partial^3 E(\tau)/\partial f_1 \partial f_2 \partial f_2$ and $\partial^3 E(\tau)/\partial f_2 \partial f_2 \partial f_2$, which will be determined in this subsection by solving a corresponding 3rd-Level Adjoint Sensitivity System (3rd-LASS). For bookkeeping purposes, the quantity $\partial^2 E(\tau)/\partial f_2 \partial f_2$ will the labeled "the *2nd* 2nd-order sensitivity" and the solution of the 3rd-LASS that corresponds to it will be labeled "the *2nd* 3rd-level adjoint function."

The expression obtained in Eq. (5.216) comprises no direct-effect term and can be evaluated after having determined the function $\delta a^{(2)}(1;2;t)$, which can be obtained as (part of) the solution of the G-differentiated 2nd-LASS defined by Eqs. (5.80) and (5.81), evaluated at the nominal values of the feature functions and state functions (i.e., dependent variables). Using the superscript "zero" to denote the respective nominal values of the various quantities and applying the definition of the G-differential to Eqs. (5.80) and (5.81) yields the following system of equations for the vector-valued function $\delta \mathbf{A}^{(2)}(2;2;t) \triangleq \left[\delta a^{(2)}(1;2;t), \delta a^{(2)}(2;2;t)\right]^{\dagger}$:

$$\left\{\frac{d}{d\varepsilon}\left[-\frac{d}{dt}+2\left(f_1^0+\varepsilon\delta f_1\right)\left(E^0+\varepsilon\delta E\right)\right]\left[a^{(2,0)}(1;2;t)+\varepsilon\delta a^{(2)}(1;2;t)\right]\right\}_{\varepsilon=0}$$
$$+2\left\{\frac{d}{d\varepsilon}\left(f_1^0+\varepsilon\delta f_1\right)\left(a^{(1,0)}+\varepsilon\delta a^{(1)}\right)\left[a^{(2,0)}(2;2;t)+\varepsilon\delta a^{(2)}(2;2;t)\right]\right\}_{\varepsilon=0}=0, \quad (5.217)$$

$$\left\{\frac{d}{d\varepsilon}\left[\frac{d}{dt}+2\left(f_1^0+\varepsilon\delta f_1\right)\left(E^0+\varepsilon\delta E\right)\right]\left[a^{(2,0)}(2;2;t)+\varepsilon\delta a^{(2)}(2;2;t)\right]\right\}_{\varepsilon=0}=0. \quad (5.218)$$

$$\left\{\delta \mathbf{B}_A^{(2)}\left[2;\delta \mathbf{A}^{(2)}(2;2;t);\mathbf{f}\right]\right\}_{\mathbf{f}^0} \triangleq \left\{\begin{pmatrix} \delta a^{(2)}(1;2;\tau) \\ \delta a^{(2)}(2;2;0) \end{pmatrix}\right\}_{\mathbf{f}^0} = \begin{pmatrix} 0 \\ 0 \end{pmatrix}. \quad (5.219)$$

Carrying out the operations with respect to $\varepsilon$ indicated in Eqs. (5.217)–(5.219) yields the following system of equations:

$$\left(2f_1^0\right)\delta E(t)+\left(2f_1^0\right)\delta a^{(1)}+\left[-\frac{d}{dt}+2f_1^0E^0(t)\right]\delta a^{(2)}(1;2;t)$$
$$+\left(2f_1^0a^{(1)}\right)\delta a^{(2)}(2;2;t)=-\left[E^0(t)a^{(2,0)}(1;2;t)+2a^{(1,0)}a^{(2,0)}(2;2;t)\right]\delta f_1; \quad (5.220)$$

$$2f_1^0a^{(2,0)}(2;2;t)\delta E+\left[\frac{d}{dt}+2f_1^0E^0(t_1)\right]\delta a^{(2)}(2;2;t)$$
$$=-2E^0(t)a^{(2,0)}(2;2;t)\delta f_1. \quad (5.221)$$

Concatenating Eqs. (5.220) and (5.221) with the equations underlying the 2nd-LVSS provided in Eqs. (5.64) and (5.65) for the 2nd-level vector-valued variational function $\mathbf{V}^{(2)}(2;t) \triangleq \left[\delta E(t), \delta a^{(1)}(t)\right]^{\dagger}$ yields the following 3rd-Level Variational Sensitivity System (3rd-LVSS) for the 3rd-level vector-valued variational function $\mathbf{V}^{(3)}(4;2;t) \triangleq \left[\mathbf{V}^{(2)}(2;t), \delta \mathbf{A}^{(2)}(2;2;t)\right]^{\dagger}$:

$$\left\{\mathbf{VM}^{(3)}[4\times 4;2;\mathbf{f}]\,\mathbf{V}^{(3)}(4;2;t)\right\}_{\mathbf{f}^0} = \left\{\mathbf{Q}_V^{(3)}[4;2;\mathbf{f};\delta\mathbf{f}]\right\}_{\mathbf{f}^0}, \quad 0<t<\tau, \quad (5.222)$$

together with the following boundary conditions:

$$\left[\delta E(0), \delta a^{(1)}(\tau), \delta a^{(2)}(1;2;\tau), \delta a^{(2)}(2;2;0)\right]^{\dagger} = [0,0,0,0]^{\dagger}. \quad (5.223)$$

The quantities which appear in Eq. (5.222) are defined as follows:

$$\mathbf{VM}^{(3)}[4\times 4;2;\mathbf{f}] \triangleq \begin{pmatrix} \mathbf{VM}^{(2)}[2\times 2;\mathbf{f}] & \mathbf{0}[2\times 2] \\ \mathbf{VM}_{21}^{(3)}[2\times 2;2;\mathbf{f}] & \mathbf{VM}_{22}^{(3)}[2\times 2;2;\mathbf{f}] \end{pmatrix}; \quad (5.224)$$

$$\mathbf{VM}_{21}^{(3)}[2\times 2;2;\mathbf{f}] \triangleq \begin{pmatrix} 2f_1(\alpha) & 2f_1(\alpha) \\ 2f_1(\alpha)a^{(2)}(2;2;t) & 0 \end{pmatrix}; \quad (5.225)$$

$$\mathbf{VM}_{22}^{(3)}[2\times 2;1;\mathbf{f}] \triangleq \mathbf{AM}^{(2)}[2\times 2;\mathbf{f}] \triangleq \begin{pmatrix} -\dfrac{d}{dt} + 2f_1(\alpha)E(t) & 2f_1(\alpha)a^{(1)}(t) \\ 0 & \dfrac{d}{dt} + 2f_1(\alpha)E(t) \end{pmatrix}; \quad (5.226)$$

$$\mathbf{Q}_V^{(3)}[4;2;\mathbf{f};\delta\mathbf{f}] \triangleq \begin{pmatrix} \mathbf{Q}_V^{(2)}[2;\mathbf{f};\delta\mathbf{f}] \\ \mathbf{Q}_2^{(3)}[2;2;\mathbf{f};\delta\mathbf{f}] \end{pmatrix};$$
$$\mathbf{Q}_2^{(3)}[2;2;\mathbf{f};\delta\mathbf{f}] \triangleq \begin{pmatrix} -(\delta f_1)\left[E(t)a^{(2)}(1;2;t) + 2a^{(1)}(t)a^{(2)}(2;2;t)\right] \\ -2(\delta f_1)E(t)a^{(2)}(2;2;t) \end{pmatrix}. \quad (5.227)$$

The notation used for the above quantities is similar to that used in Section 5.5.1.1, except for the replacement of the argument "1" with the argument "2" for the quantities that appear in Eq. (5.224), to indicate that all of the respective quantities now correspond to $\partial^2 E(\tau)/\partial f_2\,\partial f_2$, which was labeled (by convention) the "*2nd* 2nd-order sensitivity."

The need for solving the 3rd-LVSS for all parameter variations is circumvented by applying the 3rd-FASAM-N to eliminate the appearance of the variational function $\mathbf{V}^{(3)}(4;2;t)$ in Eq. (5.216), by expressing this indirect-effect term in terms of a 3rd-level adjoint sensitivity function which will be denoted as $\mathbf{A}^{(3)}(4;2;t) \triangleq \left[a^{(3)}(1;2;t), a^{(3)}(2;2;t), a^{(3)}(3;2;t), a^{(3)}(4;2;t)\right]^{\dagger}$, where: (i) the 1st argument of $\mathbf{A}^{(3)}(4;2;t)$, namely, "4," indicates that this vector-valued function has four components; (ii) the 2nd argument of $\mathbf{A}^{(3)}(4;2;t)$, namely, "2," indicates that this vector-valued function corresponds to $\partial^2 E(\tau)/\partial f_2 \partial f_2$, which was labeled (by convention) the "*2nd* 2nd-order sensitivity."

The 3rd-level adjoint function $\mathbf{A}^{(3)}(4;2;t)$ will be obtained as the solution of a 3rd-Level Adjoint Sensitivity System (3rd-LASS) which is constructed in the Hilbert space $\mathcal{H}_3$ (as introduced in Section 5.5.1.1), as follows:

1. Form the inner product of the vector $\mathbf{A}^{(3)}(4;2;t)$ with Eq. (5.222) to obtain the following relation:

$$
\begin{aligned}
&\left\{\left\langle \mathbf{A}^{(3)}(4;2;t), \mathbf{VM}^{(3)}[4\times 4;2]\mathbf{V}^{(3)}(4;2;t)\right\rangle_3\right\}_{\mathbf{f}^0} = \left\{a^{(3)}(1;2;t)\delta E(t)\right.\\
&\left. -a^{(3)}(2;2;t)\delta a^{(1)}(t) - a^{(3)}(3;2;t)\delta a^{(2)}(1;2;t) + a^{(3)}(4;2;t)\delta a^{(2)}(2;2;t)\right\}_{t=0}^{t=\tau}\\
&+\left\{\left\langle \mathbf{V}^{(3)}(4;2;t), \mathbf{AM}^{(3)}[4\times 4;2]\mathbf{A}^{(3)}(4;2;t)\right\rangle_3\right\}_{\mathbf{f}^0}\\
&=\left\{\left\langle \mathbf{A}^{(3)}(4;2;t), \mathbf{Q}_V^{(3)}[4;2;\mathbf{f};\delta\mathbf{f}]\right\rangle_3\right\}_{\mathbf{f}^0}.
\end{aligned}
\tag{5.228}
$$

2. Eliminate the boundary terms on the right side of Eq. (5.228) and require the 2nd term on the right side of the 1st equality in Eq. (5.228) to represent the G-differential defined in Eq. (5.216) by imposing the following relations:

$$
\left\{\mathbf{AM}^{(3)}[4\times 4;2]\mathbf{A}^{(3)}(4;2;t)\right\}_{\mathbf{f}^0} = \left\{\mathbf{Q}_A^{(3)}[4;2;\mathbf{f}]\right\}_{\mathbf{f}^0}, \quad 0<t<\tau, \tag{5.229}
$$

$$
\left\{\mathbf{B}_A^{(3)}\left[4;\mathbf{A}^{(3)}(4;2;t);\mathbf{f}\right]\right\}_{\mathbf{f}^0} \triangleq \begin{pmatrix} a^{(3)}(1;2;\tau) \\ a^{(3)}(2;2;0) \\ a^{(3)}(3;2;0) \\ a^{(3)}(4;2;\tau) \end{pmatrix} = \begin{pmatrix} 0 \\ 0 \\ 0 \\ 0 \end{pmatrix}; \tag{5.230}
$$

where the block-matrix $\mathbf{AM}^{(3)}[4\times 4;2] \triangleq \left[\mathbf{VM}^{(3)}[4\times 4;2]\right]^*$ is the formal adjoint of the block-matrix $\mathbf{VM}^{(3)}[4\times 4;2]$.

Replacing Eqs. (5.229), (5.230), and (5.216) into Eq. (5.228) reduces the latter to the following expression:

$$\frac{\partial^3 E(\tau)}{\partial f_1 \partial f_2 \partial f_2}\delta f_1+\frac{\partial^3 E(\tau)}{\partial f_2 \partial f_2 \partial f_2}\delta f_2=\left\{\left\langle \mathbf{A}^{(3)}(4;2;t),\mathbf{Q}_V^{(3)}[4;2;\mathbf{f};\delta\mathbf{f}]\right\rangle_3\right\}_{\mathbf{f}^0}$$
$$=\left\{\int_0^\tau a^{(3)}(1;2;t)\left[-(\delta f_1)E^2(t)+(\delta f_2)\right]dt\right\}_{\mathbf{f}^0}$$
$$+\left\{\int_0^\tau a^{(3)}(2;2;t)\left[-2(\delta f_1)a^{(1)}(t)E(t)\right]dt\right\}_{\mathbf{f}^0} \tag{5.231}$$
$$-(\delta f_1)\left\{\int_0^\tau a^{(3)}(3;2;t)\left[E(t)a^{(2)}(1;2;t)+2a^{(1)}(t)a^{(2)}(2;2;t)\right]dt\right\}_{\mathbf{f}^0}$$
$$-2(\delta f_1)\left\{\int_0^\tau a^{(3)}(4;2;t)E(t)a^{(2)}(2;2;t)dt\right\}_{\mathbf{f}^0}.$$

It follows from Eq. (5.231) that:

$$\frac{\partial^3 E(\tau)}{\partial f_1 \partial f_2 \partial f_2}=-\left\{\int_0^\tau a^{(3)}(1;2;t)E^2(t)dt\right\}_{\mathbf{f}^0}-2\left\{\int_0^\tau a^{(3)}(2;2;t)a^{(1)}(t)E(t)dt\right\}_{\mathbf{f}^0}$$
$$-\left\{\int_0^\tau a^{(3)}(3;2;t)\left[E(t)a^{(2)}(1;2;t)+2a^{(1)}(t)a^{(2)}(2;2;t)\right]dt\right\}_{\mathbf{f}^0}$$
$$-\left\{\int_0^\tau a^{(3)}(4;2;t)E(t)a^{(2)}(2;2;t)dt\right\}_{\mathbf{f}^0}. \tag{5.232}$$

$$\frac{\partial^3 E(\tau)}{\partial f_2 \partial f_2 \partial f_2}=\left\{\int_0^\tau a^{(3)}(1;2;t)dt\right\}_{\mathbf{f}^0}. \tag{5.233}$$

### 5.5.1.3 Using Sensitivities with Respect to the Feature Functions to Compute Most Efficiently 3rd-Order Response Sensitivities to Primary Model Parameters

The expression of $\partial^2 E(\tau)/\partial f_2 \partial f_1$ is provided in Eq. (5.78) in terms of the 1st-level adjoint function $\mathbf{A}^{(2)}(2;1;t)\triangleq\left[a^{(2)}(1;1;t),a^{(2)}(2;1;t)\right]^\dagger$ while the equivalent expression of $\partial^2 E(\tau)/\partial f_1 \partial f_2$ is provided in Eq. (5.85) in terms of the 1st-level adjoint function $\mathbf{A}^{(2)}(2;2;t)\triangleq\left[a^{(2)}(1;2;t),a^{(2)}(2;2;t)\right]^\dagger$. Examining these two equivalent expressions indicates that the expression for $\partial^2 E(\tau)/\partial f_2 \partial f_1$ is much simpler – and hence more convenient to use – than the expression for $\partial^2 E(\tau)/\partial f_1 \partial f_2$. The 2nd-order sensitivity $\partial^2 E(\tau)/\partial f_2 \partial f_1$ gives rise to the 3rd-order sensitivities $\partial^3 E(\tau)/\partial f_1 \partial f_2 \partial f_1$ and $\partial^3 E(\tau)/\partial f_2 \partial f_2 \partial f_1$, while the 2nd-order sensitivity $\partial^2 E(\tau)/\partial f_1 \partial f_2$ gives rise to the 3rd-order sensitivities $\partial^3 E(\tau)/\partial f_1 \partial f_1 \partial f_2$ and $\partial^3 E(\tau)/\partial f_2 \partial f_1 \partial f_2$.

In summary, one large-scale computation is needed for solving the 3rd-LASS defined by Eqs. (5.205) and (5.206), which yields the unmixed 3rd-order sensitivity $\partial^3 E(\tau)/\partial f_1 \partial f_1 \partial f_1$. A 2nd large-scale computation is needed for solving the 3rd-LASS defined by Eqs. (5.229) and (5.230), which yields the unmixed 3rd-order sensitivity $\partial^3 E(\tau)/\partial f_2 \partial f_2 \partial f_2$. These two large-scale computations also yield the mixed 3rd-order sensitivities $\partial^3 E(\tau)/\partial f_2 \partial f_1 \partial f_1$ and $\partial^3 E(\tau)/\partial f_1 \partial f_2 \partial f_2$. Thus, *the computation of all 3rd-order sensitivities necessitates just two large-scale computations.* For verification purpose, one may consider performing a 3rd and/or 4th large-scale computation for solving the 3rd-LASS that corresponds to $\partial^2 E(\tau)/\partial f_2 \partial f_1$ and/or $\partial^2 E(\tau)/\partial f_1 \partial f_2$. These additional large-scale computations yield alternative equivalent expressions/results for the mixed 3rd-order sensitivities.

The 3rd-order sensitivities of the response with respect to the primary model parameters are obtained by analytical, exact differentiation of the general expression provided in Eq. (5.87) with respect to an arbitrary primary model parameter $\alpha_k$, i.e.,

$$\frac{\partial^3 E(\tau;f_1;f_2)}{\partial \alpha_k \partial \alpha_j \partial \alpha_i} = \frac{\partial}{\partial \alpha_k}\left\{\left[\frac{\partial^2 E(\tau)}{\partial f_1 \partial f_1}\frac{\partial f_1(\boldsymbol{\alpha})}{\partial \alpha_j} + \frac{\partial^2 E(\tau)}{\partial f_2 \partial f_1}\frac{\partial f_2(\boldsymbol{\alpha})}{\partial \alpha_j}\right]\frac{\partial f_1(\boldsymbol{\alpha})}{\partial \alpha_i}\right\}$$
$$+\frac{\partial}{\partial \alpha_k}\left\{\frac{\partial E(\tau)}{\partial f_1}\frac{\partial^2 f_1(\boldsymbol{\alpha})}{\partial \alpha_j \partial \alpha_i}\right\} + \frac{\partial}{\partial \alpha_k}\left\{\frac{\partial E(\tau)}{\partial f_2}\frac{\partial^2 f_2(\boldsymbol{\alpha})}{\partial \alpha_j \partial \alpha_i}\right\}$$
$$+\frac{\partial}{\partial \alpha_k}\left\{\left[\frac{\partial^2 E(\tau)}{\partial f_1 \partial f_2}\frac{\partial f_1(\boldsymbol{\alpha})}{\partial \alpha_j} + \frac{\partial^2 E(\tau)}{\partial f_2 \partial f_2}\frac{\partial f_2(\boldsymbol{\alpha})}{\partial \alpha_j}\right]\frac{\partial f_2(\boldsymbol{\alpha})}{\partial \alpha_i}\right\}; \quad i,j,k = 1,\ldots,7. \tag{5.234}$$

The numerical result for any of the 3rd-order sensitivity of the response with respect to a primary model parameter is obtained by performing the actual differentiation in Eq. (5.234) and subsequently replacing in the resulting expression the analytical expression of the sensitivity of the respective feature-functions with respect to the parameter(s) under consideration, together with the (numerical) results for the respective sensitivities of the response with respect to the feature functions. Evidently, no additional large-scale computations are required to obtain the 3rd-order response sensitivities to the primary parameters from the response sensitivities to the feature functions. Altogether, *two large-scale computations (for solving the aforementioned 3rd-LASS) suffice for obtaining all of the 343 3rd-order sensitivities* (of which 84 are distinct from each other) of the response with respect to the primary model parameters.

### 5.5.2 Computation of 3rd-Order Sensitivities Using the 3rd-CASAM-N

Just like the 3rd-FASAM-N, the 3rd-CASAM-N methodology generically considers that the 3rd-order sensitivities are the "1st-order sensitivities of the 2nd-order sensitivities." The application of the 3rd-CASAM-N for determining the 3rd-order sensitivities of the response with respect to the primary model parameters will be generally illustrated in this subsection by considering the general functional dependence of the 2nd-order sensitivities obtained in Section 5.4.2. As has been shown in Section 5.4.2, as many large-scale computations as there are primary model parameters ($TP = 7$) are needed to solve the seven distinct 2nd-Level Adjoint Sensitivity Systems (2nd-LASS) for obtaining all $TP^2 = 49$ 2nd-order sensitivities with respect to the model parameters, of which $TP(TP+1)/2 = 28$ are distinct from each other. Examining the expressions obtained in Section 5.4.2 for the 2nd-order sensitivities of the response with respect to the primary model parameters indicates that they all have the following generic form:

$$\frac{\partial^2 E(\tau)}{\partial \alpha_j \partial \alpha_i} = \int_0^\tau F\left[j;i;E(t);a^{(1)}(t);c^{(2)}(1;i;t),c^{(2)}(2;i;t);\boldsymbol{\alpha}\right]dt, \quad i,j = 1,\dots,TP = 7. \tag{5.235}$$

The 3rd-order sensitivities of the response $E(\tau)$ with respect to the primary model parameters are obtained as the 1st-order G-differential of the expression considered in Eq. (5.235) which by definition yields the following expression:

$$\delta\left\{\frac{\partial^2 E(\tau)}{\partial \alpha_j \partial \alpha_i}\right\} \triangleq \left\{\left\{\frac{d}{d\varepsilon}\int_0^\tau F\left[j;i;E(t);a^{(1)}(t);c^{(2)}(1;i;t),c^{(2)}(2;i;t);\boldsymbol{\alpha}\right]dt\right\}_{\varepsilon=0}\right\}_{\boldsymbol{\alpha}^0}$$
$$\triangleq \left\{\delta\left[\partial^2 E(\tau)/\partial\alpha_j\partial\alpha_i\right]\right\}_{dir} + \left\{\delta\left[\partial^2 E(\tau)/\partial\alpha_j\partial\alpha_i\right]\right\}_{ind}, \quad j,i = 1,\dots,7; \tag{5.236}$$

where the direct-effect and indirect-effect terms, respectively, are defined as follows:

$$\left\{\delta\left[\partial^2 E(\tau)/\partial\alpha_j\partial\alpha_i\right]\right\}_{dir} \triangleq \sum_{k=1}^{TP=7}\left\{\int_0^\tau\left[\partial F(j;i;\dots;\boldsymbol{\alpha})/\partial\alpha_k\right]dt\right\}_{\boldsymbol{\alpha}^0}\delta\alpha_k; \tag{5.237}$$

$$\left\{\delta\left[\partial^2 E(\tau)/\partial\alpha_j\partial\alpha_i\right]\right\}_{ind} \triangleq \left\{\int_0^\tau \frac{\partial F[j;i;\dots;\boldsymbol{\alpha}]}{\partial E}\delta E(t)\,dt\right\}_{\boldsymbol{\alpha}^0}$$
$$+\left\{\int_0^\tau \frac{\partial F[j;i;\dots;\boldsymbol{\alpha}]}{\partial a^{(1)}}\delta a^{(1)}(t)\,dt\right\}_{\boldsymbol{\alpha}^0} + \left\{\int_0^\tau \frac{\partial F[j;i;\dots;\boldsymbol{\alpha}]}{\partial c^{(2)}(1;i;t)}\delta c^{(2)}(1;i;t)\,dt\right\}_{\boldsymbol{\alpha}^0}$$
$$+\left\{\int_0^\tau \frac{\partial F[j;i;\dots;\boldsymbol{\alpha}]}{\partial c^{(2)}(2;i;t)}\delta c^{(2)}(2;i;t)\,dt\right\}_{\boldsymbol{\alpha}^0}. \tag{5.238}$$

The direct-effect term defined in Eq. (5.237) can be computed already at this stage. The indirect-effect term may, in principle, be computed after having determined the variational functions $\delta E(t)$, $\delta a^{(1)}(t)$, $\delta c^{(2)}(1;i;t)$ and $\delta c^{(2)}(2;i;t)$. The variational functions $\delta E(t)$ and $\delta a^{(1)}(t)$ are the solutions of the 1st-LVSS represented by Eqs. (5.100) and (5.101).

Recall that the 2nd-level adjoint functions $c^{(2)}(1;i;t), c^{(2)}(2;i;t), i = 1,\dots,7$, are the solutions of the corresponding 2nd-LASS, namely:

1. the functions $c^{(2)}(1;1;t), c^{(2)}(2;1;t)$ are the solutions of Eqs. (5.108) and (5.109);
2. the functions $c^{(2)}(1;2;t), c^{(2)}(2;2;t)$ are the solutions of Eqs. (5.122) and (5.123);
3. the functions $c^{(2)}(1;3;t), c^{(2)}(2;3;t)$ are the solutions of Eqs. (5.134) and (5.135);
4. the functions $c^{(2)}(1;4;t), c^{(2)}(2;4;t)$ are the solutions of Eqs. (5.146) and (5.147);
5. the functions $c^{(2)}(1;5;t), c^{(2)}(2;5;t)$ are the solutions of Eqs. (5.158) and (5.159);
6. the functions $c^{(2)}(1;6;t), c^{(2)}(2;6;t)$ are the solutions of Eqs. (5.170) and (5.171);
7. the functions $c^{(2)}(1;7;t), c^{(2)}(2;7;t)$ are the solutions of Eqs. (5.182) and (5.183).

The 2nd-LASS enumerated above can be written in the following generic form for $i = 1,\dots,7 = TP$, where "$TP$" denotes the "total number of primary model parameters":

$$\left\{\mathbf{AM}^{(2)}\left[2\times 2; E(t), a^{(1)}(t); \alpha\right]\mathbf{C}^{(2)}(2;i;t)\right\}_{\alpha^0} = \left\{\mathbf{Q}_A^{(2)}\left[2;i;E(t), a^{(1)}(t);\alpha\right]\right\}_{\alpha^0}, \quad 0 < t < \tau; \tag{5.239}$$

$$\left\{\mathbf{B}_A^{(2)}\left[2;\mathbf{C}^{(2)}(2;i;t);\alpha\right]\right\}_{\alpha^0} \triangleq \begin{pmatrix} c^{(2)}(1;i;\tau) \\ c^{(2)}(2;i;0) \end{pmatrix}_{\alpha^0} = \begin{pmatrix} 0 \\ 0 \end{pmatrix}; \tag{5.240}$$

where:

$$\mathbf{Q}_A^{(2)}\left[2;i;E(t), a^{(1)}(t);\alpha\right] \triangleq \begin{pmatrix} q_A^{(2)}\left(1;i;E(t), a^{(1)}(t);\alpha\right) \\ q_A^{(2)}\left(2;i;E(t), a^{(1)}(t);\alpha\right) \end{pmatrix}. \tag{5.241}$$

The variational functions $\delta c^{(2)}(1;i;t)$ and $\delta c^{(2)}(2;i;t)$ are the solutions of the equations that result by G-differentiating Eqs. (5.239) and (5.240), which can be generically represented as follows:

$$\left\{\frac{\partial\left[\mathbf{AM}^{(2)}\mathbf{C}^{(2)}(2;i;t)-\mathbf{Q}_A^{(2)}\right]}{\partial E}\delta E(t)\right\}_{\alpha^0}+\left\{\mathbf{AM}^{(2)}\left[\delta\mathbf{C}^{(2)}(2;i;t)\right]\right\}_{\alpha^0}$$

$$+\left\{\frac{\partial\left[\mathbf{AM}^{(2)}\mathbf{C}^{(2)}(2;i;t)-\mathbf{Q}_A^{(2)}\right]}{\partial a^{(1)}}\delta a^{(1)}(t)\right\}_{\alpha^0} \tag{5.242}$$

$$=\left\{\frac{\partial\left[\mathbf{Q}_A^{(2)}-\mathbf{AM}^{(2)}\mathbf{C}^{(2)}(2;i;t)\right]}{\partial\alpha}\delta\alpha\right\}_{\alpha^0}.$$

$$\left\{\delta\mathbf{B}_A^{(2)}\left[2;\mathbf{C}^{(2)}(2;i;t);\alpha\right]\right\}_{\alpha^0}\triangleq\begin{pmatrix}\delta c^{(2)}(1;i;\tau)\\ \delta c^{(2)}(2;i;0)\end{pmatrix}_{\alpha^0}=\begin{pmatrix}0\\0\end{pmatrix};\quad i=1,\ldots,7. \tag{5.243}$$

As indicated by Eqs. (5.242) and (5.243), the variational functions $\delta c^{(2)}(1;i;t)$ and $\delta c^{(2)}(2;i;t)$ are coupled to the variational functions $\delta E(t)$ and $\delta a^{(1)}(t)$. It follows that these variational functions can be determined as the solution of the system of equations obtained by concatenating Eqs. (5.242) and (5.243) with Eqs. (5.64) and (5.65). The $4\times 4$-dimensional system thus obtained is called the 3rd-Level Variational Sensitivity System (3rd-LVSS) for the 3rd-level variational vector-valued function $\mathbf{U}^{(3)}(4;i;t)\triangleq\left[\mathbf{V}^{(2)}(2;t),\delta\mathbf{C}^{(2)}(2;i;t)\right]^{\dagger}$, where $\mathbf{V}^{(2)}(2;t)\triangleq\left[\delta E(t),\delta a^{(1)}(t)\right]^{\dagger}$ and $\mathbf{C}^{(2)}(2;i;t)\triangleq\left[\delta c^{(2)}(1;i;t),\delta c^{(2)}(2;i;t)\right]^{\dagger}$, for $i=1,\ldots,TP=7$. This 3rd-LVSS is linear in $\mathbf{U}^{(3)}(4;i;t)$ and can be generically represented in the following form:

$$\left\{\mathbf{UM}^{(3)}\left[4\times 4;i;\alpha\right]\mathbf{U}^{(3)}(4;i;t)\right\}_{\alpha^0}=\left\{\mathbf{S}_V^{(3)}\left[4;i;\alpha;\delta\alpha\right]\right\}_{\alpha^0},\quad 0<t<\tau, \tag{5.244}$$

together with the following boundary conditions:

$$\left[\delta E(0),\delta a^{(1)}(\tau),\delta a^{(2)}(1;i;\tau),\delta a^{(2)}(2;i;0)\right]^{\dagger}=\left[0,0,0,0\right]^{\dagger}. \tag{5.245}$$

The specific expressions of the $4\times 4$-dimensional matrix $\mathbf{UM}^{(3)}\left[4\times 4;i;\alpha\right]$ and 4-dimensional vector $\mathbf{S}_V^{(3)}\left[4;i;\alpha;\delta\alpha\right]$ which appear in Eq. (5.244) differ according to the value of the index $i=1,\ldots,TP=7$, but their specific expressions are not needed for the purpose of presenting the generic characteristics of computing the 3rd-order response sensitivities with respect to the primary model parameters by applying the 3rd-CASAM-N methodology. Following the principles underlying this methodology, the need for determining explicitly the variational vector-valued function $\mathbf{U}^{(3)}(4;i;t)$ is avoided by eliminating its appearance in Eq. (5.238) by recasting the indirect-effect term into an equivalent expression involving a corresponding (4-dimensional vector-valued) 3rd-level adjoint sensitivity function, which will be constructed in the

Hilbert space $\mathscr{H}_3$. The corresponding vector-valued 3rd-level adjoint function will be denoted as $\mathbf{C}^{(3)}(4;j;i;t) \triangleq \left[c^{(3)}(1;j;i;t),\ldots,c^{(3)}(4;j;i;t)\right]^{\dagger} \in \mathscr{H}_3$, where the indices "$(j;i)$" indicate that this adjoint function corresponds to the "$(j;i)$th second-order sensitivity" in Eq. (5.238). The 3rd-LASS for $\mathbf{C}^{(3)}(4;j;i;t)$ is constructed by applying the 3rd-CASAM-N, which employs the same principles as the 3rd-FASAM-N, as follows:

1. Using the definition provided by Eq. (5.203), form the inner product in $\mathscr{H}_3$ of the vector $\mathbf{C}^{(3)}(4;j;i;t)$ with Eq. (5.244) to obtain the following relation:

$$
\begin{aligned}
&\left\{\left\langle \mathbf{C}^{(3)}(4;j;i;t), \mathbf{UM}^{(3)}[4\times 4;i;\boldsymbol{\alpha}]\mathbf{U}^{(3)}(4;i;t)\right\rangle_3\right\}_{\boldsymbol{\alpha}^0} = \left\{c^{(3)}(1;j;i;t)\,\delta E(t)\right.\\
&\left.-c^{(3)}(2;j;i;t)\,\delta a^{(1)}(t) - c^{(3)}(3;j;i;t)\,\delta c^{(2)}(1;i;t) + c^{(3)}(4;j;i;t)\,\delta c^{(2)}(2;i;t)\right\}_{t=0}^{t=t_f}\\
&+\left\{\left\langle \mathbf{U}^{(3)}(4;i;t), \left[\mathbf{UM}^{(3)}(4\times 4;i;\boldsymbol{\alpha})\right]^{*}\mathbf{C}^{(3)}(4;j;i;t)\right\rangle_3\right\}_{\boldsymbol{\alpha}^0}\\
&=\left\{\left\langle \mathbf{C}^{(3)}(4;j;i;t), \mathbf{S}_V^{(3)}[4;i;\boldsymbol{\alpha};\delta\boldsymbol{\alpha}]\right\rangle_3\right\}_{\boldsymbol{\alpha}^0},
\end{aligned}
\tag{5.246}
$$

where $\left[\mathbf{UM}^{(3)}(4\times 4;i;\boldsymbol{\alpha})\right]^{*}$ denotes the formal adjoint of the operator-valued matrix $\mathbf{UM}^{(3)}[4\times 4;i;\boldsymbol{\alpha}]$.

2. Eliminate the boundary terms on the right side of Eq. (5.246) and require the term $\left\{\left\langle \mathbf{U}^{(3)}(4;i;t), \left[\mathbf{UM}^{(3)}(4\times 4;i;\boldsymbol{\alpha})\right]^{*}\mathbf{C}^{(3)}(4;j;i;t)\right\rangle_3\right\}_{\boldsymbol{\alpha}^0}$ in Eq. (5.246) to represent the indirect-effect term defined in Eq. (5.238) by imposing the following relations:

$$
\begin{aligned}
&\left\{\left[\mathbf{UM}^{(3)}(4\times 4;i;\boldsymbol{\alpha})\right]^{*}\mathbf{C}^{(3)}(4;j;i;t)\right\}_{\boldsymbol{\alpha}^0}\\
&=\left\{\left[\frac{\partial F[j;i;\ldots;\boldsymbol{\alpha}]}{\partial E}, \frac{\partial F[j;i;\ldots;\boldsymbol{\alpha}]}{\partial a^{(1)}}, \frac{\partial F[j;i;\ldots;\boldsymbol{\alpha}]}{\partial c^{(2)}(1;i;t)}, \frac{\partial F[j;i;\ldots;\boldsymbol{\alpha}]}{\partial c^{(2)}(2;i;t)}\right]^{\dagger}\right\}_{\boldsymbol{\alpha}^0}.
\end{aligned}
\tag{5.247}
$$

$$
c^{(3)}(1;j;i;\tau) = c^{(3)}(2;j;i;0) = c^{(3)}(3;j;i;0) = c^{(3)}(4;j;i;\tau) = 0. \tag{5.248}
$$

The relations represented by Eqs. (5.247) and (5.248) constitute the 3rd-LASS for the 3rd-level adjoint function $\mathbf{C}^{(3)}(4;j;i;t)$.

3. Use the relations provided in Eqs. (5.246)–(5.248) in Eq. (5.238) to obtain the following expression for the indirect-effect term $\left\{\delta\left[\partial^2 E\left(t_f\right)/\partial\gamma\,\partial l_p\right]\right\}_{ind}$:

$$\left\{\delta\left[\partial^2 E(\tau)/\partial\alpha_j\,\partial\alpha_i\right]\right\}_{ind} \triangleq \left\{\left\langle \mathbf{C}^{(3)}(4;j;i;t), \mathbf{S}_V^{(3)}[4;i;\boldsymbol{\alpha};\delta\boldsymbol{\alpha}]\right\rangle_3\right\}_{\boldsymbol{\alpha}^0}. \quad (5.249)$$

4. Finally, the 3rd-order sensitivities, $\partial^3 E(\tau)/\partial\alpha_k\,\partial\alpha_j\,\partial\alpha_i, i,j,k = 1,\ldots,7 = TP$, of the response with respect to the primary model parameters are obtained by adding the expression for the indirect-effect term obtained in Eq. (5.249) with the expression for the direct-effect term defined by Eq. (5.237) and subsequently identifying the quantities that multiply the parameter variations $\delta\alpha_k$, $k = 1,\ldots,7 = TP$.

Because of the symmetry of the mixed 2nd-order sensitivities $\partial^2 E(\tau)/\partial\alpha_j\,\partial\alpha_i = \partial^2 E(\tau)/\partial\alpha_i\,\partial\alpha_j, i,j = 1,\ldots,7 = TP$, only $TP(TP+1)/2 = 28$ of the total of 49 2nd-order response sensitivities with respect to the primary model parameters will need to be considered for providing the source terms for the right side of the 3rd-LASS defined in Eq. (5.247). This means that at most 28 "responses" would need to be considered for providing the sources on the right side of the 3rd-LASS defined in Eq. (5.247), which in turn implies that at most 28 "large-scale" computations would be needed to obtain all of the 3rd-order sensitivities, $\partial^3 E(\tau)/\partial\alpha_k\,\partial\alpha_j\,\partial\alpha_i$, $i,j,k = 1,\ldots,7 = TP$. However, if all of these large-scale computations were performed, many of the mixed 3rd-order sensitivities would be computed twice, and some of them would even be computed thrice.

Note that of the total number of $TP^3 = 343$ 3rd-order sensitivities, only $TP(TP+1)(TP+2)/6 = 84$ of them are distinct. The minimum number of "large-scale" computations needed to obtain all of these 84 distinct 3rd-order sensitivities can be deduced as follows:

a. The 2nd-order sensitivities $\partial^2 E(\tau)/\partial\alpha_i\,\partial\alpha_i$, $i = 1,\ldots,7 = TP$, are uniquely obtained when solving the 2nd-LASS, so they certainly must serve as "responses" for the "large-scale" computations needed to solve the corresponding 3rd-LASS. Performing these seven large-scale computations (i.e., solving the 3rd-LASS that correspond to $\partial^2 E(\tau)/\partial\alpha_i\,\partial\alpha_i$) will yield 49 3d-order sensitivities of the form $\partial^3 E(\tau)/\partial\alpha_k\,\partial\alpha_i\,\partial\alpha_i, i,k = 1,\ldots,7 = TP$. These seven "large-scale" computations are mandatory since these are the only ones that will produce the unique 3rd-order sensitivities of the form $\partial^3 E(\tau)/\partial\alpha_k\,\partial\alpha_k\,\partial\alpha_k, k = 1,\ldots,7 = TP$, included in the 49 sensitivities of the form $\partial^3 E(\tau)/\partial\alpha_k\,\partial\alpha_i\,\partial\alpha_i, i,k = 1,\ldots,7 = TP$, that are obtained this way.
b. Five "large-scale computations" are performed using the 2nd-order sensitivities of the form $\partial^2 E(\tau)/\partial\alpha_j\,\partial\alpha_1 = \partial^2 E(\tau)/\partial\alpha_1\,\partial\alpha_j$, $j = 2,\ldots,6 = TP-1$, as "responses" for the respective 3rd-LASS. These five "large-scale computations" will yield 35 3rd-order sensitivities of the form $\partial^3 E(\tau)/\partial\alpha_k\,\partial\alpha_j\,\partial\alpha_1$, $k = 1,\ldots,7 = TP$, of which the following 15 are distinct: (i) $\partial^3 E(\tau)/\partial\alpha_k\,\partial\alpha_2\,\partial\alpha_1$ for $k = 3,\ldots,7 = TP$; (ii) $\partial^3 E(\tau)/\partial\alpha_k\,\partial\alpha_3\,\partial\alpha_1$

for $k = 4,\ldots,7 = TP$; (iii) $\partial^3 E(\tau)/\partial\alpha_k\,\partial\alpha_4\,\partial\alpha_1$ for $k = 5,6,7 = TP$; (iv) $\partial^3 E(\tau)/\partial\alpha_k\,\partial\alpha_5\,\partial\alpha_1$ for $k = 6,7 = TP$; (v) $\partial^3 E(\tau)/\partial\alpha_7\,\partial\alpha_6\,\partial\alpha_1$.

c. Four "large-scale computations" are performed using the 2nd-order sensitivities of the form $\partial^2 E(\tau)/\partial\alpha_j\,\partial\alpha_2 = \partial^2 E(\tau)/\partial\alpha_2\,\partial\alpha_j$, $j = 3,\ldots,6 = TP-1$, as "responses" for the respective 3rd-LASS. These four "large-scale computations" will yield 28 3rd-order sensitivities of the form $\partial^3 E(\tau)/\partial\alpha_k\,\partial\alpha_j\,\partial\alpha_2$, $k = 1,\ldots,7 = TP$, of which the following ten are distinct: (i) $\partial^3 E(\tau)/\partial\alpha_k\,\partial\alpha_3\,\partial\alpha_2$ for $k = 4,\ldots,7 = TP$; (ii) $\partial^3 E(\tau)/\partial\alpha_k\,\partial\alpha_4\,\partial\alpha_2$ for $k = 5,6,7 = TP$; (iii) $\partial^3 E(\tau)/\partial\alpha_k\,\partial\alpha_5\,\partial\alpha_2$ for $k = 6,7 = TP$; (iv) $\partial^3 E(\tau)/\partial\alpha_7\,\partial\alpha_6\,\partial\alpha_2$.

d. Three "large-scale computations" are performed using the 2nd-order sensitivities of the form $\partial^2 E(\tau)/\partial\alpha_j\,\partial\alpha_3 = \partial^2 E(\tau)/\partial\alpha_3\,\partial\alpha_j$, $j = 4,5,6 = TP-1$, as "responses" for the respective 3rd-LASSs. These three "large-scale computations" will yield 21 3rd-order sensitivities of the form $\partial^3 E(\tau)/\partial\alpha_k\,\partial\alpha_j\,\partial\alpha_2$, $k = 1,\ldots,7 = TP$, of which the following six are distinct: (i) $\partial^3 E(\tau)/\partial\alpha_k\,\partial\alpha_4\,\partial\alpha_3$ for $k = 5,6,7 = TP$; (ii) $\partial^3 E(\tau)/\partial\alpha_k\,\partial\alpha_5\,\partial\alpha_3$ for $k = 6,7 = TP$; (iii) $\partial^3 E(\tau)/\partial\alpha_7\,\partial\alpha_6\,\partial\alpha_3$.

e. Two "large-scale computations" are performed using the 2nd-order sensitivities of the form $\partial^2 E(\tau)/\partial\alpha_j\,\partial\alpha_4 = \partial^2 E(\tau)/\partial\alpha_4\,\partial\alpha_j$, $j = 5,6 = TP-1$, as "responses" for the respective 3rd-LASSs. These two "large-scale computations" will yield 14 3rd-order sensitivities of the form $\partial^3 E(\tau)/\partial\alpha_k\,\partial\alpha_j\,\partial\alpha_4$, $k = 1,\ldots,7 = TP$, of which the following three are distinct: (i) $\partial^3 E(\tau)/\partial\alpha_k\,\partial\alpha_5\,\partial\alpha_4$ for $k = 6,7 = TP$; (ii) $\partial^3 E(\tau)/\partial\alpha_7\,\partial\alpha_6\,\partial\alpha_4$.

f. One "large-scale computation" is performed using the 2nd-order sensitivity $\partial^2 E(\tau)/\partial\alpha_6\,\partial\alpha_5 = \partial^2 E(\tau)/\partial\alpha_5\,\partial\alpha_6$ as the "response" for the respective 3rd-LASS. This "large-scale computation" will yield seven 3rd-order sensitivities of the form $\partial^3 E(\tau)/\partial\alpha_k\,\partial\alpha_6\,\partial\alpha_5$, $k = 1,\ldots,7 = TP$, of which the following is distinct: $\partial^3 E(\tau)/\partial\alpha_7\,\partial\alpha_6\,\partial\alpha_5$.

In summary, a minimum of 22 "large-scale" computations for solving 22 distinct 3rd-LASSs are necessary to obtain all of the distinct 3rd-order sensitivities of the response with respect to the seven primary model parameters. In addition to the distinct 3rd-order sensitivities, these 22 computations will also yield alternative expressions/results for the 3rd-order sensitivities enumerated above. On the other hand, if duplicate computations of all of the mixed 3rd-order sensitivities are desired, then all 28 distinct 2nd-order sensitivities should be used as "responses" for the corresponding 28 3rd-LASSs.

## 5.6 CHAPTER SUMMARY

Computational models of physical systems often comprise not only imprecisely known primary model parameters (e.g., geometrical dimensions, microscopic nuclear cross-sections, atomic number densities, etc.) but often comprise "feature" functions of such parameters, such as macroscopic cross-sections, Reynolds numbers, Nusselt numbers, etc. When such "feature" functions of model parameters can be identified, the "*n*th-Order Feature Sensitivity Analysis Methodology for Nonlinear Systems" (abbreviated as "*n*th-FASAM-N") presented in Chapter 4 provides the unsurpassed efficiency for deriving and computing the exact explicit expressions of the sensitivities – of arbitrarily high order – of model responses with respect to the model's parameters, by initially determining the sensitivities of the respective responses with respect to the feature functions and subsequently determining analytically the sensitivities with respect to the model's parameters. The number of large-scale computations for determining the 2nd-and higher-order sensitivities within the *n*th-FASAM-N methodology is proportional to the number of feature-functions (as opposed to the number of primary parameters). For the computation of the 1st-order sensitivities, both the *n*th-FASAM-N and the *n*th-CASAM-N methodologies require a single "large-scale" "adjoint" computation, regardless of the number of model parameters. On the other hand, the application of the *n*th-FASAM-N methodology for obtaining the higher-order sensitivities requires the least number of large-scale computations, since the number of feature-functions is much smaller than the number of model parameters.

The above characteristics of the *n*th-FASAM-N and the *n*th-CASAM-N methodologies have been comparatively illustrated in this chapter by using the well-known Nordheim-Fuchs reactor dynamics/safety model. This phenomenological model describes a short-time self-limiting power transient in a nuclear reactor system having a negative temperature coefficient in which a large amount of reactivity is suddenly inserted, either intentionally or by accident. This model is sufficiently complex to demonstrate all of the important features of applying the *n*th-FASAM-N and *n*th-CASAM-N methodologies, while admitting exact closed-form expressions for all quantities of interest. This model can be mathematically cast into a form that comprises seven uncertain parameters that can be grouped into two "feature" functions of these parameters. The expressions of the 1st, 2nd, and 3rd-order sensitivities of the "released energy" model-response with respect to the model's parameters were obtained analytically in closed form, and their respective determination revealed the following characteristics:

1. For the computation of all 1st-order response sensitivities with respect to the model's primary parameters:
   a. The 1st-FASAM-N and the 1st-CASAM-N methodologies are equally efficient; each methodology requires a single large-scale computation for solving the "First-Level Adjoint Sensitivity System" (1st-LASS) to obtain all seven 1st-order response sensitivities with respect to the seven parameters involved in the Nordheim-Fuchs model.
   b. The simplest conventional 1st-order finite difference scheme would require at least two large-scale computations per parameter, using the original model with "perturbed" parameter values, to produce an approximate value for a 1st-order sensitivity. For the seven parameters involved in the Nordheim-Fuchs model, at least 14 large-scale computations would be required to compute the 1st-order sensitivities within a 1st-order accuracy in the considered parameter perturbation. Evidently,

both the 1st-FASAM-N and the 1st-CASAM-N are vastly more efficient than finite-difference schemes. Furthermore, the finite difference-schemes produce approximate values, while the 1st-FASAM-N and the 1st-CASAM-N accurately compute exact expressions of the respective sensitivities.

2. For the computation of all 2nd-order response sensitivities with respect to the model's primary parameters:
   a. The 2nd-CASAM-N methodology requires as many large-scale computations (for solving the 2nd-LASS) as there are 1st-order sensitivities to model parameters, or, equivalently, as many as there are parameters (i.e., $TP$).
   b. The 2nd-FASAM-N methodology requires as many large-scale computations (for solving the 2nd-LASS) as there are 1st-order sensitivities with respect to the feature functions, which is equivalent to "as many computations as there are feature functions" (i.e., $TF$). Since the number of feature-functions is much smaller than the number of primary parameters, i.e., $TF \ll TP$, the 2nd-FASAM-N methodology is considerably more efficient than the 2nd-CASAM-N methodology. For the illustrative example of the Nordheim-Fuchs model, the 2nd-FASAM-N methodology required two large-scale computations to obtain all of the exact expressions of the 28 distinct 2nd-order response sensitivities with respect to the model parameters, while the 2nd-FASAM-N methodology required seven large-scale computations to obtain these 28 distinct 2nd-order sensitivities. Both the 2nd-FASAM-N and the 2nd-CASAM-N methodologies yield exact values for the expressions of the 2nd-order sensitivities.
3. For the computation of all 3rd-order response sensitivities with respect to the model's primary parameters:
   a. The 3rd-CASAM-N methodology requires at most $TP(TP+1)/2$ large-scale computations for solving the "3rd-Level Adjoint Sensitivity System" (3rd-LASS). For the illustrative example of the Nordheim-Fuchs model, 22 large-scale computations were needed by applying the 3rd-CASAM-N methodology to obtain the exact expressions of the 84 distinct 3rd-order response sensitivities with respect to the model parameters.
   b. The 3rd-FASAM-N methodology requires at most $TF(TF+1)/2$ large-scale computations for solving the "3rd-Level Adjoint Sensitivity System" (3rd-LASS). For the illustrative example of the Nordheim-Fuchs model, two large-scale computations within the 3rd-FASAM-N methodology sufficed to obtain all of the exact expressions of the 84 distinct 3rd-order response sensitivities with respect to the model parameters. Evidently, the 3rd-FASAM methodology is significantly more efficient for computing the 2nd-and higher-order sensitivities than the 3rd-CASAM-N methodology. Both the 3rd-FASAM-N and the 3rd-CASAM-N methodologies yield exact values for the expressions of the 3rd-order sensitivities.

In summary, when no feature functions of parameters can be identified, the mathematical frameworks of the *n*th-FASAM-N and the *n*th-CASAM-N methodologies coincide. When feature-functions of parameters can be identified within the model, the *n*th-FASAM-N methodology requires the least number of large-scale computations of any practical methodology for computing exact expressions of 2nd- and higher-order sensitivities. Evidently, both the *n*th-FASAM-N and the *n*th-CASAM-N are vastly more efficient than finite-difference schemes. Furthermore, the finite difference-schemes are approximate, while the *n*th-FASAM-N and the *n*th-CASAM-N accurately compute exact expressions of these sensitivities. Together, the *n*th-FASAM-N and the *n*th-CASAM-N methodologies remain the most practical methodologies for computing response sensitivities comprehensively and accurately, overcoming the curse of dimensionality in sensitivity analysis of nonlinear systems.

# 6 The *n*th-Order Features Adjoint Sensitivity Analysis Methodology for Response-Coupled Forward/Adjoint Linear Systems (*n*th-FASAM-L)

## 6.1 INTRODUCTION

Nonlinear operators do not admit adjoint operators, so responses of nonlinear systems can only depend on the system's forward state functions. On the other hand, a linear operator admits a corresponding adjoint operator, which makes it possible for linear systems to admit responses that depend *simultaneously* on both the forward and adjoint state functions that model the respective linear system. Important model responses that depend simultaneously on both the forward and adjoint state functions in linear systems are various Lagrangian functionals that play fundamental roles in the derivation of numerical methods for solving equations (differential, integral, and integro-differential) and in the optimization and control of physical systems. Particularly important examples of such Lagrangian functionals are the Rayleigh quotient for computing eigenvalues and/or separation constants when solving partial differential equations and the Schwinger functional for 1st-order "normalization-free" solutions (Lewins 1965; Williams and Engle 1977; Stacey 1974, 2001). Although the sensitivity analysis of responses that simultaneously involve both forward and adjoint state functions could conceptually be considered as a "special case" of the sensitivity analysis of a nonlinear system (which would encompass both the underlying forward and adjoint equations), such a conceptual procedure would produce results for the various levels of "adjoint sensitivity systems" that would be insufficiently specific and would therefore need subsequent refinement for applications. Consequently, Cacuci (2021b, 2022b) has developed the "*n*th-Order Comprehensive Adjoint Sensitivity Analysis Methodology for Response-Coupled Forward/Adjoint Linear Systems" (*n*th-CASAM-L), aiming at computing sensitivities of such responses directly to the *primary* parameters involved in the computational model of the linear physical system under consideration.

As has been shown in Chapters 2–5, although the sensitivities to the primary model parameters are ultimately of interest for subsequent use in predictive


DOI: 10.1201/9781003478171-6

modeling activities, all the primary parameters seldom appear explicitly in the equations underlying the model. What appear explicitly in these equations are *features* (functions) of the primary model parameters, which have motivated the development of the *n*th-FASAM-N (*n*th-Order Features Adjoint Sensitivity Analysis Methodology for Nonlinear Systems) since, as shown in Chapters 2–5, the *n*th-FASAM-N significantly reduces the computational effort required for determining efficiently and exactly sensitivities of model responses to model features. The sensitivities of responses with respect to the primary model parameters are subsequently computed analytically, at virtually no computational cost, using the sensitivities to the model features.

Paralleling the mathematical framework of the *n*th-FASAM-N, this chapter presents the "*n*th-Order Features Adjoint Sensitivity Analysis Methodology for Response-Coupled Forward/Adjoint Linear Systems" (*n*th-FASAM-L), which enables the efficient and exact computation of sensitivities of model responses to model features for response-coupled forward and adjoint linear systems. This chapter is structured as follows: Section 6.2 presents the generic mathematical modeling of a response-coupled linear forward and adjoint system. Section 6.3 presents the mathematical framework of the 1st-FASAM-L, which enables the most efficient determination and computation of the exact expressions of the 1st-order sensitivities with respect to features/functions of responses that simultaneously depend on the forward and adjoint functions of linear systems. Section 6.4 presents the framework of the 2nd-FASAM-L, which enables the most efficient determination and computation of the exact expressions of the 2nd-order sensitivities (with respect to features/functions) of responses that simultaneously depend on the forward and adjoint functions of linear systems. The mathematical framework of the *n*th-FASAM-L is established in Section 6.5 by using "mathematical induction" as follows: (i) establish the mathematical framework underlying the *n*th-CASAM-L for the lowest value of the index *n*; (ii) assume that the mathematical framework is valid for an arbitrarily high-order, *n*; (iii) prove that the mathematical framework conjectured for *n* is also valid for n + 1. Section 6.6 summarizes the highlights of the *n*th-FASAM-L methodology.

## 6.2 MATHEMATICAL MODELING OF A RESPONSE-COUPLED LINEAR FORWARD AND ADJOINT SYSTEM

A *linear* physical system can generally be modeled by a system of coupled operator equations as follows:

$$\mathbf{L}[\mathbf{x};\mathbf{g}(\alpha)]\varphi(\mathbf{x}) = \mathbf{Q}[\mathbf{x};\mathbf{g}(\alpha)], \quad \mathbf{x}\in\Omega(\alpha). \tag{6.1}$$

The symbols that appear in Eq. (6.1) have the following meanings:

i. The *primary model parameters* will be denoted as $\alpha_1,\ldots,\alpha_{TP}$, where the subscript "*TP*" indicates "*Total* number of *Primary Parameters*;" the qualifier "primary" indicates that these parameters do not depend on any other parameters within the model. These model parameters are considered to include imprecisely known geometrical parameters that characterize the physical system's boundaries in the phase space of the model's independent variables. These boundaries depend on the physical system's geometrical dimensions, which may be imprecisely known because of manufacturing tolerances. In practice, these primary model parameters are subject to uncertainties. It will be convenient to consider that these parameters are components of a "vector of *primary* parameters" denoted as $\boldsymbol{\alpha} \triangleq (\alpha_1,\dots,\alpha_{TP})^\dagger \in \mathbb{R}^{TP}$, where $\mathbb{R}^{TP}$ denotes the *TP*-dimensional subset of the set of real scalars, where the symbol "$\triangleq$" denotes "is defined as" or "is by definition equal to" and transposition is indicated by a dagger $(\dagger)$ superscript. The nominal or mean parameter values are denoted as $\boldsymbol{\alpha}^0 \triangleq \left[\alpha_1^0,\dots,\alpha_i^0,\dots,\alpha_{TP}^0\right]^\dagger$ using the superscript "0."

ii. The model is considered to comprise *TI* independent variables, which will be denoted as $x_i, i=1,\dots,TI$, and are considered to be the components of a *TI*-dimensional column vector denoted as $\mathbf{x} \triangleq (x_1,\dots,x_{TI})^\dagger \in \mathbb{R}^{TI}$, where the sub/superscript "*TI*" denotes the "*Total* number of *Independent* variables." The vector $\mathbf{x} \in \mathbb{R}^{TI}$ of independent variables is considered to be defined on a phase-space domain, denoted as $\Omega(\boldsymbol{\alpha})$, $\Omega(\boldsymbol{\alpha}) \triangleq \{-\infty \le \lambda_i(\boldsymbol{\alpha}) \le x_i \le \omega_i(\boldsymbol{\alpha}) \le \infty; i=1,\dots,TI\}$, the boundaries of which may depend on some of the model parameters $\boldsymbol{\alpha}$. The lower boundary point of an independent variable is denoted as $\lambda_i(\boldsymbol{\alpha})$ (e.g., the inner radius of a sphere or cylinder, the lower range of an energy variable, the initial time value, etc.), while the corresponding upper boundary point is denoted as $\omega_i(\boldsymbol{\alpha})$ (e.g., the outer radius of a sphere or cylinder, the upper range of an energy variable, the final time value, etc.). A typical example of boundary conditions that depend on imprecisely known parameters that pertain to the geometry of the model and also on parameters that pertain to the material properties of the respective model occurs when modeling particle diffusion within a medium, the boundaries of which are facing vacuum or air. For such models, the boundary conditions for the respective state (dependent) variables (i.e., particle flux and/or current) are imposed not on the physical boundary but on the "extrapolated boundary" of the respective spatial domain. The "extrapolated boundary" depends both on the imprecisely known physical dimensions of the medium's domain/extent and also on the medium's properties, i.e., atomic number densities and microscopic transport cross sections. The boundary of $\Omega(\boldsymbol{\alpha})$, which will be denoted as $\partial\Omega[\boldsymbol{\lambda}(\boldsymbol{\alpha});\boldsymbol{\omega}(\boldsymbol{\alpha})]$, comprises the set of all of the endpoints $\lambda_i(\boldsymbol{\alpha}),\omega_i(\boldsymbol{\alpha}), i=1,\dots,TI$, of the respective intervals on which the components of $\mathbf{x}$ are defined, i.e., $\partial\Omega[\boldsymbol{\lambda}(\boldsymbol{\alpha});\boldsymbol{\omega}(\boldsymbol{\alpha})] \triangleq \{\lambda_i(\boldsymbol{\alpha}) \cup \omega_i(\boldsymbol{\alpha}), i=1,\dots,TI\}$.

iii. The vector $\boldsymbol{\varphi}(\mathbf{x}) \triangleq [\varphi_1(\mathbf{x}),\dots,\varphi_{TD}(\mathbf{x})]^\dagger$ is a *TD*-dimensional column vector of dependent variables; the subscript "*TD*" denotes the "*Total* (number of) *Dependent* variables." The functions $\varphi_i(\mathbf{x}), i=1,\dots,TD$, denote the system's "dependent variables" (also called "state functions").

iv. The vector $\mathbf{g}(\boldsymbol{\alpha}) \triangleq [g_1(\boldsymbol{\alpha}),\dots,g_{TG}(\boldsymbol{\alpha})]$ is a *TG*-dimensional vector comprising components $g_i(\boldsymbol{\alpha}), i=1,\dots,TG$, which are real-valued functions of (some of) the primary model parameters $\boldsymbol{\alpha} \in \mathbb{R}^{TP}$. The quantity *TG* denotes the total number of such functions that appear exclusively in the definition of the model's underlying equations. Such functions customarily appear in models in the form of correlations that describe "features" of the system under consideration, such as material properties, flow regimes, etc. Usually, the number of functions $g_i(\boldsymbol{\alpha})$ is considerably smaller than the total number of model parameters, i.e., $TG \ll TP$.

v. The $TD$-dimensional column vector $\mathbf{Q}[\mathbf{x};\mathbf{g}(\boldsymbol{\alpha})] \triangleq (q_1,\ldots,q_{TD})^\dagger$, comprising components $q_i[\mathbf{x};\mathbf{g}(\boldsymbol{\alpha})], i=1,\ldots,TD$, denotes inhomogeneous source terms, which usually depend nonlinearly on the uncertain parameters $\boldsymbol{\alpha}$. Since these source terms on the right side of Eq. (6.1) may contain distributions, the equality in this equation is considered to hold in the weak (i.e., "distributional") sense. Similarly, all of the equalities that involve differential equations in this work will be considered to hold in the distributional sense.

vi. The $TD \times TD$-dimensional matrix $\mathbf{L}[\mathbf{x};\mathbf{g}(\boldsymbol{\alpha})]$ comprises components denoted as $L_{ij}[\mathbf{x};\mathbf{g}(\boldsymbol{\alpha})]$, $i,j=1,\ldots,TD$, that act linearly on the dependent variables $\varphi_j(\mathbf{x})$ and also depend (in general, nonlinearly) on the uncertain model parameters $\boldsymbol{\alpha}$.

When $\mathbf{L}[\mathbf{x};\mathbf{g}(\boldsymbol{\alpha})]$ contains differential operators, a set of boundary and initial conditions that define the domain of $\mathbf{L}[\mathbf{x};\mathbf{g}(\boldsymbol{\alpha})]$ must also be given. Since the complete mathematical model is considered to be linear in $\boldsymbol{\varphi}(\mathbf{x})$, the boundary and/or initial conditions needed to define the domain of $\mathbf{L}[\mathbf{x};\mathbf{g}(\boldsymbol{\alpha})]$ must also be a linear in $\boldsymbol{\varphi}(\mathbf{x})$. Such linear boundary and initial conditions are represented in the following operator form:

$$\mathbf{B}[\mathbf{x};\mathbf{g}(\boldsymbol{\alpha});\boldsymbol{\lambda}(\boldsymbol{\alpha});\boldsymbol{\omega}(\boldsymbol{\alpha})]\boldsymbol{\varphi}(\mathbf{x}) = \mathbf{C}[\mathbf{x};\mathbf{g}(\boldsymbol{\alpha});\boldsymbol{\lambda}(\boldsymbol{\alpha});\boldsymbol{\omega}(\boldsymbol{\alpha})], \quad \mathbf{x}\in\partial\Omega[\boldsymbol{\lambda}(\boldsymbol{\alpha});\boldsymbol{\omega}(\boldsymbol{\alpha})]. \tag{6.2}$$

In Eq. (6.2), the quantity $\mathbf{B}[\mathbf{x};\mathbf{g}(\boldsymbol{\alpha});\boldsymbol{\lambda}(\boldsymbol{\alpha});\boldsymbol{\omega}(\boldsymbol{\alpha})]$ denotes a matrix of dimensions $N_B \times TD$ having components denoted as $B_{ij}(\mathbf{x};\boldsymbol{\alpha}); i=1,\ldots,N_B; j=1,\ldots,TD$, which are operators that act linearly on $\boldsymbol{\varphi}(\mathbf{x})$ and nonlinearly on the components of $\mathbf{g}(\boldsymbol{\alpha})$. The quantity $N_B$ denotes the total number of boundary and initial conditions. The $N_B$-dimensional column vector $\mathbf{C}[\mathbf{x};\mathbf{g}(\boldsymbol{\alpha});\boldsymbol{\lambda}(\boldsymbol{\alpha});\boldsymbol{\omega}(\boldsymbol{\alpha})]$ comprises components that are operators which, in general, act nonlinearly on the components of $\mathbf{g}(\boldsymbol{\alpha})$.

Physical problems modeled by linear systems and/or operators are naturally defined in Hilbert spaces. The dependent variables $\varphi_i(\mathbf{x}), i=1,\ldots,TD$, for the physical system represented by Eqs. (6.1) and (6.2) are considered to be square-integrable functions of the independent variables and are considered to belong to a Hilbert space, which will be denoted as $\mathcal{H}_0(\Omega)$, where the subscript "zero" denotes "zeroth-level" or "original." Higher-level Hilbert spaces, which will be denoted as $\mathcal{H}_1(\Omega)$ and $\mathcal{H}_2(\Omega)$, will also be used in this work. The Hilbert space $\mathcal{H}_0(\Omega)$ is considered to be endowed with the following inner product, denoted as $\langle\boldsymbol{\varphi}(\mathbf{x}),\boldsymbol{\psi}(\mathbf{x})\rangle_0$, between two elements $\boldsymbol{\varphi}(\mathbf{x})\in\mathcal{H}_0(\Omega)$ and $\boldsymbol{\psi}(\mathbf{x})\in\mathcal{H}_0(\Omega)$:

$$\langle \boldsymbol{\varphi}(\mathbf{x}), \boldsymbol{\psi}(\mathbf{x}) \rangle_0 \triangleq \prod_{i=1}^{TI} \int_{\lambda_i(\alpha)}^{\omega_i(\alpha)} \boldsymbol{\varphi}(\mathbf{x}) \bullet \boldsymbol{\psi}(\mathbf{x}) d\mathbf{x}$$
$$= \sum_{j=1}^{TD} \int_{\lambda_1(\alpha)}^{\omega_1(\alpha)} \dots \int_{\lambda_i(\alpha)}^{\omega_i(\alpha)} \dots \int_{\lambda_{TI}(\alpha)}^{\omega_{TI}(\alpha)} \varphi_j(\mathbf{x}) \psi_j(\mathbf{x}) dx_1 \dots dx_i \dots dx_{TI}. \tag{6.3}$$

The "dot" in Eq. (6.3) indicates the "scalar product of two vectors," which is defined as follows:

$$\boldsymbol{\varphi}(\mathbf{x}) \bullet \boldsymbol{\psi}(\mathbf{x}) \triangleq \sum_{i=1}^{TD} \varphi_i(\mathbf{x}) \psi_i(\mathbf{x}) \tag{6.4}$$

The product notation $\prod_{i=1}^{TI} \int_{\lambda_i(\alpha)}^{\omega_i(\alpha)} [\,] dx_i$ in Eq. (6.3) denotes the respective multiple integrals.

The linear operator $\mathbf{L}[\mathbf{x}; \mathbf{g}(\boldsymbol{\alpha})]$ is considered to admit an adjoint operator, which will be denoted as $\mathbf{L}^*[\mathbf{x}; \mathbf{g}(\boldsymbol{\alpha})]$ and which is defined through the following relation for a vector $\boldsymbol{\psi}(\mathbf{x}) \in \mathcal{H}_0$:

$$\langle \boldsymbol{\psi}(\mathbf{x}), \mathbf{L}[\mathbf{x}; \mathbf{g}(\boldsymbol{\alpha})] \boldsymbol{\varphi}(\mathbf{x}) \rangle_0 = \langle \mathbf{L}^*[\mathbf{x}; \mathbf{g}(\boldsymbol{\alpha})] \boldsymbol{\psi}(\mathbf{x}), \boldsymbol{\varphi}(\mathbf{x}) \rangle_0 . \tag{6.5}$$

In Eq. (6.5), the formal adjoint operator $\mathbf{L}^*[\mathbf{x}; \mathbf{g}(\boldsymbol{\alpha})]$ is the $TD \times TD$ matrix comprising elements $L^*_{ji}[\mathbf{x}; \mathbf{g}(\boldsymbol{\alpha})]$ which are obtained by transposing the formal adjoints of the forward operators $L_{ij}[\mathbf{x}; \mathbf{g}(\boldsymbol{\alpha})]$. Hence, the system adjoint to the linear system represented by (6.1) and (6.2) can generally be represented as follows:

$$\mathbf{L}^*[\mathbf{x}; \mathbf{g}(\boldsymbol{\alpha})] \boldsymbol{\psi}(\mathbf{x}) = \mathbf{Q}^*[\mathbf{x}; \mathbf{g}(\boldsymbol{\alpha})], \quad \mathbf{x} \in \Omega(\boldsymbol{\alpha}), \tag{6.6}$$

$$\mathbf{B}^*[\mathbf{x}; \mathbf{g}(\boldsymbol{\alpha}); \boldsymbol{\lambda}(\boldsymbol{\alpha}); \boldsymbol{\omega}(\boldsymbol{\alpha})] \boldsymbol{\psi}(\mathbf{x}) = \mathbf{C}^*[\mathbf{x}; \mathbf{g}(\boldsymbol{\alpha}); \boldsymbol{\lambda}(\boldsymbol{\alpha}); \boldsymbol{\omega}(\boldsymbol{\alpha})],$$
$$\mathbf{x} \in \partial\Omega[\boldsymbol{\lambda}(\boldsymbol{\alpha}); \boldsymbol{\omega}(\boldsymbol{\alpha})]. \tag{6.7}$$

When the forward operator $\mathbf{L}[\mathbf{x}; \mathbf{g}(\boldsymbol{\alpha})]$ comprises differential operators, the operations (e.g., integration by parts) that implement the transition from the left side to the right side of Eq. (6.5) give rise to boundary terms, which are collectively called the "bilinear concomitant." The domain of $\mathbf{L}^*[\mathbf{x}; \mathbf{g}(\boldsymbol{\alpha})]$ is determined by selecting adjoint boundary and/or initial conditions so as to ensure that the adjoint system is well-posed mathematically. It is also desirable that the selected adjoint boundary conditions should cause the bilinear concomitant to vanish when implemented in Eq. (6.5) together with the forward boundary conditions given in Eq. (6.2). The adjoint boundary conditions thus selected (to ensure that the respective bilinear concomitant vanishes) are represented in operator form by Eq. (6.7).

The relationship shown in Eq. (6.5), which is the basis for defining the adjoint operator, also provides the following fundamental "reciprocity-like" relation between the sources of the forward and the adjoint equations, i.e., Eqs. (6.1) and (6.6), respectively:

$$\left\langle \psi(\mathbf{x}), \mathbf{Q}\left[\mathbf{x};\mathbf{g}(\alpha)\right]\right\rangle_0 = \left\langle \mathbf{Q}^*\left[\mathbf{x};\mathbf{g}(\alpha)\right], \varphi(\mathbf{x})\right\rangle_0. \tag{6.8}$$

The functional on the right side of Eq. (6.8) represents a "detector response," i.e., a reaction rate between the particles and the medium represented by $\mathbf{Q}^*\left[\mathbf{x};\mathbf{g}(\alpha)\right]$ which is equivalent to the "number of counts" of particles incident on a detector of particles that "measures" the particle flux $\varphi(\mathbf{x})$. In view of the relation provided in (Eq. 6.8), the vector-valued source term $\mathbf{Q}^*\left[\mathbf{x};\mathbf{g}(\alpha)\right] \triangleq \left\{q_1^*\left[\mathbf{x};\mathbf{g}(\alpha)\right],\ldots,q_{TD}^*\left[\mathbf{x};\mathbf{g}(\alpha)\right]\right\}^\dagger$ in the adjoint equation Eq. (6.6) is usually associated with the "result of interest" to be measured and/or computed, which is customarily called the system's "response." In particular, if $q_i^*\left[\mathbf{x};\mathbf{g}(\alpha)\right] = \delta(\mathbf{x}-\mathbf{x}_d)$ and $q_{j\neq i}^*\left[\mathbf{x};\mathbf{g}(\alpha)\right] = 0$, then $\left\langle \mathbf{Q}^*\left[\mathbf{x};\mathbf{g}(\alpha)\right], \varphi(\mathbf{x})\right\rangle_0 = \varphi_i(\mathbf{x}_d)$, which means that, in such a case, the right side of Eq. (6.8) provides the value of the $i$th-dependent variable (particle flux, temperature, velocity, etc.) at the point in phase space where the respective measurement is performed.

The results computed using a mathematical model are customarily called "model responses" (or "system responses" or "objective functions" or "indices of performance"). For linear physical systems, the system's response may depend not only on the model's state functions and on the system parameters but may simultaneously also depend on the adjoint state function. As has been discussed by Cacuci (2022b, 2023a), any response of a linear system can be formally represented (using expansions or interpolations, if necessary) and fundamentally analyzed in terms of the following generic integral representation:

$$R\left[\varphi(\mathbf{x}),\psi(\mathbf{x});\mathbf{f}(\alpha)\right] \triangleq$$

$$\int_{\lambda_1(\alpha)}^{\omega_1(\alpha)} \cdots \int_{\lambda_{TI}(\alpha)}^{\omega_{TI}(\alpha)} S\left[\varphi(\mathbf{x}),\psi(\mathbf{x});\mathbf{g}(\alpha);\mathbf{h}(\alpha);\mathbf{x}\right] dx_1 \ldots dx_{TI}, \tag{6.9}$$

where $S\left[\varphi(\mathbf{x}),\psi(\mathbf{x});\mathbf{g}(\alpha);\mathbf{h}(\alpha);\mathbf{x}\right]$ is a suitably differentiable nonlinear function of $\varphi(\mathbf{x}),\psi(\mathbf{x})$, and $\alpha$. The integral representation of the model response provided in Eq. (6.9) can represent "averaged" and/or "point-valued" quantities in the phase space of independent variables. For example, if $R\left[\varphi(\mathbf{x}),\psi(\mathbf{x});\mathbf{f}(\alpha)\right]$ represents the computation or the measurement (which would be a "detector-response") of a quantity of interest at a point $\mathbf{x}_d$ in the phase-space of independent variables, then $S\left[\varphi(\mathbf{x}),\psi(\mathbf{x});\mathbf{g}(\alpha);\mathbf{h}(\alpha);\mathbf{x}\right]$ would contain a Dirac-delta functional of the form $\delta(\mathbf{x}-\mathbf{x}_d)$. Responses that represent "differentials/derivatives of quantities" would contain derivatives of Dirac-delta functionals in the definition of $S\left[\varphi(\mathbf{x}),\psi(\mathbf{x});\mathbf{g}(\alpha);\mathbf{h}(\alpha);\mathbf{x}\right]$. The vector $\mathbf{h}(\alpha) \triangleq \left[h_1(\alpha),\ldots,h_{TH}(\alpha)\right]$, having components $h_i(\alpha), i=1,\ldots,TH$, which appears among the arguments of the function $S\left[\varphi(\mathbf{x}),\psi(\mathbf{x});\mathbf{g}(\alpha);\mathbf{h}(\alpha);\mathbf{x}\right]$, represents functions of primary parameters that often appear solely in the definition of the response but do not appear in the mathematical definition of the model [i.e., $\mathbf{h}(\alpha)$ does not appear in Eqs. (6.1), (6.2), (6.6), and (6.7)]. The quantity $TH$ denotes the total number of such functions that appear exclusively in the definition of the model's response. Evidently, the response will depend directly and/or indirectly (through the "feature" functions) on all of the primary model parameters. This fact has been indicated in Eq. (6.9) by using the vector-valued function $\mathbf{f}(\boldsymbol{\alpha})$ as an argument in the definition of the response $R[\boldsymbol{\varphi}(\mathbf{x}),\boldsymbol{\psi}(\mathbf{x});\mathbf{f}(\boldsymbol{\alpha})]$ to represent the concatenation of all of the "features" of the model and response under consideration. The vector $\mathbf{f}(\boldsymbol{\alpha})$ of "model features" is thus defined as follows:

$$\begin{aligned}\mathbf{f}(\boldsymbol{\alpha}) &\triangleq [\mathbf{g}(\boldsymbol{\alpha});\mathbf{h}(\boldsymbol{\alpha});\boldsymbol{\lambda}(\boldsymbol{\alpha});\boldsymbol{\omega}(\boldsymbol{\alpha})]^{\dagger} \\ &\triangleq [f_1(\boldsymbol{\alpha}),\ldots,f_{TF}(\boldsymbol{\alpha})]^{\dagger}; \quad TF \triangleq TG+TH+2TI.\end{aligned} \tag{6.10}$$

As defined in Eq. (6.10), the quantity $TF$ denotes the total number of "feature functions of the model's parameters" that appear in the definition of the nonlinear model's underlying equations and response.

Solving Eqs. (6.1) and (6.2), at the nominal (or mean) values, denoted as $\boldsymbol{\alpha}^0 \triangleq [\alpha_1^0,\ldots,\alpha_i^0,\ldots,\alpha_{TP}^0]^{\dagger}$, of the model parameters, yields the nominal forward solution, which will be denoted as $\boldsymbol{\varphi}^0(\mathbf{x})$. Solving Eqs. (6.6) and (6.7) at the nominal values, $\boldsymbol{\alpha}^0$, of the model parameters yields the nominal adjoint solution, which will be denoted as $\boldsymbol{\psi}^0(\mathbf{x})$. The nominal value of the response, $R[\boldsymbol{\varphi}^0(\mathbf{x}),\boldsymbol{\psi}^0(\mathbf{x});\mathbf{f}(\boldsymbol{\alpha}^0)]$, is determined by using the nominal parameter values $\boldsymbol{\alpha}^0$, the nominal value $\boldsymbol{\varphi}^0(\mathbf{x})$ of the forward state function, and the nominal value $\boldsymbol{\psi}^0(\mathbf{x})$ of the adjoint state function.

The definition provided by Eq. (6.9) implies that the model response $R[\boldsymbol{\varphi}(\mathbf{x}),\boldsymbol{\psi}(\mathbf{x});\mathbf{f}(\boldsymbol{\alpha})]$ depends on the components of the feature function $\mathbf{f}(\boldsymbol{\alpha})$, and would therefore admit a Taylor-series expansion around the nominal value $\mathbf{f}^0 \triangleq \mathbf{f}(\boldsymbol{\alpha}^0)$ having the following form:

$$R[\mathbf{f}(\boldsymbol{\alpha})] = R(\mathbf{f}^0) + \sum_{j_1=1}^{TF}\left\{\frac{\partial R(\mathbf{f})}{\partial f_{j_1}}\right\}_{\mathbf{f}^0}\delta f_{j_1} + \frac{1}{2}\sum_{j_1=1}^{TF}\sum_{j_2=1}^{TF}\left\{\frac{\partial^2 R(\mathbf{f})}{\partial f_{j_1}\partial f_{j_2}}\right\}_{\mathbf{f}^0}\delta f_{j_1}\delta f_{j_2} + \cdots \tag{6.11}$$

where $\delta f_j \triangleq [f_j(\boldsymbol{\alpha}) - f_j^0]$; $f_j^0 \triangleq f_j(\boldsymbol{\alpha}^0)$; $j = 1,\ldots,TF$. The "sensitivities of the model response with respect to the (feature) functions" are naturally defined as being the functional derivatives of $R[\mathbf{f}(\boldsymbol{\alpha})]$ with respect to the components ("features") $f_j(\boldsymbol{\alpha})$ of $\mathbf{f}(\boldsymbol{\alpha})$. The notation $\{\ \}_{\mathbf{f}^0}$ indicates that the quantity enclosed within the braces is to be evaluated at the nominal values $\mathbf{f}^0 \triangleq \mathbf{f}(\boldsymbol{\alpha}^0)$. Since $TF \ll TP$, the computations of the functional derivatives of $R_k[\mathbf{f}(\boldsymbol{\alpha})]$ with respect to the functions $f_j(\boldsymbol{\alpha})$, which appear in Eq. (6.11), will be considerably less expensive computationally than the computation of the functional derivatives involved in the Taylor series of the response with respect to the model parameters. The functional derivatives of the response with respect to the parameters can be obtained from the functional derivatives of the response with respect to the "feature" functions $f_j(\boldsymbol{\alpha})$ by simply using the chain rule, i.e.:

$$\begin{aligned}\left\{\frac{\partial R(\boldsymbol{\alpha})}{\partial \alpha_{j_1}}\right\}_{\boldsymbol{\alpha}^0} &= \sum_{i_1=1}^{TF}\left\{\frac{\partial R(\mathbf{f})}{\partial f_{i_1}}\frac{\partial f_{i_1}(\boldsymbol{\alpha})}{\partial \alpha_{j_1}}\right\}_{\boldsymbol{\alpha}^0}; \\ \left\{\frac{\partial^2 R(\boldsymbol{\alpha})}{\partial \alpha_{j_1}\partial \alpha_{j_2}}\right\}_{\boldsymbol{\alpha}^0} &= \frac{\partial}{\partial \alpha_{j_2}}\sum_{i_1=1}^{TF}\left\{\frac{\partial R(\mathbf{f})}{\partial f_{i_1}}\frac{\partial f_{i_1}(\boldsymbol{\alpha})}{\partial \alpha_{j_1}}\right\}_{\boldsymbol{\alpha}^0};\end{aligned} \tag{6.12}$$

and so on. The evaluation/computation of the functional derivatives $\frac{\partial f_{i1}(\alpha)}{\partial \alpha_{j1}}$, $\frac{\partial^2 f_{i1}(\alpha)}{\partial \alpha_{j1} \partial \alpha_{j2}}$, etc., does not require computations involving the model and is therefore computationally trivial by comparison to the evaluation of the functional derivatives ("sensitivities") of the response with respect to either the functions ("features") $f_j(\alpha)$ or the model parameters $\alpha_i, i = 1,\ldots,TP$.

The range of validity of the Taylor series shown in Eq. (6.11) is defined by its radius of convergence. The accuracy – as opposed to the "validity" – of the Taylor series in predicting the value of the response at an arbitrary point in the phase space of model parameters depends on the order of sensitivities retained in the Taylor expansion: the higher the respective order, the more accurate the respective response value predicted by the Taylor series. In the particular case when the response happens to be a polynomial function of the "feature" functions $f_j(\alpha)$, the Taylor series represented by Eq. (6.11) is finite and exactly represents the respective model response.

In turn, the functions $f_i(\alpha)$ can also be formally expanded in a multivariate Taylor series around the nominal (mean) parameter values $\alpha^0$, namely:

$$f_i(\alpha) = f_i(\alpha^0) + \sum_{j1=1}^{TP} \left\{ \frac{\partial f_i(\alpha)}{\partial \alpha_{j1}} \right\}_{\alpha^0} \delta\alpha_{j1}$$
$$+ \frac{1}{2} \sum_{j1=1}^{TP} \sum_{j2=1}^{TP} \left\{ \frac{\partial^2 f_i(\alpha)}{\partial \alpha_{j1} \partial \alpha_{j2}} \right\}_{\alpha^0} \delta\alpha_{j1} \delta\alpha_{j2} \tag{6.13}$$
$$+ \frac{1}{3!} \sum_{j1=1}^{TP} \sum_{j2=1}^{TP} \sum_{j3=1}^{TP} \left\{ \frac{\partial^3 f_i(\alpha)}{\partial \alpha_{j1} \partial \alpha_{j2} \partial \alpha_{j3}} \right\}_{\alpha^0} \delta\alpha_{j1} \delta\alpha_{j2} \delta\alpha_{j3} + \cdots,$$

The choice of feature functions $f_i(\alpha)$ is not unique but can be tailored by the user to the problem at hand. The two most important guiding principles for constructing the feature functions $f_i(\alpha)$ based on the primary parameters are as follows:

i. As will be shown below in Section 6.4 while establishing the mathematical framework underlying the 2nd-FASAM-L, the number of large-scale computations needed to determine the numerical value of the second-order sensitivities is proportional to the number of 1st-order sensitivities of the model's response with respect to the feature functions $f_i(\alpha)$. Consequently, it is important to minimize the number of feature functions $f_i(\alpha)$, while ensuring that all of the primary model parameters are considered within the expressions constructed for the feature functions $f_i(\alpha)$. In the extreme case, when some primary parameters, $\alpha_j$, cannot be grouped into the expressions of the feature functions $f_i(\alpha)$, each of the respective primary model parameters $\alpha_j$ becomes a feature function $f_i(\alpha)$.

ii. The expressions of the feature functions $f_i(\boldsymbol{\alpha})$ must be independent of the model's state functions; they must be exact, closed-form, scalar-valued functions of the primary model parameters $\alpha_j$, so the exact expressions of the derivatives of $f_i(\boldsymbol{\alpha})$ with respect to the primary model parameters $\alpha_j$ can be obtained analytically (with "pencil and paper") and, hence, inexpensively from a computational standpoint. The motivation for this requirement is to ensure that the subsequent numerical computation of the derivatives of the feature functions $f_i(\boldsymbol{\alpha})$ with respect to the primary model parameters $\alpha_j$ becomes trivial computationally.

The domain of validity of the Taylor series in Eq. (6.13) is defined by its own radius of convergence. Of course, in the extreme case when no feature function can be constructed, the feature functions will be the primary parameters themselves, in which case the *n*th-FASAM-L methodology becomes identical to the previously established *n*th-CASAM-L methodology conceived by Cacuci (2021b, 2022b).

## 6.3 THE 1ST-ORDER FEATURE ADJOINT SENSITIVITY ANALYSIS METHODOLOGY FOR RESPONSE-COUPLED FORWARD AND ADJOINT LINEAR SYSTEMS (1ST-FASAM-L)

The "First-Order Feature Adjoint Sensitivity Analysis Methodology for Response-Coupled Forward/Adjoint Linear Systems" (1st-FASAM-L) aims at enabling the most efficient computation of the 1st-order sensitivities of a generic model response of the form $R[\boldsymbol{\varphi}(\mathbf{x}),\boldsymbol{\psi}(\mathbf{x});\boldsymbol{\alpha}]$ with respect to the components of the "features" function $\mathbf{f}(\boldsymbol{\alpha})$. In preparation for subsequent generalizations toward establishing the generic pattern for computing sensitivities of arbitrarily high-order, the function $\mathbf{u}^{(1)}(2;\mathbf{x}) \triangleq [\boldsymbol{\varphi}(\mathbf{x}),\boldsymbol{\psi}(\mathbf{x})]^{\dagger}$ will be called the "1st-level forward/adjoint function," and the system of equations satisfied by this function (which is obtained by concatenating the original forward and adjoint equations together with their respective boundary/initial conditions) will be called "the 1st-Level Forward/Adjoint System (1st-LFAS)" and will be rewritten in the following concatenated matrix form:

$$\mathbf{F}^{(1)}[2\times 2;\mathbf{x};\mathbf{f}]\,\mathbf{u}^{(1)}(2;\mathbf{x}) = \mathbf{q}_F^{(1)}(2;\mathbf{x};\mathbf{f}); \quad \mathbf{x}\in\Omega(\boldsymbol{\alpha}); \tag{6.14}$$

$$\mathbf{b}_F^{(1)}\left[\mathbf{u}^{(1)}(2;\mathbf{x});\mathbf{f}\right] = \mathbf{0}; \quad \mathbf{x}\in\partial\Omega[\boldsymbol{\lambda}(\boldsymbol{\alpha});\boldsymbol{\omega}(\boldsymbol{\alpha})]; \tag{6.15}$$

where the following definitions were used:

$$\begin{aligned} &\mathbf{F}^{(1)}[2\times 2;\mathbf{x};\mathbf{f}] \triangleq \begin{pmatrix} \mathbf{L}(\mathbf{x};\mathbf{f}) & \mathbf{0} \\ \mathbf{0} & \mathbf{L}^{*}(\mathbf{x};\mathbf{f}) \end{pmatrix}; \\ &\mathbf{u}^{(1)}(2;\mathbf{x}) \triangleq [\boldsymbol{\varphi}(\mathbf{x}),\boldsymbol{\psi}(\mathbf{x})]^{\dagger}; \end{aligned} \tag{6.16}$$

$$\mathbf{q}_F^{(1)}(2;\mathbf{x};\mathbf{f}) \triangleq \begin{pmatrix} \mathbf{Q}(\mathbf{x};\mathbf{g}) \\ \mathbf{Q}^*(\mathbf{x};\mathbf{g}) \end{pmatrix};$$

$$\mathbf{b}_F^{(1)}\left[2;\mathbf{u}^{(1)}(2;\mathbf{x});\mathbf{f}\right] \triangleq \begin{pmatrix} \mathbf{B}(\mathbf{x};\mathbf{f})\varphi(\mathbf{x}) - \mathbf{C}(\mathbf{f}) \\ \mathbf{B}^*(\mathbf{x};\mathbf{f})\psi(\mathbf{x}) - \mathbf{C}^*(\mathbf{f}) \end{pmatrix}. \tag{6.17}$$

In the list of arguments of the matrix $\mathbf{F}^{(1)}[2\times 2;\mathbf{x};\mathbf{f}]$, the argument "$2\times 2$" indicates that this square matrix comprises four component sub-matrices, as indicated in Eq. (6.16). Similarly, the argument "2" that appears in the block vectors $\mathbf{u}^{(1)}(2;\mathbf{x})$, $\mathbf{q}_F^{(1)}(2;\mathbf{x};\mathbf{f})$, and $\mathbf{b}_F^{(1)}\left[2;\mathbf{u}^{(1)}(2;\mathbf{x});\mathbf{f}\right]$ indicates that each of these column block-vectors comprises two sub-vectors as components. Also, throughout this chapter, the quantity "$\mathbf{0}$" will be used to denote either a vector or a matrix with zero-valued components, depending on the context. For example, the vector "$\mathbf{0}$" in Eq. (6.15) is considered to have as many components as the vector $\mathbf{b}_F^{(1)}\left[\mathbf{u}^{(1)}(2;\mathbf{x});\mathbf{f}\right]$. On the other hand, the quantity "$\mathbf{0}$" that appears in Eq. (6.16) may represent either a (sub) matrix or a vector of the requisite dimensions.

The nominal (or mean) parameter values, $\alpha^0$, are considered to be known, but these values will differ from the true values $\alpha$, which are unknown, by variations $\delta\alpha \triangleq (\delta\alpha_1,\ldots,\delta\alpha_{TP})^\dagger$, where $\delta\alpha_i \triangleq \alpha_i - \alpha_i^0$. The parameter variations $\delta\alpha$ will induce variations $\delta\mathbf{f}(\alpha) \triangleq [\delta f_1(\alpha),\ldots,\delta f_{TF}(\alpha)]^\dagger$ in the vector-valued function $\mathbf{f}(\alpha)$, around the nominal value $\mathbf{f}^0 \triangleq \mathbf{f}(\alpha^0)$, and will also induce variations $\delta\varphi(\mathbf{x})$ and $\delta\psi(\mathbf{x})$, respectively, around the nominal solution $(\varphi^0,\psi^0)$ through the equations underlying the model. All of these variations will induce variations in the model response $R\left[\mathbf{u}^{(1)}(2;\mathbf{x});\mathbf{f}\right] \equiv R[\varphi(\mathbf{x}),\psi(\mathbf{x});\mathbf{f}(\alpha)]$, in a neighborhood $\left[\varphi^0(\mathbf{x})+\varepsilon\delta\varphi(\mathbf{x}),\psi^0(\mathbf{x})+\varepsilon\delta\psi(\mathbf{x});\mathbf{f}^0+\varepsilon\delta\mathbf{f}\right]$ around $(\varphi^0,\psi^0;\mathbf{f}^0)$, where $\varepsilon$ is a real-valued scalar.

Formally, the 1st-order sensitivities of the response $R\left[\mathbf{u}^{(1)}(2;\mathbf{x});\mathbf{f}\right]$ with respect to the components of the feature function $\mathbf{f}(\alpha)$ are provided by the 1st-order Gateaux (G-)variation of $R(\varphi,\psi,\mathbf{f})$ at the phase-space point $(\varphi^0,\psi^0,\mathbf{f}^0)$, which is defined as follows:

$$\delta R(\varphi^0,\psi^0,\mathbf{f}^0;\delta\varphi,\delta\psi,\delta\mathbf{f}) \triangleq \left\{\frac{d}{d\varepsilon} R\left[\varphi^0(\mathbf{x})+\varepsilon\delta\varphi(\mathbf{x}),\psi^0(\mathbf{x})+\varepsilon\delta\psi(\mathbf{x});\mathbf{f}^0+\varepsilon\delta\mathbf{f}\right]\right\}_{\varepsilon=0}$$

$$\equiv \left\{\frac{d}{d\varepsilon} R\left[\mathbf{u}^{(1,0)}(2;\mathbf{x})+\varepsilon\mathbf{v}^{(1)}(2;\mathbf{x});\mathbf{f}^0+\varepsilon\delta\mathbf{f}\right]\right\}_{\varepsilon=0} \equiv \delta R\left[\mathbf{u}^{(1,0)}(2;\mathbf{x});\mathbf{f}^0;\mathbf{v}^{(1)}(2;\mathbf{x}),\delta\mathbf{f}\right], \tag{6.18}$$

where the following definitions were used:

$$\mathbf{u}^{(1,0)}(2;\mathbf{x}) \triangleq \left[\varphi^0(\mathbf{x}),\psi^0(\mathbf{x})\right]^\dagger; \quad \mathbf{v}^{(1)}(2;\mathbf{x}) \triangleq [\delta\varphi(\mathbf{x}),\delta\psi]^\dagger. \tag{6.19}$$

In general, the G-variation $\delta R\left(\varphi^0,\psi^0,\mathbf{f}^0;\delta\varphi,\delta\psi,\delta\mathbf{f}\right)$ is nonlinear in the variations $\delta\varphi(\mathbf{x})$, $\delta\psi(\mathbf{x})$, and $\delta\mathbf{f}(\alpha)$. In such cases, the partial functional Gateaux-(G-) derivatives of the response $R(\varphi,\psi,\mathbf{f})$ with respect to the functions $\varphi,\psi,\mathbf{f}$ do not exist, which implies that the response sensitivities to the model parameters do not exist, either. Therefore, it will be henceforth assumed in this chapter that $\delta R\left(\varphi^0,\psi^0,\mathbf{f}^0;\delta\varphi,\delta\psi,\delta\mathbf{f}\right)$ is *linear* in the respective variations, so the corresponding partial G-derivatives exist and $\delta R\left(\varphi^0,\psi^0,\mathbf{f}^0;\delta\varphi,\delta\psi,\delta\mathbf{f}\right)$ is actually the 1st-order *G-differential* of the response. The usual numerical methods (e.g., Newton's method and variants thereof) for solving the equations underlying the model also require the existence of the 1st-order G-derivatives of the original model equations; these will also be assumed to exist. When the 1st-order G-derivatives exist, the G-differential $\delta R\left[\mathbf{u}^{(1,0)}(2;\mathbf{x});\mathbf{f}^0;\mathbf{v}^{(1)}(2;\mathbf{x}),\delta\mathbf{f}\right]$ can be written as follows:

$$\delta R\left[\mathbf{u}^{(1,0)}(2;\mathbf{x});\mathbf{f}^0;\mathbf{v}^{(1)}(2;\mathbf{x}),\delta\mathbf{f}\right]=\left\{\delta R\left[\mathbf{u}^{(1)}(2;\mathbf{x});\mathbf{f};\delta\mathbf{f}\right]\right\}_{dir} \\ +\left\{\delta R\left[\mathbf{u}^{(1)}(2;\mathbf{x});\mathbf{f};\mathbf{v}^{(1)}(2;\mathbf{x})\right]\right\}_{ind}. \tag{6.20}$$

In Eq. (6.20), the "direct-effect" term $\left\{\delta R\left[\mathbf{u}^{(1)}(2;\mathbf{x});\mathbf{f};\delta\mathbf{f}\right]\right\}_{dir}$ comprises only dependencies on $\delta\mathbf{f}(\alpha)$ and is defined as follows:

$$\left\{\delta R\left[\mathbf{u}^{(1)}(2;\mathbf{x});\mathbf{f};\delta\mathbf{f}\right]\right\}_{dir}\triangleq\left\{\frac{\partial R\left(\mathbf{u}^{(1)};\mathbf{f}\right)}{\partial\mathbf{f}}\delta\mathbf{f}\right\}_{\alpha^0}. \tag{6.21}$$

The following convention/definition was used in Eq. (6.21):

$$\frac{\partial[\;]}{\partial\mathbf{f}}\delta\mathbf{f}\triangleq\sum_{i=1}^{TF}\frac{\partial[\;]}{\partial f_i}\delta f_i \\ =\sum_{i=1}^{TG}\frac{\partial[\;]}{\partial g_i}\delta g_i+\sum_{i=1}^{TH}\frac{\partial[\;]}{\partial h_i}\delta h_i \\ +\sum_{i=1}^{TI}\frac{\partial[\;]}{\partial\omega_i}\delta\omega_i+\sum_{i=1}^{TI}\frac{\partial[\;]}{\partial\lambda_i}\delta\lambda_i. \tag{6.22}$$

The above convention implies that:

a. For $j=1,\ldots,TG$:

$$\frac{\partial R\left(\mathbf{u}^{(1)};\mathbf{f}\right)}{\partial f_j}\delta f_j\triangleq\left\{\int_{\lambda_1(\alpha)}^{\omega_1(\alpha)}\cdots\int_{\lambda_{TI}(\alpha)}^{\omega_{TI}(\alpha)}\frac{\partial S(\varphi,\psi;\mathbf{g};\mathbf{h})}{\partial g_i}dx_1\ldots dx_{TI}\right\}_{\alpha^0}\delta g_i; \tag{6.23}$$

$i=1,\ldots,TG$;

b. For $j=TG+1,\ldots,TG+TH$:

$$\frac{\partial R\left(\mathrm{u}^{(1)};\mathrm{f}\right)}{\partial f_j}\delta f_j \triangleq \left\{\int\limits_{\lambda_1(\alpha)}^{\omega_1(\alpha)}\cdots\int\limits_{\lambda_{TI}(\alpha)}^{\omega_{TI}(\alpha)}\frac{\partial S(\varphi,\psi;\mathbf{g};\mathbf{h})}{\partial h_i}dx_1\ldots dx_{TI}\right\}_{\alpha^0}\delta h_i; \tag{6.24}$$

$i = 1,\ldots,TH;$

c. For $j = TG + TH + 1,\ldots,TG + TH + TI$:

$$\frac{\partial R\left(\mathbf{u}^{(1)};\mathbf{f}\right)}{\partial f_j}\delta f_j \triangleq \left\{\frac{\partial}{\partial \omega_i}\int\limits_{\lambda_1}^{\omega_1}dx_1\ldots\int\limits_{\lambda_{TI}}^{\omega_{TI}}dx_{TI}\int_{\lambda_{TI}}^{\omega_{TI}}dx_{TI}S(\varphi,\psi;\mathbf{g};\mathbf{h})\right\}_{\alpha^0}\delta\omega_i, \tag{6.25}$$

$i = 1,\ldots,TI;$

d. For $j = TG + TH + TI + 1,\ldots,TG + TH + 2TI$:

$$\frac{\partial R\left(\mathbf{u}^{(1)};\mathbf{f}\right)}{\partial f_j}\delta f_j \triangleq \left\{\frac{\partial}{\partial \lambda_i}\int\limits_{\lambda_1}^{\omega_1}dx_1\ldots\int\limits_{\lambda_{TI}}^{\omega_{TI}}dx_{TI}S(\varphi,\psi;\mathbf{g};\mathbf{h})\right\}_{\alpha^0} \tag{6.26}$$

$\delta\lambda_i, \quad i = 1,\ldots,TI.$

The notation on the left side of Eq. (6.22) represents the inner product between two vectors, but the symbol "(†)," which indicates "transposition" has been omitted in order to keep the notation as simple as possible. "Daggers" indicating transposition will also be omitted in other inner products, whenever possible, while avoiding ambiguities.

In Eq. (6.20), the "indirect-effect" term $\left\{\delta R\left[\mathbf{u}^{(1)}(2;\mathbf{x});\mathbf{f};\mathbf{v}^{(1)}(2;\mathbf{x})\right]\right\}_{ind}$ depends only on the variations $\mathbf{v}^{(1)}(2;\mathbf{x}) \triangleq \left[\delta\varphi(\mathbf{x}),\delta\psi\right]^{\dagger}$ in the state functions and is defined as follows:

$$\begin{aligned}\left\{\delta R\left[\mathbf{u}^{(1)}(2;\mathbf{x});\mathbf{f};\mathbf{v}^{(1)}(2;\mathbf{x})\right]\right\}_{ind} &\triangleq \left\{\int\limits_{\lambda_1(\alpha)}^{\omega_1(\alpha)}dx_1\ldots\int\limits_{\lambda_{TI}(\alpha)}^{\omega_{TI}(\alpha)}dx_{TI}\frac{\partial S(\varphi,\psi;\mathbf{g};\mathbf{h})}{\partial \mathbf{u}^{(1)}(2;\mathbf{x})}\mathbf{v}^{(1)}(2;\mathbf{x})\right\}_{\alpha^0} \\ &\triangleq \left\{\int\limits_{\lambda_1(\alpha)}^{\omega_1(\alpha)}dx_1\ldots\int\limits_{\lambda_{TI}(\alpha)}^{\omega_{TI}(\alpha)}dx_{TI}\frac{\partial S(\varphi,\psi;\mathbf{g};\mathbf{h})}{\partial \varphi}\delta\varphi\right\}_{\alpha^0} \\ &+\left\{\int\limits_{\lambda_1(\alpha)}^{\omega_1(\alpha)}dx_1\ldots\int\limits_{\lambda_{TI}(\alpha)}^{\omega_{TI}(\alpha)}dx_{TI}\,dx_{TI}\frac{\partial S(\varphi,\psi;\mathbf{g};\mathbf{h})}{\partial \psi}\delta\psi\right\}_{\alpha^0}.\end{aligned} \tag{6.27}$$

In Eqs. (6.21) and (6.27), the notation $\{\}_{\alpha^0}$ was used to indicate that the quantity within the brackets is to be evaluated at the nominal values of the parameters and state functions. This simplified notation is justified by the fact that when the parameters take on their nominal values, it implicitly means that the corresponding state functions also take on their corresponding nominal values. This simplified notation will be used throughout this chapter.

The direct-effect term can be computed after having solved the forward system modeled by Eqs. (6.1) and (6.2), as well as the adjoint system modeled by Eqs. (6.6) and (6.7), to obtain the nominal values $\boldsymbol{\varphi}^0, \boldsymbol{\psi}^0$ of the forward and adjoint dependent variables.

On the other hand, the indirect-effect term $\left\{\delta R\left[\mathbf{u}^{(1)}(2;\mathbf{x});\mathbf{f};\mathbf{v}^{(1)}(2;\mathbf{x})\right]\right\}_{ind}$ defined in Eq. (6.27) can be quantified only after having determined the variations $\mathbf{v}^{(1)}(2;\mathbf{x}) \triangleq \left[\delta\boldsymbol{\varphi}(\mathbf{x}), \delta\boldsymbol{\psi}\right]^{\dagger}$ in the state functions $(\boldsymbol{\varphi}, \boldsymbol{\psi})$ of the 1st-Level Forward/Adjoint System (1st-LFAS). The variations $\mathbf{v}^{(1)}(2;\mathbf{x})$ are obtained as the solutions of the system of equations obtained by taking the 1st-order G-differentials of the 1st-LFAS defined by Eqs. (6.14) and (6.15), which are obtained by definition as follows:

$$\begin{gathered}\left\{\frac{d}{d\varepsilon}\mathbf{F}^{(1)}\left[2\times 2;\mathbf{x};\mathbf{f}^0+\varepsilon\delta\mathbf{f}\right]\left[\mathbf{u}^{(1,0)}(2;\mathbf{x})+\varepsilon\mathbf{v}^{(1)}(2;\mathbf{x})\right]\right\}_{\varepsilon=0} \\ =\left\{\frac{d}{d\varepsilon}\mathbf{q}_F^{(1)}\left[2;\mathbf{x};\mathbf{f}^0+\varepsilon\delta\mathbf{f}\right]\right\}_{\varepsilon=0},\end{gathered} \tag{6.28}$$

$$\left\{\frac{d}{d\varepsilon}\mathbf{b}_F^{(1)}\left[2;\mathbf{u}^{(1,0)}(2;\mathbf{x})+\varepsilon\mathbf{v}^{(1)}(2;\mathbf{x});\mathbf{f}^0+\varepsilon\delta\mathbf{f}\right]\right\}_{\varepsilon=0}=\mathbf{0}[2]. \tag{6.29}$$

Carrying out the differentiations with respect to $\varepsilon$ in the above equations and setting $\varepsilon = 0$ in the resulting expressions yields the following matrix-vector equations:

$$\begin{gathered}\left\{\mathbf{V}^{(1)}[2\times 2;\mathbf{x};\mathbf{f}]\mathbf{v}^{(1)}(2;\mathbf{x})\right\}_{\alpha^0}=\left\{\mathbf{q}_V^{(1)}\left[2;\mathbf{u}^{(1)}(2;\mathbf{x});\mathbf{f};\delta\mathbf{f}\right]\right\}_{\alpha^0}; \\ \mathbf{x}\in\Omega\left(\boldsymbol{\alpha}^0\right);\end{gathered} \tag{6.30}$$

$$\left\{\mathbf{b}_V^{(1)}\left(\mathbf{u}^{(1)};\mathbf{v}^{(1)};\mathbf{f};\delta\mathbf{f}\right)\right\}_{\alpha^0}=\mathbf{0};\quad \mathbf{x}\in\partial\Omega\left[\boldsymbol{\lambda}\left(\boldsymbol{\alpha}^0\right);\boldsymbol{\omega}\left(\boldsymbol{\alpha}^0\right)\right]; \tag{6.31}$$

where:

$$\mathbf{V}^{(1)}[2\times 2;\mathbf{x};\mathbf{f}]\triangleq\begin{pmatrix}\mathbf{L}(\mathbf{x};\mathbf{f}) & \mathbf{0}\\ \mathbf{0} & \mathbf{L}^*(\mathbf{x};\mathbf{f})\end{pmatrix}=\mathbf{F}^{(1)}[2\times 2;\mathbf{x};\mathbf{f}] \tag{6.32}$$

$$\begin{gathered}\mathbf{q}_V^{(1)}\left[2;\mathbf{u}^{(1)};\mathbf{f};\delta\mathbf{f}\right]\triangleq\begin{pmatrix}\mathbf{q}_1^{(1)}(\boldsymbol{\varphi};\mathbf{f};\delta\mathbf{f})\\ \mathbf{q}_2^{(1)}(\boldsymbol{\psi};\mathbf{f};\delta\mathbf{f})\end{pmatrix}; \\ \mathbf{b}_V^{(1)}\left(\mathbf{u}^{(1)};\mathbf{v}^{(1)};\mathbf{f};\delta\mathbf{f}\right)\triangleq\begin{pmatrix}\mathbf{b}_1^{(1)}(\boldsymbol{\varphi};\delta\boldsymbol{\varphi};\mathbf{f};\delta\mathbf{f})\\ \mathbf{b}_2^{(1)}(\boldsymbol{\psi};\delta\boldsymbol{\psi};\mathbf{f};\delta\mathbf{f})\end{pmatrix};\end{gathered} \tag{6.33}$$

$$\mathbf{q}_1^{(1)}(\varphi;\mathbf{f};\delta\mathbf{f}) \triangleq \frac{\partial[\mathbf{Q}-\mathbf{L}\varphi(\mathbf{x})]}{\partial\mathbf{f}}\delta\mathbf{f} \triangleq \sum_{j_1=1}^{TF}\mathbf{s}_1^{(1)}(j_1;\varphi;\mathbf{f})\delta f_{j_1}, \tag{6.34}$$

$$\mathbf{q}_2^{(1)}(\psi,\mathbf{f};\delta\mathbf{f}) \triangleq \frac{\partial\left[\mathbf{Q}^*-\mathbf{L}^*\psi(\mathbf{x})\right]}{\partial\mathbf{f}}\delta\mathbf{f} \triangleq \sum_{j_1=1}^{TF}\mathbf{s}_2^{(1)}(j_1;\psi;\mathbf{f})\delta f_{j_1}, \tag{6.35}$$

$$\mathbf{b}_1^{(1)}(\varphi;\delta\varphi;\mathbf{f};\delta\mathbf{f}) \triangleq \mathbf{B}\delta\varphi + \frac{\partial(\mathbf{B}\varphi-\mathbf{C})}{\partial\mathbf{f}}\delta\mathbf{f}; \tag{6.36}$$

$$\mathbf{b}_2^{(1)}(\psi;\delta\psi;\mathbf{f};\delta\mathbf{f}) \triangleq \mathbf{B}^*\delta\psi + \frac{\partial(\mathbf{B}^*\psi-\mathbf{C}^*)}{\partial\mathbf{f}}\delta\mathbf{f}. \tag{6.37}$$

In order to keep the notation as simple as possible in Eqs. (6.30)-(6.37), the differentials with respect to the various components of the feature function $\mathbf{f}(\alpha)$ have all been written in the form $\left(\frac{\partial[\;]}{\partial\mathbf{f}}\right)\delta\mathbf{f}$, keeping in mind the convention/notation introduced in Eq. (6.22). The system of equations comprising Eqs. (6.30) and (6.31) will be called the "*1st-Level Variational Sensitivity System* (1st-LVSS)," and its solution, $\mathbf{v}^{(1)}(2;\mathbf{x})$, will be called the "1st-level variational sensitivity function," which is indicated by the superscript "(1)." The solution, $\mathbf{v}^{(1)}(2;\mathbf{x})$, of the 1st-LVSS will be a function of the components of the vector of variations $\delta\mathbf{f}$. In principle, therefore, if the response sensitivities with respect to the components of the feature function $\mathbf{f}(\alpha)$ are of interest, then the 1st-LVSS would need to be solved as many times as there are components in the variational feature function $\delta\mathbf{f}$. On the other hand, if the response sensitivities with respect to the primary parameters are of interest, then the 1st-LVSS would need to be solved as many times as there are primary parameters. Solving the 1st-LVSS involves "large-scale computations."

On the other hand, the need for solving repeatedly the 1st-LVSS for every variation in the components of the feature function (or for every variation in the model's parameters) can be avoided by expressing the indirect-effect term $\left\{\delta R\left[\mathbf{u}^{(1)}(2;\mathbf{x});\mathbf{f};\mathbf{v}^{(1)}(2;\mathbf{x})\right]\right\}_{ind}$ defined in Eq. (6.27) in terms of the solutions of the "1st-Level Adjoint Sensitivity System" (1st-LASS), which will be constructed by implementing the following sequence of steps:

1. Introduce a Hilbert space, denoted as $\mathcal{H}_1$, comprising vector-valued elements of the form $\chi^{(1)}(2;\mathbf{x}) \triangleq \left[\chi_1^{(1)}(\mathbf{x}),\chi_2^{(1)}(\mathbf{x})\right]^\dagger$, where the components $\chi_i^{(1)}(\mathbf{x}) \triangleq \left[\chi_{i,1}^{(1)}(\mathbf{x}),\ldots,\chi_{i,j}^{(1)}(\mathbf{x}),\ldots,\chi_{i,TD}^{(1)}(\mathbf{x})\right]^\dagger$, $i=1,2,$ are square-integrable functions. Consider further that this Hilbert space is endowed with an inner product denoted as $\left\langle\chi^{(1)}(2;\mathbf{x}),\theta^{(1)}(2;\mathbf{x})\right\rangle_1$ between two elements, $\chi^{(1)}(2;\mathbf{x})\in\mathcal{H}_1$, $\theta^{(1)}(2;\mathbf{x})\in\mathcal{H}_1$, which is defined as follows:

$$\left\langle\chi^{(1)}(2;\mathbf{x}),\theta^{(1)}(2;\mathbf{x})\right\rangle_1 \triangleq \sum_{i=1}^{2}\left\langle\chi_i^{(1)}(\mathbf{x}),\theta_i^{(1)}(\mathbf{x})\right\rangle_0 \tag{6.38}$$

2. In the Hilbert $\mathscr{H}_1$, form the inner product of Eq. (6.30) with a yet undefined vector-valued function $\mathbf{a}^{(1)}(2;\mathbf{x}) \triangleq \left[\mathbf{a}_1^{(1)}(\mathbf{x}), \mathbf{a}_2^{(1)}(\mathbf{x})\right]^{\dagger} \in \mathscr{H}_1$ to obtain the following relation:

$$\left\{\left\langle \mathbf{a}^{(1)}(2;\mathbf{x}), \mathbf{V}^{(1)}\left[2\times 2;\mathbf{x};\mathbf{f}^0\right]\mathbf{v}^{(1)}(2;\mathbf{x})\right\rangle_1\right\}_{\alpha^0}$$
$$= \left\{\left\langle \mathbf{a}^{(1)}(2;\mathbf{x}), \mathbf{q}_V^{(1)}\left[2;\mathbf{u}^{(1)}(2;\mathbf{x});\mathbf{f};\delta\mathbf{f}\right]\right\rangle_1\right\}_{\alpha^0}. \tag{6.39}$$

3. Using the definition of the adjoint operator in the Hilbert space $\mathscr{H}_1$, recast the left side of Eq. (6.39) as follows:

$$\left\{\left\langle \mathbf{a}^{(1)}(2;\mathbf{x}), \mathbf{V}^{(1)}\left[2\times 2;\mathbf{x};\mathbf{f}\right]\mathbf{v}^{(1)}(2;\mathbf{x})\right\rangle_1\right\}_{\alpha^0}$$
$$= \left\{\left\langle \mathbf{v}^{(1)}(2;\mathbf{x}), \mathbf{A}^{(1)}\left[2\times 2;\mathbf{x};\mathbf{f}\right]\mathbf{a}^{(1)}(2;\mathbf{x})\right\rangle_1\right\}_{\alpha^0} \tag{6.40}$$
$$+ \left\{P^{(1)}\left[\mathbf{v}^{(1)}(2;\mathbf{x});\mathbf{a}^{(1)}(2;\mathbf{x});\mathbf{f};\delta\mathbf{f}\right]\right\}_{\alpha^0},$$

where $\left\{P^{(1)}\left[\mathbf{v}^{(1)}(2;\mathbf{x});\mathbf{a}^{(1)}(2;\mathbf{x});\mathbf{f};\delta\mathbf{f}\right]\right\}_{\alpha^0}$ denotes the bilinear concomitant defined on the phase-space boundary $\mathbf{x} \in \partial\Omega\left(\alpha^0\right)$, and where $\mathbf{A}^{(1)}\left[2\times 2;\mathbf{x};\mathbf{f}\right]$ is the operator formally adjoint to $\mathbf{V}^{(1)}\left[2\times 2;\mathbf{x};\mathbf{f}\right]$, i.e.,

$$\mathbf{A}^{(1)}\left[2\times 2;\mathbf{x};\mathbf{f}\right] \triangleq \left\{\mathbf{V}^{(1)}\left[2\times 2;\mathbf{x};\mathbf{f}\right]\right\}^* = \begin{pmatrix} \mathbf{L}^*(\mathbf{x};\mathbf{f}) & \mathbf{0} \\ \mathbf{0} & \mathbf{L}(\mathbf{x};\mathbf{f}) \end{pmatrix} \tag{6.41}$$

4. Require the first term on the right side of Eq. (6.40) to represent the indirect-effect term defined in Eq. (6.27), by imposing the following relation:

$$\mathbf{A}^{(1)}\left[2\times 2;\mathbf{x};\mathbf{f}\right]\mathbf{a}^{(1)}(2;\mathbf{x}) = \mathbf{q}_A^{(1)}\left[2;\mathbf{u}^{(1)}(2;\mathbf{x});\mathbf{f}\right], \quad \mathbf{x} \in \Omega\left(\alpha^0\right); \tag{6.42}$$

where:

$$\mathbf{q}_A^{(1)}\left[2;\mathbf{u}^{(1)}(2;\mathbf{x});\mathbf{f}\right] \triangleq \left[\frac{\partial S\left(\mathbf{u}^{(1)};\mathbf{f}\right)}{\partial \mathbf{u}^{(1)}(2;\mathbf{x})}\right]^{\dagger} \triangleq \begin{pmatrix} \left[\dfrac{\partial S\left(\mathbf{u}^{(1)};\mathbf{f}\right)}{\partial \varphi}\right]^{\dagger} \\ \left[\dfrac{\partial S\left(\mathbf{u}^{(1)};\mathbf{f}\right)}{\partial \psi}\right]^{\dagger} \end{pmatrix}. \tag{6.43}$$

5. Implement the boundary conditions represented by Eq. (6.31) into Eq. (6.40) and eliminate the remaining unknown boundary values of the function $\mathbf{v}^{(1)}(2;\mathbf{x})$ from the expression of the bilinear concomitant $\left\{P^{(1)}\left[\mathbf{v}^{(1)}(2;\mathbf{x});\mathbf{a}^{(1)}(2;\mathbf{x});\mathbf{f};\delta\mathbf{f}\right]\right\}_{\alpha^0}$ by selecting boundary conditions for the function $\mathbf{a}^{(1)}(2;\mathbf{x}) \triangleq \left[\mathbf{a}_1^{(1)}(\mathbf{x}),\mathbf{a}_2^{(1)}(\mathbf{x})\right]^\dagger$ so as to ensure that Eq. (6.42) is well-posed while being independent of *unknown* values of $\mathbf{v}^{(1)}(2;\mathbf{x})$ and of $\delta\mathbf{f}$. The boundary conditions thus chosen for the function $\mathbf{a}^{(1)}(2;\mathbf{x}) \triangleq \left[\mathbf{a}_1^{(1)}(\mathbf{x}),\mathbf{a}_2^{(1)}(\mathbf{x})\right]^\dagger$ can be represented in operator form as follows:

$$\left\{\mathbf{b}_A^{(1)}\left[\mathbf{u}^{(1)}(2;\mathbf{x});\mathbf{a}^{(1)}(2;\mathbf{x});\mathbf{f}\right]\right\}_{\alpha^0} = \mathbf{0}, \quad \mathbf{x} \in \partial\Omega\left[\lambda(\alpha^0);\omega(\alpha^0)\right]. \tag{6.44}$$

The selection of the boundary conditions for $\mathbf{a}^{(1)}(2;\mathbf{x}) \triangleq \left[\mathbf{a}_1^{(1)}(\mathbf{x}),\mathbf{a}_2^{(1)}(\mathbf{x})\right]^\dagger$ represented by Eq. (6.44) eliminates the appearance of the *unknown* values of $\mathbf{v}^{(1)}(2;\mathbf{x})$ in $\left\{P^{(1)}\left[\mathbf{v}^{(1)}(2;\mathbf{x});\mathbf{a}^{(1)}(2;\mathbf{x});\mathbf{f};\delta\mathbf{f}\right]\right\}_{\alpha^0}$ and reduces this bilinear concomitant to a residual quantity that contains boundary terms involving only known values of $\mathbf{u}^{(1)}(2;\mathbf{x})$, $\mathbf{a}^{(1)}(2;\mathbf{x})$, $\boldsymbol{f}$, and $\delta\mathbf{f}$. This residual quantity will be denoted as $\left\{\hat{P}^{(1)}\left[\mathbf{u}^{(1)}(2;\mathbf{x});\mathbf{a}^{(1)}(2;\mathbf{x});\mathbf{f};\delta\mathbf{f}\right]\right\}_{\alpha^0}$. In general, this residual quantity does not automatically vanish, although it may do so occasionally.

6. The system of equations comprising Eq. (6.42) together with the boundary conditions represented by Eq. (6.44) will be called the 1st-Level Adjoint Sensitivity System (1st-LASS). The solution $\mathbf{a}^{(1)}(2;\mathbf{x}) \triangleq \left[\mathbf{a}_1^{(1)}(\mathbf{x}),\mathbf{a}_2^{(1)}(\mathbf{x})\right]^\dagger$ of the 1st-LASS will be called the 1st-level adjoint sensitivity function. The 1st-LASS is called "first-level" (as opposed to "1st-order") because it does not contain any differential or functional derivatives, but its solution, $\mathbf{a}^{(1)}(2;\mathbf{x})$, will be used below to compute the 1st-order sensitivities of the response with respect to the components of the feature function $\mathbf{f}(\alpha)$.
7. Using Eq. (6.39) together with the forward and adjoint boundary conditions represented by Eqs. (6.31) and (6.44) in Eq. (6.40) reduce the latter to the following relation:

$$\begin{aligned}&\left\{\left\langle \mathbf{a}^{(1)}(2;\mathbf{x}),\mathbf{q}_V^{(1)}\left[2;\mathbf{u}^{(1)}(2;\mathbf{x});\mathbf{f};\delta\mathbf{f}\right]\right\rangle_1\right\}_{\alpha^0} \\ &= \left\{\left\langle \mathbf{v}^{(1)}(2;\mathbf{x}),\mathbf{A}^{(1)}[2\times 2;\mathbf{x};\mathbf{f}]\mathbf{a}^{(1)}(2;\mathbf{x})\right\rangle_1\right\}_{\alpha^0} \\ &+\left\{\hat{P}^{(1)}\left[\mathbf{u}^{(1)}(2;\mathbf{x});\mathbf{a}^{(1)}(2;\mathbf{x});\mathbf{f};\delta\mathbf{f}\right]\right\}_{\alpha^0}.\end{aligned} \tag{6.45}$$

8. In view of Eqs. (6.27) and (6.42), the 1st term on the right side of Eq. (6.45) represents the indirect-effect term $\left\{\delta R\left[\mathbf{u}^{(1)}(2;\mathbf{x});\mathbf{f};\mathbf{v}^{(1)}\right]\right\}_{ind}$. It therefore follows from Eq. (6.45) that the indirect-effect term can be expressed in terms of the 1st-level adjoint sensitivity function $\mathbf{a}^{(1)}(2;\mathbf{x}) \triangleq \left[\mathbf{a}_1^{(1)}(\mathbf{x}),\mathbf{a}_2^{(1)}(\mathbf{x})\right]^\dagger$ as follows:

$$\left\{\delta R\left[\mathbf{u}^{(1)}(2;\mathbf{x});\mathbf{f};\mathbf{v}^{(1)}(2;\mathbf{x})\right]\right\}_{ind}=\left\{\left\langle \mathbf{a}^{(1)}(2;\mathbf{x}),\mathbf{q}_V^{(1)}\left[2;\mathbf{u}^{(1)}(2;\mathbf{x});\mathbf{f};\delta\mathbf{f}\right]\right\rangle_1\right\}_{\alpha^0}$$
$$-\left\{\hat{P}^{(1)}\left[\mathbf{u}^{(1)}(2;\mathbf{x});\mathbf{a}^{(1)}(2;\mathbf{x});\mathbf{f};\delta\mathbf{f}\right]\right\}_{\alpha^0}\equiv\left\{\delta R\left[\mathbf{u}^{(1)}(2;\mathbf{x});\mathbf{a}^{(1)}(2;\mathbf{x});\mathbf{f};\delta\mathbf{f}\right]\right\}_{ind}. \tag{6.46}$$

As indicated by the identity shown in Eq. (6.46), the variations $\delta\boldsymbol{\varphi}$ and $\delta\boldsymbol{\psi}$ have been eliminated from the original expression of the indirect-effect term, which now depends on the 1st-level adjoint sensitivity function $\mathbf{a}^{(1)}(2;\mathbf{x})\triangleq\left[\mathbf{a}_1^{(1)}(\mathbf{x}),\mathbf{a}_2^{(1)}(\mathbf{x})\right]^\dagger$. Adding the expression obtained in Eq. (6.46) with the expression for the direct-effect term defined in Eq. (6.21) yields, according to Eq. (6.20), the following expression for the total 1st-order sensitivity $\left\{\delta R\left(\boldsymbol{\varphi},\boldsymbol{\psi},\mathbf{f};\delta\boldsymbol{\varphi},\delta\boldsymbol{\psi},\delta\mathbf{f}\right)\right\}_{\alpha^0}$ of the response $R\left[\boldsymbol{\varphi}(\mathbf{x}),\boldsymbol{\psi}(\mathbf{x});\mathbf{f}\right]$ with respect to the components of the feature function $\mathbf{f}(\boldsymbol{\alpha})$:

$$\left\{\delta R\left(\boldsymbol{\varphi},\boldsymbol{\psi},\mathbf{f};\delta\boldsymbol{\varphi},\delta\boldsymbol{\psi},\delta\mathbf{f}\right)\right\}_{\alpha^0}=\left\{\left\langle \mathbf{a}^{(1)}(2;\mathbf{x}),\mathbf{q}_V^{(1)}\left[2;\mathbf{u}^{(1)}(2;\mathbf{x});\mathbf{f};\delta\mathbf{f}\right]\right\rangle_1\right\}_{\alpha^0}$$
$$-\left\{\hat{P}^{(1)}\left[\mathbf{u}^{(1)}(2;\mathbf{x});\mathbf{a}^{(1)}(2;\mathbf{x});\mathbf{f};\delta\mathbf{f}\right]\right\}_{\alpha^0}+\left\{\frac{\partial\mathbf{R}\left(\mathbf{u}^{(1)};\mathbf{f}\right)}{\partial\mathbf{f}}\delta\mathbf{f}\right\}_{\alpha^0} \tag{6.47}$$
$$\equiv\sum_{j_1=1}^{TF}\left\{R^{(1)}\left[j_1;\mathbf{u}^{(1)}(2;\mathbf{x});\mathbf{a}^{(1)}(2;\mathbf{x});\mathbf{f}(\boldsymbol{\alpha})\right]\delta f_{j_1}\right\}_{\alpha^0}.$$

The identity that appears in Eq. (6.47) emphasizes the fact that the variations $\delta\boldsymbol{\varphi}$ and $\delta\boldsymbol{\psi}$, which are expensive to compute, have been eliminated from the final expressions of the 1st-order sensitivities $R^{(1)}\left[j_1;\mathbf{u}^{(1)}(2;\mathbf{x});\mathbf{a}^{(1)}(2;\mathbf{x});\mathbf{f}(\boldsymbol{\alpha})\right]$ of the response with respect to the components $f_{j_1}(\boldsymbol{\alpha})$, $j_1=1,\ldots,TF$, of the feature functions. The dependence on the variations $\delta\boldsymbol{\varphi}$ and $\delta\boldsymbol{\psi}$ has been replaced in the expression of $R^{(1)}\left[j_1;\mathbf{u}^{(1)}(2;\mathbf{x});\mathbf{a}^{(1)}(2;\mathbf{x});\mathbf{f}(\boldsymbol{\alpha})\right]$ by the dependence on the 1st-level adjoint sensitivity function $\mathbf{a}^{(1)}(2;\mathbf{x})\triangleq\left[\mathbf{a}_1^{(1)}(\mathbf{x}),\mathbf{a}_2^{(1)}(\mathbf{x})\right]^\dagger$. It is very important to note that the 1st-LASS is independent of variations $\delta\mathbf{f}(\boldsymbol{\alpha})$ in the components of the feature function and is consequently also independent of any variations $\delta\boldsymbol{\alpha}$ in the primary model parameters. Hence, the 1st-LASS needs to be solved only once to determine the 1st-level adjoint sensitivity function $\mathbf{a}^{(1)}(2;\mathbf{x})\triangleq\left[\mathbf{a}_1^{(1)}(\mathbf{x}),\mathbf{a}_2^{(1)}(\mathbf{x})\right]^\dagger$. Subsequently, the "indirect-effect term" is computed efficiently and exactly by simply performing the integrations required to compute the inner product over the adjoint function $\mathbf{a}^{(1)}(2;\mathbf{x})\triangleq\left[\mathbf{a}_1^{(1)}(\mathbf{x}),\mathbf{a}_2^{(1)}(\mathbf{x})\right]^\dagger$, as indicated on the right side of Eq. (6.47). Solving the 1st-Level Adjoint Sensitivity System (1st-LASS) requires the same computational effort as solving the original coupled linear system, entailing the following operations: (i) inverting (i.e., solving) the left side of the original adjoint equation with the source $\left[\frac{\partial S\left(\mathbf{u}^{(1)};\boldsymbol{\alpha}\right)}{\partial\boldsymbol{\varphi}}\right]^\dagger$ to obtain the 1st-level adjoint sensitivity function $\mathbf{a}_1^{(1)}(\mathbf{x})$;

and (ii) inverting the left side of the original forward equation with the source $\left[\dfrac{\partial S\left(\mathbf{u}^{(1)};\boldsymbol{\alpha}\right)}{\partial \boldsymbol{\psi}}\right]^{\dagger}$ to obtain the 1st-level adjoint sensitivity function $\mathbf{a}_2^{(1)}(\mathbf{x})$.

The 1st-order sensitivities $R^{(1)}\left[j_1;\mathbf{u}^{(1)}(2;\mathbf{x});\mathbf{a}^{(1)}(2;\mathbf{x});\mathbf{f}(\boldsymbol{\alpha})\right]$, $j_1 = 1,\dots,TF$, can be expressed as an integral over the independent variables as follows:

$$R^{(1)}\left[j_1;\mathbf{u}^{(1)}(2;\mathbf{x});\mathbf{a}^{(1)}(2;\mathbf{x});\mathbf{f}(\boldsymbol{\alpha})\right] \triangleq \int_{\lambda_1(\boldsymbol{\alpha})}^{\omega_1(\boldsymbol{\alpha})} dx_1 \dots \int_{\lambda_{TI}(\boldsymbol{\alpha})}^{\omega_{TI}(\boldsymbol{\alpha})} dx_{TI} S^{(1)}\left[j_1;\mathbf{u}^{(1)}(2;\mathbf{x});\mathbf{a}^{(1)}(2;\mathbf{x});\mathbf{f}(\boldsymbol{\alpha})\right]. \tag{6.48}$$

In particular, if the residual bilinear concomitant is non-zero, the functions $S^{(1)}\left[j_1;\mathbf{u}^{(1)}(2;\mathbf{x});\mathbf{a}^{(1)}(2;\mathbf{x});\mathbf{f}(\boldsymbol{\alpha})\right]$ would contain suitably defined Dirac delta-functionals for expressing the respective non-zero boundary terms as volume integrals over the phase space of the independent variables. Dirac-delta functionals would also be used in the expression of $S^{(1)}\left[j_1;\mathbf{u}^{(1)}(2;\mathbf{x});\mathbf{a}^{(1)}(2;\mathbf{x});\mathbf{f}(\boldsymbol{\alpha})\right]$ to represent terms containing the derivatives of the boundary end-points with respect to the model and/or response parameters.

The response sensitivities with respect to the primary model parameters would be obtained by using the expression obtained in Eq. (6.48) in conjunction with the "chain rule" of differentiation provided in Eq. (6.12).

It is important to compare the results produced by the 1st-FASAM-L (for obtaining the sensitivities of the model response with respect to the model's features) with the 1st-CASAM-L (the 1st-Order Comprehensive Adjoint Sensitivity Analysis Methodology for Response-Coupled Forward/Adjoint Linear Systems) methodology, which provides the expressions of the response sensitivities directly with respect to the model's primary parameters. Recall that the 1st-CASAM-L (Cacuci 2021b, 2022b) yields the following expression for the 1st-order sensitivities of the response with respect to the primary model parameters:

$$\begin{aligned}
\left\{\frac{\partial R\left[j_1;\mathbf{u}^{(1)}(2;\mathbf{x});\mathbf{a}^{(1)}(2;\mathbf{x});\boldsymbol{\alpha}\right]}{\partial \alpha_{j_1}}\right\}_{\boldsymbol{\alpha}^0} &= \left\{\int_{\lambda_1(\boldsymbol{\alpha})}^{\omega_1(\boldsymbol{\alpha})} dx_1 \dots \int_{\lambda_{TI}(\boldsymbol{\alpha})}^{\omega_{TI}(\boldsymbol{\alpha})} dx_{TI} \frac{\partial S\left[\mathbf{u}^{(1)}(2;\mathbf{x});\boldsymbol{\alpha}\right]}{\partial \alpha_{j_1}}\right\}_{\boldsymbol{\alpha}^0} \\
&+ \sum_{k=1}^{TI} \prod_{m=1,k\neq j}^{TI} \left\{\int_{\lambda_m(\boldsymbol{\alpha})}^{\omega_m(\boldsymbol{\alpha})} dx_m S\left[\mathbf{u}^{(1)}(2;\dots,\omega_k,\dots);\boldsymbol{\alpha}\right]\frac{\partial \omega_k(\boldsymbol{\alpha})}{\partial \alpha_{j_1}} - S\left[\mathbf{u}^{(1)}(2;\dots,\lambda_k,\dots);\boldsymbol{\alpha}\right]\frac{\partial \lambda_k(\boldsymbol{\alpha})}{\partial \alpha_{j_1}}\right\}_{\boldsymbol{\alpha}^0} \\
&+ \left\{\left\langle \mathbf{a}^{(1)}(2;\mathbf{x}), \frac{\partial}{\partial \alpha_{j_1}} \mathbf{q}^{(1)}\left[\mathbf{u}^{(1)}(2;\mathbf{x});\boldsymbol{\alpha}\right]\right\rangle_1\right\}_{\boldsymbol{\alpha}^0} - \left\{\frac{\partial}{\partial \alpha_{j_1}} \hat{P}^{(1)}\left[\mathbf{u}^{(1)};\mathbf{a}^{(1)};\boldsymbol{\alpha}\right]\right\}_{\boldsymbol{\alpha}^0}; \quad j_1 = 1,\dots,TP.
\end{aligned} \tag{6.49}$$

The same 1st-level adjoint sensitivity function, denoted as $\mathbf{a}^{(1)}(2;\mathbf{x})$, appears both in Eq. (6.49) and in Eq. (6.48). Therefore, the same number of "large-scale computations" (which are needed to solve the 1st-LASS to determine the 1st-level adjoint sensitivity function) is needed for obtaining either the response sensitivities with respect to the components, $f_j(\boldsymbol{\alpha})$, $j=1,\ldots,TF$, of the feature function $\mathbf{f}(\boldsymbol{\alpha})$ using the 1st-FASAM-L or for obtaining the response sensitivities directly with respect to the primary model parameters $\alpha_j$, $j=1,\ldots,TP$, by using the 1st-CASAM-L. The use of the 1st-CASAM-L would also require performing a number of $TP$ integrations to compute all of the response sensitivities with respect to the primary parameters. In contradistinction, the use of the 1st-FASAM-L would require only $TF$ integrations ($TF<TP$) to compute all of the response sensitivities with respect to the components $f_j(\boldsymbol{\alpha})$ of the feature function. Since integrations using quadrature schemes are significantly less expensive computationally by comparison to solving systems of equations (e.g., the original equations underlying the model and the 1st-LASS), the computational savings provided by the use of the 1st-FASAM-L are small by comparison to using the 1st-CASAM-L. However, this conclusion is valid only for the computation of 1st-order sensitivities. As will be shown in Section 6.4, the computational savings are significantly larger when computing the 2nd-order sensitivities by using the 2nd-FASAM-L rather than using the 2nd-CASAM-L (or any other method).

## 6.4 THE 2ND-ORDER FEATURE ADJOINT SENSITIVITY ANALYSIS METHODOLOGY FOR RESPONSE-COUPLED FORWARD AND ADJOINT LINEAR SYSTEMS (2ND-FASAM-L)

The "Second-Order Function/Feature Adjoint Sensitivity Analysis Methodology for Response-Coupled Forward/Adjoint Linear Systems" (2nd-FASAM-L) determines the 2nd-order sensitivities $\dfrac{\partial^2 R\left[\mathbf{u}^{(1)}(2;\mathbf{x});\mathbf{f}(\boldsymbol{\alpha})\right]}{\partial f_{j_2}\,\partial f_{j_1}}$ of the model's response with respect to the components of the "feature" function $\mathbf{f}(\boldsymbol{\alpha})$ by conceptually considering that the first-order sensitivities $R^{(1)}\left[j_1;\mathbf{u}^{(1)}(2;\mathbf{x});\mathbf{a}^{(1)}(2;\mathbf{x});\mathbf{f}(\boldsymbol{\alpha})\right]\triangleq\dfrac{\partial R\left[\mathbf{u}^{(1)}(2;\mathbf{x});\mathbf{f}(\boldsymbol{\alpha})\right]}{\partial f_{j_1}}$, which were obtained in Eq. (6.48), are "model responses." Consequently, the 2nd-order sensitivities are obtained as the "1st-order sensitivities of the 1st-order sensitivities" by applying the concepts underlying 1st-FASAM to each 1st-order sensitivity $R^{(1)}\left[j_1;\mathbf{u}^{(1)}(2;\mathbf{x});\mathbf{a}^{(1)}(2;\mathbf{x});\mathbf{f}(\boldsymbol{\alpha})\right]$, $j_1=1,\ldots,TF$; each of these 1st-order sensitivities depends on both the vector $\mathbf{u}^{(1)}(2;\mathbf{x})$, which comprises the original state variables, as well as on the 1st-level adjoint function $\mathbf{a}^{(1)}(2;\mathbf{x})$.

To establish the pattern underlying the computation of sensitivities of arbitrarily high-order, it is useful to introduce a systematic classification of the systems of equations that will underlie the computation of the sensitivities of various orders. As has been shown in Section 6.3, the 1st-order response sensitivities $R^{(1)}\left[j_1;\mathbf{u}^{(1)}(2;\mathbf{x});\mathbf{a}^{(1)}(2;\mathbf{x});\mathbf{f}(\boldsymbol{\alpha})\right]$ depend on the original state functions $\mathbf{u}^{(1)}(2;\mathbf{x})\triangleq\left[\varphi(\mathbf{x}),\psi(\mathbf{x})\right]^{\dagger}$ and on the 1st-level adjoint sensitivity function

$\mathbf{a}^{(1)}(2;\mathbf{x}) \triangleq \left[\mathbf{a}_1^{(1)}(\mathbf{x}),\mathbf{a}_2^{(1)}(\mathbf{x})\right]^{\dagger}$. The system of equations satisfied by these functions will be called "the 2nd-Level Forward/Adjoint System" (2nd-LFAS) and will be rewritten in the following concatenated form:

$$\mathbf{F}^{(2)}\left[2^2\times 2^2;\mathbf{f}(\alpha)\right]\mathbf{u}^{(2)}\left(2^2;\mathbf{x}\right)=\mathbf{q}_F^{(2)}\left[2^2;\mathbf{u}^{(1)}(2;\mathbf{x});\mathbf{f}(\alpha)\right];\quad \mathbf{x}\in\Omega(\alpha); \tag{6.50}$$

$$\mathbf{b}_F^{(2)}\left(2^2;\mathbf{u}^{(2)};\mathbf{f}\right)\triangleq\left(\mathbf{b}_F^{(1)},\mathbf{b}_A^{(1)}\right)^{\dagger}=\mathbf{0};\quad \mathbf{x}\in\partial\Omega\left[\boldsymbol{\lambda}(\alpha);\boldsymbol{\omega}(\alpha)\right]; \tag{6.51}$$

where the following definitions were used:

$$\mathbf{F}^{(2)}\left[2^2\times 2^2;\mathbf{f}(\alpha)\right]\triangleq diag\left(\mathbf{F}^{(1)},\mathbf{A}^{(1)}\right);\mathbf{u}^{(2)}\left(2^2;\mathbf{x}\right)\triangleq\left[\mathbf{u}^{(1)}(2;\mathbf{x}),\mathbf{a}^{(1)}(2;\mathbf{x})\right]^{\dagger}; \tag{6.52}$$

$$\mathbf{q}_F^{(2)}\left[2^2;\mathbf{u}^{(1)}(2;\mathbf{x});\mathbf{f}(\alpha)\right]\triangleq\left[\mathbf{q}_F^{(1)}(2;\mathbf{x};\mathbf{f}),\mathbf{q}_A^{(1)}\left[2;\mathbf{u}^{(1)};\mathbf{f}\right]\right]^{\dagger}. \tag{6.53}$$

The notation used for the matrix $\mathbf{F}^{(2)}\left[2^2\times 2^2;\mathbf{f}(\alpha)\right]$ indicates the following characteristics: (i) the superscript "2" indicates "2nd-level;" (ii) the argument "$2^2\times 2^2$" indicates that this square matrix comprises $4\times 4=16$ component sub-matrices. Similarly, the argument "$2^2$" that appears in the block vectors $\mathbf{u}^{(2)}\left(2^2;\mathbf{x}\right)$, $\mathbf{q}_F^{(2)}\left[2^2;\mathbf{u}^{(1)}(2;\mathbf{x});\mathbf{f}(\alpha)\right]$ and $\mathbf{b}_F^{(2)}\left(2^2;\mathbf{u}^{(2)};\alpha\right)$ indicates that each of these column block-vectors comprises four sub-vectors as components.

The 1st-order G-differential of a first-order sensitivity $R^{(1)}\left[j_1;\mathbf{u}^{(2)}\left(2^2;\mathbf{x}\right);\mathbf{f}(\alpha)\right]$, $j_1=1,\ldots,TF$, is obtained by definition as follows:

$$\begin{aligned}&\left\{\delta R^{(1)}\left[j_1;\mathbf{u}^{(2)}\left(2^2;\mathbf{x}\right);\mathbf{v}^{(2)}\left(2^2;\mathbf{x}\right);\mathbf{f};\delta\mathbf{f}\right]\right\}_{\alpha^0}\\&\triangleq\left\{\frac{d}{d\varepsilon}\delta R^{(1)}\left[j_1;\mathbf{u}^{(1)}(2;\mathbf{x})+\varepsilon\mathbf{v}^{(1)}(2;\mathbf{x});\mathbf{a}^{(1)}(2;\mathbf{x})+\varepsilon\delta\mathbf{a}^{(1)}(2;\mathbf{x});\mathbf{f}+\varepsilon\delta\mathbf{f}\right]\right\}_{\varepsilon=0}\\&=\left\{\delta R^{(1)}\left[j_1;\mathbf{u}^{(2)}\left(2^2;\mathbf{x}\right);\mathbf{v}^{(2)}\left(2^2;\mathbf{x}\right);\mathbf{f}\right]\right\}_{ind}+\left\{\delta R^{(1)}\left[j_1;\mathbf{u}^{(2)}\left(2^2;\mathbf{x}\right);\delta\mathbf{f}\right]\right\}_{dir}.\end{aligned} \tag{6.54}$$

The direct-effect term $\left\{\delta R^{(1)}\left[j_1;\mathbf{u}^{(2)}\left(2^2;\mathbf{x}\right);\delta\mathbf{f}\right]\right\}_{dir}$ in Eq. (6.54) is defined as follows:

$$\begin{aligned}&\left\{\delta R^{(1)}\left[j_1;\mathbf{u}^{(2)}\left(2^2;\mathbf{x}\right);\delta\mathbf{f}\right]\right\}_{dir}\\&\triangleq\sum_{j_2=1}^{TF}\left\{\frac{\partial}{\partial f_{j_2}}\int_{\lambda_1(\alpha)}^{\omega_1(\alpha)}dx_1\ldots\int_{\lambda_{TI}(\alpha)}^{\omega_{TI}(\alpha)}dx_{TI}S^{(1)}\left[j_1;\mathbf{u}^{(2)}\left(2^2;\mathbf{x}\right);\mathbf{f}(\alpha)\right]\right\}_{\alpha^0}\delta f_{j_2},\end{aligned} \tag{6.55}$$

and can be computed immediately. The indirect-effect term $\left\{\delta R^{(1)}\left[j_1;\mathbf{u}^{(2)}\left(2^2;\mathbf{x}\right);\mathbf{v}^{(2)}\left(2^2;\mathbf{x}\right);\mathbf{f}\right]\right\}_{ind}$ in Eq. (6.54) depends on the 2nd-level variational sensitivity function $\mathbf{v}^{(2)}\left(2^2;\mathbf{x}\right)\triangleq\left[\mathbf{v}^{(1)}(2;\mathbf{x}),\delta\mathbf{a}^{(1)}(2;\mathbf{x})\right]$ and is defined as follows:

$$\left\{\delta R^{(1)}\left[j_1;\mathbf{u}^{(2)}\left(2^2;\mathbf{x}\right);\mathbf{v}^{(2)}\left(2^2;\mathbf{x}\right);\mathbf{f}\right]\right\}_{ind}$$
$$\triangleq\int_{\lambda_1(\alpha)}^{\omega_1(\alpha)}dx_1\ldots\int_{\lambda_{TI}(\alpha)}^{\omega_{TI}(\alpha)}dx_{TI}\left[\mathbf{s}^{(2)}\left(2^2;j_1;\mathbf{u}^{(2)};\mathbf{f}\right)\bullet\mathbf{v}^{(2)}\left(2^2;\mathbf{x}\right)\right], \tag{6.56}$$

where:

$$\mathbf{s}^{(2)}\left(2^2;j_1;\mathbf{u}^{(2)};\mathbf{f}\right)\triangleq\frac{\partial R^{(1)}\left[j_1;\mathbf{u}^{(2)}\left(2^2;\mathbf{x}\right);\mathbf{v}^{(2)}\left(2^2;\mathbf{x}\right);\mathbf{f}\right]}{\partial\mathbf{u}^{(2)}}. \tag{6.57}$$

Evidently, the functions $\mathbf{v}^{(1)}(2;\mathbf{x})$ and $\delta\mathbf{a}^{(1)}(2;\mathbf{x})$ are needed in order to evaluate the above indirect-effect term. These functions are the solutions of the system of equations obtained by taking the first-G-differential of the 2nd-LFAS defined by Eqs. (6.52) and (6.53). Applying the definition of the first G-differential, the 2nd-LFAS yields the following "2nd-Level Variational Sensitivity System" (2nd-LVSS) for the 2nd-Level variational sensitivity function $\mathbf{v}^{(2)}\left(2^2;\mathbf{x}\right)\triangleq\left[\mathbf{v}^{(1)}(2;\mathbf{x}),\delta\mathbf{a}^{(1)}(2;\mathbf{x})\right]^{\dagger}$:

$$\left\{\frac{d}{d\varepsilon}\mathbf{F}^{(2)}\left[2^2\times2^2;\mathbf{f}^0+\varepsilon\delta\mathbf{f}\right]\left[\mathbf{u}^{(2,0)}\left(2^2;\mathbf{x}\right)+\varepsilon\mathbf{v}^{(2)}\left(2^2;\mathbf{x}\right)\right]\right\}_{\varepsilon=0}$$
$$=\left\{\frac{d}{d\varepsilon}\mathbf{q}_F^{(2)}\left[2^2;\mathbf{u}^{(1,0)}(2;\mathbf{x})+\varepsilon\mathbf{v}^{(1)}(2;\mathbf{x});\mathbf{f}^0+\varepsilon\delta\mathbf{f}\right]\right\}_{\varepsilon=0};\quad\mathbf{x}\in\Omega\left(\boldsymbol{\alpha}^0\right); \tag{6.58}$$

$$\left\{\frac{d}{d\varepsilon}\mathbf{b}_F^{(2)}\left[\mathbf{u}^{(2,0)}\left(2^2;\mathbf{x}\right)+\varepsilon\mathbf{v}^{(2)}\left(2^2;\mathbf{x}\right);\mathbf{f}^0+\varepsilon\delta\mathbf{f}\right]\right\}_{\varepsilon=0}=\mathbf{0};$$
$$\mathbf{x}\in\partial\Omega\left[\boldsymbol{\lambda}\left(\boldsymbol{\alpha}^0\right);\boldsymbol{\omega}\left(\boldsymbol{\alpha}^0\right)\right]. \tag{6.59}$$

Carrying out the differentiation with respect to $\varepsilon$ in Eqs. (6.58) and (6.59), and setting $\varepsilon=0$ in the resulting expressions yields the following 2nd-LVSS:

$$\left\{\mathbf{V}^{(2)}\left[2^2\times2^2;\mathbf{x};\mathbf{f}\right]\mathbf{v}^{(2)}\left(2^2;\mathbf{x}\right)\right\}_{\boldsymbol{\alpha}^0}$$
$$=\left\{\mathbf{q}_V^{(2)}\left[2^2;\mathbf{u}^{(2)}\left(2^2;\mathbf{x}\right);\mathbf{f};\delta\mathbf{f}\right]\right\}_{\boldsymbol{\alpha}^0};\quad\mathbf{x}\in\Omega\left(\boldsymbol{\alpha}^0\right); \tag{6.60}$$

$$\left\{\mathbf{b}_V^{(2)}\left(\mathbf{u}^{(2)};\mathbf{v}^{(2)};\mathbf{f};\delta\mathbf{f}\right)\right\}_{\boldsymbol{\alpha}^0}=\mathbf{0};\quad\mathbf{x}\in\partial\Omega\left[\boldsymbol{\lambda}\left(\boldsymbol{\alpha}^0\right);\boldsymbol{\omega}\left(\boldsymbol{\alpha}^0\right)\right]; \tag{6.61}$$

where the following definitions were used:

$$\mathbf{V}^{(2)}\left[2^2\times 2^2;\mathbf{x};\mathbf{f}\right]=\begin{pmatrix} \mathbf{V}^{(1)}\left[2\times 2;\mathbf{x};\mathbf{f}\right] & \mathbf{0} \\ \mathbf{V}_{21}^{(2)}\left(2\times 2;\mathbf{u}^{(1)};\mathbf{f}\right) & \mathbf{A}^{(1)}\left[2\times 2;\mathbf{x};\mathbf{f}\right] \end{pmatrix};$$

$$\mathbf{V}_{21}^{(2)}\left(2\times 2;\mathbf{u}^{(1)};\mathbf{f}\right)\triangleq\begin{pmatrix} -\dfrac{\partial^2 S\left(\mathbf{u}^{(1)};\mathbf{f}\right)}{\partial\varphi\,\partial\varphi} & -\dfrac{\partial^2 S\left(\mathbf{u}^{(1)};\mathbf{f}\right)}{\partial\varphi\,\partial\psi} \\ -\dfrac{\partial^2 S\left(\mathbf{u}^{(1)};\mathbf{f}\right)}{\partial\psi\,\partial\varphi} & -\dfrac{\partial^2 S\left(\mathbf{u}^{(1)};\mathbf{f}\right)}{\partial\psi\,\partial\psi} \end{pmatrix}; \tag{6.62}$$

$$\mathbf{q}_V^{(2)}\left[2^2;\mathbf{u}^{(2)}\left(2^2;\mathbf{x}\right);\mathbf{f};\delta\mathbf{f}\right]\triangleq\begin{pmatrix} \mathbf{q}_V^{(1)}\left[2;\mathbf{u}^{(1)};\mathbf{f};\delta\mathbf{f}\right] \\ \mathbf{p}_A^{(1)}\left[2;\mathbf{u}^{(1)};\mathbf{a}_1^{(1)};\mathbf{f};\delta\mathbf{f}\right] \end{pmatrix};$$

$$\mathbf{p}_A^{(1)}\left[2;\mathbf{u}^{(1)};\mathbf{a}_1^{(1)};\mathbf{f};\delta\mathbf{f}\right]\triangleq\begin{pmatrix} \mathbf{p}_1^{(1)}\left(\mathbf{u}^{(1)};\mathbf{a}_1^{(1)};\delta\mathbf{f}\right) \\ \mathbf{p}_2^{(1)}\left(\mathbf{u}^{(1)};\mathbf{a}_1^{(1)};\delta\mathbf{f}\right) \end{pmatrix}; \tag{6.63}$$

$$\mathbf{p}_1^{(1)}\left(\mathbf{u}^{(1)};\mathbf{a}_1^{(1)};\mathbf{f};\delta\mathbf{f}\right)\triangleq\frac{\partial^2 S\left(\mathbf{u}^{(1)};\mathbf{f}\right)}{\partial\mathbf{f}\,\partial\varphi}\delta\mathbf{f}-\frac{\partial\left[\mathbf{L}^*(\mathbf{f})\mathbf{a}_1^{(1)}\right]}{\partial\mathbf{f}}\delta\mathbf{f}; \tag{6.64}$$

$$\mathbf{p}_2^{(1)}\left(\mathbf{u}^{(1)};\mathbf{a}_2^{(1)};\mathbf{f};\delta\mathbf{f}\right)\triangleq\frac{\partial^2 S\left(\mathbf{u}^{(1)};\mathbf{f}\right)}{\partial\mathbf{f}\,\partial\psi}\delta\mathbf{f}-\frac{\partial\left[\mathbf{L}(\mathbf{f})\mathbf{a}_2^{(1)}\right]}{\partial\mathbf{f}}\delta\mathbf{f}; \tag{6.65}$$

$$\mathbf{b}_V^{(2)}\left(\mathbf{u}^{(2)};\mathbf{v}^{(2)};\mathbf{f};\delta\mathbf{f}\right)\triangleq\begin{pmatrix} \mathbf{b}_V^{(1)}\left(\mathbf{u}^{(1)};\mathbf{v}^{(1)};\mathbf{f};\delta\mathbf{f}\right) \\ \delta\mathbf{b}_A^{(1)}\left(\mathbf{u}^{(2)};\mathbf{v}^{(2)};\mathbf{f};\delta\mathbf{f}\right) \end{pmatrix}. \tag{6.66}$$

The matrix $\mathbf{V}_{21}^{(2)}\left(2\times 2;\mathbf{u}^{(1)};\mathbf{f}\right)$ depends on the system's response and is responsible for coupling the forward and adjoint systems. Although the forward and adjoint systems are coupled, they could nevertheless be solved successively rather than simultaneously because the matrix $\mathbf{V}^{(2)}\left[2^2\times 2^2;\mathbf{x};\mathbf{f}\right]$ is block-diagonal. All of the components of the matrices and vectors underlying the 2nd-LVSS are to be computed at nominal parameter and state function values, as indicated in Eqs. (6.60) and (6.61).

Computing the indirect-effect term $\left\{\delta R^{(1)}\left[j_1;\mathbf{u}^{(2)}\left(2^2;\mathbf{x}\right);\mathbf{v}^{(2)}\left(2^2;\mathbf{x}\right);\mathbf{f}\right]\right\}_{ind}$ by solving the 2nd-LVSS would require at least $2TF(TF+1)$ large-scale computations (to solve the 2nd-LVSS) for every component of the feature function $\mathbf{f}(\alpha)$.

The need for solving the 2nd-LVSS will be circumvented by deriving an alternative expression for the indirect-effect term $\left\{\delta R^{(1)}\left[j_1;\mathbf{u}^{(2)}\left(2^2;\mathbf{x}\right);\mathbf{v}^{(2)}\left(2^2;\mathbf{x}\right);\mathbf{f}\right]\right\}_{ind}$, as defined in Eq. (6.65), in which the second-level variational function $\mathbf{v}^{(2)}\left(2^2;\mathbf{x}\right)$ will be replaced by a 2nd-level adjoint function that is independent of variations in the model parameter and state functions. This 2nd-level adjoint function will be the solution of a 2nd-Level Adjoint Sensitivity System (2nd-LASS), which will be constructed by using the same principles as employed for deriving the 1st-LASS. The 2nd-LASS is constructed in a Hilbert space, denoted as $\mathcal{H}_2$, which will comprise as elements block-vectors of the same form as $\mathbf{v}^{(2)}\left(2^2;\mathbf{x}\right)$, i.e., a vector in $\mathcal{H}_2$ has the generic structure $\boldsymbol{\chi}^{(2)}\left(2^2;\mathbf{x}\right)\triangleq\left[\boldsymbol{\chi}_1^{(2)}(\mathbf{x}),\boldsymbol{\chi}_2^{(2)}(\mathbf{x}),\boldsymbol{\chi}_3^{(2)}(\mathbf{x}),\boldsymbol{\chi}_4^{(2)}(\mathbf{x})\right]^\dagger$, comprising four vector-valued components of the form $\boldsymbol{\chi}_i^{(2)}(\mathbf{x})\triangleq\left[\chi_{i,1}^{(2)}(\mathbf{x}),\ldots,\chi_{i,j}^{(2)}(\mathbf{x}),\ldots,\chi_{i,TD}^{(2)}(\mathbf{x})\right]^\dagger$, $i=1,2,3,4=2^2$. The inner product between two elements, $\boldsymbol{\chi}^{(2)}\left(2^2;\mathbf{x}\right)\in\mathcal{H}_2$ and $\boldsymbol{\theta}^{(2)}\left(2^2;\mathbf{x}\right)\in\mathcal{H}_2$, of this Hilbert space, will be denoted as $\boldsymbol{\chi}^{(2)}\left(2^2;\mathbf{x}\right),\boldsymbol{\theta}^{(2)}\left(2^2;\mathbf{x}\right)_{2^2}$ and is defined as follows:

$$\left\langle\boldsymbol{\chi}^{(2)}\left(2^2;\mathbf{x}\right),\boldsymbol{\theta}^{(2)}\left(2^2;\mathbf{x}\right)\right\rangle_2\triangleq\sum_{i=1}^{2^2}\left\langle\boldsymbol{\chi}_i^{(2)}(\mathbf{x}),\boldsymbol{\theta}_i^{(2)}(\mathbf{x})\right\rangle_0 \tag{6.67}$$

Note that there are $j_1=1,\ldots,TF$ distinct indirect-effect terms $\left\{\delta R^{(1)}\left[j_1;\mathbf{u}^{(2)}\left(2^2;\mathbf{x}\right);\mathbf{v}^{(2)}\left(2^2;\mathbf{x}\right);\mathbf{f}\right]\right\}_{ind}$. Each of these indirect-effect terms will serve as a "source" for a "2nd-Level Adjoint Sensitivity System" (2nd-LASS) that will be constructed by applying the same sequence of steps that were used in Section 6.3 to construct the 1st-LASS. This implies that a distinct 2nd-level adjoint sensitivity function, of the form $\mathbf{a}^{(2)}\left(2^2;j_1;\mathbf{x}\right)\triangleq\left[\mathbf{a}_1^{(2)}(j_1;\mathbf{x}),\mathbf{a}_2^{(2)}(j_1;\mathbf{x}),\mathbf{a}_3^{(2)}(j_1;\mathbf{x}),\mathbf{a}_4^{(2)}(j_1;\mathbf{x})\right]^\dagger\in\mathcal{H}_2$, $j_1=1,\ldots,TF$, corresponding to each distinct indirect-effect term, will be needed for constructing each of the corresponding 2nd-LASS, as follows:

1. For each $j_1=1,\ldots,TP$, form the inner product in the Hilbert space $\mathcal{H}_2$ of Eq. (6.60) with a yet undefined function $\mathbf{a}^{(2)}\left(2^2;j_1;\mathbf{x}\right)$ to obtain the following relation:

$$\begin{aligned}&\left\{\left\langle\mathbf{a}^{(2)}\left(2^2;j_1;\mathbf{x}\right),\mathbf{V}^{(2)}\left[2^2\times2^2;\mathbf{x};\mathbf{f}\right]\mathbf{v}^{(2)}\left(2^2;\mathbf{x}\right)\right\rangle_{2^2}\right\}_{\alpha^0}\\&=\left\{\left\langle\mathbf{a}^{(2)}\left(2^2;j_1;\mathbf{x}\right),\mathbf{q}_V^{(2)}\left[2^2;\mathbf{u}^{(2)}\left(2^2;\mathbf{x}\right);\mathbf{f};\delta\mathbf{f}\right]\right\rangle_{2^2}\right\}_{\alpha^0};\quad\mathbf{x}\in\Omega\left(\alpha^0\right).\end{aligned} \tag{6.68}$$

2. Using the definition of the adjoint operator in the Hilbert space $\mathcal{H}_2$, recast the left side of Eq. (6.68) as follows:

$$\left\{\left\langle \mathbf{a}^{(2)}\left(2^2;j_1;\mathbf{x}\right),\mathbf{V}^{(2)}\left[2^2\times 2^2;\mathbf{x};\mathbf{f}\right]\mathbf{v}^{(2)}\left(2^2;\mathbf{x}\right)\right\rangle_{2^2}\right\}_{\alpha^0}$$
$$=\left\{\left\langle \mathbf{v}^{(2)}\left(2^2;\mathbf{x}\right),\mathbf{A}^{(2)}\left[2^2\times 2^2;\mathbf{x};\mathbf{f}\right]\mathbf{a}^{(2)}\left(2^2;j_1;\mathbf{x}\right)\right\rangle_{2^2}\right\}_{\alpha^0} \tag{6.69}$$
$$+\left\{P^{(2)}\left[\mathbf{v}^{(2)}\left(2^2;\mathbf{x}\right);\mathbf{a}^{(2)}\left(2^2;j_1;\mathbf{x}\right);\mathbf{f};\delta\mathbf{f}\right]\right\}_{\alpha^0},$$

where $\left\{P^{(2)}\left[\mathbf{v}^{(2)}\left(2^2;\mathbf{x}\right);\mathbf{a}^{(2)}\left(2^2;j_1;\mathbf{x}\right);\mathbf{f};\delta\mathbf{f}\right]\right\}_{\alpha^0}$ denotes the bilinear concomitant defined on the phase-space boundary $\mathbf{x}\in\partial\Omega_x\left(\alpha^0\right)$ and where $\mathbf{A}^{(2)}\left[2^2\times 2^2;\mathbf{x};\mathbf{f}\right]\triangleq\left[\mathbf{V}^{(2)}\left[2^2\times 2^2;\mathbf{x};\mathbf{f}\right]\right]^*$ is the operator formally adjoint to $\mathbf{V}^{(2)}\left[2^2\times 2^2;\mathbf{x};\mathbf{f}\right]$.

3. The 1st term on the right side of Eq. (6.69) is now required to represent the indirect-effect term $\left\{\delta R^{(1)}\left[j_1;\mathbf{u}^{(2)}\left(2^2;\mathbf{x}\right);\mathbf{v}^{(2)}\left(2^2;\mathbf{x}\right);\mathbf{f}\right]\right\}_{ind}$ defined in Eq. (6.56). This requirement is satisfied by recalling Eq. (6.57) and imposing the following relation on each function $\mathbf{a}^{(2)}\left(2^2;j_1;\mathbf{x}\right)$, $j_1=1,\dots,TF$:

$$\left\{\mathbf{A}^{(2)}\left[2^2\times 2^2;\mathbf{x};\mathbf{f}\right]\mathbf{a}^{(2)}\left(2^2;j_1;\mathbf{x}\right)\right\}_{\alpha^0}$$
$$=\left\{\mathbf{s}^{(2)}\left(2^2;j_1;\mathbf{u}^{(2)};\mathbf{f}\right)\right\}_{\alpha^0},\quad j_1=1,\dots,TF, \tag{6.70}$$

4. The definition of the vector $\mathbf{a}^{(2)}\left(2^2;j_1;\mathbf{x}\right)$ will now be completed by selecting boundary conditions, which will be represented in operator form as follows:

$$\left\{\mathbf{b}_A^{(2)}\left[\mathbf{u}^{(2)}\left(2^2;\mathbf{x}\right);\mathbf{a}^{(2)}\left(2^2;j_1;\mathbf{x}\right);\mathbf{f}\right]\right\}_{\alpha^0}=0,$$
$$\mathbf{x}\in\partial\Omega\left(\alpha^0\right),\quad j_1=1,\dots,TF. \tag{6.71}$$

5 The boundary conditions represented by Eq. (6.71) are selected so as to satisfy the following requirements: (i) these boundary conditions together with Eq. (6.70) constitute a well-posed problem for the functions $\mathbf{a}^{(2)}\left(2^2;j_1;\mathbf{x}\right)$; (ii) the implementation in Eq. (6.69) of these boundary conditions together with those provided in Eq. (6.61) eliminates all of the unknown values of the functions $\mathbf{v}^{(2)}\left(2^2;\mathbf{x}\right)$ and $\mathbf{a}^{(2)}\left(2^2;j_1;\mathbf{x}\right)$ in the expression of the bilinear concomitant $\left\{P^{(2)}\left[\mathbf{v}^{(2)}\left(2^2;\mathbf{x}\right);\mathbf{a}^{(2)}\left(2^2;j_1;\mathbf{x}\right);\mathbf{f};\delta\mathbf{f}\right]\right\}_{\alpha^0}$. This bilinear concomitant may vanish after these boundary conditions are implemented, but if it does not, it will be reduced to a residual quantity, which will be denoted as $\hat{P}^{(2)}\left[\mathbf{u}^{(2)}\left(2^2;\mathbf{x}\right);\mathbf{a}^{(2)}\left(2^2;j_1;\mathbf{x}\right);\mathbf{f};\delta\mathbf{f}\right]$ and which will comprise only known values of $\mathbf{u}^{(2)}\left(2^2;\mathbf{x}\right)$, $\mathbf{a}^{(2)}\left(2^2;j_1;\mathbf{x}\right)$, $\mathbf{f}$ and $\delta\mathbf{f}$.

The system of equations represented by Eq. (6.70) together with the boundary conditions represented by Eq. (6.71) constitute the *2nd-Level Adjoint Sensitivity System* (2nd-LASS). The solution of the 2nd-LASS, i.e., the four-component vector $\mathbf{a}^{(2)}\left(2^2;j_1;\mathbf{x}\right)$, $j_1,\ldots,TP$, will be called the *2nd-level adjoint sensitivity function.* It is important to note that the 2nd-LASS is independent of any variations, $\delta\mathbf{f}$, in the components of the feature function and, hence, is independent of any parameter variations, $\delta\boldsymbol{\alpha}$, as well.

The equations underlying the 2nd-LASS, represented by Eqs. (6.70) and (6.71), together with the equations underlying the 2nd-LVSS, represented by Eqs. (6.60) and (6.61), are now employed in Eq. (6.69) in conjunction with Eq. (6.56) to obtain the following expression for the indirect-effect term $\left\{\delta R^{(1)}\left[j_1;\mathbf{u}^{(2)}\left(2^2;\mathbf{x}\right);\mathbf{v}^{(2)}\left(2^2;\mathbf{x}\right);\mathbf{f}\right]\right\}_{ind}$ in terms of the 2nd-level adjoint sensitivity functions $\mathbf{a}^{(2)}\left(2^2;j_1;\mathbf{x}\right)$, for $j_1=1,\ldots,TP$:

$$
\begin{aligned}
&\left\{\delta R^{(1)}\left[j_1;\mathbf{u}^{(2)}\left(2^2;\mathbf{x}\right);\mathbf{v}^{(2)}\left(2^2;\mathbf{x}\right);\mathbf{f}\right]\right\}_{ind} \\
&=\left\{\left\langle \mathbf{a}^{(2)}\left(2^2;j_1;\mathbf{x}\right),\mathbf{q}_V^{(2)}\left[2^2;\mathbf{u}^{(2)}\left(2^2;\mathbf{x}\right);\mathbf{f};\delta\mathbf{f}\right]\right\rangle_2\right\}_{\alpha^0} \\
&-\left\{\hat{P}^{(2)}\left[\mathbf{u}^{(2)};\mathbf{a}^{(2)};\mathbf{f};\delta\mathbf{f}\right]\right\}_{\alpha^0} \\
&\equiv\left\{\delta R^{(1)}\left[j_1;\mathbf{u}^{(2)}\left(2^2;\mathbf{x}\right);\mathbf{a}^{(2)}\left(2^2;j_1;\mathbf{x}\right);\mathbf{f};\delta\mathbf{f}\right]\right\}_{ind}.
\end{aligned}
\tag{6.72}
$$

As the last equality (identity) in Eq. (6.72) indicates, the 2nd-level variational sensitivity function $\mathbf{v}^{(2)}\left(2^2;\mathbf{x}\right)$ has been eliminated from appearing in the expression of the indirect-effect term, having been replaced by the 2nd-level adjoint sensitivity function $\mathbf{a}^{(2)}\left(2^2;j_1;\mathbf{x}\right)$, for each $j_1=1,\ldots,TF$.

Inserting the expressions that define the vector $\mathbf{q}_V^{(2)}\left[2^2;\mathbf{u}^{(2)}\left(2^2;\mathbf{x}\right);\mathbf{f};\delta\mathbf{f}\right]$ from Eq. (6.63)-(6.65) into Eq. (6.72) and adding the resulting expression for the indirect-effect term to the expression of the direct-effect term given in Eq. (6.54) yields the following expression for the total 2nd-order G-differential of the response $R\left[\boldsymbol{\varphi}(\mathbf{x}),\boldsymbol{\psi}(\mathbf{x});\mathbf{f}\right]$:

$$
\begin{aligned}
&\left\{\delta R^{(1)}\left[j_1;\mathbf{u}^{(2)}\left(2^2;\mathbf{x}\right);\mathbf{a}^{(2)}\left(j_1;2^2;\mathbf{x}\right);\mathbf{f};\delta\mathbf{f}\right]\right\}_{\alpha^0} \\
&=\sum_{j_2=1}^{TF}\left\{R^{(2)}\left[j_2;j_1;\mathbf{u}^{(2)}\left(2^2;\mathbf{x}\right);\mathbf{a}^{(2)}\left(j_1;2^2;\mathbf{x}\right);\mathbf{f}\right]\right\}_{\alpha^0}\delta f_{j_2},
\end{aligned}
\tag{6.73}
$$

where $R^{(2)}\left[j_2;j_1;\mathbf{u}^{(2)}(\mathbf{x});\mathbf{a}^{(2)}(j_1;\mathbf{x});\mathbf{f}\right]\equiv\dfrac{\partial^2 R\left[\boldsymbol{\varphi}(\mathbf{x}),\boldsymbol{\psi}(\mathbf{x});\mathbf{f}\right]}{\partial f_{j_1}\,\partial f_{j_2}}$ denotes the *2nd-order partial sensitivity* of the response $R\left[\boldsymbol{\varphi}(\mathbf{x}),\boldsymbol{\psi}(\mathbf{x});\mathbf{f}\right]$ with respect to the components $f_{j_2}(\boldsymbol{\alpha})$ of the feature function $\mathbf{f}(\boldsymbol{\alpha})$, evaluated at the nominal parameter values $\boldsymbol{\alpha}^0$ and having the following expression for $j_1,j_2=1,\ldots,TP$:

$$
\begin{aligned}
&R^{(2)}\left[j_2;j_1;\mathbf{u}^{(2)}\left(2^2;\mathbf{x}\right);\mathbf{a}^{(2)}\left(j_1;2^2;\mathbf{x}\right);\mathbf{f}\right] \\
&=\left\{\frac{\partial}{\partial f_{j2}}\int_{\lambda_1(\alpha)}^{\omega_1(\alpha)}dx_1\ldots\int_{\lambda_{TI}(\alpha)}^{\omega_{TI}(\alpha)}dx_{TI}S^{(1)}\left[j_1;\mathbf{u}^{(2)}\left(2^2;\mathbf{x}\right);\mathbf{f}(\alpha)\right]\right\}_{\alpha^0} \\
&+\left\{\left\langle \mathbf{a}_1^{(2)}(j_1;\mathbf{x}),\ \frac{\partial\left[\mathbf{Q}(\mathbf{f})-\mathbf{L}(\mathbf{f})\boldsymbol{\varphi}(\mathbf{x})\right]}{\partial f_{j2}}\right\rangle_0\right\}_{\alpha^0} \\
&+\left\{\left\langle \mathbf{a}_2^{(2)}(j_1;\mathbf{x}),\ \frac{\partial\left[\mathbf{Q}^*(\mathbf{f})-\mathbf{L}^*(\mathbf{f})\boldsymbol{\psi}(\mathbf{x})\right]}{\partial f_{j2}}\right\rangle_0\right\}_{\alpha^0} \qquad (6.74)\\
&+\left\{\left\langle \mathbf{a}_3^{(2)}(j_1;\mathbf{x}),\ \frac{\partial^2 S\left(\mathbf{u}^{(1)}(\mathbf{x});\mathbf{f}\right)}{\partial f_{j2}\,\partial\boldsymbol{\varphi}}-\frac{\partial\left[\mathbf{L}^*(\mathbf{f})\mathbf{a}_1^{(1)}\right]}{\partial f_{j2}}\right\rangle_0\right\}_{\alpha^0} \\
&+\left\{\left\langle \mathbf{a}_4^{(2)}(j_1;\mathbf{x}),\ \frac{\partial^2 S\left(\mathbf{u}^{(1)}(\mathbf{x});\mathbf{f}\right)}{\partial f_{j2}\,\partial\boldsymbol{\psi}}-\frac{\partial\left[\mathbf{L}(\mathbf{f})\mathbf{a}_2^{(1)}\right]}{\partial f_{j2}}\right\rangle_0\right\}_{\alpha^0} \\
&-\left\{\frac{\partial}{\partial f_{j2}}\hat{P}^{(2)}\left[\mathbf{u}^{(2)}(\mathbf{x});\mathbf{a}^{(2)}(j_1;\mathbf{x});\mathbf{f}(\alpha)\right]\right\}_{\alpha^0}.
\end{aligned}
$$

Since the 2nd-LASS is independent of variations in the components of the feature-functions (and, hence, variations in the model parameters), the exact computation of all partial 2nd-order sensitivities $R^{(2)}\left[j_2;j_1;\mathbf{u}^{(2)}\left(2^2;\mathbf{x}\right);\mathbf{a}^{(2)}\left(j_1;2^2;\mathbf{x}\right);\mathbf{f}\right]\equiv\dfrac{\partial^2 R\left[\boldsymbol{\varphi}(\mathbf{x}),\boldsymbol{\psi}(\mathbf{x});\mathbf{f}\right]}{\partial f_{j1}\,\partial f_{j2}}$ requires at most $TF$ large-scale (adjoint) computations using the 2nd-LASS. When the 2nd-LASS is solved $TF$-times, the "off-diagonal" 2nd-order mixed sensitivities $\dfrac{\partial^2 R}{\partial f_{j1}\,\partial f_{j2}}$ will be computed twice, in two different ways, using two distinct 2nd-level adjoint sensitivity functions, thereby providing an independent intrinsic (numerical) verification that the 1st- and 2nd-order response sensitivities with respect to the components of the feature functions are computed accurately.

In component form, the equations comprising the 2nd-LASS are solved, for each $j_1=1,\ldots,TF$, in the following order:

$$
\mathbf{L}(\mathbf{f})\mathbf{a}_3^{(2)}(j_1;\mathbf{x})=\frac{\partial S^{(1)}\left(j_1;\mathbf{u}^{(2)};\mathbf{f}\right)}{\partial\mathbf{a}_1^{(1)}}, \qquad (6.75)
$$

$$
\mathbf{L}^*(\mathbf{f})\mathbf{a}_4^{(2)}(j_1;\mathbf{x})=\frac{\partial S^{(1)}\left(j_1;\mathbf{u}^{(2)};\mathbf{f}\right)}{\partial\mathbf{a}_2^{(1)}}, \qquad (6.76)
$$

$$\mathbf{L}^{*}(\mathbf{f})\mathbf{a}_{1}^{(2)}(j_1;\mathbf{x})=\frac{\partial^2 S\left(\mathbf{u}^{(1)};\mathbf{f}\right)}{\partial\varphi\,\partial\varphi}\mathbf{a}_{3}^{(2)}(j_1;\mathbf{x})+\frac{\partial^2 S\left(\mathbf{u}^{(1)};\mathbf{f}\right)}{\partial\psi\,\partial\varphi}\mathbf{a}_{4}^{(2)}(j_1;\mathbf{x})+\frac{\partial S^{(1)}\left(j_1;\mathbf{u}^{(2)};\mathbf{f}\right)}{\partial\varphi},\tag{6.77}$$

$$\mathbf{L}(\mathbf{f})\mathbf{a}_{2}^{(2)}(j_1;\mathbf{x})=\frac{\partial^2 S\left(\mathbf{u}^{(1)};\mathbf{f}\right)}{\partial\varphi\,\partial\psi}\mathbf{a}_{3}^{(2)}(j_1;\mathbf{x})+\frac{\partial^2 S\left(\mathbf{u}^{(1)};\mathbf{f}\right)}{\partial\psi\,\partial\psi}\mathbf{a}_{4}^{(2)}(j_1;\mathbf{x})+\frac{\partial S^{(1)}\left(j_1;\mathbf{u}^{(2)};\mathbf{f}\right)}{\partial\psi}.\tag{6.78}$$

Dirac delta-functionals may need to be used in Eq. (6.74) in order to express in integral- form the non-zero residual terms in the residual bilinear concomitant and/or the terms containing derivatives with respect to the lower- and upper-boundary points. Ultimately, the expression of the partial 2nd-order sensitivities $R^{(2)}\left[j_2;j_1;\mathbf{u}^{(2)}\left(2^2;\mathbf{x}\right);\mathbf{a}^{(2)}\left(2^2;j_1;\mathbf{x}\right);\mathbf{f}\right]$ obtained in Eq. (6.74) is written in the following integral form, which mirrors Eq. (6.48):

$$R^{(2)}\left[j_2;j_1;\mathbf{u}^{(2)}\left(2^2;\mathbf{x}\right);\mathbf{a}^{(2)}\left(2^2;j_1;\mathbf{x}\right);\mathbf{f}(\boldsymbol{\alpha})\right]\triangleq\int_{\lambda_1(\boldsymbol{\alpha})}^{\omega_1(\boldsymbol{\alpha})}dx_1\ldots\int_{\lambda_{TI}(\boldsymbol{\alpha})}^{\omega_{TI}(\boldsymbol{\alpha})}dx_{TI}S^{(2)}\left[j_2;j_1;\mathbf{u}^{(2)}\left(2^2;\mathbf{x}\right);\mathbf{a}^{(2)}\left(2^2;j_1;\mathbf{x}\right);\mathbf{f}(\boldsymbol{\alpha})\right].\tag{6.79}$$

The computation of the partial 2nd-order sensitivities $R^{(2)}\left[j_2;j_1;\mathbf{u}^{(2)}\left(2^2;\mathbf{x}\right);\mathbf{a}^{(2)}\left(2^2;j_1;\mathbf{x}\right);\mathbf{f}\right]$ using Eq. (6.74) requires quadratures for performing the integrations over the four components of the 2nd-level adjoint sensitivity function $\mathbf{a}^{(2)}\left(2^2;j_1;\mathbf{x}\right)$, which are obtained by solving the 2nd-LASS for $j_1=1,\ldots,TF$. Thus, obtaining all of the 2nd-order sensitivities $R^{(2)}\left[j_2;j_1;\mathbf{u}^{(2)}\left(2^2;\mathbf{x}\right);\mathbf{a}^{(2)}\left(2^2;j_1;\mathbf{x}\right);\mathbf{f}\right]\equiv\dfrac{\partial^2 R}{\partial f_{j_1}\,\partial f_{j_2}}$ with respect to the components $f_{j_1}$ of the feature function $\mathbf{f}(\boldsymbol{\alpha})$ requires performing at most $TF$ large-scale computations for solving the 2nd-LASS.

By comparison, if the 2nd-CASAM-L (Cacuci 2021b, 2022b) were applied to compute the 2nd-order sensitivities of the response directly with respect to the model parameters, $TP$ (instead of $TF$) large-scale computations for solving the corresponding 2nd-LASS would have been required, where $TP$ denotes the total number of primary model parameters. Since $TF < TP$, fewer large-scale computations are needed when using the 2nd-FASAM-L rather than the 2nd-CASAM-L. Notably, the left sides of the 2nd-LASS to be solved within the 2nd-FASAM-L are the same as those to be solved within the 2nd-CASAM-L. However, the source terms on the right sides of these 2nd-LASS are different from each other: there are as many source terms on the right sides as there are components of the feature function within the

2nd-FASAM-L, and there are as many right-side sources as there are primary model parameters within the 2nd-CASAM-L.

## 6.5 THE *N*TH-ORDER FEATURE ADJOINT SENSITIVITY ANALYSIS METHODOLOGY FOR RESPONSE-COUPLED FORWARD AND ADJOINT LINEAR SYSTEMS (*N*TH-FASAM-L)

The validity of the mathematical framework underlying the "*n*th-Order Feature Adjoint Sensitivity Analysis Methodology for Response-Coupled Forward and Adjoint Linear Systems" (*n*th-FASAM-L) methodology will be established in this section by using the "proof by mathematical induction" comprising the usual steps, as follows:

1. Conjecture the pattern underlying the *n*th-FASAM-L, for arbitrary $n$, based on prior experience.
2. Prove that the conjectured pattern for arbitrary $n$ is valid for the lowest value of $n$, i.e., for $n = 2$.
3. Assuming that the pattern underlying the *n*th-FASAM-L is valid for an arbitrarily high-order $n$, prove that this pattern is also valid for $n \to n+1$, i.e., for the $(n + 1)$th-FASAM-L.

As has been noted in Section 6.3, the roles played by the components $f_i(\boldsymbol{\alpha}), i = 1,\dots,TF$, of the "feature function" $\mathbf{f}(\boldsymbol{\alpha}) \triangleq \left[f_1(\boldsymbol{\alpha}),\dots,f_{TF}(\boldsymbol{\alpha})\right]^\dagger$ within the framework of the 1st-FASAM-L correspond to the roles played by the components $\alpha_j, j = 1,\dots,TP$, of the "vector of primary model parameters" $\boldsymbol{\alpha} \triangleq (\alpha_1,\dots,\alpha_{TP})^\dagger$ within the framework of the 1st-CASAM-L (Cacuci 2021b, 2022b). The same correspondence holds between the frameworks of the 2nd-FASAM-L and the 2nd-CASAM-L, as noted in Section 6.4. It can therefore be conjectured that the same correspondence would be expected to hold in general between the general frameworks of the *n*th-FASAM-L and the *n*th-CASAM-L (Cacuci 2021b, 2022b) methodologies. As will be demonstrated in this section, this conjecture is indeed correct.

Considering the analogy with the framework of the *n*th-CASAM-L methodology (Cacuci 2021b, 2022b), it is conjectured that the G-differential of the $(n-1)$th-order sensitivity of the model's response $R\left[\mathbf{u}(\mathbf{x});\mathbf{f}(\boldsymbol{\alpha})\right]$ with respect to the components $f_1,\dots,f_{TF}$ of the "feature" function $\mathbf{f}(\boldsymbol{\alpha}) \triangleq \left[f_1(\boldsymbol{\alpha}),\dots,f_{TF}(\boldsymbol{\alpha})\right]^\dagger$ will have the following form:

$$
\begin{aligned}
&\left\{\delta R^{(n-1)}\left[j_{n-1};\dots;j_1;\mathbf{u}^{(n)};\mathbf{f}\right]\right\}_{\alpha^0} = -\left\{\hat{P}^{(n)}\left(\mathbf{a}^{(n)};\mathbf{u}^{(n)};\mathbf{f};\delta\mathbf{f}\right)\right\}_{\alpha^0} \\
&+\left\{\left\langle \mathbf{a}^{(n)}\left(j_{n-1},\dots,j_1;\mathbf{x}\right),\mathbf{q}_V^{(n)}\left[2^n;\mathbf{u}^{(n)}\left(2^n;\mathbf{x}\right);\mathbf{f};\delta\mathbf{f}\right]\right\rangle_n\right\}_{\alpha^0} \\
&+\sum_{j_n=1}^{TF}\left\{\frac{\partial}{\partial f_{j_n}}\int_{\lambda_1(\alpha)}^{\omega_1(\alpha)} dx_1\dots\int_{\lambda_{TI}(\alpha)}^{\omega_{TI}(\alpha)} dx_{TI}\,S^{(n-1)}\left(j_{n-1},\dots,j_1;\mathbf{u}^{(n)};\boldsymbol{\alpha}\right)\delta f_{j_n}\right\}_{\alpha^0} \\
&\equiv \sum_{j_n=1}^{TF}\left\{R^{(n)}\left(j_n,\dots,j_1;\mathbf{u}^{(n)};\mathbf{a}^{(n)};\mathbf{f}\right)\right\}_{(\alpha^0)}\delta f_{j_n},
\end{aligned}
\tag{6.80}
$$

such that the *nth-order sensitivity* of the model's response $R[\mathbf{u}(\mathbf{x});\mathbf{f}(\alpha)]$ with respect to the components $f_{j1},\ldots,f_{jn}$ of the "feature" function $\mathbf{f}(\alpha)\triangleq[f_{j1}(\alpha),\ldots,f_{jn}(\alpha)]^{\dagger}$ is expected to have the following functional form:

$$R^{(n)}\left[j_n;\ldots;j_1;\mathbf{u}^{(n)}\left(2^n;j_{n-2},\ldots,j_1;\mathbf{x}\right);\mathbf{a}^{(n)}\left(2^n;j_{n-1},\ldots,j_1;\mathbf{x}\right);\mathbf{f}(\alpha)\right]$$

$$\triangleq\int_{\lambda_1(\alpha)}^{\omega_1(\alpha)}dx_1\ldots\int_{\lambda_{TI}(\alpha)}^{\omega_{TI}(\alpha)}dx_{TI}S^{(n)}\left[j_n;\ldots;j_1;\mathbf{u}^{(n)};\mathbf{a}^{(n)};\mathbf{f}(\alpha)\right] \tag{6.81}$$

$$\triangleq\frac{\partial^n R[\mathbf{u}(\mathbf{x});\mathbf{f}(\alpha)]}{\partial f_{jn}\ldots\partial f_{j1}};\quad j_1=1,\ldots,TF;\quad j_n=1,\ldots,j_{n-1};\quad n=2,3,\ldots,$$

where $TF$ denotes the "total number of features," i.e., functions of the primary model parameters. It is also conjectured that the $n$th-order functions $\mathbf{u}^{(n)}\left(2^n;j_{n-2},\ldots,j_1;\mathbf{x}\right)$ and $\mathbf{a}^{(n)}\left(j_{n-1},\ldots,j_1;2^n;\mathbf{x}\right)$, which are needed to compute the $n$th-order sensitivities defined in Eq. (6.81), are recursively defined as follows:

i. $\mathbf{u}^{(n)}\left(2^n;j_{n-2},\ldots,j_1;\mathbf{x}\right)\triangleq\left[\mathbf{u}^{(n-1)}\left(2^{n-1};j_{n-3},\ldots,j_1;\mathbf{x}\right),\mathbf{a}^{(n-1)}\left(2^{n-1};j_{n-2},\ldots,j_1;\mathbf{x}\right)\right]^{\dagger}$ are the solutions of the following *nth-Level Forward/Adjoint System (nth-LFAS)* for $j_n=1,\ldots,j_{n-1};\ n\geq 3$:

$$\mathbf{F}^{(n)}\left[2^n\times 2^n;\mathbf{f}(\alpha)\right]\mathbf{u}^{(n)}\left(2^n;j_{n-2},\ldots,j_1;\mathbf{x}\right)$$
$$=\mathbf{q}_F^{(n)}\left[2^n;\mathbf{u}^{(n-1)};\mathbf{f}(\alpha)\right];\quad \mathbf{x}\in\Omega; \tag{6.82}$$

$$\mathbf{b}_F^{(n)}\left(2^n;\mathbf{u}^{(n)};\mathbf{f}\right)=\mathbf{0};\quad \mathbf{x}\in\partial\Omega; \tag{6.83}$$

ii. $\mathbf{a}^{(n)}\left(j_{n-1},\ldots,j_1;2^n;\mathbf{x}\right)$ are the solutions of the following *nth-Level Adjoint Sensitivity System (nth-LASS)* for $j_n=1,\ldots,j_{n-1};\ n\geq 3$:

$$\mathbf{A}^{(n)}\left[2^n\times 2^n;\mathbf{f}(\alpha)\right]\mathbf{a}^{(n)}\left(2^n;j_{n-1},\ldots,j_1;\mathbf{x}\right)=\mathbf{s}^{(n)}\left(2^n;\mathbf{u}^{(n)};\mathbf{f}\right);\quad \mathbf{x}\in\Omega; \tag{6.84}$$

$$\left\{\mathbf{b}_A^{(n)}\left[\mathbf{u}^{(n)}\left(2^n;j_{n-1},\ldots,j_1;\mathbf{x}\right);\mathbf{a}^{(n)}\left(2^n;j_{n-1},\ldots,j_1;\mathbf{x}\right);\mathbf{f}\right]\right\}_{\alpha^0}=\mathbf{0},\quad \mathbf{x}\in\partial\Omega. \tag{6.85}$$

Through their implicit dependence on lower-level forward and adjoint functions, the block-matrix valued operators $\mathbf{F}^{(n)}\left[2^n\times 2^n;\mathbf{f}(\alpha)\right]$ and $\mathbf{A}^{(n)}\left[2^n\times 2^n;\mathbf{f}(\alpha)\right]$, as well as the source terms $\mathbf{q}_F^{(n)}\left[2^n;\mathbf{u}^{(n-1)};\mathbf{f}(\alpha)\right]$ and $\mathbf{s}^{(n)}\left(2^n;\mathbf{u}^{(n)};\mathbf{f}\right)$, also depend on lower-level indices $j_k,k<n$, but this dependence is not material to establishing the general framework of the $n$th-FASAM-L and has therefore been omitted to keep the notation as simple as possible.

The proof that the framework conjectured for the nth-FASAM-L methodology is correct/valid for the lowest value $n = 2$ of the index $n$ can be readily performed by setting $n = 2$ in Eqs. (6.81) - (6.85) and verifying that these equations indeed become identical to the corresponding equations and expressions derived in Section 6.4.

The 3rd (and last) step of the proof by mathematical induction to establish the validity of the nth-FASAM-L Framework conjectured in Section 6.5.1 is to show that the formalism assumed to be correct for the computation of the nth-order sensitivities also holds true for the computation of the $(n+1)$th-order sensitivities. This proof entails showing that the formulas obtained by computing the $(n+1)$th-order sensitivities using Eqs. (6.81) - (6.85) as the starting point will be the same as would be obtained by replacing "$n$" with "$(n+1)$" in Eqs. (6.81) - (6.85).

The nth-order response sensitivity defined in Eq. (6.81) can be considered to be a function of the $(n+1)$th-level function $\mathbf{u}^{(n+1)}\left(2^{(n+1)};\mathbf{x}\right) \triangleq \left[\mathbf{u}^{(n)}\left(2^n;\mathbf{x}\right),\mathbf{a}^{(n)}\left(2^n;\mathbf{x}\right)\right]^\dagger$, which is the solution of the $(n+1)$th-Level Forward/Adjoint System, abbreviated as "$(n+1)$th-LFAS," which is obtained by concatenating Eqs. (6.82) - (6.85) and is written as follows:

$$\mathbf{F}^{(n+1)}\left[2^{n+1}\times 2^{n+1};\mathbf{f}(\alpha)\right]\mathbf{u}^{(n+1)}\left(2^{n+1};\mathbf{x}\right)$$
$$= \mathbf{q}_F^{(n+1)}\left[2^{n+1};\mathbf{u}^{(n)}\left(2^n;\mathbf{x}\right);\mathbf{f}(\alpha)\right]; \quad \mathbf{x}\in\Omega; \tag{6.86}$$

$$\mathbf{b}_F^{(n+1)}\left(2^{(n+1)};\mathbf{u}^{(n+1)};\mathbf{f}\right) \triangleq \left(\mathbf{b}_F^{(n)},\mathbf{b}_A^{(n)}\right)^\dagger = \mathbf{0}; \quad \mathbf{x}\in\partial\Omega. \tag{6.87}$$

The following definitions were used in Eqs. (6.86) and (6.87), where the explicit dependence on the indices $j_k, k = 1,\dots,n$, has been omitted for simplicity:

$$\mathbf{F}^{(n+1)}\left[2^{n+1}\times 2^{n+1};\mathbf{f}(\alpha)\right] \triangleq diag\left(\mathbf{F}^{(n)},\mathbf{A}^{(n)}\right);$$
$$\mathbf{u}^{(n+1)}\left(2^{n+1};\mathbf{x}\right) \triangleq \left[\mathbf{u}^{(n)}\left(2^n;\mathbf{x}\right),\mathbf{a}^{(n)}\left(2^n;\mathbf{x}\right)\right]^\dagger; \tag{6.88}$$

$$\mathbf{q}_F^{(n+1)}\left[2^{n+1};\mathbf{u}^{(n+1)}\left(2^{n+1};\mathbf{x}\right);\mathbf{f}(\alpha)\right]$$
$$\triangleq \left[\mathbf{q}_F^{(n)}\left(2^n;\mathbf{x};\mathbf{f}\right),\mathbf{q}_A^{(n)}\left(2^n;\mathbf{u}^{(n)};\mathbf{f}\right)\right]^\dagger. \tag{6.89}$$

Next, it will be assumed that, for each index $j_1,\dots,j_n$, the 1st-order total G-differential of the nth-order sensitivities $R^{(n)}\left[j_n;\dots;j_1;\mathbf{u}^{(n+1)}\left(2^{n+1};\mathbf{x}\right);\mathbf{f}(\alpha)\right]$ will exist and will be linear in the variational functions $\mathbf{v}^{(n+1)}\left(2^{n+1};j_{n-1},\dots,j_1;\mathbf{x}\right) \triangleq \left[\mathbf{v}^{(n)}\left(2^n;\mathbf{x}\right),\delta\mathbf{a}^{(n)}\left(2^n;\mathbf{x}\right)\right]^\dagger$ and $\delta\mathbf{f}$ in a neighborhood around the nominal values of the respective state functions and components of the feature function. In this case, the 1st-order total G-differential of $R^{(n)}\left[j_n;\dots;j_1;\mathbf{u}^{(n+1)};\mathbf{f}\right]$ is by definition obtained as follows:

$$\left\{\delta R^{(n)}\left[j_n;\ldots;j_1;\mathbf{u}^{(n+1)};\mathbf{f}\right]\right\}_{\alpha^0}$$
$$\triangleq\left\{\frac{d}{d\varepsilon}R^{(n)}\left[j_n;\ldots;j_1;\mathbf{u}^{(n+1)}+\varepsilon\mathbf{v}^{(n+1)};\mathbf{f}+\varepsilon\delta\mathbf{f}\right]\right\}_{\varepsilon=0}$$
$$\triangleq\sum_{j_{n+1}=1}^{TF}\left\{\frac{\partial R^{(n)}\left[\ldots;\mathbf{u}^{(n+1)};\mathbf{f}\right]}{\partial f_{j_{n+1}}}\right\}_{\alpha^0}\delta f_{j_{n+1}}+\left\{\delta R^{(n)}\left[j_n;\ldots;j_1;\mathbf{u}^{(n+1)};\mathbf{v}^{(n+1)};\mathbf{f}\right]\right\}_{ind},\tag{6.90}$$

where the quantity $\left\{\delta R^{(n)}\left[j_n;\ldots;j_1;\mathbf{u}^{(n+1)};\mathbf{v}^{(n+1)};\mathbf{f}\right]\right\}_{ind}$ denotes the "indirect-effect term" and is defined as follows:

$$\left\{\delta R^{(n)}\left[j_n;\ldots;j_1;\mathbf{u}^{(n+1)};\mathbf{v}^{(n+1)};\mathbf{f}\right]\right\}_{ind}$$
$$\triangleq\int_{\lambda_1(\alpha)}^{\omega_1(\alpha)}dx_1\ldots\int_{\lambda_{TI}(\alpha)}^{\omega_{TI}(\alpha)}dx_{TI}\left\{\frac{\partial S^{(n)}}{\partial\mathbf{u}^{(n+1)}(\mathbf{x})}\mathbf{v}^{(n+1)}(\mathbf{x})\right\}_{\alpha^0}.\tag{6.91}$$

The vector $\mathbf{v}^{(n+1)}\left(2^{n+1};j_{n-1},\ldots,j_1;\mathbf{x}\right)$, which is needed to evaluate the indirect-effect term $\left\{\delta R^{(n)}\left[j_n;\ldots;j_1;\mathbf{u}^{(n+1)};\mathbf{v}^{(n+1)};\mathbf{f}\right]\right\}_{ind}$, is the solution of the $(n+1)$th-Level Variational Sensitivity System, abbreviated as "$(n+1)$th-LVSS," which is obtained by taking the (1st-order) G-differential of the $(n+1)$th-LFAS defined by Eqs. (6.88) and (6.89). Performing this G-differentiation yields the following relations, which define the $(n+1)$th-LVSS:

$$\left\{\frac{d}{d\varepsilon}\mathbf{F}^{(n+1)}\left[2^{n+1}\times2^{n+1};\mathbf{f}^0+\varepsilon\delta\mathbf{f}\right]\left[\mathbf{u}^{(n+1,0)}\left(2^{n+1};\mathbf{x}\right)+\varepsilon\mathbf{v}^{(n+1)}\left(2^{n+1};\mathbf{x}\right)\right]\right\}_{\varepsilon=0}$$
$$=\left\{\frac{d}{d\varepsilon}\mathbf{q}_F^{(n+1)}\left[2^{n+1};\mathbf{u}^{(n,0)}\left(2^n;\mathbf{x}\right)+\varepsilon\mathbf{v}^{(n)}\left(2^n;\mathbf{x}\right);\mathbf{f}^0+\varepsilon\delta\mathbf{f}\right]\right\}_{\varepsilon=0},\quad\mathbf{x}\in\Omega;\tag{6.92}$$

$$\left\{\frac{d}{d\varepsilon}\mathbf{b}_F^{(n+1)}\left[2^{n+1};\mathbf{u}^{(n+1,0)}\left(2^{n+1};\mathbf{x}\right)+\varepsilon\mathbf{v}^{(n+1)}\left(2^{n+1};\mathbf{x}\right);\mathbf{f}^0+\varepsilon\delta\mathbf{f}\right]\right\}_{\varepsilon=0}=\mathbf{0};\quad\mathbf{x}\in\partial\Omega;\tag{6.93}$$

Carrying out the differentiation with respect to $\varepsilon$ in Eqs. (6.92) and (6.93), and setting $\varepsilon=0$ in the resulting expressions yields the following $(n+1)$th-LVSS for the $(n+1)$th-level variational function $\mathbf{v}^{(n+1)}\left(2^{n+1};j_{n-1},\ldots,j_1;\mathbf{x}\right)$:

$$\left\{\mathbf{V}^{(n+1)}\left[2^{n+1}\times2^{n+1};\mathbf{x};\mathbf{f}\right]\mathbf{v}^{(n+1)}\left(2^{n+1};\mathbf{x}\right)\right\}_{\alpha^0}$$
$$=\left\{\mathbf{q}_V^{(n+1)}\left[2^{n+1};\mathbf{u}^{(n+1)}\left(2^{n+1};\mathbf{x}\right);\mathbf{f};\delta\mathbf{f}\right]\right\}_{\alpha^0};\quad\mathbf{x}\in\Omega;\tag{6.94}$$

$$\left\{\mathbf{b}_V^{(n+1)}\left(\mathbf{u}^{(n+1)};\mathbf{v}^{(n+1)};\mathbf{f};\delta\mathbf{f}\right)\right\}_{\alpha^0}=\mathbf{0};\quad \mathbf{x}\in\partial\Omega. \tag{6.95}$$

Solving the $(n+1)$th-LVSS is prohibitive computationally. Therefore, the need for solving the $(n+1)$th-LVSS will be avoided by expressing the indirect-effect term $\left\{\delta R^{(n)}\left[j_n;\ldots;j_1;\mathbf{u}^{(n+1)};\mathbf{v}^{(n+1)};\mathbf{f}\right]\right\}_{ind}$ in an alternative way, which eliminates the appearance of the variational function $\mathbf{v}^{(n+1)}\left(2^{n+1};j_{n-1},\ldots,j_1;\mathbf{x}\right)$, which is achieved by replacing it with the solution of the "$(n+1)$th-Level Adjoint Sensitivity System," abbreviated as "$(n+1)$th-LASS." This $(n+1)$th-LASS will be constructed below by implementing the same sequence of logical steps as were followed when constructing the lower-level adjoint sensitivity systems, namely:

i. The $(n+1)$th-LASS is constructed in a Hilbert space, denoted as $\mathcal{H}_{n+1}$, comprising block-vectors of the form $\boldsymbol{\chi}^{(n+1)}\left(2^{n+1};\mathbf{x}\right)\triangleq\left[\ldots,\boldsymbol{\chi}_k^{(n+1)}(\mathbf{x}),\ldots\right]^\dagger\in\mathcal{H}_{n+1}$, for $k=1,\ldots,2^{n+1}$, comprising elements having the following structure: $\boldsymbol{\chi}_k^{(n+1)}(\mathbf{x})\triangleq\left[\chi_{k,1}^{(n+1)}(\mathbf{x}),\ldots,\chi_{k,TD}^{(n+1)}(\mathbf{x})\right]^\dagger$. The inner product in $\mathcal{H}_{n+1}$ between two elements $\boldsymbol{\chi}^{(n+1)}(\mathbf{x})\in\mathcal{H}_{n+1}$ and $\boldsymbol{\theta}^{(n+1)}(\mathbf{x})\in\mathcal{H}_{n+1}$ will be denoted as $\left\langle\boldsymbol{\chi}^{(n+1)}(\mathbf{x}),\boldsymbol{\theta}^{(n+1)}(\mathbf{x})\right\rangle_{n+1}$ and is defined as follows:

$$\left\langle\boldsymbol{\chi}^{(n+1)}\left(2^{n+1};\mathbf{x}\right),\boldsymbol{\theta}^{(n+1)}\left(2^{n+1};\mathbf{x}\right)\right\rangle_{(n+1)}\triangleq\sum_{i=1}^{2^{n+1}}\left\langle\boldsymbol{\chi}_i^{(n+1)}(\mathbf{x}),\boldsymbol{\theta}_i^{(n+1)}(\mathbf{x})\right\rangle_0 \tag{6.96}$$

ii. Using the definition provided in Eq. (6.96), form the inner product in $\mathcal{H}_{n+1}$ of Eq. (6.94) with a yet undefined vector-valued function $\mathbf{a}^{(n+1)}(j_n,\ldots,j_1;\mathbf{x})\triangleq\left[\ldots,\mathbf{a}_k^{(n+1)}(j_n,\ldots,j_1;\mathbf{x}),\ldots\right]^\dagger\in\mathcal{H}_{n+1};\quad k=1,\ldots,2^{n+1}$, $j_1=1,\ldots,TF$, $j_2=1,\ldots,j_1$; $j_{n+1}=1,\ldots,j_n$, to obtain the following relation:

$$\begin{aligned}
&\left\{\left\langle\mathbf{a}^{(n+1)}(j_n,\ldots,j_1;\mathbf{x}),\mathbf{V}^{(n+1)}\left[2^{n+1}\times2^{n+1};\mathbf{x};\mathbf{f}\right]\mathbf{v}^{(n+1)}\left(2^{n+1};\mathbf{x}\right)\right\rangle_{n+1}\right\}_{\alpha^0}\\
&=\left\{\left\langle\mathbf{a}^{(n+1)}(j_n,\ldots,j_1;\mathbf{x}),\mathbf{q}_V^{(n+1)}\left[2^{n+1};\mathbf{u}^{(n+1)}\left(2^{n+1};\mathbf{x}\right);\mathbf{f};\delta\mathbf{f}\right]\right\rangle_{n+1}\right\}_{\alpha^0}\\
&=\left\{\left\langle\mathbf{v}^{(n+1)}\left(2^{n+1};\mathbf{x}\right),\mathbf{A}^{(n+1)}\left[2^{n+1}\times2^{n+1};\mathbf{x};\mathbf{f}\right]\mathbf{a}^{(n+1)}(j_n,\ldots,j_1;\mathbf{x})\right\rangle_{n+1}\right\}_{\alpha^0}\\
&\quad+\left\{P^{(n+1)}\left[\mathbf{v}^{(n+1)};\mathbf{a}^{(n+1)};\mathbf{f};\delta\mathbf{f}\right]\right\}_{\alpha^0},
\end{aligned} \tag{6.97}$$

where $\left\{P^{(n+1)}\left[\mathbf{v}^{(n+1)};\mathbf{a}^{(n+1)};\mathbf{f};\delta\mathbf{f}\right]\right\}_{\alpha^0}$ denotes the bilinear concomitant defined on the phase-space boundary $\mathbf{x}\in\partial\Omega$, evaluated at the nominal values of the model parameter and respective functions, and where $\mathbf{A}^{(n+1)}\left[2^{n+1}\times2^{n+1};\mathbf{x};\mathbf{f}\right]$ is the formal adjoint of the matrix-valued operator $\mathbf{V}^{(n+1)}\left[2^{n+1}\times2^{n+1};\mathbf{x};\mathbf{f}\right]$, i.e.,

$$\mathbf{A}^{(n+1)}\left[2^{n+1}\times2^{n+1};\mathbf{x};\mathbf{f}\right]\triangleq\left\{\mathbf{V}^{(n+1)}\left[2^{n+1}\times2^{n+1};\mathbf{x};\mathbf{f}\right]\right\}^*. \tag{6.98}$$

iii. The 1st term on right side of the 2nd equality in Eq. (6.97) is now required to represent the indirect-effect term $\left\{\delta R^{(n)}\left[j_n;\ldots;j_1;\mathbf{u}^{(n+1)};\mathbf{v}^{(n+1)};\mathbf{f}\right]\right\}_{ind}$. This is achieved by requiring that the $(n+1)$th-level adjoint sensitivity function $\mathbf{a}^{(n+1)}(j_n,\ldots,j_1;\mathbf{x})\triangleq\left[\ldots,\mathbf{a}_k^{(n+1)}(j_n,\ldots,j_1;\mathbf{x}),\ldots\right]^{\dagger}\in\mathcal{H}_{n+1};\ k=1,\ldots,2^{n+1}$, be the solution of the following $(n+1)$th-LASS, for $j_1=1,\ldots,TP;\ j_2=1,\ldots,j_1;\ \ldots\ j_n=1,\ldots,j_{n-1}$:

$$\mathbf{A}^{(n+1)}\left[2^{n+1}\times 2^{n+1};\mathbf{x};\mathbf{f}\right]\mathbf{a}^{(n+1)}(j_n,\ldots,j_1;\mathbf{x})=\mathbf{s}_A^{(n+1)}(j_n,\ldots,j_1;\mathbf{f}), \tag{6.99}$$

subject to the following boundary conditions:

$$\left\{\mathbf{b}_A^{(n+1)}\left[\mathbf{a}^{(n+1)}(j_n,\ldots,j_1;\mathbf{x});\mathbf{u}^{(n+1)}(j_{n-1},\ldots,j_1;\mathbf{x});\mathbf{f}\right]\right\}_{\alpha^0}=\mathbf{0},\quad \mathbf{x}\in\partial\Omega, \tag{6.100}$$

where the vector $\mathbf{s}_A^{(n+1)}(j_n,\ldots,j_1;\mathbf{f})\triangleq\left[\ldots,\mathbf{s}_k^{(n+1)}(j_n,\ldots,j_1;\mathbf{f}),\ldots\right]^{\dagger},\ k=1,\ldots,2^{n+1}$, comprises $2^{n+1}$ components defined as follows, for each $j_1=1,\ldots,TP;\ j_2=1,\ldots,j_1;\ \ldots;\ j_n=1,\ldots,j_{n-1}$:

$$\mathbf{s}_A^{(n+1)}(j_n,\ldots,j_1;\mathbf{f})\triangleq\frac{\partial S^{(n)}}{\partial\mathbf{u}^{(n+1)}(\mathbf{x})} \tag{6.101}$$

iv. The $(n+1)$th-level adjoint boundary conditions represented by Eq. (6.100) are selected so as to eliminate, in conjunction with the boundary conditions represented by Eq. (6.95), all of the unknown values of the functions $\mathbf{v}^{(n+1)}\left(2^{n+1};j_{n-1},\ldots,j_1;\mathbf{x}\right)$ in the expression of the bilinear concomitant $\left\{P^{(n+1)}\left[\mathbf{v}^{(n+1)};\mathbf{a}^{(n+1)};\mathbf{f};\delta\mathbf{f}\right]\right\}_{\alpha^0}$. This bilinear concomitant may vanish after implementing the boundary conditions represented by Eqs. (6.95) and (6.100). However, if it does not vanish, this bilinear concomitant will be reduced to a residual quantity that will comprise only known values of $\mathbf{a}^{(n+1)}(j_n,\ldots,j_1;\mathbf{x})$, $\mathbf{u}^{(n+1)}(j_{n-1}\ldots j_1;\mathbf{x})$, $\mathbf{f}(\alpha)$ and $\delta\mathbf{f}(\alpha)$, and that will be denoted as $\left\{\hat{P}^{(n+1)}\left(\mathbf{a}^{(n+1)};\mathbf{u}^{(n+1)};\mathbf{f};\delta\mathbf{f}\right)\right\}_{\alpha^0}$.

v. Using in Eq. (6.91) the equations underlying the $(n+1)$th-LASS together with the relation provided in Eq. (6.97) yields the following expression for the indirect-effect term $\left\{\delta R^{(n)}\left[j_n;\ldots;j_1;\mathbf{u}^{(n+1)};\mathbf{v}^{(n+1)};\mathbf{f}\right]\right\}_{ind}$ in terms of the $(n+1)$th-level adjoint sensitivity functions $\mathbf{a}^{(n+1)}(j_n,\ldots,j_1;\mathbf{x})$, for each $j_1=1,\ldots,TP;\ j_2=1,\ldots,j_1;\ \ldots;\ j_n=1,\ldots,j_{n-1}$:

$$\begin{aligned}\left\{\delta R^{(n)}\left[j_n;\ldots;j_1;\mathbf{u}^{(n+1)};\mathbf{v}^{(n+1)};\mathbf{f}\right]\right\}_{ind}&=-\left\{\hat{P}^{(n+1)}\left(\mathbf{a}^{(n+1)};\mathbf{u}^{(n+1)};\mathbf{f};\delta\mathbf{f}\right)\right\}_{\alpha^0}\\&+\left\{\left\langle\mathbf{a}^{(n+1)}(j_n,\ldots,j_1;\mathbf{x}),\mathbf{q}_V^{(n+1)}\left[2^{n+1};\mathbf{u}^{(n+1)}\left(2^{n+1};\mathbf{x}\right);\mathbf{f};\delta\mathbf{f}\right]\right\rangle_{n+1}\right\}_{\alpha^0}.\end{aligned} \tag{6.102}$$

Adding the result obtained in Eq. (6.102) for the indirect effect term to the result provided in Eq. (6.90) for the direct effect term yields the following expression for the total $n$th-order G-variation of the response:

$$\left\{\delta R^{(n)}\left[j_n;...;j_1;\mathbf{u}^{(n+1)};\mathbf{f}\right]\right\}_{\alpha^0} = -\left\{\hat{P}^{(n+1)}\left(\mathbf{a}^{(n+1)};\mathbf{u}^{(n+1)};\mathbf{f};\delta\mathbf{f}\right)\right\}_{\alpha^0}$$

$$+\left\{\left\langle \mathbf{a}^{(n+1)}\left(j_n,...,j_1;\mathbf{x}\right),\mathbf{q}_V^{(n+1)}\left[2^{n+1};\mathbf{u}^{(n+1)}\left(2^{n+1};\mathbf{x}\right);\mathbf{f};\delta\mathbf{f}\right]\right\rangle_{n+1}\right\}_{\alpha^0}$$

$$+\sum_{j_{n+1}=1}^{TF}\left\{\frac{\partial}{\partial f_{j_{n+1}}}\int\limits_{\lambda_1(\alpha)}^{\omega_1(\alpha)} dx_1 ... \int\limits_{\lambda_{TI}(\alpha)}^{\omega_{TI}(\alpha)} dx_{TI} S^{(n)}\left(j_n,...,j_1;\mathbf{u}^{(n+1)};\alpha\right)\delta f_{j_{n+1}}\right\}_{\alpha^0}$$

$$\equiv \sum_{j_{n+1}=1}^{TF}\left\{R^{(n+1)}\left(j_{n+1},...,j_1;\mathbf{u}^{(n+1)};\mathbf{a}^{(n+1)};\mathbf{f}\right)\right\}_{\left(\alpha^0\right)}\delta f_{j_{n+1}},$$

where $R^{(n+1)}\left(j_{n+1},...,j_1;\mathbf{u}^{(n+1)};\mathbf{a}^{(n+1)};\mathbf{f}\right)$ denotes the $(n+1)$th-order partial sensitivity of the response $R\left[\mathbf{u}^{(1)}(\mathbf{x});\alpha\right]$ with respect to the components of the feature function $\mathbf{f}(\alpha)$, evaluated at the nominal parameter values $\alpha^0$.

The result obtained in Eq. (6.103) for the expression of the $(n+1)$th-order sensitivity, which was obtained by determining the 1st-order differential of the $n$th-order sensitivity, is identical to the expression that would be obtained by advancing the index, from $n$ to $(n+1)$, in the expression of the $n$th-order sensitivity that was conjectured in Eq. (6.80). Thus, the proof by mathematical induction of the general mathematical framework underlying the $n$th-CASAM-L is thereby completed.

The highlights of the 1st-FASAM-L, 2nd-FASAM-L, $n$th-FASAM-L, and $(n+1)$th-FASAM-L, respectively, are tabularized in Tables 6.1–6.4. An overview, in tabular form, of the computational frameworks of the $n$th-CASAM-L, $n$th-CASAM-N, $n$th-FASAM-L, and $n$th-FASAM-N methodologies, highlighting their objectives, characteristics, and interrelationships, is presented in Table 6.5.

The summary presented in Table 6.5 highlights the conceptual parallelisms between the $n$th-FASAM-L and the $n$th-CASAM-L (as well as the parallelisms between the $n$th-FASAM-N and the $n$th-CASAM-N). Formally, the results produced by the $n$th-FASAM-L can be written in the same mathematical forms as those produced by the $n$th-CASAM-L, with the fundamental difference that the number of large-scale computations needed within the $n$th-FASAM-L is dictated by the number *TF* of "feature function components," whereas the number of large-scale computations needed within the $n$th-CASAM-L is dictated by the number *TP* of primary model parameters. In particular, a single large-scale adjoint computation is needed to solve the 1st-LASS (which is the same for both the 1st-FASAM-L and the 1st-CASAM-L) to obtain the 1st-order sensitivities with respect to the model parameters. Obtaining the 2nd-order sensitivities of the response with respect to the primary model parameters requires at most *TP* large-scale computations (to solve the 2nd-LASS) within the 2nd-CASAM-L, while the computation of the 2nd-order sensitivities within the 2nd-FASAM-L requires only at most TF large-scale

## TABLE 6.1
## 1st-FASAM-L: 1st-Order ($n=1$) Sensitivities of Response to Model Features

| | |
|---|---|
| 1st-LFAS | $\mathbf{F}^{(1)}[2\times 2;\mathbf{x};\mathbf{f}]\mathbf{u}^{(1)}(2;\mathbf{x})=\mathbf{q}_F^{(1)}(2;\mathbf{x};\mathbf{f});\quad \mathbf{x}\in\Omega;$<br>$\mathbf{b}_F^{(1)}\left[\mathbf{u}^{(1)}(2;\mathbf{x});\mathbf{f}\right]=\mathbf{0};\quad \mathbf{x}\in\partial\Omega[\boldsymbol{\lambda}(\boldsymbol{\alpha});\boldsymbol{\omega}(\boldsymbol{\alpha})];$<br>$\mathbf{u}^{(1)}(2;\mathbf{x})\triangleq[\boldsymbol{\varphi}(\mathbf{x}),\boldsymbol{\psi}(\mathbf{x})]^{\dagger}.$ |
| 1st-LVSS | $\mathbf{V}^{(1)}[2\times 2;\mathbf{x};\mathbf{f}]\mathbf{v}^{(1)}(2;\mathbf{x})=\mathbf{q}_V^{(1)}\left[2;\mathbf{u}^{(1)}(2;\mathbf{x});\mathbf{f};\delta\mathbf{f}\right];\quad \mathbf{x}\in\Omega;$<br>$\left\{\mathbf{b}_v^{(1)}\left(\mathbf{u}^{(1)};\mathbf{v}^{(1)};\mathbf{f};\delta\mathbf{f}\right)\right\}_{\alpha^0}=\mathbf{0};\quad \mathbf{x}\in\partial\Omega\left[\boldsymbol{\lambda}(\boldsymbol{\alpha}^0);\boldsymbol{\omega}(\boldsymbol{\alpha}^0)\right];$<br>$\mathbf{v}^{(1)}(2;\mathbf{x})\triangleq[\delta\boldsymbol{\varphi}(\mathbf{x}),\delta\boldsymbol{\psi}]^{\dagger}.$ |
| 1st-level Hilbert space | $\mathcal{H}_1:\left\langle\boldsymbol{\chi}^{(1)}(2;\mathbf{x}),\boldsymbol{\theta}^{(1)}(2;\mathbf{x})\right\rangle_1\triangleq\sum_{i=1}^{2}\left\langle\chi_i^{(1)}(\mathbf{x}),\boldsymbol{\theta}_i^{(1)}(\mathbf{x})\right\rangle_0$<br>$\left\langle\boldsymbol{\varphi}(\mathbf{x}),\boldsymbol{\psi}(\mathbf{x})\right\rangle_0\triangleq\sum_{j=1}^{TD}\int_{\lambda_1(\alpha)}^{\omega_1(\alpha)}\ldots\int_{\lambda_i(\alpha)}^{\omega_i(\alpha)}\ldots\int_{\lambda_{TI}(\alpha)}^{\omega_{TI}(\alpha)}\varphi_j(\mathbf{x})\psi_j(\mathbf{x})dx_1\ldots dx_i\ldots dx_{TI}.$ |
| 1st-LASS | $\mathbf{A}^{(1)}[2\times 2;\mathbf{x};\mathbf{f}]\mathbf{a}^{(1)}(2;\mathbf{x})=\mathbf{q}_A^{(1)}\left[2;\mathbf{u}^{(1)}(2;\mathbf{x});\mathbf{f}\right],\quad \mathbf{x}\in\Omega;$<br>$\mathbf{b}_A^{(1)}\left[\mathbf{u}^{(1)}(2;\mathbf{x});\mathbf{a}^{(1)}(2;\mathbf{x});\mathbf{f}\right]=0,\quad \mathbf{x}\in\partial\Omega;\ \mathbf{a}^{(1)}(2;\mathbf{x})\triangleq\left[\mathbf{a}_1^{(1)}(\mathbf{x}),\mathbf{a}_2^{(1)}(\mathbf{x})\right]^{\dagger};$ |
| 1st-order resp. sensitivities to model features | $R^{(1)}\left[j_1;\mathbf{u}^{(1)}(2;\mathbf{x});\mathbf{a}^{(1)}(2;\mathbf{x});\mathbf{f}(\times)\right];\quad j_1=1,\ldots,TF.$ |

## TABLE 6.2
## 2nd-FASAM-L: 2nd-Order ($n=2$) Sensitivities of Response to Model Features

| | |
|---|---|
| 2nd-LFAS = 1st-LFAS + 1st-LASS | $\mathbf{F}^{(2)}\left[2^2\times 2^2;\mathbf{f}(\boldsymbol{\alpha})\right]\mathbf{u}^{(2)}\left(2^2;\mathbf{x}\right)=\mathbf{q}_F^{(2)}\left[2^2;\mathbf{u}^{(1)}(2;\mathbf{x});\mathbf{f}(\boldsymbol{\alpha})\right];\quad \mathbf{x}\in\Omega;$<br>$\mathbf{b}_F^{(2)}\left(2^2;\mathbf{u}^{(2)};\mathbf{f}\right)\triangleq\left(\mathbf{b}_F^{(1)},\mathbf{b}_A^{(1)}\right)^{\dagger}=\mathbf{0};\quad \mathbf{x}\in\partial\Omega;$<br>$\mathbf{u}^{(2)}\left(2^2;\mathbf{x}\right)\triangleq\left[\mathbf{u}^{(1)}(2;\mathbf{x}),\mathbf{a}^{(1)}(2;\mathbf{x})\right]^{\dagger};$ |
| 2nd-LVSS | $\mathbf{V}^{(2)}\left[2^2\times 2^2;\mathbf{x};\mathbf{f}\right]\mathbf{v}^{(2)}\left(2^2;\mathbf{x}\right)=\mathbf{q}_V^{(2)}\left[2^2;\mathbf{u}^{(2)}\left(2^2;\mathbf{x}\right);\mathbf{f};\delta\mathbf{f}\right];\quad \mathbf{x}\in\Omega;$<br>$\mathbf{b}_v^{(2)}\left(\mathbf{u}^{(2)};\mathbf{v}^{(2)};\mathbf{f};\delta\mathbf{f}\right)=\mathbf{0};\quad \mathbf{x}\in\partial\Omega;$<br>$\mathbf{v}^{(2)}\left(2^2;\mathbf{x}\right)\triangleq\left[\mathbf{v}^{(1)}(2;\mathbf{x}),\delta\mathbf{a}^{(1)}(2;\mathbf{x})\right];$ |
| 2nd-level Hilbert space | $\mathcal{H}_2:\left\langle\boldsymbol{\chi}^{(2)}\left(2^2;\mathbf{x}\right),\boldsymbol{\theta}^{(2)}\left(2^2;\mathbf{x}\right)\right\rangle_2\triangleq\sum_{i=1}^{2^2}\left\langle\chi_i^{(2)}(\mathbf{x}),\boldsymbol{\theta}_i^{(2)}(\mathbf{x})\right\rangle_0$<br>$\left\langle\boldsymbol{\chi}^{(2)}(2;\mathbf{x}),\boldsymbol{\theta}^{(2)}(2;\mathbf{x})\right\rangle_{2^2}\triangleq\sum_{i=1}^{2^2}\left\langle\chi_i^{(2)}(\mathbf{x}),\boldsymbol{\theta}_i^{(2)}(\mathbf{x})\right\rangle_0$ |

*(Continued)*

**TABLE 6.2 (*Continued*)**
**2nd-FASAM-L: 2nd-Order ($n=2$) Sensitivities of Response to Model Features**

| | |
|---|---|
| 2nd-LASS | $\mathbf{A}^{(2)}\left[2^2\times 2^2;\mathbf{x};\mathbf{f}\right]\mathbf{a}^{(2)}\left(2^2;j_1;\mathbf{x}\right)=\mathbf{s}^{(2)}\left(2^2;j_1;\mathbf{u}^{(2)};\mathbf{f}\right);\quad \mathbf{x}\in\Omega;\quad j_1=1,\ldots,TF;$ <br> $\left\{\mathbf{b}_A^{(2)}\left[\mathbf{u}^{(2)}\left(2^2;\mathbf{x}\right);\mathbf{a}^{(2)}\left(2^2;j_1;\mathbf{x}\right);\mathbf{f}\right]\right\}_{\alpha^0}=\mathbf{0},\quad \mathbf{x}\in\partial\Omega,\quad j_1=1,\ldots,TF.$ <br> $\mathbf{a}^{(2)}\left(2^2;j_1;\mathbf{x}\right)\triangleq\left[\mathbf{a}_1^{(2)}(j_1;\mathbf{x}),\mathbf{a}_2^{(2)}(j_1;\mathbf{x}),\mathbf{a}_3^{(2)}(j_1;\mathbf{x}),\mathbf{a}_4^{(2)}(j_1;\mathbf{x})\right]^\dagger$ <br> $=\left[\ldots,\mathbf{a}_k^{(2)}(j_1;\mathbf{x}),\ldots\right]^\dagger;\quad k=1,\ldots,2^2.$ |
| 2nd-order resp. sensitivities to model features | $R^{(2)}\left[j_2;j_1;\mathbf{u}^{(2)}\left(2^2;\mathbf{x}\right);\mathbf{a}^{(2)}\left(2^2;j_1;\mathbf{x}\right);\mathbf{f}(\alpha)\right];\quad j_1=1,\ldots,TF;\quad j_2=1,\ldots,j_1.$ <br> Distinct sensitivities: $\dfrac{TF(TF+1)}{2!}$ |

**TABLE 6.3**
**$n$th-FASAM-L: $n$th-Order Sensitivities of Response to Model Features**

| | |
|---|---|
| $n$th-LFAS $=(n-1)$th-LFAS $+(n-1)$th-LASS: | $\mathbf{F}^{(n)}\left[2^n\times 2^n;\mathbf{f}(\alpha)\right]\mathbf{u}^{(n)}\left(2^n;\mathbf{x}\right)=\mathbf{q}_F^{(n)}\left[2^n;\mathbf{u}^{(n-1)}\left(2^{n-1};\mathbf{x}\right);\mathbf{f}(\alpha)\right];$ <br> $\mathbf{b}_F^{(n)}\left(2^{(n)};\mathbf{u}^{(n)};\mathbf{f}\right)\triangleq\left(\mathbf{b}_F^{(n-1)},\mathbf{b}_A^{(n-1)}\right)^\dagger=\mathbf{0};\quad \mathbf{x}\in\partial\Omega;$ <br> $\mathbf{F}^{(n)}\left[2^n\times 2^n;\mathbf{f}\right]\triangleq diag\left(\mathbf{F}^{(n-1)},\mathbf{A}^{(n-1)}\right);$ <br> $\mathbf{u}^{(n)}\left(2^n;j_{n-2},\ldots,j_1;\mathbf{x}\right)=\left[\mathbf{u}^{(n-1)}\left(2^{n-1};j_{n-3},\ldots,j_1;\mathbf{x}\right),\mathbf{a}^{(n-1)}\left(2^{n-1};j_{n-2},\ldots,j_1;\mathbf{x}\right)\right]^\dagger;$ <br> $\mathbf{q}_F^{(n)}\left[2^n;\mathbf{u}^{(n)}\left(2^n;\mathbf{x}\right);\mathbf{f}(\alpha)\right]\triangleq\left[\mathbf{q}_F^{(n-1)}\left(2^{n-1};\mathbf{x};\mathbf{f}\right),\mathbf{q}_A^{(n-1)}\left(2^{n-1};\mathbf{x};\mathbf{f}\right)\right]^\dagger;$ |
| $n$th-LVSS | $\mathbf{V}^{(n)}\left[2^n\times 2^n;\mathbf{x};\mathbf{f}\right]\mathbf{v}^{(n)}\left(2^n;\mathbf{x}\right)=\mathbf{q}_V^{(n)}\left[2^n;\mathbf{u}^{(n)}\left(2^n;\mathbf{x}\right);\mathrm{f};\delta\mathbf{f}\right];\quad \mathbf{x}\in\Omega;$ <br> $\mathbf{v}^{(n)}\left(2^n;j_{n-2},\ldots,j_1;\mathbf{x}\right)\triangleq\left[\mathbf{v}^{(n-1)}\left(2^{n-1};\mathbf{x}\right),\delta\mathbf{a}^{(n-1)}\left(2^{n-1};\mathbf{x}\right)\right]^\dagger$ <br> $\mathbf{b}_v^{(n)}\left(\mathbf{u}^{(n)};\mathbf{v}^{(n)};\mathbf{f};\delta\mathbf{f}\right)\triangleq\left[\mathbf{b}_V^{(n-1)},\delta\mathbf{b}_A^{(n-1)}\right]^\dagger=\mathbf{0};\quad \mathbf{x}\in\partial\Omega.$ |
| $n$th-level Hilbert space | $\mathcal{H}_n$: $\left\langle\chi^{(n)}\left(2^n;\mathbf{x}\right),\theta^{(n)}\left(2^n;\mathbf{x}\right)\right\rangle_n\triangleq\sum_{i=1}^{2^n}\left\langle\chi_i^{(n)}(\mathbf{x}),\theta_i^{(n)}(\mathbf{x})\right\rangle_0;$ |
| $n$th-LASS | $\mathbf{A}^{(n)}\left[2^n\times 2^n;\mathbf{x};\mathbf{f}\right]\mathbf{a}^{(n)}\left(2^n;j_{n-1},\ldots,j_1;\mathbf{x}\right)=\mathbf{s}_A^{(n)}\left(2^n;j_{n-1},\ldots,j_1;\mathbf{f}\right);$ <br> $\mathbf{b}_A^{(n)}\left[\mathbf{a}^{(n)}\left(j_{n-1},\ldots,j_1;\mathbf{x}\right);\mathbf{u}^{(n)}\left(j_{n-2},\ldots,j_1;\mathbf{x}\right);\mathbf{f}\right]=\mathbf{0},\quad \mathbf{x}\in\partial\Omega;$ <br> $\mathbf{A}^{(n)}\left[2^n\times 2^n;\mathbf{x};\mathbf{f}\right]\triangleq\left\{\mathbf{V}^{(n)}\left[2^n\times 2^n;\mathbf{x};\mathbf{f}\right]\right\}^*;$ |
| $n$th-order resp. sensitivities to model features | $R^{(n)}\left[j_n;\ldots;j_1;\mathbf{u}^{(n)}\left(2^n;j_{n-2},\ldots,j_1;\mathbf{x}\right);\mathbf{a}^{(n)}\left(2^n;j_{n-1},\ldots,j_1;\mathbf{x}\right);\mathbf{f}(\alpha)\right]$ <br> $\triangleq\dfrac{\partial^n R[\varphi(\mathbf{x}),\psi(\mathbf{x});\alpha]}{\partial f_{j_1}\ldots\partial f_{j_n};j_1=1,\ldots,TF;};\ j_2=1,\ldots,j_1;\ldots j_n=1,\ldots,j_{n-1};$ <br> Distinct sensitivities: $\dfrac{TF(TF+1)(TF+2)\ldots(TF+n-1)}{n!}$ |

**TABLE 6.4**
**(*n* + 1)th-FASAM-L: (*n* + 1)th-Order Sensitivities of Response to Model Features**

| | |
|---|---|
| $(n+1)$th-LASS = $n$th-LFAS + $n$th-LASS: | $\mathbf{F}^{(n+1)}\left[2^{n+1}\times 2^{n+1};\mathbf{f}(\alpha)\right]\mathbf{u}^{(n+1)}\left(2^{n+1};\mathbf{x}\right)=\mathbf{q}_F^{(n+1)}\left[2^{n+1};\mathbf{u}^{(n)}\left(2^{n};\mathbf{x}\right);\mathbf{f}(\alpha)\right];$ <br> $\mathbf{F}^{(n+1)}\left[2^{n+1}\times 2^{n+1};\mathbf{f}(\alpha)\right]\triangleq diag\left(\mathbf{F}^{(n)},\mathbf{A}^{(n)}\right);$ <br> $\mathbf{u}^{(n+1)}\left(2^{n+1};j_{n-1},\ldots,j_1;\mathbf{x}\right)=\left[\mathbf{u}^{(n)}\left(2^{n};j_{n-2},\ldots,j_1;\mathbf{x}\right),\mathbf{a}^{(n)}\left(2^{n};j_{n-1},\ldots,j_1;\mathbf{x}\right)\right]^{\dagger}$ <br> $\mathbf{q}_F^{(n+1)}\left[2^{n+1};\mathbf{u}^{(n+1)}\left(2^{n+1};\mathbf{x}\right);\mathbf{f}(\alpha)\right]\triangleq\left[\mathbf{q}_F^{(n)}\left(2^{n};\mathbf{x};\mathbf{f}\right),\mathbf{q}_A^{(n)}\left(2^{n};\mathbf{u}^{(n)};\mathbf{f}\right)\right]^{\dagger}$ <br> $\mathbf{b}_F^{(n+1)}\left(2^{(n+1)};\mathbf{u}^{(n+1)};\mathbf{f}\right)\triangleq\left(\mathbf{b}_F^{(n)},\mathbf{b}_A^{(n)}\right)^{\dagger}=\mathbf{0};\quad \mathbf{x}\in\partial\Omega.$ |
| $(n+1)$th-LVSS | $\mathbf{V}^{(n+1)}\left[2^{n+1}\times 2^{n+1};\mathbf{x};\mathbf{f}\right]\mathbf{v}^{(n+1)}\left(2^{n+1};\mathbf{x}\right)=\mathbf{q}_V^{(n+1)}\left[2^{n+1};\mathbf{u}^{(n+1)}\left(2^{n+1};\mathbf{x}\right);\mathbf{f};\delta\mathbf{f}\right];\quad \mathbf{x}\in\Omega;$ <br> $\mathbf{v}^{(n+1)}\left(2^{n+1};j_{n-1},\ldots,j_1;\mathbf{x}\right)\triangleq\left[\mathbf{v}^{(n)}\left(2^{n};\mathbf{x}\right),\delta\mathbf{a}^{(n)}\left(2^{n};\mathbf{x}\right)\right]^{\dagger}$ <br> $\mathbf{b}_V^{(n+1)}\left(\mathbf{u}^{(n+1)};\mathbf{v}^{(n+1)};\mathbf{f};\delta\mathbf{f}\right)\triangleq\left[\mathbf{b}_V^{(n)},\delta\mathbf{b}_A^{(n)}\right]^{\dagger}=\mathbf{0};\quad \mathbf{x}\in\partial\Omega.$ |
| $(n+1)$th-level Hilbert space | $\mathcal{H}_{n+1}:\left\langle\chi^{(n+1)}\left(2^{n+1};\mathbf{x}\right),\theta^{(n+1)}\left(2^{n+1};\mathbf{x}\right)\right\rangle_{(n+1)}\triangleq\sum_{i=1}^{2^{n+1}}\left\langle\chi_i^{(n+1)}(\mathbf{x}),\theta_i^{(n+1)}(\mathbf{x})\right\rangle_0$ |
| $(n+1)$th-LASS | $\mathbf{A}^{(n+1)}\left[2^{n+1}\times 2^{n+1};\mathbf{x};\mathbf{f}\right]\mathbf{a}^{(n+1)}\left(2^{n+1};j_n,\ldots,j_1;\mathbf{x}\right)=\mathbf{s}_A^{(n+1)}\left(2^{n+1};j_n,\ldots,j_1;\mathbf{f}\right),$ <br> $\mathbf{A}^{(n+1)}\left[2^{n+1}\times 2^{n+1};\mathbf{x};\mathbf{f}\right]\triangleq\left\{\mathbf{V}^{(n+1)}\left[2^{n+1}\times 2^{n+1};\mathbf{x};\mathbf{f}\right]\right\}^{*};$ <br> $\mathbf{b}_A^{(n+1)}\left[\mathbf{a}^{(n+1)}\left(j_n,\ldots,j_1;\mathbf{x}\right);\mathbf{u}^{(n+1)}\left(j_{n-1},\ldots,j_1;\mathbf{x}\right);\mathbf{f}\right]=\mathbf{0},\quad \mathbf{x}\in\partial\Omega;$ |
| $(n+1)$th-Resp. sensitivities to model features | $R^{(n+1)}\left(j_{n+1},\ldots,j_1;\mathbf{u}^{(n+1)};\mathbf{a}^{(n+1)};\mathbf{f}\right)\triangleq\frac{\partial^{n+1}R\left[\varphi(\mathbf{x}),\psi(\mathbf{x});\alpha\right]}{\partial f_{j_1}\ldots\partial f_{j_{n+1}}};$ <br> $j_1=1,\ldots,TF;\ldots j_{n+1}=1,\ldots,j_n;$ <br> Distinct sensitivities: $\frac{TF(TF+1)(TF+2)\ldots(TF+n)}{(n+1)!}$ |

computations. Obtaining the 3rd-order sensitivities of the response with respect to the primary model parameters requires at most $\frac{TP(TP+1)}{2}$ large-scale computations (to solve the 3rd-LASS) within the 3rd-CASAM-L. Obtaining the 3rd-order sensitivities using the 3rd-FASAM-L requires at most $\frac{TF(TF+1)}{2}$ large-scale computations (to solve the respective 3rd-LASS) followed by analytical derivations to obtain the 3rd-order sensitivities with respect to the model parameters from the 3rd-order sensitivities with respect to the components of the feature function produced by the 2nd-FASAM-L. The same parallel holds for the computation of all of the higher-order sensitivities: the computation of the 4th-order sensitivities with respect to the primary model parameters requires at most $\frac{TP(TP+1)(TP+2)}{3!}$ computations if using the 4th-CASAM-L, as opposed to at most $\frac{TF(TF+1)(TF+2)}{3!}$

**TABLE 6.5**
**Overview of the Computational Frameworks of the *n*th-CASAM-L, *n*th-CASAM-N, *n*th-FASAM-L, and *n*th-FASAM-N Methodologies**

| Methodology | Objective | Characteristics | Interrelationships |
|---|---|---|---|
| *n*th-FASAM-L | Develop forward and adjoint operators in linearly increasing Hilbert spaces to enable the most efficient computation of exact expressions of any-order sensitivities of responses to features/ functions of primary model parameters. | Especially applicable to response-coupled forward/adjoint linear models. Also applicable to responses that depend just on the forward or just the adjoint state functions in linear systems. | Reduces to the *n*th-CASAM-L in the absence of "feature functions," i.e., when the feature functions coincide with the primary parameters. |
| *n*th-CASAM-L | Develop forward and adjoint operators in linearly increasing Hilbert spaces to enable the most efficient computation of exact expressions of any-order sensitivities of responses to primary model parameters. | Same characteristics as *n*th-FASAM-L but directly considering the primary model parameters. | Becomes identical to the *n*th-FASAM-L in the absence of "feature functions" of parameters. |
| *n*th-FASAM-N | Same objective as the *n*th-FASAM-L, but for nonlinear models. | Subsumes the *n*th-FASAM-L if the responses depend just on the forward state functions. | Reduces to the *n*th-CASAM-N in the absence of "feature functions," i.e., when the feature functions coincide with the primary parameters. |
| *n*th-CASAM-N | Same objective as the *n*th-CASAM-L, but for nonlinear models. | Subsumes the *n*th-CASAM-L if the responses depend just on the forward state functions. | Becomes identical to the *n*th-FASAM-N in the absence of "feature functions" of parameters. |

large-scale computations plus analytical derivations if using the 4th-FASAM-L and so on. Since $TF \ll TP$, it is evident that the *n*th-FASAM-L methodology becomes increasingly more efficient than the *n*th-CASAM-L as the order of computed sensitivities increases.

## 6.6 CHAPTER SUMMARY

This chapter has presented the "*n*th-Order Feature Adjoint Sensitivity Analysis Methodology for Response-Coupled Forward/Adjoint Linear Systems" (abbreviated as "*n*th-FASAM-L"), which is the most efficient methodology for computing exact expressions of sensitivities of model responses to features of model parameters and, subsequently, to the model parameters themselves for such linear systems. This efficiency stems from the maximal reduction of the number of "large-scale" computations, by comparison to any other method for computing high-order sensitivities. Specific details are as follows:

i. Comparing the mathematical framework of the *n*th-FASAM-N methodology to the framework of the *n*th-CASAM-N methodology indicates that the components $f_i(\boldsymbol{\alpha}), i = 1,\ldots,TF$, of the "feature function" $\mathbf{f}(\boldsymbol{\alpha}) \triangleq [f_1(\boldsymbol{\alpha}),\ldots,f_{TF}(\boldsymbol{\alpha})]^\dagger$ play within the *n*th-FASAM-N the same role as played by the components $\alpha_j, j = 1,\ldots,TP$, of the "vector of primary model parameters" $\boldsymbol{\alpha} \triangleq (\alpha_1,\ldots,\alpha_{TP})^\dagger$ within the framework of the *n*th-CASAM-N. It is important to note that the total number of model parameters is always larger (usually by a wide margin) than the total number of components of the feature function $\mathbf{f}(\boldsymbol{\alpha})$, i.e., $TP \gg TF$.

ii. The 1st-FASAM-N and the 1st-CASAM-N methodologies require a *single* large-scale "adjoint" computation for solving the 1st-LASS (1st-Level Adjoint Sensitivity System), so they are comparably efficient for computing the exact expressions of the *1st-order* sensitivities of a model response to the model's uncertain parameters, boundaries, and internal interfaces. The 1st-FASAM-N enjoys a slight computational advantage over the 1st-CASAM-N, since it requires only $TF$ quadratures, as opposed to $TP$ quadratures required by the 1st-CASAM-N methodology. The same conclusion can be drawn when comparing the 1st-FASAM-L and the 1st-CASAM-L methodologies.

iii. For computing the exact expressions of the 2nd-order response sensitivities with respect to the primary model's parameters, the 2nd-FASAM-N methodology requires, at most, as many large-scale "adjoint" computations as there are "feature functions of parameters" $f_i(\boldsymbol{\alpha}), i = 1,\ldots,TF$ (where $TF$ denotes the total number of feature functions) for solving the left side of the 2nd-LASS with $TF$ distinct sources on its right side. By comparison, the 2nd-CASAM-N methodology requires at most $TP$ (where $TP$ denotes the total number of model parameters) large-scale computations for solving the same left side of the 2nd-LASS but with $TP$ distinct sources. Since $TF \ll TP$, the 2nd-FASAM-N methodology is considerably more efficient than the 2nd-CASAM-N methodology for computing the exact expressions of the

2nd-order sensitivities of a model response to the model's uncertain parameters, boundaries, and internal interfaces. The same comparison can be drawn between the 2nd-CASAM-L and the 2nd-FASAM-L methodologies.

iv. For computing the exact expressions of the 3rd-order response sensitivities with respect to the primary model's parameters, the 3rd-FASAM-N requires at most $TF\frac{(TF+1)}{2}$ large-scale "adjoint" computations for solving the 3rd-LASS with $TF\frac{(TF+1)}{2}$ distinct sources, while the 3rd-CASAM-N methodology requires at most $TP\frac{(TP+1)}{2}$ large-scale computations for solving the 3rd-LASS with $TP\frac{(TP+1)}{2}$ distinct sources. The same computational count of "large-scale computations" carries over when computing the higher-order sensitivities, i.e., the formula for calculating the "number of large-scale adjoint computations" is formally the same for both the *n*th-FASAM-N and the *n*th-CASAM-N methodologies, but the "variable" in the formula for determining the number of adjoint computations for the *n*th-FASAM-N methodology is $TF$ (i.e., total number of feature functions) while the counterpart for the formula for determining the number of adjoint computations for the *n*th-CASAM-N methodology is $TP$ (i.e., total number of model parameters). Since $TF \ll TP$, it follows that the higher the order of computed sensitivities, the more efficient the *n*th-FASAM-N methodology becomes by comparison to the *n*th-CASAM-N methodology. The same comparison can be drawn between the *n*th-CASAM-L and the *n*th-FASAM-L methodologies.

v. When a model has no "feature" functions of parameters but only comprises primary parameters, the *n*th-FASAM-N methodology becomes identical to the *n*th-CASAM-N methodology. Similarly, under these conditions, the *n*th-FASAM-L methodology becomes identical to the *n*th-CASAM-L methodology.

vi. Both the *n*th-FASAM-N and the *n*th-CASAM-N methodologies are formulated in linearly increasing higher-dimensional Hilbert spaces – as opposed to exponentially increasing parameter-dimensional spaces – thus overcoming the curse of dimensionality in sensitivity analysis of nonlinear systems. Both the *n*th-FASAM-N and the *n*th-CASAM-N methodologies are incomparably more efficient and more accurate than any other methods (statistical, finite differences, etc.) for computing exact expressions of response sensitivities (of any order) with respect to the model's uncertain parameters, boundaries, and internal interfaces. The same conclusion holds for the *n*th-FASAM-L and the *n*th-CASAM-L methodologies.

# 7 Illustrative Application of the *n*th-FASAM-L

## 7.1 INTRODUCTION

This chapter presents an illustrative application of the "*n*th-Order Feature Adjoint Sensitivity Analysis Methodology for Response-Coupled Forward/Adjoint Linear Systems" (*n*th-FASAM-L) to an energy-dependent neutron slowing-down model of fundamental importance to reactor physics. The physical considerations underlying this model are presented in Section 7.2, which briefly reviews the concept of "contributon-flux density response" and particularizes this concept within the modeling of neutron slowing down in a mixture of materials. This physical model enables the derivation of exact closed-form results for the application of the *n*th-FASAM-L methodology. Section 7.2 also defines the "features" (functions of primary model parameters) inherent to this model, which enable the advantageous application of the *n*th-FASAM-L methodology.

Section 7.3 presents the application of the 1st-FASAM-L to determine the 1st-order sensitivities of the contributon-flux with respect to the slowing-down model's features and primary model parameters and compares the application of the 1st-FASAM-L versus the 1st-CASAM-L methodology. Using either the 1st-FASAM-L or the 1st-CASAM-L methodologies involves solving the same operator equations and boundary conditions within the respective 1st-LASS, but with differing source terms. For the computation of the 1st-order sensitivities, the 1st-FASAM-L enjoys a slight computational advantage since it requires only one quadrature per component of the feature function, whereas the 1st-CASAM-L requires one quadrature per primary model parameter.

Section 7.4 presents the 2nd-order adjoint sensitivity analysis of the contributon-flux with respect to the slowing-down model's features and primary model parameters, comparing the application of the 2nd-FASAM-L versus the 2nd-CASAM-L methodology. It is shown that the 2nd-FASAM-L requires as many large-scale "adjoint" computations as there are nonvanishing 1st-order response sensitivities with respect to the components of the feature functions. In contradistinction, the 2nd-CASAM-L methodology requires as many large-scale computations as there are non-nonvanishing 1st-order response sensitivities with respect to the primary model parameters. Hence, the 2nd-FASAM-L methodology is inherently more efficient than the 2nd-CASAM-L methodology. In particular, one of the three distinct 2nd-order sensitivities with respect to the model's features vanishes identically within the 2nd-FASAM-L but none of the many (tens of) 2nd-order sensitivities with respect to the primary model parameters vanishes within the 2nd-CASAM-L. The identically vanishing 2nd-order sensitivity within the 2nd-FASAM-L framework will evidently *not* give rise to higher-order sensitivities, thus ensuring that the nth-FASAM-L methodology will require the fewest number of computations for higher-order sensitivity analysis.


DOI: 10.1201/9781003478171-7

Section 7.5 presents the 3rd-order adjoint sensitivity analysis of the contributon-flux with respect to the slowing-down model's features and primary model parameters, comparing the application of the 3rd-FASAM-L versus the 3rd-CASAM-L methodology. For computing the exact expressions of the 3rd-order contributon response sensitivities, the 3rd-FASAM-L requires just *two* large-scale computations, whereas the 3rd-CASAM-L methodology would require hundreds of large-scale computations for computing the 3rd-order sensitivities.

The concluding discussion presented in Section 7.6 emphasizes the fact that the unparalleled efficiency of the *n*th-FASAM-N methodology increases as the order of computed sensitivities increases, and the probability of encountering vanishing sensitivities is much higher when using the *n*th-FASAM-L methodology rather than any other methodology. Both the *n*th-FASAM-L and the *n*th-CASAM-L methodologies overcome the curse of dimensionality in sensitivity analysis of linear systems, being incomparably more efficient and more accurate than any other method (statistical, finite differences, etc.) for computing exact expressions of response sensitivities (of any order) with respect to the model's uncertain parameters, boundaries, and internal interfaces.

## 7.2 CONTRIBUTON-FLUX RESPONSE IN A PARADIGM NEUTRON SLOWING-DOWN MODEL

Fundamentally important responses of linear models depend *simultaneously* on both the forward and adjoint state functions governing the respective linear model. Typical examples of such responses arise in the modeling of self-diffusion processes in which the interaction mean-free path is independent of the phase-space density. Such processes are modeled by linear equations of Lorentz-Boltzmann type, and they occur in neutron, electron, and photon transport through media, as well as certain types of transport processes in gas or plasma dynamics. Even though the time-dependent integro-differential equations that model these processes are linear, solving them numerically is representative of "large-scale" computations and will be used in the sequel for illustrating the application of the *n*th-FASAM-L methodology. For example, the distribution of neutrons in a medium is modeled by the following standard form of the linear Boltzmann equation:

$$L(\mathbf{r},E,\mathbf{\Omega},t)\varphi(\mathbf{r},E,\mathbf{\Omega},t)=Q(\mathbf{r},E,\mathbf{\Omega},t), \tag{7.1}$$

where the linear integro-differential operator $L(\mathbf{r},E,\mathbf{\Omega},t)$ is defined below:

$$\begin{aligned} L(\mathbf{r},E,\mathbf{\Omega},t)\varphi(\mathbf{r},E,\mathbf{\Omega},t) &\triangleq \frac{1}{\mathrm{v}}\frac{\partial\varphi(\mathbf{r},E,\mathbf{\Omega},t)}{\partial t}+\mathbf{\Omega}\bullet\nabla\varphi(\mathbf{r},E,\mathbf{\Omega},t)+\Sigma_t(\mathbf{r},E)\varphi(\mathbf{r},E,\mathbf{\Omega},t) \\ &-\int_0^{E_f}dE'\int_{4\pi}d\Omega'\Sigma_s(\mathbf{r},E'\to E,\mathbf{\Omega}'\to\mathbf{\Omega})\varphi(\mathbf{r},E',\mathbf{\Omega}',t) \\ &-\int_0^{E_f}dE'\int_{4\pi}\chi(\mathbf{r},E'\to E)\nu\Sigma(\mathbf{r},E')\varphi(\mathbf{r},E',\mathbf{\Omega}',t)d\Omega'. \end{aligned} \tag{7.2}$$

The quantities which appear in the standard notation used in Eq. (7.2) are defined as follows:

i. $\mathbf{r}$ denotes the three-dimensional position vector in space; $E$ denotes the energy independent variable; the directional vector $\mathbf{\Omega}$ denotes the scattering solid angle; $t$ denotes the time independent variable; v denotes the neutron particle speed.
ii. $\varphi(\mathbf{r},E,\mathbf{\Omega},t)$ denotes the flux of particles (i.e., particle number density multiplied by the particle speed) in the energy range $dE$ about $E$, in the volume element $d\mathbf{r}$ about $\mathbf{r}$, with directions of motion in the solid angle element $d\mathbf{\Omega}$ about $\mathbf{\Omega}$.
iii. $Q(\mathbf{r},E,\mathbf{\Omega},t)$ denotes the rate at which particles are produced in the same element of phase space from sources that are independent of the flux.
iv. $\Sigma_t(\mathbf{r},E)$ denotes the macroscopic total cross section.
v. $\Sigma_s(\mathbf{r},E'\to E,\mathbf{\Omega}'\to\mathbf{\Omega})$ denotes the macroscopic scattering transfer cross section from energy $E'$ to energy $E$, and from scattering angle $\mathbf{\Omega}'$ to scattering angle $\mathbf{\Omega}$.
vi. $\nu$ denotes the number of particles emitted isotropically $(1/4\pi)$ per fission.
vii. $\Sigma_f(\mathbf{r},E)$ denotes the macroscopic fission cross section.
viii. $\chi(\mathbf{r},E'\to E)$ denotes the fraction of fission particles appearing in energy $dE$ about $E$ from fissions in $dE'$ about $E'$.

The adjoint Boltzmann transport equation is constructed in a Hilbert space, denoted as $\mathcal{H}_B$ and endowed with the following inner product, denoted as $\langle\varphi(\mathbf{r},E,\mathbf{\Omega},t),\psi(\mathbf{r},E,\mathbf{\Omega},t)\rangle_B$, between two elements $\varphi(\mathbf{r},E,\mathbf{\Omega},t)\in\mathcal{H}_B$ and $\psi(\mathbf{r},E,\mathbf{\Omega},t)\in\mathcal{H}_B$:

$$\langle\varphi,\psi\rangle_B \triangleq \int_0^{t_f} dt \int_0^{\infty} dE \int_{4\pi} d\Omega \int_V dV \varphi(\mathbf{r},E,\mathbf{\Omega},t)\psi(\mathbf{r},E,\mathbf{\Omega},t). \tag{7.3}$$

In the Hilbert space $\mathcal{H}_B$, the generic adjoint Boltzmann transport equation has the following form:

$$L^*(\mathbf{r},E,\mathbf{\Omega},t)\psi(\mathbf{r},E,\mathbf{\Omega},t) = Q^*(\mathbf{r},E,\mathbf{\Omega},t) \tag{7.4}$$

where the (adjoint) linear integro-differential operator $L^*(\mathbf{r},E,\mathbf{\Omega},t)$ is defined below:

$$\begin{aligned} L^*(\mathbf{r},E,\mathbf{\Omega},t)\psi(\mathbf{r},E,\mathbf{\Omega},t) \triangleq & -\frac{1}{\mathrm{v}}\frac{\partial\psi(\mathbf{r},E,\mathbf{\Omega},t)}{\partial t} - \mathbf{\Omega}\bullet\nabla\psi(\mathbf{r},E,\mathbf{\Omega},t) \\ & +\Sigma_t(\mathbf{r},E)\psi(\mathbf{r},E,\mathbf{\Omega},t) \\ & -\int_0^{E_f} dE' \int_{4\pi} d\Omega'\Sigma_s(\mathbf{r},E\to E',\mathbf{\Omega}\to\mathbf{\Omega}')\psi(\mathbf{r},E',\mathbf{\Omega}',t) \\ & -\nu\Sigma(\mathbf{r},E)\int_0^{E_f} dE' \int_{4\pi} \chi(\mathbf{r},E\to E')\psi(\mathbf{r},E',\mathbf{\Omega}',t)d\Omega'. \end{aligned} \tag{7.5}$$

By construction, the forward and adjoint transport equations satisfy the following relation:

$$\langle \varphi, L^{*}\psi \rangle_{B} - \langle \psi, L\varphi \rangle_{B} = P[\varphi,\psi] = \langle \varphi, Q^{*} \rangle_{B} - \langle \psi, Q \rangle_{B}, \tag{7.6}$$

where $P[\varphi,\psi]$ denotes the bilinear concomitant evaluated on the boundary of the phase-space domain under consideration. The "generalized reciprocity relation" expressed by Eq. (7.6) relates the bilinear concomitant, which is a functional of the forward and adjoint fluxes at the initial and final times along incoming and outgoing directions at the surface of the medium, to the fluxes in the interior of the medium comprising fixed sources. This reciprocity relation provides a physical interpretation of the adjoint flux as an "importance function" which quantifies the contribution of a source to a detector and enables transport problems to be posed either in the forward or the adjoint representations. This reciprocity relation also restricts the combination of forward and adjoint boundary conditions to those that render both the forward and the adjoint formulations to be "well-posed" mathematically. The reciprocity relation expressed by Eq. (7.6) is extensively used in the so-called "source-detector" problems in steady-state subcritical systems, where $Q^{*}(\mathbf{r},E,\boldsymbol{\Omega})$ models the detector's properties (cross section) in the sub-region occupied by the respective detector.

When the boundary conditions for Eq. (7.1) are homogeneous and when there is no external source, i.e., when $Q(r,E,\Omega,t) = 0$, the stationary neutron transport problem becomes an eigenvalue problem. The largest (i.e., fundamental) eigenvalue in such a case is called the "effective multiplication factor" and, depending on its value, corresponds to a critical, subcritical, or supercritical physical system (e.g., nuclear reactor). This eigenvalue (multiplication factor) is an important system (model) response, and its mathematical expression is a functional ("Raleigh quotient") of the forward and the adjoint fluxes. Additional important model responses that are functionals of both the forward and adjoint fluxes include the system's reactivity, generation time, lifetime, along with several other Lagrangian functionals used in variational principles for developing efficient Raleigh-Ritz type numerical methods (see, e.g., Lewins 1965; Stacey 1974, 2001). Perhaps the simplest quantity that depends on both the forward and adjoint fluxes, and which has important applications in particle transport (particularly in particle shielding) is the so-called "contributon-flux" (Williams and Engle 1977), which arises as follows:

i. Multiplying the stationary form of Eq. (7.1) by $\psi(r,E,\Omega)$, multiplying the stationary form of Eq. (7.4) by $\varphi(r,E,\Omega)$, subtracting the resulting equations from each other and integrating the resulting equation over only the energy and solid angle-independent variables yields the following relation:

$$\nabla \bullet \mathbf{v}_{c} R_{c}(\mathbf{r}) = S_{c}(\mathbf{r}) - S_{c}^{*}(\mathbf{r}), \tag{7.7}$$

where:

$$R_{c}(\mathbf{r}) \triangleq \frac{1}{\mathrm{V}} \int_{0}^{E_f} dE \int_{4\pi} d\Omega \varphi(r,E,\Omega) \psi(r,E,\Omega), \tag{7.8}$$

$$\mathbf{v}_c \triangleq \frac{\int_0^{E_f} dE \int_{4\pi} d\Omega \left[ \Omega \varphi(r,E,\Omega) \psi(r,E,\Omega) \right]}{\frac{1}{\mathrm{v}} \int_0^{E_f} dE \int_{4\pi} d\Omega \varphi(r,E,\Omega) \psi(r,E,\Omega)}, \tag{7.9}$$

$$S_c(\mathbf{r}) \triangleq \int_0^{E_f} dE \int_{4\pi} d\Omega \left[ Q(r,E,\Omega) \psi(r,E,\Omega) \right]. \tag{7.10}$$

$$S_c^*(\mathbf{r}) \triangleq \int_0^{E_f} dE \int_{4\pi} d\Omega \left[ Q^*(r,E,\Omega) \varphi(r,E,\Omega) \right]. \tag{7.11}$$

ii. The form of Eq. (7.7) is the same as the mass continuity balance/equation for compressible flow, indicating that the "contributon response density" $R_c(\mathbf{r})$ is conserved as it flows from the "contributon response source" $S_c(\mathbf{r})$ toward the "contributon response sink" $S_c^*(\mathbf{r})$, with a "contributon response mean velocity" $\mathbf{v}_c$ corresponding to the neutron speed $v$.

The application of the *n*th-FASAM-L methodology presented in Chapter 6 will be illustrated in this chapter by considering the simplified model of the distribution, in the asymptotic energy range, of neutrons produced by a source placed in an isotropic medium comprising a homogeneous mixture of "$M$" non-fissionable materials having constant (i.e., energy-, space- and angle-independent) properties. For simplicity, but without diminishing the applicability of the *n*th-FASAM-L methodology, this medium is considered to be infinitely large. The simplified form of the Boltzmann neutron transport equation, cf., Eq. (7.1), that models the energy distributions of neutrons in such a mixture of materials is called the "neutron slowing-down equation" and is written using the neutron *lethargy* (rather than the neutron energy) as the independent variable. The neutron lethargy is customarily denoted using the variable/letter "$u$" and is defined as follows: $u \triangleq \ln(E_0/E)$, where $E$ denotes the energy variable and $E_0$ denotes the highest energy in the system. Thus, the neutron slowing-down model (see, e.g., Lamarsh 1966; Stacey 1974, 2001) for the energy distribution of the neutron flux in a homogeneous mixture of non-fissionable materials of infinite extent takes on the following simplified form of Eq. (7.1):

$$\frac{d\varphi(u)}{du} + \frac{\Sigma_a}{\xi \Sigma_t} \varphi(u) = \frac{S(u)}{\xi \Sigma_t}; \quad 0 < u \le u_{th}; \tag{7.12}$$

$$\varphi(0) = 0; \quad \text{at} \quad u = 0. \tag{7.13}$$

The quantities which appear in Eq. (7.12) are defined as follows:

i. The lethargy-dependent neutron flux is denoted as $\varphi(u)$; $u_{th}$ denotes a cut-off lethargy, usually taken to be the lethargy that corresponds to the thermal neutron energy (ca. 0.0024 electron-volts).

ii. The macroscopic elastic scattering cross section for the homogeneous mixture of "*M*" materials is denoted as $\Sigma_s$ and is defined as follows:

$$\Sigma_s \triangleq \sum_{i=1}^{M} N_m^{(i)} \sigma_s^{(i)}; \tag{7.14}$$

where $\sigma_s^{(i)}, i = 1,\ldots,M$, denotes the elastic scattering cross section of material "*i*," and where the atomic or molecular number density of material "*i*" is denoted as $N_m^{(i)}, i = 1,\ldots,M$, and is defined as follows: $N_m^{(i)} \triangleq \rho_i N_A / A_i$, where $N_A$ is Avogadro's number $\left(0.602 \times 10^{24} \text{ nuclei/mol}\right)$, while $A_i$ and $\rho_i$ denote the respective material's mass number and density.

iii. The average gain in lethargy of a neutron per collision is denoted as $\overline{\xi}$ and is defined as follows for the homogeneous mixture:

$$\overline{\xi} \triangleq \frac{1}{\Sigma_s} \sum_{i=1}^{M} \xi_i N_m^{(i)} \sigma_s^{(i)}; \quad \xi_i \triangleq 1 + \frac{a_i \ln a_i}{1 - a_i}; \quad a_i \triangleq \left( \frac{A_i - 1}{A_i + 1} \right)^2. \tag{7.15}$$

iv. The macroscopic absorption cross section is denoted as $\Sigma_a$ and is defined as follows for the homogeneous mixture:

$$\Sigma_a \triangleq \sum_{i=1}^{M} N_m^{(i)} \sigma_\gamma^{(i)}, \tag{7.16}$$

where $\sigma_\gamma^{(i)}, i = 1,\ldots,M$, denotes the microscopic radiative-capture cross section of material "*i*."

v. The macroscopic total cross section is denoted as $\Sigma_t$ and is defined as follows for the homogeneous mixture:

$$\Sigma_t \triangleq \Sigma_a + \Sigma_s. \tag{7.17}$$

vi. The source $S(u)$ is considered to be a simplified "spontaneous fission" source stemming from fissionable actinides, such as $^{239}Pu$ and $^{240}Pu$, emitting monoenergetic neutrons at the highest energy (i.e., zero lethargy). Such a source is comprised within the OECD/NEA polyethylene-reflected plutonium (PERP) OECD/NEA reactor physics benchmark (Valentine 2006; Cacuci and Fang 2023) which can be modeled by the following simplified expression:

$$S(u) = S_0 \delta(u); \quad S_0 \triangleq \sum_{k=1}^{2} \lambda_k^S N_k^S F_k^S \nu_k^S W_k^S, \tag{7.18}$$

where the superscript "$S$" indicates "source;" the subscript index $k=1$ indicates material properties pertaining to the isotope $^{239}$Pu; the subscript index $k=2$ indicates material properties pertaining to the isotope $^{240}$Pu; $\lambda_k^S$ denotes the decay constant; $N_k^S$ denotes the atomic density of the respective actinide; $F_k^S$ denotes the spontaneous fission branching ratio; $\nu_k^S$ denotes the average number of neutrons per spontaneous fission; $W_k^S$ denotes a function of parameters used in a Watt's fission spectrum to approximate the spontaneous fission neutron spectrum of the respective actinide. The detailed forms of the parameters $W_k^S$ are unimportant for illustrating the application of the $n$th-FASAM-L methodology. The nominal values for these imprecisely known parameters are available from a library file contained in SOURCES4C (Wilson et al. 2002).

Mirroring the considerations for the Boltzmann transport equation presented in Eqs. (7.1)–(7.6), the "adjoint slowing-down model" is constructed in the Hilbert space $\mathcal{H}_B$ of square-integrable functions $\varphi(u)\in\mathcal{H}_B$, $\psi(u)\in\mathcal{H}_B$, endowed with the following inner product, denoted as $\langle\varphi(u),\psi(u)\rangle_B$:

$$\langle\varphi(u),\psi(u)\rangle_B \triangleq \int_0^{u_{th}} \varphi(u)\psi(u)\,du. \tag{7.19}$$

Using the inner product $\langle\varphi(u),\psi(u)\rangle_B$ defined in Eq. (7.19), the adjoint slowing-down model is constructed by the usual procedure, namely: (i) construct the inner product of Eq. (7.12) with a function $\psi(u)\in\mathcal{H}_B$; (ii) integrate by parts the resulting relation so as to transfer the differential operation from the forward function $\varphi(u)$ onto the adjoint function $\psi(u)$; (iii) use the initial condition provided in Eq. (7.13) and eliminate the unknown function $\varphi(u_{th})$ by choosing the final-value condition $\psi(u_{th})=0$; (iv) choose the source for the resulting adjoint slowing-down model so as to satisfy the generalized reciprocity relation shown in Eq. (7.6). The result of these operations is the following adjoint slowing-down model for the adjoint slowing-down function $\psi(u)$:

$$-\frac{d\psi(u)}{du}+f_1(\alpha)\psi(u)=\delta(u-u_d), \tag{7.20}$$

$$\psi(u_{th})=0, \quad \text{at} \quad u=u_{th}. \tag{7.21}$$

The "contributon-flux response density" $R_c(\varphi,\psi)$ was generally defined in Eq. (7.8). Specializing this definition to the neutron slowing-down model described in the foregoing happens to coincide with the inner product for this model, i.e.,

$$R_c(\varphi,\psi)\triangleq\int_0^{u_{th}}\varphi(u)\psi(u)\,du\equiv\langle\varphi(u),\psi(u)\rangle_B. \tag{7.22}$$

It is important to note that $R_c(\varphi,\psi)$ does *not* depend *explicitly* on either the feature function $\mathbf{f}(\boldsymbol{\alpha})$ or on any primary model parameter. Therefore, the G-differential of $R_c(\varphi,\psi)$ will not comprise a direct-effect term but will consist entirely of the indirect-effect term.

For this "contributon-flux response density" model, the following primary model parameters are subject to experimental uncertainties:

i. For each material "$i$," $i=1,\ldots,M$, included in the homogeneous mixture, the following are primary model parameters subject to experimental uncertainties: the atomic number densities $N_m^{(i)}$; the microscopic radiative-capture cross section $\sigma_\gamma^{(i)}$; the scattering cross section $\sigma_s^{(i)}$;
ii. The source parameters $\lambda_k^S$, $N_k^S$, $F_k^S$, $\nu_k^S$, $W_k^S$, for $k=1,2$, are also primary model parameters subject to experimental uncertainties

The above primary parameters are considered to constitute the components of a "vector of primary model parameters" defined as follows:

$$\boldsymbol{\alpha} \triangleq \left(N_m^{(1)},\sigma_\gamma^{(1)},\sigma_s^{(1)},\ldots,N_m^{(M)},\sigma_\gamma^{(M)},\sigma_s^{(M)},\lambda_1^S,\lambda_2^S,N_1^S,N_2^S,F_1^S,F_2^S,\nu_1^S,\nu_2^S,W_1^S,W_2^S\right)^\dagger$$
$$\triangleq (\alpha_1,\ldots,\alpha_{TP})^\dagger; \quad TP \triangleq 3M+10. \tag{7.23}$$

The 1st-Level Forward/Adjoint System (1st-LFAS) for the "1st-level forward/adjoint function" $\mathbf{u}^{(1)}(2;u) \triangleq [\varphi(u),\psi(u)]^\dagger$ comprises Eqs. (7.12), (7.13), (7.20), and (7.21). The structure of the 1st-LFAS suggests that the components $f_i(\boldsymbol{\alpha})$ of the feature function $\mathbf{f}(\boldsymbol{\alpha})$ can be defined as follows:

$$\mathbf{f}(\boldsymbol{\alpha}) \triangleq [f_1(\boldsymbol{\alpha}),f_2(\boldsymbol{\alpha})]^\dagger; \quad f_1(\boldsymbol{\alpha}) \triangleq \frac{\Sigma_a(\boldsymbol{\alpha})}{\bar{\xi}(\boldsymbol{\alpha})\Sigma_t(\boldsymbol{\alpha})}; \quad f_2(\boldsymbol{\alpha}) \triangleq \frac{S_0(\boldsymbol{\alpha})}{\bar{\xi}(\boldsymbol{\alpha})\Sigma_t(\boldsymbol{\alpha})}. \tag{7.24}$$

Solving Eqs. (7.12) and (7.13) while using the definitions introduced in Eq. (7.24) yields the following expression for the flux $\varphi(u)$ in terms of the components $f_i(\boldsymbol{\alpha})$ of the feature function $\mathbf{f}(\boldsymbol{\alpha})$:

$$\varphi(u) = H(u)f_2(\boldsymbol{\alpha})\exp[-uf_1(\boldsymbol{\alpha})]; \quad H(0)=0; \quad H(u)=1, \quad \text{if} \quad u>0. \tag{7.25}$$

Solving the above adjoint slowing-down model yields the following closed-form expression for the adjoint slowing-down function $\psi(u)$:

$$\psi(u) = H(u_d-u)\exp\left[(u-u_d)f_1(\boldsymbol{\alpha})\right]. \tag{7.26}$$

In terms of the components $f_i(\boldsymbol{\alpha})$ of the feature function $\mathbf{f}(\boldsymbol{\alpha})$, the closed-form expression of the "contributon response density" is obtained by substituting the expressions provided in Eqs. (7.25) and (7.26) into Eq. (7.22) and performing the integration over lethargy, which yields

$$R_c(\varphi,\psi)=\int_0^{u_{th}} H(u)f_2(\alpha)\exp[-uf_1(\alpha)]H(u_d-u)\exp[(u-u_d)f_1(\alpha)]\,du \tag{7.27}$$
$$=u_d f_2(\alpha)\exp[-u_d f_1(\alpha)].$$

In terms of the primary model parameters, the closed-form expression of the "contributon response density" is:

$$R_c(\varphi,\psi)=u_d\frac{S_0(\alpha)}{\xi(\alpha)\Sigma_t(\alpha)}\exp\left[-u_d\frac{\Sigma_a(\alpha)}{\xi(\alpha)\Sigma_t(\alpha)}\right]. \tag{7.28}$$

As Eq. (7.28) indicates, the model response can be considered to depend directly on $TP \triangleq 3M+10$ primary model parameters. In view of Eq. (7.27), however, the model response can be alternatively considered to depend directly on the two feature functions denoted as $f_1(\alpha)$ and $f_2(\alpha)$, while depending only indirectly (through the two feature functions) on the primary model parameters. In the former consideration/interpretation, the response sensitivities with respect to the primary model parameters will be obtained by applying the *n*th-CASAM-L methodology. In the later consideration/interpretation, the response sensitivities to the primary model parameters will be obtained by applying the *n*th-FASAM-L methodology, which will involve two stages: (a) the response sensitivities with respect to the feature functions will be obtained in the 1st stage; (b) the subsequent computation of the response sensitivities to the primary model parameters will be performed in the 2nd stage by using the response sensitivities with respect to the feature functions obtained in the 1st stage. The computational distinctions that stem from these differing considerations/interpretations underlying the *n*th-CASAM-L methodology versus the *n*th-FASAM-L methodology will become evident in Sections 7.3–7.5, by using a paradigm neutron slowing-down model, which is representative of the general situation for any linear system.

## 7.3 FIRST-ORDER ADJOINT SENSITIVITY ANALYSIS OF THE CONTRIBUTON-FLUX RESPONSE

The 1st-order sensitivities of the response $R_c\left[\mathbf{u}^{(1)}(2;u)\right]$, where $\mathbf{u}^{(1)}(2;u)\triangleq[\varphi(u),\psi(u)]^\dagger$, are obtained by determining the first-order Gateaux- (G-) differential, denoted as $\left\{\delta R_c\left[\mathbf{u}^{(1)}(2;u),\mathbf{v}^{(1)}(2;u)\right]\right\}_{\alpha^0}$, of this response for variations $\mathbf{v}^{(1)}(2;u)\triangleq[\delta\varphi(u),\delta\psi(u)]^\dagger$ around the phase-space point $(\varphi^0,\psi^0)$. By definition, the 1st-order G-differential $\left\{\delta R_c\left[\mathbf{u}^{(1)}(2;u),\mathbf{v}^{(1)}(2;u)\right]\right\}_{\alpha^0}$ is obtained as follows:

$$
\begin{aligned}
&\left\{\delta R_c\left[\mathbf{u}^{(1)}(2;u),\mathbf{v}^{(1)}(2;u)\right]\right\}_{\alpha^0} \\
&\triangleq\left\{\frac{d}{d\varepsilon}\int_0^{u_{th}}\left[\varphi^0(u)+\varepsilon v^{(1)}(u)\right]\left[\psi^0(u)+\varepsilon\delta\psi(u)\right]du\right\}_{\varepsilon=0} \qquad (7.29)\\
&=\left\{\int_0^{u_{th}}\left[v^{(1)}(u)\psi(u)+\varphi(u)\delta\psi(u)\right]du\right\}_{\alpha^0}.
\end{aligned}
$$

The sensitivities of $R_c\left[\mathbf{u}^{(1)}(2;u)\right]$ with respect to the feature functions (and subsequently with respect to the primary model parameters) will be determined in Section 7.3.1 by applying the 1st-FASAM-L. Alternatively, the expressions of the sensitivities of $R_c\left[\mathbf{u}^{(1)}(2;u)\right]$ with respect to the primary model parameters will be directly determined in Section 7.3.2 by applying the 1st-CASAM-L.

### 7.3.1 Application of the 1st-FASAM-L Methodology

The 1st-level variational sensitivity function $\mathbf{v}^{(1)}(2;u)\triangleq\left[v^{(1)}(u),\delta\psi(u)\right]^{\dagger}$ is the solution of the 1st-Level Variational Sensitivity System (1st-LVSS) obtained by differentiating the 1st-Level Forward/Adjoint System (1st-LFAS). The function $v^{(1)}(u)$ is obtained by taking the 1st-order G-differentials of Eqs. (7.12) and (7.13), to obtain the following equations:

$$
\left\{\frac{d}{d\varepsilon}\left[\frac{d\left(\varphi^0+\varepsilon v^{(1)}\right)}{du}+\left(f_1^0+\varepsilon\delta f_1\right)\left(\varphi^0+\varepsilon v^{(1)}\right)\right]\right\}_{\varepsilon=0}=\delta(u)\left\{\frac{d}{d\varepsilon}\left(f_2^0+\varepsilon\delta f_2\right)\right\}_{\varepsilon=0}, \qquad (7.30)
$$

$$
\left\{\frac{d}{d\varepsilon}\left[\varphi^0(u)+\varepsilon v^{(1)}(u)\right]\right\}_{\varepsilon=0}=0;\quad \text{at}\quad u=0. \qquad (7.31)
$$

Carrying out the differentiations with respect to $\varepsilon$ in the above equations and setting $\varepsilon=0$ in the resulting expressions yields the following relations:

$$
\frac{dv^{(1)}(u)}{du}+f_1\left(\alpha^0\right)v^{(1)}(u)=\left(\delta f_2\right)\delta(u)-\left(\delta f_1\right)\varphi^0(u), \qquad (7.32)
$$

$$
v^{(1)}(u)=0;\quad \text{at}\quad u=0. \qquad (7.33)
$$

The equations satisfied by the variational function $\delta\psi(u)$ are obtained by G-differentiating Eqs. (7.20) and (7.21) to obtain the equations below:

$$
-\frac{d}{du}\left[\delta\psi(u)\right]+f_1\left(\alpha^0\right)\left[\delta\psi(u)\right]=-\left(\delta f_1\right)\psi(u), \qquad (7.34)
$$

$$\delta\psi(u_{th}) = 0, \quad \text{at} \quad u = u_{th}. \tag{7.35}$$

Concatenating Eqs. (7.32)–(7.35) yields the following 1st-Level Variational Sensitivity System (1st-LVSS) for the 1st-level variational sensitivity function $\mathbf{v}^{(1)}(2;u) \triangleq [\delta\varphi(u), \delta\psi(u)]^{\dagger}$:

$$\left\{\mathbf{V}^{(1)}[2\times 2;u;\mathbf{f}]\mathbf{v}^{(1)}(2;u)\right\}_{\alpha^0} = \left\{\mathbf{q}_V^{(1)}\left[2;\mathbf{u}^{(1)}(2;u);\mathbf{f};\delta\mathbf{f}\right]\right\}_{\alpha^0}, \tag{7.36}$$

$$\left\{\mathbf{b}_v^{(1)}\left(\mathbf{v}^{(1)};\mathbf{f};\delta\mathbf{f}\right)\right\}_{\alpha^0} = \mathbf{0}, \tag{7.37}$$

where:

$$\mathbf{V}^{(1)}[2\times 2;u;\mathbf{f}] \triangleq \begin{pmatrix} d/du + f_1 & 0 \\ 0 & -d/du + f_1 \end{pmatrix}, \quad \mathbf{b}_v^{(1)}\left(\mathbf{v}^{(1)};\mathbf{f};\delta\mathbf{f}\right) \triangleq \begin{pmatrix} v^{(1)}(0) \\ \delta\psi(u_{th}) \end{pmatrix}, \tag{7.38}$$

$$\mathbf{q}_V^{(1)}\left[2;\mathbf{u}^{(1)};\mathbf{f};\delta\mathbf{f}\right] \triangleq \begin{pmatrix} (\delta f_2)\delta(u) - (\delta f_1)\varphi(u) \\ -(\delta f_1)\psi(u) \end{pmatrix}; \tag{7.39}$$

Rather than repeatedly solving the 1st-LVSS for every possible variations $\delta f_i, i = 1,2$, the appearance of the 1st-level variational sensitivity function $\mathbf{v}^{(1)}(2;u) \triangleq [\delta\varphi(u), \delta\psi(u)]^{\dagger}$ will be eliminated from the expression of the G-differential of the response $\left\{\delta R_c\left[\mathbf{u}^{(1)}(2;u), \mathbf{v}^{(1)}(2;u)\right]\right\}_{\alpha^0}$, defined in Eq. (7.29), by applying the principles of the 1st-FASAM-L outlined in Chapter 6. The specific steps are as follows:

1. Introduce a Hilbert space, denoted as $\mathcal{H}_1$, endowed with the following inner product denoted as $\left\langle \boldsymbol{\chi}^{(1)}(2;u), \boldsymbol{\theta}^{(1)}(2;u)\right\rangle_1$, between two elements, $\boldsymbol{\chi}^{(1)}(2;u) \triangleq \left[\chi_1^{(1)}(u), \chi_2^{(1)}(u)\right]^{\dagger} \in \mathcal{H}_1$, and $\boldsymbol{\theta}^{(1)}(2;u) \triangleq \left[\theta_1^{(1)}(u), \theta_2^{(1)}(u)\right]^{\dagger} \in \mathcal{H}_1$:

$$\left\langle \boldsymbol{\chi}^{(1)}(2;u), \boldsymbol{\theta}^{(1)}(2;u)\right\rangle_1 \triangleq \sum_{i=1}^{2} \int_0^{u_{th}} \chi_i^{(1)}(u)\theta_i^{(1)}(u)\, du. \tag{7.40}$$

2. In the Hilbert-space $\mathcal{H}_1$, form the inner product of Eq. (7.36) with a yet undefined vector-valued function $\mathbf{a}^{(1)}(2;u) \triangleq \left[a_1^{(1)}(u), a_2^{(1)}(u)\right]^{\dagger} \in \mathcal{H}_1$ to obtain the following relation:

$$\left\{\left\langle \mathbf{a}^{(1)}(2;u), \mathbf{V}^{(1)}\left[2\times 2;u;\mathbf{f}^{0}\right]\mathbf{v}^{(1)}(2;u)\right\rangle_{1}\right\}_{\alpha^{0}}$$
$$=\left\{\left\langle \mathbf{a}^{(1)}(2;u), \mathbf{q}_{V}^{(1)}\left[2;\mathbf{u}^{(1)}(2;u);\mathbf{f};\delta\mathbf{f}\right]\right\rangle_{1}\right\}_{\alpha^{0}}. \tag{7.41}$$

3. Integrate by parts the left side of Eq. (7.41) to obtain the following relation, where the specification $\{\ \}_{\alpha^0}$ is omitted to simplify the notation:

$$\int_{0}^{u_{th}} a_1^{(1)}(u)\left[\frac{dv^{(1)}}{du}+f_1 v^{(1)}\right]du+\int_{0}^{u_{th}} a_2^{(1)}(u)\left[-\frac{d}{du}\delta\psi+f_1\delta\psi\right]du$$

$$=\int_{0}^{u_{th}} v^{(1)}\left[-\frac{d}{du}a_1^{(1)}(u)+f_1 a_1^{(1)}(u)\right]du+\int_{0}^{u_{th}} \delta\psi(u)\left[\frac{d}{du}a_2^{(1)}(u)+f_1 a_2^{(1)}(u)\right]du \tag{7.42}$$

$$+a_1^{(1)}(u_{th})v^{(1)}(u_{th})-a_1^{(1)}(0)v^{(1)}(0)-a_2^{(1)}(u_{th})\delta\psi(u_{th})+a_2^{(1)}(0)\delta\psi(0).$$

4. Require the first and second terms on the right side of Eq. (7.42) to represent the G-differentiated response defined in Eq. (7.29), and eliminate the unknown boundary values of the function $\mathbf{v}^{(1)}(2;u)$ from the bilinear concomitant on the right side of Eq. (7.42) to obtain the following 1st-Level Adjoint Sensitivity System (1st-LASS) for the 1st-level adjoint sensitivity function $\mathbf{a}^{(1)}(2;u)\triangleq\left[a_1^{(1)}(u),a_2^{(1)}(u)\right]^{\dagger}$:

$$\mathbf{A}^{(1)}\left[2\times 2;\mathbf{x};\mathbf{f}\right]\mathbf{a}^{(1)}(2;\mathbf{x})=\mathbf{q}_{A}^{(1)}\left[2;\mathbf{u}^{(1)}(2;\mathbf{x});\mathbf{f}\right], \tag{7.43}$$

$$\left\{\mathbf{b}_{A}^{(1)}\left[\mathbf{u}^{(1)}(2;u);\mathbf{a}^{(1)}(2;u);\mathbf{f}\right]\right\}_{\alpha^{0}}\triangleq\begin{pmatrix} a_1^{(1)}(u_{th}) \\ a_2^{(1)}(0)\end{pmatrix}=\mathbf{0}, \tag{7.44}$$

where:

$$\mathbf{A}^{(1)}\left[2\times 2;u;\mathbf{f}\right]\triangleq\begin{pmatrix} -d/du+f_1 & 0 \\ 0 & d/du+f_1\end{pmatrix}=\left\{\mathbf{V}^{(1)}\left[2\times 2;u;\mathbf{f}\right]\right\}^{*}, \tag{7.45}$$

$$\mathbf{q}_{A}^{(1)}\left[2;\mathbf{u}^{(1)}(2;\mathbf{x});\mathbf{f}\right]\triangleq\begin{pmatrix} \psi(u) \\ \varphi(u)\end{pmatrix}. \tag{7.46}$$

5. It follows from Eqs. (7.29), (7.41)–(7.44) that the G-differentiated response defined in Eq. (7.29) takes on the following expression in terms of the 1st-level adjoint sensitivity function $\mathbf{a}^{(1)}(2;u)\triangleq\left[a_1^{(1)}(u),a_2^{(1)}(u)\right]^{\dagger}$:

$$\left\{\delta R_c\left[\mathbf{u}^{(1)}(2;u),\mathbf{a}^{(1)}(2;u)\right]\right\}_{\alpha^0} = \left\{\int_0^{u_{th}} a_1^{(1)}(u)\left[(\delta f_2)\delta(u)-(\delta f_1)\varphi(u)\right]du\right\}_{\alpha^0} + \left\{\int_0^{u_{th}} a_2^{(1)}(u)\left[-(\delta f_1)\psi(u)\right]du\right\}_{\alpha^0}, \tag{7.47}$$

The expressions of the sensitivities of the response $R_c(\varphi,\psi)$ with respect to the components of the feature function $\mathbf{f}(\alpha)$ are given by the expressions that multiply the respective components of $\mathbf{f}(\alpha)$ in Eq. (7.47), namely:

$$\frac{\partial R_c(\varphi,\psi)}{\partial f_1} = -\int_0^{u_{th}}\left[a_1^{(1)}(u)\varphi(u)+a_2^{(1)}(u)\psi(u)\right]du; \tag{7.48}$$

$$\frac{\partial R_c(\varphi,\psi)}{\partial f_2} = \int_0^{u_{th}} a_1^{(1)}(u)\,\delta(u)\,du. \tag{7.49}$$

The above expressions are to be evaluated at the nominal parameter values $\alpha^0$ but the indication $\{\ \}_{\alpha^0}$ has been omitted for simplicity.

The 1st-order sensitivities of the response $R_c(\varphi,\psi)$ with respect to the primary model parameters are obtained by using the results obtained in Eqs. (7.48) and (7.49), respectively, in conjunction with the "chain rule" of differentiating the components of the feature function $\mathbf{f}(\alpha)$ with respect to the primary model parameters defined in Eq. (7.29) to obtain the following expressions:

$$\begin{aligned}\frac{\partial R_c(\varphi,\psi)}{\partial \alpha_i} &= \frac{\partial R_c(\varphi,\psi)}{\partial f_1}\frac{\partial f_1}{\partial \alpha_i} + \frac{\partial R_c(\varphi,\psi)}{\partial f_2}\frac{\partial f_2}{\partial \alpha_i} \\ &= -\left(\frac{\partial f_1}{\partial \alpha_i}\right)\int_0^{u_{th}}\left[a_1^{(1)}(u)\varphi(u)+a_2^{(1)}(u)\psi(u)\right]du \\ &\quad + \left(\frac{\partial f_2}{\partial \alpha_i}\right)\int_0^{u_{th}} a_1^{(1)}(u)\delta(u)\,du.\end{aligned} \tag{7.50}$$

Solving the 1st-LASS defined by Eqs. (7.43) and (7.44) yields the following closed-form expressions for the components of the 1st-level adjoint sensitivity function $\mathbf{a}^{(1)}(2;u) \triangleq \left[a_1^{(1)}(u), a_2^{(1)}(u)\right]^{\dagger}$:

$$a_1^{(1)}(u) = (u_d - u)H(u_d - u)\exp\left[(u-u_d)f_1(\alpha)\right], \tag{7.51}$$

$$a_2^{(1)}(u) = uf_2(\alpha)\exp\left[-uf_1(\alpha)\right]. \tag{7.52}$$

Using the above expressions in Eqs. (7.48) and (7.49) yields the following closed-form expressions for the respective sensitivities:

$$\frac{\partial R_c(\varphi,\psi)}{\partial f_1} = -(u_d)^2 f_2(\boldsymbol{\alpha})\exp\left[-u_d f_1(\boldsymbol{\alpha})\right]; \tag{7.53}$$

$$\frac{\partial R_c(\varphi,\psi)}{\partial f_2} = u_d \exp\left[-u_d f_1(\boldsymbol{\alpha})\right]. \tag{7.54}$$

The correctness of the expressions obtained in Eqs. (7.53) and (7.54) can be verified by differentiating accordingly the closed-form expression given in Eq. (7.27).

### 7.3.2 Application of the 1st-CASAM-L Methodology

The 1st-CASAM-L methodology delivers the 1st-order sensitivities of the response directly with respect to the primary model parameters. The expression of the G-differentiated response is as shown in Eq. (7.29), but the source term on the right side of the 1st-LVSS takes on the following form:

$$\mathbf{q}_V^{(1)}\left[2;\mathbf{u}^{(1)};\mathbf{f};\delta\mathbf{f}\right] \triangleq \begin{pmatrix} \delta(u)\sum_{i=1}^{TP}\frac{\partial f_2}{\partial \alpha_i}\delta\alpha_i - \varphi(u)\sum_{i=1}^{TP}\frac{\partial f_1}{\partial \alpha_i}\delta\alpha_i \\ -\psi(u)\sum_{i=1}^{TP}\frac{\partial f_1}{\partial \alpha_i}\delta\alpha_i \end{pmatrix}. \tag{7.55}$$

If one were to actually solve the 1st-LVSS to obtain the 1st-level variational function and subsequently use the respective variational function to compute each sensitivity, one would need to solve the 1st-LVSS *TP-times*, using each time a source that would correspond to the *i*th-primary parameter, of the form $\mathbf{q}_V^{(1)}\left[i;2;\mathbf{u}^{(1)};\mathbf{f};\delta\mathbf{f}\right] \triangleq \left[\delta(u)\partial f_2/\partial\alpha_i - \varphi(u)\partial f_1/\partial\alpha_i, -\psi(u)\partial f_1/\partial\alpha_i\right]^\dagger$, for each primary parameter $i = 1,\ldots,TP$.

Since the left side of the 1st-LVSS remains the same as in Eq. (7.36), and the boundary conditions also remain the same as obtained in Eq. (7.37), it follows that the 1st-LASS and its solution $\mathbf{a}^{(1)}(2;u) \triangleq \left[a_1^{(1)}(u), a_2^{(1)}(u)\right]^\dagger$ remain unchanged. It therefore follows that the counterpart of the expression of the G-differential obtained in Eq. (7.47) has the following form:

$$\begin{aligned}\left\{\delta R_c\left[\mathbf{u}^{(1)}(2;u),\mathbf{a}^{(1)}(2;u)\right]\right\}_{\alpha^0} &= -\left\{\sum_{i=1}^{TP}\frac{\partial f_1}{\partial \alpha_i}\delta\alpha_i \int_0^{u_{th}} a_2^{(1)}(u)\psi(u)\,du\right\}_{\alpha^0} \\ &+ \left\{\int_0^{u_{th}} a_1^{(1)}(u)\left[\delta(u)\sum_{i=1}^{TP}\frac{\partial f_2}{\partial \alpha_i}\delta\alpha_i - \varphi(u)\sum_{i=1}^{TP}\frac{\partial f_1}{\partial \alpha_i}\delta\alpha_i\right]du\right\}_{\alpha^0}.\end{aligned} \tag{7.56}$$

The 1st-order sensitivities of the response $R_c(\varphi,\psi)$ with respect to the primary model parameters $\alpha_i, i=1,\dots,TP$ are obtained by identifying the expressions that multiply the respective variations $\delta\alpha_i$ in Eq. (7.47), which yields the following result:

$$\frac{\partial R_c(\varphi,\psi)}{\partial \alpha_i} = -\left(\frac{\partial f_1}{\partial \alpha_i}\right)\int_0^{u_{th}}\left[a_1^{(1)}(u)\varphi(u)+a_2^{(1)}(u)\psi(u)\right]du+\left(\frac{\partial f_2}{\partial \alpha_i}\right)\int_0^{u_{th}} a_1^{(1)}(u)\,\delta(u)\,du. \tag{7.57}$$

As expected, the result obtained in Eq. (7.57) is identical to the result produced in Eq. (7.50) by using the 1st-FASAM-L methodology. Both the 1st-FASAM-L and the 1st-CASAM-L methodologies require "one large-scale computation" for solving the 1st-LASS represented by Eqs. (7.43) and (7.44).

## 7.4 SECOND-ORDER ADJOINT SENSITIVITY ANALYSIS OF THE CONTRIBUTON-FLUX RESPONSE

In practice, closed-form expressions such as those shown in Eqs. (7.53) and (7.54) are unavailable. The 1st-FASAM-L methodology yields the expressions provided in Eqs. (7.48) and (7.49), while the 1st-CASAM-L methodology yields the expressions provided in Eq. (7.57). Hence, these expressions will provide the starting points for obtaining the 2nd-order sensitivities that stem from the respective 1st-order sensitivities. As has been outlined within the general frameworks of both the *n*th-FASAM-L and the *n*th-CASAM-L methodologies, the 2nd-order sensitivities are obtained by conceptually considering them to arise as the "1st-order sensitivities of the 1st-order sensitivities."

### 7.4.1 Application of the 2nd-FASAM-L Methodology

The 2nd-FASAM-L methodology uses the 1st-order sensitivities obtained using the 1st-FASAM-L methodology provided in Eqs. (7.48) and (7.49) to obtain the respective 2nd-order sensitivities, as presented below in Sections 7.4.1.1 and 7.4.1.2.

#### 7.4.1.1 Second-Order Sensitivities Stemming from the 1st-Order Sensitivity $\partial R_c/\partial f_1$

The 2nd-order sensitivities which stem from the 1st-order sensitivity $\partial R_c/\partial f_1$ are obtained by determining the G-differential of $\partial R_c/\partial f_1$. For subsequent "bookkeeping" purposes, this 1st-order sensitivity will be denoted as $R^{(1)}\left[1;\mathbf{u}^{(2)}(2^2;u);\mathbf{f}(\alpha)\right] \triangleq \partial R_c/\partial f_1$, where the superscript "(1)" denotes "1st-order" (sensitivity), and the argument "1" indicates that this sensitivity is with respect to the *1st component*, namely $f_1(\alpha)$, of the feature function $\mathbf{f}(\alpha)$. The function $\mathbf{u}^{(2)}(2^2;u) \triangleq \left[\mathbf{u}^{(1)}(2;u),\mathbf{a}^{(1)}(2;u)\right]^\dagger$ is the solution of the "2nd-Level Forward/Adjoint System (2nd-LFAS)" obtained by concatenating the 1st-Level Forward/Adjoint System (1st-LFAS) with the 1st-Level Adjoint Sensitivity System (1st-LASS), comprising Eqs. (7.12), (7.13), (7.20), (7.21), (7.43), and (7.44).

Applying the definition of the G-differential to Eq. (7.48) yields the following expression for the G-differential $\left\{\delta R^{(1)}\left[1;\mathbf{u}^{(2)}\left(2^2;u\right);\mathbf{v}^{(2)}\left(2^2;u\right);\mathbf{f}(\boldsymbol{\alpha})\right]\right\}_{\boldsymbol{\alpha}^0}$:

$$\left\{\delta R^{(1)}\left[1;\mathbf{u}^{(2)}\left(2^2;u\right);\mathbf{v}^{(2)}\left(2^2;u\right);\mathbf{f}(\boldsymbol{\alpha})\right]\right\}_{\boldsymbol{\alpha}^0}$$

$$\triangleq -\left\{\frac{d}{d\varepsilon}\int_0^{u_{th}}\left[a_1^{(1)}(u)+\varepsilon\delta a_1^{(1)}(u)\right]\left[\varphi(u)+\varepsilon v^{(1)}(u)\right]du\right\}_{\boldsymbol{\alpha}^0,\varepsilon=0}$$

$$-\left\{\frac{d}{d\varepsilon}\int_0^{u_{th}}\left[a_2^{(1)}(u)+\varepsilon\delta a_2^{(1)}(u)\right]\left[\psi(u)+\varepsilon\delta\psi(u)\right]du\right\}_{\boldsymbol{\alpha}^0,\varepsilon=0} \tag{7.58}$$

$$= -\int_0^{u_{th}}\varphi(u)\left[\delta a_1^{(1)}(u)\right]du - \int_0^{u_{th}} a_1^{(1)}(u)\,v^{(1)}(u)\,du - \int_0^{u_{th}}\psi(u)\left[\delta a_2^{(1)}(u)\right]du$$

$$-\int_0^{u_{th}} a_2^{(1)}(u)\left[\delta\psi(u)\right]du \equiv \sum_{j=1}^{2}\frac{\partial^2 R(\varphi;\mathbf{f})}{\partial f_j\,\partial f_1}\left(\delta f_j\right).$$

The components $v^{(1)}(u)$, $\delta\psi(u)$, $\delta a_1^{(1)}(u)$ and $\delta a_2^{(1)}(u)$ of the 2nd-level variational sensitivity function $\mathbf{v}^{(2)}\left(2^2;u\right)\triangleq\left[v^{(1)}(u),\delta\psi(u),\delta a_1^{(1)}(u),\delta a_2^{(1)}(u)\right]^{\dagger}$ are the solutions of the 2nd-Level Variational Sensitivity System (2nd-LVSS), which is obtained by G-differentiating the 2nd-LFAS. Thus, performing the G-differentiation of Eqs. (7.12), (7.13), (7.20), (7.21), (7.43), and (7.44) yields the following 2nd-Level Variational Sensitivity System (2nd-LVSS) for the 2nd-level variational sensitivity function $\mathbf{v}^{(2)}\left(2^2;u\right)\triangleq\left[v^{(1)}(u),\delta\psi(u),\delta a_1^{(1)}(u),\delta a_2^{(1)}(u)\right]^{\dagger}$:

$$\left\{\mathbf{V}^{(2)}\left[2^2\times 2^2;u;\mathbf{f}\right]\mathbf{v}^{(2)}\left(2^2;u\right)\right\}_{\boldsymbol{\alpha}^0} = \left\{\mathbf{q}_V^{(2)}\left[2^2;u;\mathbf{f};\delta\mathbf{f}\right]\right\}_{\boldsymbol{\alpha}^0}, \tag{7.59}$$

$$\left\{\mathbf{b}_v^{(2)}\left(u;\mathbf{f};\delta\mathbf{f}\right)\right\}_{\boldsymbol{\alpha}^0} = \mathbf{0}, \tag{7.60}$$

where:

$$\mathbf{V}^{(2)}\left[2^2\times 2^2;u;\mathbf{f}\right]\triangleq\begin{pmatrix} d/du+f_1 & 0 & 0 & 0 \\ 0 & -d/du+f & 0 & 0 \\ 0 & -1 & -d/du+f_1 & 0 \\ -1 & 0 & 0 & d/du+f_1 \end{pmatrix}; \tag{7.61}$$

$$\mathbf{q}_V^{(2)}\left[2^2;u;\mathbf{f};\delta\mathbf{f}\right]\triangleq\begin{pmatrix}(\delta f_2)\delta(u)-(\delta f_1)\varphi(u)\\-(\delta f_1)\psi(u)\\-(\delta f_1)a_1^{(1)}(u)\\-(\delta f_1)a_2^{(1)}(u)\end{pmatrix};\quad \mathbf{b}_V^{(2)}(u;\mathbf{f};\delta\mathbf{f})\triangleq\begin{pmatrix}v^{(1)}(0)\\\delta\psi(u_{th})\\\delta a_1^{(1)}(u_{th})\\\delta a_2^{(1)}(0)\end{pmatrix}. \tag{7.62}$$

The 2nd-level variational sensitivity function $\mathbf{v}^{(2)}(2^2;u)$ will be eliminated from appearing in the expression of $\left\{\delta R^{(1)}\left[1;\mathbf{u}^{(2)}(2^2;u);\mathbf{v}^{(2)}(2^2;u);\mathbf{f}(\alpha)\right]\right\}_{\alpha^0}$ by constructing the 2nd-Level Adjoint Sensitivity System (2nd-LASS) corresponding to the above 2nd-LVSS. The solution of the 2nd-LASS will be used in Eq. (7.58) to construct the alternative expression for $\left\{\delta R^{(1)}\left[1;\mathbf{u}^{(2)}(2^2;u);\mathbf{v}^{(2)}(2^2;u);\mathbf{f}(\alpha)\right]\right\}_{\alpha^0}$ that will not depend on $\mathbf{v}^{(2)}(2^2;u)$. This 2nd-LASS will be constructed in a Hilbert space denoted as $\mathcal{H}_2$, comprising as elements four-component vector-valued functions of the form $\chi^{(2)}(2^2;1;u)\triangleq\left[\chi_1^{(2)}(1;u),\chi_2^{(2)}(1;u),\chi_3^{(2)}(1;u),\chi_4^{(2)}(1;u)\right]^\dagger\in\mathcal{H}_2$. The Hilbert space $\mathcal{H}_2$ is endowed with the following inner product between two vectors $\chi^{(2)}(2^2;1;u)$ and $\theta^{(2)}(2^2;1;u)$:

$$\left\langle\chi^{(2)}(2^2;u),\theta^{(2)}(2^2;u)\right\rangle_2\triangleq\sum_{i=1}^{2^2}\int_0^{u_{th}}\chi_i^{(2)}(1;u)\theta_i^{(2)}(1;u)\,du. \tag{7.63}$$

The inner product defined in Eq. (7.63) will be used to construct the inner product of Eq. (7.59) with a function denoted as $\mathbf{a}^{(2)}(2^2;1;u)\triangleq\left[a_1^{(2)}(1;u),a_2^{(2)}(1;u),a_3^{(2)}(1;u),a_4^{(2)}(1;u)\right]^\dagger\in\mathcal{H}_2$, where the argument "1" of the function $\mathbf{a}^{(2)}(2^2;1;u)$ indicates that this (adjoint) function corresponds to the first-order sensitivity of the response with respect to the "1st" component, namely $f_1(\alpha)$, of the feature function $\mathbf{f}(\alpha)$. Constructing this inner product yields the following relation, where the specification $\{\ \}_{\alpha^0}$ has been omitted, to simplify the notation:

$$\begin{aligned}&\left\langle\mathbf{a}^{(2)}(2^2;1;\mathbf{x}),\mathbf{V}^{(2)}\left[2^2\times2^2;u;\mathbf{f}\right]\mathbf{v}^{(2)}(2^2;u)\right\rangle_{2^2}\\&=\int_0^{u_{th}}a_1^{(2)}(1;u)\left[dv^{(1)}/du+f_1v^{(1)}\right]du+\int_0^{u_{th}}a_2^{(2)}(1;u)\left[-d(\delta\psi)/du+f_1(\delta\psi)\right]du\\&+\int_0^{u_{th}}a_3^{(2)}(1;u)\left[-\delta\psi-d\left(\delta a_1^{(1)}\right)/du+f_1\left(\delta a_1^{(1)}\right)\right]du\\&+\int_0^{u_{th}}a_4^{(2)}(1;u)\left[-v^{(1)}(u)+d\left(\delta a_2^{(1)}\right)/du+f_1\left(\delta a_2^{(1)}\right)\right]du\end{aligned} \tag{7.64}$$

$$
= \int_0^{u_{th}} a_1^{(2)}(1;u)\left[(\delta f_2)\delta(u)-(\delta f_1)\varphi(u)\right]du + \int_0^{u_{th}} a_2^{(2)}(1;u)\left[-(\delta f_1)\psi(u)\right]du
$$

$$
+ \int_0^{u_{th}} a_3^{(2)}(1;u)\left[-(\delta f_1)a_1^{(1)}(u)\right]du + \int_0^{u_{th}} a_4^{(2)}(1;u)\left[-(\delta f_1)a_2^{(1)}(u)\right]du.
$$

Integrating by parts the left-side of Eq. (7.64) yields the following relation:

$$
\begin{aligned}
&\int_0^{u_{th}} a_1^{(2)}(1;u)\left[dv^{(1)}/du + f_1 v^{(1)}\right]du + \int_0^{u_{th}} a_2^{(2)}(1;u)\left[-d(\delta\psi)/du + f_1(\delta\psi)\right]du \\
&\quad + \int_0^{u_{th}} a_3^{(2)}(1;u)\left[-\delta\psi - d\left(\delta a_1^{(1)}\right)/du + f_1\left(\delta a_1^{(1)}\right)\right]du \\
&\quad + \int_0^{u_{th}} a_4^{(2)}(1;u)\left[-v^{(1)}(u) + d\left(\delta a_2^{(1)}\right)/du + f_1\left(\delta a_2^{(1)}\right)\right]du \\
&= a_1^{(2)}(1;u_{th})v^{(1)}(u_{th}) - a_1^{(2)}(1;0)v^{(1)}(0) \\
&\quad + \int_0^{u_{th}} v^{(1)}(u)\left[-da_1^{(2)}(1;u)/du + f_1 a_1^{(2)}(1;u)\right]du \\
&\quad - a_2^{(2)}(1;u_{th})\delta\psi(u_{th}) + a_2^{(2)}(1;0)\delta\psi(0) \\
&\quad + \int_0^{u_{th}} (\delta\psi)\left[da_2^{(2)}(1;u)/du + f_1 a_2^{(2)}(1;u)\right]du \\
&\quad - a_3^{(2)}(1;u_{th})\delta a_1^{(1)}(u_{th}) + a_3^{(2)}(1;0)\delta a_1^{(1)}(0) - \int_0^{u_{th}} (\delta\psi)a_3^{(2)}(1;u)\,du \\
&\quad + \int_0^{u_{th}} \delta a_1^{(1)}(u)\left[da_3^{(2)}(1;u)/du + f_1 a_3^{(2)}(1;u)\right]du - \int_0^{u_{th}} v^{(1)}(u)a_4^{(2)}(1;u)\,du \\
&\quad + a_4^{(2)}(1;u_{th})\delta a_2^{(1)}(u_{th}) - a_4^{(2)}(1;0)\delta a_2^{(1)}(0) \\
&\quad + \int_0^{u_{th}} \delta a_2^{(1)}(u)\left[-da_4^{(2)}(1;u)/du + f_1 a_4^{(2)}(1;u)\right]du.
\end{aligned}
\tag{7.65}
$$

The right side of Eq. (7.65) is now tailored to represent the G-differential $\left\{\delta R^{(1)}\left[1;\mathbf{u}^{(2)}\left(2^2;u\right);\mathbf{v}^{(2)}\left(2^2;u\right);\mathbf{f}(\alpha)\right]\right\}_{\alpha^0}$ expressed by Eq. (7.58) by requiring the 2nd-level adjoint sensitivity function $\mathbf{a}^{(2)}\left(2^2;1;u\right)$ to be the solution of the following 2nd-Level Adjoint Sensitivity System (2nd-LASS):

$$\left\{\mathbf{A}^{(2)}\left[2^2\times 2^2;u;\mathbf{f}\right]\mathbf{a}^{(2)}\left(2^2;1;u\right)\right\}_{\alpha^0}=\left\{\mathbf{s}^{(2)}\left(2^2;1;u;\mathbf{f}\right)\right\}_{\alpha^0}, \tag{7.66}$$

$$\left\{\mathbf{b}_A^{(2)}\left(u;\mathbf{f}\right)\right\}_{\alpha^0}=\mathbf{0}, \tag{7.67}$$

where:

$$\mathbf{A}^{(2)}\left[2^2\times 2^2;u;\mathbf{f}\right]\triangleq\begin{pmatrix} -d/du+f_1 & 0 & 0 & -1 \\ 0 & d/du+f_1 & -1 & 0 \\ 0 & 0 & d/du+f_1 & 0 \\ 0 & 0 & 0 & -d/du+f_1 \end{pmatrix}; \tag{7.68}$$

$$\mathbf{s}^{(2)}\left(2^2;1;u;\mathbf{f}\right)\triangleq\begin{pmatrix} -a_1^{(1)}(u) \\ -a_2^{(1)}(u) \\ -\varphi(u) \\ -\psi(u) \end{pmatrix};\quad \mathbf{b}_A^{(2)}\left(u;\mathbf{f}\right)\triangleq\begin{pmatrix} a_1^{(2)}(1;u_{th}) \\ a_2^{(2)}(1;0) \\ a_3^{(2)}(1;0) \\ a_4^{(2)}(1;u_{th}) \end{pmatrix}. \tag{7.69}$$

Implementing the equations underlying the 2nd-LVSS, the 2nd-LASS, and Eq. (7.58) into Eq. (7.64) provides the following alternative expression for the G-differential $\left\{\delta R^{(1)}\left[1;\mathbf{u}^{(2)}\left(2^2;u\right);\mathbf{v}^{(2)}\left(2^2;u\right);\mathbf{f}(\alpha)\right]\right\}_{\alpha^0}$:

$$\begin{aligned}&\left\{\delta R^{(1)}\left[1;\mathbf{u}^{(2)}\left(2^2;u\right);\mathbf{v}^{(2)}\left(2^2;u\right);\mathbf{f}(\alpha)\right]\right\}_{\alpha^0}\\ &=\left\{\int_0^{u_{th}} a_1^{(2)}(1;u)\left[(\delta f_2)\delta(u)-(\delta f_1)\varphi(u)\right]du\right\}_{\alpha^0}+\left\{\int_0^{u_{th}} a_2^{(2)}(1;u)\left[-(\delta f_1)\psi(u)\right]du\right\}_{\alpha^0}\\ &+\left\{\int_0^{u_{th}} a_3^{(2)}(1;u)\left[-(\delta f_1)a_1^{(1)}(u)\right]du\right\}_{\alpha^0}+\left\{\int_0^{u_{th}} a_4^{(2)}(1;u)\left[-(\delta f_1)a_2^{(1)}(u)\right]du\right\}_{\alpha^0}.\end{aligned} \tag{7.70}$$

The expressions that multiply the respective components of $\mathbf{f}(\alpha)$ in Eq. (7.70) are the expressions of the 2nd-order sensitivities $\partial^2 R_c(\varphi,\psi)/\partial f_1\,\partial f_j$ (stemming from the

1st-order sensitivity $\partial R_c/\partial f_1$) of the response $R_c(\varphi,\psi)$ with respect to the components of the feature function $\mathbf{f}(\alpha)$. Thus, identifying in Eq. (7.70) the expressions that multiply the respective variations in the components of the feature function $\mathbf{f}(\alpha)$ yields the following expressions:

$$\frac{\partial^2 R_c(\varphi,\psi)}{\partial f_1 \partial f_1} = -\int_0^{u_{th}} a_1^{(2)}(1;u)\varphi(u)\,du - \int_0^{u_{th}} a_2^{(2)}(1;u)\psi(u)\,du - \int_0^{u_{th}} a_3^{(2)}(1;u)\,a_1^{(1)}(u)\,du - \int_0^{u_{th}} a_4^{(2)}(1;u)\,a_2^{(1)}(u)\,du; \tag{7.71}$$

$$\frac{\partial R_c(\varphi,\psi)}{\partial f_2 \partial f_1} = \int_0^{u_{th}} a_1^{(2)}(1;u)\delta(u)du; \tag{7.72}$$

Solving the 2nd-LASS represented by Eqs. (7.66) and (7.67) yields the following closed-form expressions for the components of the 2nd-level adjoint sensitivity function $\mathbf{a}^{(2)}(2^2;1;u)$:

$$a_1^{(2)}(1;u) = -(u_d-u)^2 H(u_d-u)\exp[(u-u_d)f_1(\alpha)], \tag{7.73}$$

$$a_2^{(2)}(1;u) = -f_2(\alpha)u^2\exp[-uf_1(\alpha)], \tag{7.74}$$

$$a_3^{(2)}(1;u) = -f_2(\alpha)u\exp[-uf_1(\alpha)], \tag{7.75}$$

$$a_4^{(2)}(1;u) = -(u_d-u)H(u_d-u)\exp[(u-u_d)f_1(\alpha)]. \tag{7.76}$$

Using the explicit closed-form expressions obtained in Eqs. (7.73)–(7.76) in Eqs. (7.71) and (7.72) yields the following closed-form explicit expressions for the respective 2nd-order sensitivities:

$$\frac{\partial^2 R_c(\varphi,\psi)}{\partial f_1 \partial f_1} = (u_d)^3 f_2(\alpha)\exp[-u_d f_1(\alpha)]; \tag{7.77}$$

$$\frac{\partial R_c(\varphi,\psi)}{\partial f_2 \partial f_1} = -(u_d)^2\exp[-u_d f_1(\alpha)]. \tag{7.78}$$

The correctness of the expressions obtained in Eqs. (7.77) and (7.78) can be verified by differentiating accordingly the closed-form expression given in Eq. (7.53).

#### 7.4.1.2 Second-Order Sensitivities Stemming from the 1st-Order Sensitivity $\partial R_c/\partial f_2$

The 2nd-order sensitivities that stem from the 1st-order sensitivity $\partial R_c/\partial f_2$ are obtained by determining the G-differential of $\partial R_c/\partial f_2$. For subsequent "bookkeeping" purposes, this 1st-order sensitivity will be denoted as $R^{(1)}\left[2;\mathbf{u}^{(2)}\left(2^2;u\right);\mathbf{f}(\boldsymbol{\alpha})\right] \triangleq \partial R_c/\partial f_2$, where the superscript "(1)" denotes "1st-order" (sensitivity), the argument "2" indicates that this sensitivity is with respect to the *2nd-component*, namely $f_2(\boldsymbol{\alpha})$, of the feature function $\mathbf{f}(\boldsymbol{\alpha})$, while also depending on the function $\mathbf{u}^{(2)}\left(2^2;u\right) \triangleq \left[\mathbf{u}^{(1)}(2;u),\mathbf{a}^{(1)}(2;u)\right]^{\dagger}$. Applying the definition of the G-differential to the expression provided in Eq. (7.49) yields the result below for the G-differential $\left\{\delta R^{(1)}\left[2;\mathbf{u}^{(2)}\left(2^2;u\right);\mathbf{v}^{(2)}\left(2^2;u\right);\mathbf{f}(\boldsymbol{\alpha})\right]\right\}_{\boldsymbol{\alpha}^0}$:

$$\left\{\delta R^{(1)}\left[2;\mathbf{u}^{(2)}\left(2^2;u\right);\mathbf{v}^{(2)}\left(2^2;u\right);\mathbf{f}(\boldsymbol{\alpha})\right]\right\}_{\boldsymbol{\alpha}^0} = \int_0^{u_{th}} \delta a_1^{(1)}(u)\,\delta(u)\,du \equiv \sum_{j=1}^{2} \frac{\partial^2 R(\varphi;\mathbf{f})}{\partial f_j\,\partial f_2}\left(\delta f_j\right). \tag{7.79}$$

The function $\delta a_1^{(1)}(u)$, which appears in Eq. (7.79), is the component of the 2nd-level variational sensitivity function $\mathbf{v}^{(2)}\left(2^2;u\right) \triangleq \left[v^{(1)}(u),\delta\psi(u),\delta a_1^{(1)}(u),\delta a_2^{(1)}(u)\right]^{\dagger}$, which is the solution of the 2nd-LVSS comprising Eqs. (7.59) and (7.60). The component $\delta a_1^{(1)}(u)$ will be eliminated from the expression of $\left\{\delta R^{(1)}\left[2;\mathbf{u}^{(2)}\left(2^2;u\right);\mathbf{v}^{(2)}\left(2^2;u\right);\mathbf{f}(\boldsymbol{\alpha})\right]\right\}_{\boldsymbol{\alpha}^0}$ by following the same procedure as in Section 4.1.1, to construct a 2nd-Level Adjoint Sensitivity System (2nd-LASS), the solution of which will be denoted as $\mathbf{a}^{(2)}\left(2^2;2;u\right) \triangleq \left[a_1^{(2)}(2;u),a_2^{(2)}(2;u),a_3^{(2)}(2;u),a_4^{(2)}(2;u)\right]^{\dagger} \in \mathcal{H}_2$ and will be used in Eq. (7.79) to eliminate $\delta a_1^{(1)}(u)$. The argument "2" in $\mathbf{a}^{(2)}\left(2^2;2;u\right)$ indicates that this 2nd-level adjoint sensitivity function corresponds to the 1st-order sensitivity of the response with respect to the "2nd" component, $f_2(\boldsymbol{\alpha})$, of the feature function $\mathbf{f}(\boldsymbol{\alpha})$. The 2nd-LASS for the function $\mathbf{a}^{(2)}\left(2^2;2;u\right)$ will have the same left side and boundary conditions as obtained in Eqs. (7.66) and (7.67), but the right side of this 2nd-LASS will correspond to the G-differential obtained in Eq. (7.79), which leads to the following 2nd-LASS:

$$\left\{\mathbf{A}^{(2)}\left[2^2\times 2^2;u;\mathbf{f}\right]\mathbf{a}^{(2)}\left(2^2;2;u\right)\right\}_{\boldsymbol{\alpha}^0} = \left\{\mathbf{s}^{(2)}\left(2^2;2;u;\mathbf{f}\right)\right\}_{\boldsymbol{\alpha}^0}, \tag{7.80}$$

$$\left\{\mathbf{b}_A^{(2)}(u;\mathbf{f})\right\}_{\boldsymbol{\alpha}^0} = \mathbf{0}, \tag{7.81}$$

where:

$$\mathbf{s}^{(2)}\left(2^2;1;u;\mathbf{f}\right) \triangleq \left[0,0,\delta(u),0\right]^{\dagger}. \tag{7.82}$$

The alternative expression for the G-differential $\left\{\delta R^{(1)}\left[2;\mathbf{u}^{(2)}\left(2^2;u\right);\mathbf{v}^{(2)}\left(2^2;u\right);\mathbf{f}(\alpha)\right]\right\}_{\alpha^0}$ in terms of the components of $\mathbf{a}^{(2)}\left(2^2;2;u\right)$ has the same formal expression as shown in Eq. (7.70), but with the components of the function $\mathbf{a}^{(2)}\left(2^2;1;u\right)$ being replaced by the components of $\mathbf{a}^{(2)}\left(2^2;2;u\right)$, namely:

$$\delta\left(\partial R_c/\partial f_2\right)=\int_0^{u_{th}} a_1^{(2)}(2;u)\left[(\delta f_2)\delta(u)-(\delta f_1)\varphi(u)\right]du+\int_0^{u_{th}} a_2^{(2)}(2;u)\left[-(\delta f_1)\psi(u)\right]du$$
$$+\int_0^{u_{th}} a_3^{(2)}(2;u)\left[-(\delta f_1)a_1^{(1)}(u)\right]du+\int_0^{u_{th}} a_4^{(2)}(2;u)\left[-(\delta f_1)a_2^{(1)}(u)\right]du. \tag{7.83}$$

Solving the 2nd-LASS represented by Eqs. (7.80) and (7.81) yields the following expressions:

$$a_1^{(2)}(2;u)=0, \tag{7.84}$$

$$a_2^{(2)}(2;u)=u\exp\left[-uf_1(\alpha)\right], \tag{7.85}$$

$$a_3^{(2)}(2;u)=H(u)\exp\left[-uf_1(\alpha)\right], \tag{7.86}$$

$$a_4^{(2)}(2;u)=0. \tag{7.87}$$

Identifying in Eq. (7.83) the expressions that multiply the respective variations $\delta f_i$, $i=1,2$, in the components of the feature function $\mathbf{f}(\alpha)$ and using the closed-form expressions obtained in Eqs. (7.84)–(7.87), (7.26), and (7.51) yields the following closed-form explicit expressions for the respective 2nd-order sensitivities:

$$\frac{\partial^2 R_c(\varphi,\psi)}{\partial f_1\,\partial f_2}=-\int_0^{u_{th}} a_2^{(2)}(2;u)\psi(u)du-\int_0^{u_{th}} a_3^{(2)}(2;u)a_1^{(1)}(u)du$$
$$=-(u_d)^2\exp\left[-u_d f_1(\alpha)\right]; \tag{7.88}$$

$$\frac{\partial R_c(\varphi,\psi)}{\partial f_2\,\partial f_2}=0; \tag{7.89}$$

The correctness of the expressions obtained in Eqs. (7.88) and (7.89) can be verified by differentiating accordingly the closed-form expression given in Eq. (7.54).

Notably, due to the symmetry of the mixed 2nd-order sensitivities, the expressions obtained in Eqs. (7.88) and (7.72) provide an intrinsic mutual verification mechanism of the accuracy of the computations of the 2nd-level adjoint sensitivity functions $\mathbf{a}^{(2)}\left(2^2;1;u\right)$ and $\mathbf{a}^{(2)}\left(2^2;2;u\right)$.

### 7.4.2 Application of the 2nd-CASAM-L Methodology

The starting point for the application of the 2nd-CASAM-L methodology is to determine the G-differential of the *TP* 1st-order sensitivities represented by Eq. (7.57). For "bookkeeping" purposes, it is convenient to designate these *TP* 1st-order sensitivities as follows:

$$R_c^{(1)}\left[i;\mathbf{u}^{(2)}\left(2^2;u\right);\boldsymbol{\alpha}\right] \triangleq \partial R_c\left(\varphi,\psi\right)/\partial\alpha_i$$
$$= -g_1\left(i;\boldsymbol{\alpha}\right)\int_0^{u_{th}}\left[a_1^{(1)}(u)\varphi(u)+a_2^{(1)}(u)\psi(u)\right]du + g_2\left(i;\boldsymbol{\alpha}\right)\int_0^{u_{th}} a_1^{(1)}(u)\,\delta(u)\,du, \tag{7.90}$$

where:

$$g_1\left(i;\boldsymbol{\alpha}\right) \triangleq \partial f_1/\partial\alpha_i;\quad g_2\left(i;\boldsymbol{\alpha}\right) \triangleq \partial f_2/\partial\alpha_i;\quad i=1,\ldots,TP. \tag{7.91}$$

The G-differential of the expression in Eq. (7.90) is obtained, by definition, as follows:

$$\left\{\delta R_c^{(1)}\left[i;\mathbf{u}^{(2)}\left(2^2;u\right);\mathbf{v}^{(2)}\left(2^2;u\right);\boldsymbol{\alpha};\delta\boldsymbol{\alpha}\right]\right\}_{\boldsymbol{\alpha}^0}$$
$$\triangleq -\left\{\int_0^{u_{th}}\left[a_1^{(1)}(u)\varphi(u)+a_2^{(1)}(u)\psi(u)\right]du\left[\frac{d}{d\varepsilon}g_1\left(i;\boldsymbol{\alpha}+\varepsilon\delta\boldsymbol{\alpha}\right)\right]\right\}_{\boldsymbol{\alpha}^0,\varepsilon=0}$$
$$-\left\{g_1\left(i;\boldsymbol{\alpha}\right)\frac{d}{d\varepsilon}\int_0^{u_{th}}\left[a_1^{(1)}(u)+\varepsilon\delta a_1^{(1)}(u)\right]\left[\varphi(u)+\varepsilon v^{(1)}(u)\right]du\right\}_{\boldsymbol{\alpha}^0,\varepsilon=0}$$
$$-\left\{g_1\left(i;\boldsymbol{\alpha}\right)\frac{d}{d\varepsilon}\int_0^{u_{th}}\left[a_2^{(1)}(u)+\varepsilon\delta a_2^{(1)}(u)\right]\left[\psi(u)+\varepsilon\delta\psi(u)\right]du\right\}_{\boldsymbol{\alpha}^0,\varepsilon=0} \tag{7.92}$$
$$+\left\{\int_0^{u_{th}} a_1^{(1)}(u)\delta(u)\,du\left[\frac{d}{d\varepsilon}g_2\left(i;\boldsymbol{\alpha}+\varepsilon\delta\boldsymbol{\alpha}\right)\right]\right\}_{\boldsymbol{\alpha}^0,\varepsilon=0}$$
$$+\left\{g_2\left(i;\boldsymbol{\alpha}\right)\frac{d}{d\varepsilon}\int_0^{u_{th}}\left[a_1^{(1)}(u)+\varepsilon\delta a_1^{(1)}(u)\right]\delta(u)\,du\right\}_{\boldsymbol{\alpha}^0,\varepsilon=0}$$
$$\triangleq\left\{\delta R_c^{(1)}\left[i;\mathbf{u}^{(2)}\left(2^2;u\right);\mathbf{v}^{(2)}\left(2^2;u\right);\boldsymbol{\alpha}\right]\right\}_{ind}+\left\{\delta R_c^{(1)}\left[i;\mathbf{u}^{(2)}\left(2^2;u\right);\boldsymbol{\alpha};\delta\boldsymbol{\alpha}\right]\right\}_{dir},$$

where the direct-effect and indirect-effect terms are defined, respectively, as follows:

$$\left\{\delta R_c^{(1)}\left[i;\mathbf{u}^{(2)}\left(2^2;u\right);\boldsymbol{\alpha};\delta\boldsymbol{\alpha}\right]\right\}_{dir} \triangleq \left\{\left[\sum_{j=1}^{TP}\frac{\partial g_2(i;\boldsymbol{\alpha})}{\partial \alpha_j}\delta\alpha_j\right]\int_0^{u_{th}} a_1^{(1)}(u)\delta(u)\,du\right\}_{\boldsymbol{\alpha}^0}$$
$$-\left\{\left[\sum_{j=1}^{TP}\frac{\partial g_1(i;\boldsymbol{\alpha})}{\partial \alpha_j}\delta\alpha_j\right]\int_0^{u_{th}}\left[a_1^{(1)}(u)\varphi(u)+a_2^{(1)}(u)\psi(u)\right]du\right\}_{\boldsymbol{\alpha}^0}, \tag{7.93}$$

$$\left\{\delta R_c^{(1)}\left[i;\mathbf{u}^{(2)}\left(2^2;u\right);\mathbf{v}^{(2)}\left(2^2;u\right);\boldsymbol{\alpha}\right]\right\}_{ind} \triangleq \left\{g_2(i;\boldsymbol{\alpha})\int_0^{u_{th}}\delta a_1^{(1)}(u)\delta(u)\,du\right\}_{\boldsymbol{\alpha}^0}$$
$$-\left\{g_1(i;\boldsymbol{\alpha})\int_0^{u_{th}}\left[a_1^{(1)}(u)v^{(1)}(u)+\varphi(u)\delta a_1^{(1)}(u)\right]du\right\}_{\boldsymbol{\alpha}^0} \tag{7.94}$$
$$-\left\{g_1(i;\boldsymbol{\alpha})\int_0^{u_{th}}\left[a_2^{(1)}(u)\delta\psi(u)+\psi(u)\delta a_2^{(1)}(u)\right]du\right\}_{\boldsymbol{\alpha}^0}.$$

The direct-effect term can be evaluated/computed already at this stage. On the other hand, the indirect-effect depends on the 2nd-level variational function $\mathbf{v}^{(2)}\left(2^2;u\right)\triangleq\left[v^{(1)}(u),\delta\psi(u),\delta a_1^{(1)}(u),\delta a_2^{(1)}(u)\right]^\dagger$, which is the solution of the counterpart of 2nd-LVSS defined by Eqs. (7.59) and (7.60), with the same boundary conditions and left sides, but with distinct source terms, each source term involving the quantities $\partial g_1(i;\boldsymbol{\alpha})/\partial\alpha_j$ and $\partial g_2(i;\boldsymbol{\alpha})/\partial\alpha_j$, for $i,j=1,\ldots,TP$. If this path were chosen to compute the 2nd-order sensitivities, the 2nd-LVSS would need to be solved $TP^2$ times, with $TP^2$ different sources on the respective right sides, albeit with the same left side and boundary conditions.

The components $v^{(1)}(u),\delta\psi(u),\delta a_1^{(1)}(u),\delta a_2^{(1)}(u)$ are eliminated from the expression of the indirect-effect term $\left\{\delta R_c^{(1)}\left[i;\mathbf{u}^{(2)}\left(2^2;u\right);\mathbf{v}^{(2)}\left(2^2;u\right);\boldsymbol{\alpha}\right]\right\}_{ind}$ defined in Eq. (7.94) by constructing a corresponding 2nd-LASS in the Hilbert space $\mathcal{H}_2$ by following the same sequence of steps as in Section 7.4.1. The formal expression of the 2nd-LASS thus obtained will have the same left side and boundary conditions as in Section 7.4.1, but the right side of this formal 2nd-LASS will have a source-term that will correspond to the indirect-effect term defined in Eq. (7.94) and hence will be different for each $i=1,\ldots,TP$, namely:

$$\left\{\mathbf{A}^{(2)}\left[2^2\times 2^2;u;\boldsymbol{\alpha}\right]\mathbf{a}^{(2)}\left(2^2;i;2;u\right)\right\}_{\boldsymbol{\alpha}^0}=\left\{\mathbf{s}^{(2)}\left(2^2;i;u;\boldsymbol{\alpha}\right)\right\}_{\boldsymbol{\alpha}^0},\quad i=1,\ldots,TP; \tag{7.95}$$

$$\left\{\mathbf{b}_A^{(2)}(u;\boldsymbol{\alpha})\right\}_{\boldsymbol{\alpha}^0}=\mathbf{0};\quad i=1,\ldots,TP; \tag{7.96}$$

where:

$$\mathbf{s}^{(2)}\left(2^2;i;u;\boldsymbol{\alpha}\right) \triangleq \left[-g_1(i;\boldsymbol{\alpha})a_1^{(1)}, -g_1(i;\boldsymbol{\alpha})a_2^{(1)}, g_2(i;\boldsymbol{\alpha})\delta(u) - g_1(i;\boldsymbol{\alpha})\varphi, -g_1(i;\boldsymbol{\alpha})\psi\right]^{\dagger}. \tag{7.97}$$

In terms of the solution $\mathbf{a}^{(2)}\left(2^2;i;2;u\right)$ of the 2nd-LASS represented by Eqs. (7.95) and (7.96), the indirect-effect term $\left\{\delta R_c^{(1)}\left[i;\mathbf{u}^{(2)}\left(2^2;u\right);\mathbf{v}^{(2)}\left(2^2;u\right);\boldsymbol{\alpha}\right]\right\}_{ind}$ defined in Eq. (7.94) will have a representation that will formally resemble the expressions provided Section 7.4.1, such as in Eq. (7.83), but with the 2nd-level adjoint function(s) from Section 7.4.1 being replaced by the 2nd-level adjoint sensitivity function $\mathbf{a}^{(2)}\left(2^2;i;2;u\right)$. Finally, the total G-differential $\left\{\delta R_c^{(1)}\left[i;\mathbf{u}^{(2)}\left(2^2;u\right);\mathbf{v}^{(2)}\left(2^2;u\right);\boldsymbol{\alpha};\delta\boldsymbol{\alpha}\right]\right\}_{\boldsymbol{\alpha}^0}$ will be obtained, as shown in Eq. (7.92), by adding the expression of the indirect-effect term obtained in terms of the 2nd-level adjoint sensitivity function $\mathbf{a}^{(2)}\left(2^2;i;2;u\right)$ and the expression of the direct-effect term provided in Eq. (7.93). The expression of the individual 2nd-order sensitivities, $\partial^2 R_c(\varphi,\psi)/\partial\alpha_i\,\partial\alpha_j$, $i,j=1,\ldots,TP$, will subsequently be obtained by identifying in the final expression of the total G-differential $\left\{\delta R_c^{(1)}\left[i;\mathbf{u}^{(2)}\left(2^2;u\right);\mathbf{v}^{(2)}\left(2^2;u\right);\boldsymbol{\alpha};\delta\boldsymbol{\alpha}\right]\right\}_{\boldsymbol{\alpha}^0}$ those terms that multiply the parameter variations $\partial\alpha_j$, $j=1,\ldots,TP$.

### 7.4.3 Comparison: 2nd-FASAM-L versus 2nd-CASAM-L

The computational savings provided by using, whenever possible, the 2nd-FASAM-L methodology rather than the 2nd-CASAM-L methodology are evident by comparing the results obtained in Section 7.4.1 versus the results obtained in Section 7.4.2. The feature function $\mathbf{f}(\boldsymbol{\alpha})$ comprises two components $f_i(\boldsymbol{\alpha})$, $i=1,2$; consequently, the 2nd-FASAM-L methodology requires two large-scale computations (to solve the corresponding 2nd-LASS) to obtain the 2nd-order response sensitivities with respect to the components of the feature function. Subsequently, the 2nd-order response sensitivities with respect to the primary model parameters are obtained analytically by using the chain rule of differentiation.

In contradistinction, there are $TP \triangleq 3M+10$ primary model parameters, where the number ($M$) of materials in the medium can easily exceed two dozen. Consequently, the 2nd-CASAM-L methodology requires $TP$ large-scale computations (to solve the corresponding 2nd-LASS) to obtain the 2nd-order response sensitivities with respect to the primary model parameters. The boundary conditions and the operators on the left sides for all of the 2nd-LASS, for both the 2nd-FASAM-L and the 2nd-CASAM-L methodologies, are the same; only the source terms on the left sides of these 2nd-LASS differ from each other. It is therefore computationally advantageous if the inverse operators of the left sides of these 2nd-LASS could be computed just once and stored for subsequent use, in which case the computational advantage of using the 2nd-FASAM-L methodology would not be massive. Such a procedure could be feasible for relatively small models but would be impractical for large-scale problems, for which the advantage of using the 2nd-FASAM-L rather than the 2nd-CASAM-L methodology increases as the number of model parameters increases.

## 7.5 THIRD-ORDER ADJOINT SENSITIVITY ANALYSIS OF THE CONTRIBUTON-FLUX

The 3rd-FASAM-L methodology determines the 3rd-order sensitivities by applying the principles of the 1st-FASAM to the 2nd-order sensitivities, i.e., considering that the 3rd-order sensitivities are "the 1st-order sensitivities of the 2nd-order sensitivities." The unmixed 2nd-order sensitivity $\partial^2 R_c(\varphi,\psi)/\partial f_2\,\partial f_2$ is identically zero. The two non-zero 2nd-order sensitivities of the model response with respect to the components of the feature function $\mathbf{f}(\boldsymbol{\alpha})$ are as follows: (i) the unmixed 2nd-order sensitivity $\partial^2 R_c(\varphi,\psi)/\partial f_1\,\partial f_1$, expressed by Eq. (7.71); (ii) the mixed 2nd-order sensitivity $\partial^2 R_c(\varphi,\psi)/\partial f_1\,\partial f_2 = \partial^2 R_c(\varphi,\psi)/\partial f_2\,\partial f_1$, expressed by either Eq. (7.72) or Eq. (7.88), which are equivalent, in view of the symmetry property of the mixed 2nd-order sensitivities. Therefore, either the expression obtained in Eq. (7.88) or the expression obtained in Eq. (7.72) can be used as the starting point for obtaining the 3rd-order sensitivities stemming from this mixed 2nd-order sensitivity. It appears that the expression provided in Eq. (7.72) is the simpler of the two, so it will be used as the starting point for obtaining the corresponding 3rd-order sensitivities.

The 2nd-order sensitivity $\partial^2 R_c(\varphi,\psi)/\partial f_1\,\partial f_1$ expressed by Eq. (7.71) depends on the components of the 3rd-level forward/adjoint function, denoted as $\mathbf{u}^{(3)}(2^3;1;1;u) = \left[\mathbf{u}^{(2)}(2^2;u), \mathbf{a}^{(2)}(2^2;1;u)\right]^{\dagger}$, which is the solution of the 3rd-Level Forward/Adjoint System (3rd-LFAS) obtained by concatenating the 2nd-Level Forward/Adjoint System (2nd-LFAS) with the 2nd-LASS, thus comprising Eqs. (7.12), (7.13), (7.20), (7.21), (7.43), (7.44), (7.66), and (7.67). The arguments "1;1" of $\mathbf{u}^{(3)}(2^3;1;1;u)$ indicate that this 3rd-level function corresponds to the (unmixed) 2nd-order sensitivity $\partial^2 R_c(\varphi,\psi)/\partial f_1\,\partial f_1$ of the response with respect to the "1st" feature function, $f_1$. Therefore, the 2nd-order sensitivity $\partial^2 R_c(\varphi,\psi)/\partial f_1\,\partial f_1$ is denoted as follows: $R^{(2)}\left[1;1;\mathbf{u}^{(3)};\mathbf{f}(\boldsymbol{\alpha})\right] \triangleq \partial^2 R_c(\varphi,\psi)/\partial f_1\,\partial f_1$, where the arguments "1;1" indicate that this 3rd-level function corresponds to the (unmixed) 2nd-order sensitivity $\partial^2 R_c(\varphi,\psi)/\partial f_1\,\partial f_1$ and where the arguments of the function $\mathbf{u}^{(3)}(2^3;1;1;u)$ were omitted, for simplicity. Similarly, the mixed 2nd-order sensitivity $\partial^2 R_c(\varphi,\psi)/\partial f_1\,\partial f_2$ depends on the components of the same function $\mathbf{u}^{(3)}(2^3;1;1;u)$ and will therefore be denoted as $R^{(2)}\left[2;1;\mathbf{u}^{(3)};\mathbf{f}(\boldsymbol{\alpha})\right] \triangleq \partial^2 R_c(\varphi,\psi)/\partial f_2\,\partial f_1$, where the argument "2,1" indicates that this 2nd-order sensitivity is with respect to the components $(f_2, f_1)$ of $\mathbf{f}(\boldsymbol{\alpha})$.

### 7.5.1 Application of the 3rd-FASAM-L Methodology to Compute the 3rd-Order Sensitivities Stemming from $\partial^2 R_c(\varphi,\psi)/\partial f_1\,\partial f_1$

The 3rd-order sensitivities stemming from $R^{(2)}\left[1;1;\mathbf{u}^{(3)};\mathbf{f}(\boldsymbol{\alpha})\right] \triangleq \partial^2 R_c(\varphi,\psi)/\partial f_1\,\partial f_1$ are obtained from the G-differential of Eq. (7.71), which will be denoted as

$\left\{\delta R^{(2)}\left[1;1;\mathbf{u}^{(3)};\mathbf{v}^{(3)};\mathbf{f}(\alpha)\right]\right\}_{\alpha^0} \triangleq \left\{\delta\left[\partial^2 R_c(\varphi,\psi)/\partial f_1\,\partial f_1\right]\right\}_{\alpha^0}$, and which is determined (by definition) as follows:

$$\begin{aligned}
&\left\{\delta R^{(2)}\left[1;1;\mathbf{u}^{(3)};\mathbf{v}^{(3)};\mathbf{f}(\alpha)\right]\right\}_{\alpha^0} \\
&\triangleq -\left\{\frac{d}{d\varepsilon}\int_0^{u_{th}}\left[a_1^{(2)}(1;u)+\varepsilon\delta a_1^{(2)}(1;u)\right]\left[\varphi(u)+\varepsilon v^{(1)}(u)\right]du\right\}_{\alpha^0,\varepsilon=0} \\
&-\left\{\frac{d}{d\varepsilon}\int_0^{u_{th}}\left[a_2^{(2)}(1;u)+\varepsilon\delta a_2^{(2)}(1;u)\right]\left[\psi(u)+\varepsilon\delta\psi(u)\right]du\right\}_{\alpha^0,\varepsilon=0} \\
&-\left\{\frac{d}{d\varepsilon}\int_0^{u_{th}}\left[a_3^{(2)}(1;u)+\varepsilon\delta a_3^{(2)}(1;u)\right]\left[a_1^{(1)}(u)+\varepsilon\delta a_1^{(1)}(u)\right]du\right\}_{\alpha^0,\varepsilon=0} \\
&-\left\{\frac{d}{d\varepsilon}\int_0^{u_{th}}\left[a_4^{(2)}(1;u)+\varepsilon\delta a_4^{(2)}(1;u)\right]\left[a_2^{(1)}(u)+\varepsilon\delta a_2^{(1)}(u)\right]du\right\}_{\alpha^0,\varepsilon=0}.
\end{aligned} \tag{7.98}$$

Performing the differentiation with respect to $\varepsilon$ in Eq. (7.98) and setting $\varepsilon = 0$ in the resulting expression yields:

$$\begin{aligned}
\left\{\delta R^{(2)}\left[1;1;\mathbf{u}^{(3)};\mathbf{v}^{(3)};\mathbf{f}(\alpha)\right]\right\}_{\alpha^0} &= -\left\{\int_0^{u_{th}}\left[a_1^{(2)}(1;u)v^{(1)}(u)+\varphi(u)\delta a_1^{(2)}(1;u)\right]du\right\}_{\alpha^0} \\
&-\left\{\int_0^{u_{th}}\left[a_2^{(2)}(1;u)\delta\psi(u)+\psi(u)\delta a_2^{(2)}(1;u)\right]du\right\}_{\alpha^0} \\
&-\left\{\int_0^{u_{th}}\left[a_3^{(2)}(1;u)\delta a_1^{(1)}(u)+a_1^{(1)}(u)\delta a_3^{(2)}(1;u)\right]du\right\}_{\alpha^0} \\
&-\left\{\int_0^{u_{th}}\left[a_4^{(2)}(1;u)\delta a_2^{(1)}(u)+a_2^{(1)}(u)\delta a_4^{(2)}(1;u)\right]du\right\}_{\alpha^0}.
\end{aligned} \tag{7.99}$$

The 3rd-level variational function $\mathbf{v}^{(3)} \triangleq \mathbf{v}^{(3)}\left(2^3;1;1;u\right) \triangleq \left[\mathbf{v}^{(2)}\left(2^2;u\right),\boldsymbol{\delta}\mathbf{a}^{(2)}\left(2^2;1;u\right)\right]^\dagger$, where $\boldsymbol{\delta}\mathbf{a}^{(2)}\left(2^2;1;u\right) \triangleq \left[\delta a_1^{(2)}(1;u),\delta a_2^{(2)}(1;u),\delta a_3^{(2)}(1;u),\delta a_4^{(2)}(1;u)\right]^\dagger$, is the solution of the 3rd-LVSS obtained by concatenating the 2nd-LVSS, i.e., Eqs. (7.59) and (7.60), with the equations obtained by G-differentiating the 2nd-LASS for the function $\mathbf{a}^{(2)}\left(2^2;1;u\right)$, namely Eqs. (7.66) and (7.67). The resulting 3rd-LVSS for the 3rd-level variational function $\mathbf{v}^{(3)}\left(2^3;1;1;u\right)$ comprises the following matrix-equation, where the dots are used to denote zero-elements, for better visibility of the structure:

$$\begin{pmatrix} L & \bullet & \bullet & \bullet & \bullet & \bullet & \bullet & \bullet \\ \bullet & M & \bullet & \bullet & \bullet & \bullet & \bullet & \bullet \\ \bullet & -1 & M & \bullet & \bullet & \bullet & \bullet & \bullet \\ -1 & \bullet & \bullet & L & \bullet & \bullet & \bullet & \bullet \\ \bullet & \bullet & 1 & \bullet & M & \bullet & \bullet & -1 \\ \bullet & \bullet & \bullet & 1 & \bullet & L & -1 & \bullet \\ 1 & \bullet & \bullet & \bullet & \bullet & \bullet & L & \bullet \\ \bullet & 1 & \bullet & \bullet & \bullet & \bullet & \bullet & M \end{pmatrix} \begin{pmatrix} v^{(1)}(u) \\ \delta\psi(u) \\ \delta a_1^{(1)}(u) \\ \delta a_2^{(1)}(u) \\ \delta a_1^{(2)}(1;u) \\ \delta a_2^{(2)}(1;u) \\ \delta a_3^{(2)}(1;u) \\ \delta a_4^{(2)}(1;u) \end{pmatrix} = \begin{pmatrix} (\delta f_2)\delta(u)-(\delta f_1)\varphi(u) \\ -(\delta f_1)\psi(u) \\ -(\delta f_1)a_1^{(1)}(u) \\ -(\delta f_1)a_2^{(1)}(u) \\ -(\delta f_1)a_1^{(2)}(1;u) \\ -(\delta f_1)a_2^{(2)}(1;u) \\ -(\delta f_1)a_3^{(2)}(1;u) \\ -(\delta f_1)a_4^{(2)}(1;u) \end{pmatrix}, \tag{7.100}$$

$$L(u) \triangleq \frac{d}{du} + f_1(\boldsymbol{\alpha});\ M(u) \triangleq -\frac{d}{du} + f_1(\boldsymbol{\alpha});\ M(u) = L^*(u);$$

$$\begin{gathered} v^{(1)}(0)=0;\quad \delta\psi(u_{th})=0;\quad \delta a_1^{(1)}(u_{th})=0;\quad \delta a_2^{(1)}(0)=0; \\ \delta a_1^{(2)}(1;u_{th})=0;\quad \delta a_2^{(2)}(1;0)=0;\quad \delta a_3^{(2)}(1;0)=0;\quad \delta a_4^{(2)}(1;u_{th})=0. \end{gathered} \tag{7.101}$$

The 3rd-LVSS comprising Eqs. (7.100) and (7.101) can be formally expressed in the following $2^3 \times 2^3$-matrix form:

$$\mathbf{V}^{(3)}\left[2^3\times 2^3;u;\mathbf{f}\right]\mathbf{v}^{(3)}\left(2^3;1;1;u\right) = \mathbf{q}_V^{(3)}\left[2^3;\mathbf{u}^{(3)}\left(2^3;u\right);\mathbf{f};\delta\mathbf{f}\right]; \tag{7.102}$$

$$\mathbf{b}_V^{(3)}\left[\mathbf{v}^{(3)}\left(2^3;1;1;u\right)\right] = \mathbf{0}. \tag{7.103}$$

The above matrix form of the 3rd-LVSS will be used as a "condensed notation" to construct the 3rd-LASS, the solution of which will be used to derive the alternative expression for the G-differential $\left\{\delta R^{(2)}\left[1;1;\mathbf{u}^{(3)}\left(2^3;1;1;u\right);\mathbf{v}^{(3)}\left(2^3;1;1;u\right);\mathbf{f}(\boldsymbol{\alpha})\right]\right\}_{\alpha^0}$. This 3rd-LASS will be constructed in a Hilbert space denoted as $\mathcal{H}_3$, comprising as elements eight-component vector-valued functions of the form $\boldsymbol{\chi}^{(3)}\left(2^3;1;1;u\right) \triangleq \left[\chi_1^{(2)}(1;1;u),\dots,\chi_8^{(2)}(1;1;u)\right]^\dagger \in \mathcal{H}_3$, and endowed with the following inner product between two vectors $\boldsymbol{\chi}^{(3)}\left(2^3;1;1;u\right)$ and $\boldsymbol{\theta}^{(3)}\left(2^3;1;1;u\right)$:

$$\left\langle \boldsymbol{\chi}^{(3)}\left(2^3;1;1;u\right),\boldsymbol{\theta}^{(3)}\left(2^3;1;1;u\right)\right\rangle_3 \triangleq \sum_{i=1}^{2^3}\int_0^{u_{th}} \chi_i^{(3)}(1;1;u)\theta_i^{(3)}(1;1;u)du. \tag{7.104}$$

The inner product defined in Eq. (7.104) will be used to construct the inner product of Eq. (7.102) with a function denoted as $\mathbf{a}^{(3)}\left(2^3;1;1;u\right)\triangleq\left[a_1^{(3)}(1;1;u),\ldots,a_8^{(3)}(1;1;u)\right]^\dagger \in \mathcal{H}_3$, where the argument "1,1" of this (3rd-level adjoint) function indicates that it corresponds to the unmixed 2nd-order sensitivity of the response with respect to the "1st" component, $f_1(\boldsymbol{\alpha})$, of the feature function $\mathbf{f}(\boldsymbol{\alpha})$. Constructing this inner product yields the following relation, where the specification $\{\ \}_{\alpha^0}$ has been omitted, to simplify the notation:

$$\begin{aligned}&\left\langle \mathbf{a}^{(3)}\left(2^3;1;1;u\right),\mathbf{V}^{(3)}\left[2^3\times 2^3;u;\mathbf{f}\right]\mathbf{v}^{(3)}\left(2^3;1;1;u\right)\right\rangle_3\\&=\left\langle \mathbf{a}^{(3)}\left(2^3;1;1;u\right),\mathbf{q}_V^{(3)}\left[2^3;\mathbf{u}^{(3)}\left(2^3;u\right);\mathbf{f};\delta\mathbf{f}\right]\right\rangle_3,\end{aligned} \tag{7.105}$$

where:

$$\begin{aligned}&\left\langle \mathbf{a}^{(3)}\left(2^3;1;1;u\right),\mathbf{q}_V^{(3)}\left[2^3;\mathbf{u}^{(3)}\left(2^3;u\right);\mathbf{f};\delta\mathbf{f}\right]\right\rangle_3\\&\equiv\int_0^{u_{th}} a_1^{(3)}(1;1;u)\left[(\delta f_2)\delta(u)-(\delta f_1)\varphi(u)\right]du\\&+\int_0^{u_{th}} a_2^{(3)}(1;1;u)\left[-(\delta f_1)\psi(u)\right]du+\int_0^{u_{th}} a_3^{(3)}(1;1;u)\left[-(\delta f_1)a_1^{(1)}(u)\right]du\\&+\int_0^{u_{th}} a_4^{(3)}(1;1;u)\left[-(\delta f_1)a_2^{(1)}(u)\right]du+\int_0^{u_{th}} a_5^{(3)}(1;1;u)\left[-(\delta f_1)a_1^{(2)}(1;u)\right]du\\&+\int_0^{u_{th}} a_6^{(3)}(1;1;u)\left[-(\delta f_1)a_2^{(2)}(1;u)\right]du+\int_0^{u_{th}} a_7^{(3)}(1;1;u)\left[-(\delta f_1)a_3^{(2)}(1;u)\right]du\\&+\int_0^{u_{th}} a_8^{(3)}(1;1;u)\left[-(\delta f_1)a_4^{(2)}(1;u)\right]du.\end{aligned} \tag{7.106}$$

and where:

$$
\begin{aligned}
&\left\langle \mathbf{a}^{(3)}\left(2^3;1;1;u\right), \mathbf{V}^{(3)}\left[2^3\times 2^3;u;\mathbf{f}\right]\mathbf{v}^{(3)}\left(2^3;1;1;u\right)\right\rangle_3 \\
&\equiv \int_0^{u_{th}} a_1^{(3)}(1;1;u)\left[\frac{d}{du}v^{(1)}(u)+f_1 v^{(1)}(u)\right]du \\
&+\int_0^{u_{th}} a_2^{(3)}(1;1;u)\left[-\frac{d}{du}\delta\psi(u)+f_1\delta\psi(u)\right]du \\
&+\int_0^{u_{th}} a_3^{(3)}(1;1;u)\left[-\delta\psi(u)-\frac{d}{du}\delta a_1^{(1)}(u)+f_1\delta a_1^{(1)}(u)\right]du \\
&+\int_0^{u_{th}} a_4^{(3)}(1;1;u)\left[-v^{(1)}(u)+\frac{d}{du}\delta a_2^{(1)}(u)+f_1\delta a_2^{(1)}(u)\right]du \qquad (7.107)\\
&+\int_0^{u_{th}} a_5^{(3)}(1;1;u)\left[-\frac{d}{du}\delta a_1^{(2)}(1;u)+f_1\delta a_1^{(2)}(1;u)-\delta a_4^{(2)}(u)+\delta a_1^{(1)}(u)\right]du \\
&+\int_0^{u_{th}} a_6^{(3)}(1;1;u)\left[\frac{d}{du}\delta a_2^{(2)}(1;u)+f_1\delta a_2^{(2)}(1;u)-\delta a_3^{(2)}(u)+\delta a_2^{(1)}(u)\right]du \\
&+\int_0^{u_{th}} a_7^{(3)}(1;1;u)\left[\frac{d}{du}\delta a_3^{(2)}(1;u)+f_1\delta a_3^{(2)}(1;u)+v^{(1)}(u)\right]du \\
&+\int_0^{u_{th}} a_8^{(3)}(1;1;u)\left[-\frac{d}{du}\delta a_4^{(2)}(1;u)+f_1\delta a_4^{(2)}(1;u)+\delta\psi(u)\right]du.
\end{aligned}
$$

The right side of Eq. (7.107) is integrated by parts to obtain the relation below, in which the arguments "1;1" have been omitted when writing the components $a_i^{(3)}(1;1;u)$, $i=1,\dots,8$, to simplify the notation:

$$
\begin{aligned}
&\left\langle \mathbf{a}^{(3)}\left(2^3;1;1;u\right), \mathbf{V}^{(3)}\left[2^3\times 2^3;u;\mathbf{f}\right]\mathbf{v}^{(3)}\left(2^3;1;1;u\right)\right\rangle_3 \\
&= P\left[\mathbf{v}^{(3)};\mathbf{a}^{(3)};\mathbf{f};\delta\mathbf{f}\right]+\left\langle \mathbf{v}^{(3)}\left(2^3;1;1;u\right), \mathbf{A}^{(3)}\left[2^3\times 2^3;u;\mathbf{f}\right]\mathbf{a}^{(3)}\left(2^3;1;1;u\right)\right\rangle_3,
\end{aligned}
\qquad (7.108)
$$

where the bilinear concomitant $P\left[\mathbf{v}^{(3)};\mathbf{a}^{(3)};\mathbf{f};\delta\mathbf{f}\right]$ is defined as follows:

$$P\left[\mathbf{v}^{(3)};\mathbf{a}^{(3)};\mathbf{f};\delta\mathbf{f}\right]\triangleq a_1^{(3)}(u_{th})v^{(1)}(u_{th})-a_1^{(3)}(0)v^{(1)}(0)$$
$$-a_2^{(3)}(u_{th})\delta\psi(u_{th})+a_2^{(3)}(0)\delta\psi(0)-a_3^{(3)}(u_{th})\delta a_1^{(1)}(u_{th})$$
$$+a_3^{(3)}(0)\delta a_1^{(1)}(0)+a_4^{(3)}(u_{th})\delta a_2^{(1)}(u_{th})-a_4^{(3)}(0)\delta a_2^{(1)}(0)$$
$$-a_5^{(3)}(u_{th})\delta a_1^{(2)}(1;u_{th})+a_5^{(3)}(0)\delta a_1^{(2)}(1;0)+a_6^{(3)}(u_{th})\delta a_2^{(2)}(1;u_{th})$$
$$-a_6^{(3)}(0)\delta a_2^{(2)}(1;0)+a_7^{(3)}(u_{th})\delta a_3^{(2)}(1;u_{th})-a_7^{(3)}(0)\delta a_3^{(2)}(1;0)$$
$$-a_8^{(3)}(u_{th})\delta a_4^{(2)}(1;u_{th})+a_8^{(3)}(0)\delta a_4^{(2)}(1;0), \tag{7.109}$$

and where $\mathbf{A}^{(3)}\left[2^3\times 2^3;u;\mathbf{f}\right]\triangleq\left\{\mathbf{V}^{(3)}\left[2^3\times 2^3;u;\mathbf{f}\right]\right\}^*$ denotes the formal adjoint of $\mathbf{V}^{(3)}\left[2^3\times 2^3;u;\mathbf{f}\right]$, satisfying identically the following relation:

$$\left\langle \mathbf{v}^{(3)}(2^3;1;1;u),\mathbf{A}^{(3)}\left[2^3\times 2^3;u;\mathbf{f}\right]\mathbf{a}^{(3)}(2^3;1;1;u)\right\rangle_3$$
$$\equiv\int_0^{u_{th}} v^{(1)}(u)\left[-\frac{d}{du}a_1^{(3)}(u)+f_1a_1^{(3)}(u)\right]du+\int_0^{u_{th}}\delta\psi(u)\left[\frac{d}{du}a_2^{(3)}(u)+f_1a_2^{(3)}(u)\right]du$$
$$+\int_0^{u_{th}}\delta a_1^{(1)}(u)\left[\frac{d}{du}a_3^{(3)}(u)+f_1a_3^{(3)}(u)\right]du-\int_0^{u_{th}}a_3^{(3)}(u)\delta\psi(u)du-\int_0^{u_{th}}a_4^{(3)}(u)v^{(1)}(u)du$$
$$+\int_0^{u_{th}}\delta a_2^{(1)}(u)\left[-\frac{d}{du}a_4^{(3)}(u)+f_1a_4^{(3)}(u)\right]du+\int_0^{u_{th}}\delta a_1^{(2)}(1;u)\left[\frac{d}{du}a_5^{(3)}(u)+f_1a_5^{(3)}(u)\right]du$$
$$+\int_0^{u_{th}}a_5^{(3)}(u)\left[-\delta a_4^{(2)}(u)+\delta a_1^{(1)}(u)\right]du+\int_0^{u_{th}}\delta a_2^{(2)}(1;u)\left[\frac{d}{du}a_6^{(3)}(u)+f_1a_6^{(3)}(u)\right]du$$
$$+\int_0^{u_{th}}a_6^{(3)}(u)\left[-\delta a_3^{(2)}(u)+\delta a_2^{(1)}(u)\right]du+\int_0^{u_{th}}\delta a_3^{(2)}(1;u)\left[-\frac{d}{du}a_7^{(3)}(u)+f_1a_7^{(3)}(u)\right]du$$
$$+\int_0^{u_{th}}a_7^{(3)}(u)v^{(1)}(u)du+\int_0^{u_{th}}\delta a_4^{(2)}(1;u)\left[\frac{d}{du}a_8^{(3)}(u)+f_1a_8^{(3)}(u)\right]du+\int_0^{u_{th}}a_8^{(3)}(u)\delta\psi(u)du. \tag{7.110}$$

The boundary terms which appear in the bilinear concomitant defined in Eq. (7.109) will vanish by using Eq. (7.101) and by imposing the following boundary conditions on the components $a_i^{(3)}(1;1;u)$, $i = 1,\dots,8$ of the 3rd-level adjoint sensitivity function $\mathbf{a}^{(3)}(2^3;1;1;u)$:

$$\begin{aligned} &a_1^{(3)}(1;1;u_{th}) = 0; \quad a_2^{(3)}(1;1;0) = 0; \quad a_3^{(3)}(1;1;0) = 0; \quad a_4^{(3)}(1;1;u_{th}) = 0; \\ &a_5^{(3)}(1;1;0) = 0; \quad a_6^{(3)}(1;1;u_{th}) = 0; \quad a_7^{(3)}(1;1;u_{th}) = 0; \quad a_8^{(3)}(1;1;0) = 0. \end{aligned} \tag{7.111}$$

The right side of Eq. (7.110) is now required to represent the G-differential $\left\{\delta R^{(2)}\left[1;1;\mathbf{u}^{(3)}(2^3;1;1;u);\mathbf{v}^{(3)}(2^3;1;1;u);\mathbf{f}(\alpha)\right]\right\}_{\alpha^0}$ defined in Eq. (7.99) by imposing the following relation:

$$\mathbf{A}^{(3)}\left[2^3 \times 2^3;u;\mathbf{f}\right]\mathbf{a}^{(3)}(2^3;1;1;u) = -\mathbf{s}_A^{(3)}(2^3;1;1;\mathbf{f}), \tag{7.112}$$

where the eight-component vector $\mathbf{s}_A^{(3)}(2^3;1;1;\mathbf{f})$ is defined as follows:

$$\begin{aligned} &\mathbf{s}_A^{(3)}(2^3;1;1;\mathbf{f}) \\ &\triangleq \left[a_1^{(2)}(1;u),a_2^{(2)}(1;u),a_3^{(2)}(1;u),a_4^{(2)}(1;u);\varphi(u),\psi(u),a_1^{(1)}(u),a_2^{(1)}(u)\right]^\dagger. \end{aligned} \tag{7.113}$$

The relations provided in Eqs. (7.111) and (7.112) constitute the 3rd-LASS for the 3rd-level adjoint sensitivity function $\mathbf{a}^{(3)}(2^3;1;1;u)$. In component form, Eq. (7.112) has the following expression, where the dots are used to denote zero-elements, for better visibility of the structure:

$$\begin{pmatrix} M & \bullet & \bullet & -1 & \bullet & \bullet & 1 & \bullet \\ \bullet & L & -1 & \bullet & \bullet & \bullet & \bullet & 1 \\ \bullet & \bullet & L & \bullet & 1 & \bullet & \bullet & \bullet \\ \bullet & \bullet & \bullet & M & \bullet & 1 & \bullet & \bullet \\ \bullet & \bullet & \bullet & \bullet & L & \bullet & \bullet & \bullet \\ \bullet & \bullet & \bullet & \bullet & \bullet & M & \bullet & \bullet \\ \bullet & \bullet & \bullet & \bullet & \bullet & -1 & M & \bullet \\ \bullet & \bullet & \bullet & \bullet & -1 & \bullet & \bullet & L \end{pmatrix} \begin{pmatrix} a_1^{(3)}(1;1;u) \\ a_2^{(3)}(1;1;u) \\ a_3^{(3)}(1;1;u) \\ a_4^{(3)}(1;1;u) \\ a_5^{(3)}(1;1;u) \\ a_6^{(3)}(1;1;u) \\ a_7^{(3)}(1;1;u) \\ a_8^{(3)}(1;1;u) \end{pmatrix} = \begin{pmatrix} -a_1^{(2)}(1;u) \\ -a_2^{(2)}(1;u) \\ -a_3^{(2)}(1;u) \\ -a_4^{(2)}(1;u) \\ -\varphi(u) \\ -\psi(u) \\ -a_1^{(1)}(u) \\ -a_2^{(1)}(u) \end{pmatrix}. \tag{7.114}$$

Using the relations in Eqs. (7.99), (7.102), (7.103), (7.111), and (7.112) yields the following alternative expression for $\left\{\delta R^{(2)}\left[1;1;\mathbf{u}^{(3)}(2^3;1;1;u);\mathbf{v}^{(3)}(2^3;1;1;u);\mathbf{f}(\alpha)\right]\right\}_{\alpha^0}$:

$$\begin{aligned}
&\left\{\delta R^{(2)}\left[1;1;\mathbf{u}^{(3)}\left(2^3;1;1;u\right);\mathbf{v}^{(3)}\left(2^3;1;1;u\right);\mathbf{f}(\alpha)\right]\right\}_{\alpha^0} \\
&= \left\{\int_0^{u_{th}} a_1^{(3)}(1;1;u)\left[(\delta f_2)\delta(u)-(\delta f_1)\varphi(u)\right]du\right\}_{\alpha^0} \\
&-\left\{\int_0^{u_{th}} a_2^{(3)}(1;1;u)(\delta f_1)\psi(u)\,du\right\}_{\alpha^0} -\left\{\int_0^{u_{th}} a_3^{(3)}(1;1;u)(\delta f_1)a_1^{(1)}(u)\,du\right\}_{\alpha^0} \\
&-\left\{\int_0^{u_{th}} a_4^{(3)}(1;1;u)(\delta f_1)a_2^{(1)}(u)\,du\right\}_{\alpha^0} -\left\{\int_0^{u_{th}} a_5^{(3)}(1;1;u)(\delta f_1)a_1^{(2)}(1;u)\,du\right\}_{\alpha^0} \\
&-\left\{\int_0^{u_{th}} a_6^{(3)}(1;1;u)(\delta f_1)a_2^{(2)}(1;u)\,du\right\}_{\alpha^0} -\left\{\int_0^{u_{th}} a_7^{(3)}(1;1;u)(\delta f_1)a_3^{(2)}(1;u)\,du\right\}_{\alpha^0} \\
&-\left\{\int_0^{u_{th}} a_8^{(3)}(1;1;u)(\delta f_1)a_4^{(2)}(1;u)\,du\right\}_{\alpha^0}.
\end{aligned} \tag{7.115}$$

The 3rd-order sensitivities stemming from the relation obtained in Eq. (7.115) are the expressions that multiply the respective variations $\delta f_1$, $\delta f_2$, and are as follows:

$$\begin{aligned}
\partial^3 R_c(\varphi,\psi)/\partial f_1\,\partial f_1\,\partial f_1 &= -\int_0^{u_{th}} a_1^{(3)}(1;1;u)\varphi(u)\,du - \int_0^{u_{th}} a_2^{(3)}(1;1;u)\psi(u)\,du \\
&-\int_0^{u_{th}} a_3^{(3)}(1;1;u)a_1^{(1)}(u)\,du - \int_0^{u_{th}} a_4^{(3)}(1;1;u)a_2^{(1)}(u)\,du \\
&-\int_0^{u_{th}} a_5^{(3)}(1;1;u)a_1^{(2)}(1;u)\,du - \int_0^{u_{th}} a_6^{(3)}(1;1;u)a_2^{(2)}(1;u)\,du \\
&-\int_0^{u_{th}} a_7^{(3)}(1;1;u)a_3^{(2)}(1;u)\,du - \int_0^{u_{th}} a_8^{(3)}(1;1;u)a_4^{(2)}(1;u)\,du;
\end{aligned} \tag{7.116}$$

$$\partial^3 R_c(\varphi,\psi)/\partial f_1\,\partial f_1\,\partial f_2 = \int_0^{u_{th}} a_1^{(3)}(1;1;u)\delta(u)\,du. \tag{7.117}$$

The expressions obtained in Eqs. (7.116) and (7.117) are to be evaluated at the nominal values of parameters and state functions, but the notation $\{\ \}_{\alpha^0}$ has been omitted for simplicity.

Solving Eqs. (7.114) and (7.111) yields the following expressions for the components of the 3rd-level adjoint sensitivity function $\mathbf{a}^{(3)}(2^3;1;1;u)$:

$$a_1^{(3)}(1;1;u) = (u_d - u)^3 H(u_d - u)\exp[(u - u_d) f_1(\alpha)]; \tag{7.118}$$

$$a_2^{(3)}(1;1;u) = f_2(\alpha) u^3 \exp[-u f_1(\alpha)]; \tag{7.119}$$

$$a_3^{(3)}(1;1;u) = f_2(\alpha) u^2 \exp[-u f_1(\alpha)]; \tag{7.120}$$

$$a_4^{(3)}(1;1;u) = (u_d - u)^2 H(u_d - u)\exp[(u - u_d) f_1(\alpha)]; \tag{7.121}$$

$$a_5^{(3)}(1;1;u) = -f_2(\alpha) u \exp[-u f_1(\alpha)]; \tag{7.122}$$

$$a_6^{(3)}(1;1;u) = -(u_d - u) H(u_d - u)\exp[(u - u_d) f_1(\alpha)]; \tag{7.123}$$

$$a_7^{(3)}(1;1;u) = -(u_d - u)^2 H(u_d - u)\exp[(u - u_d) f_1(\alpha)]; \tag{7.124}$$

$$a_8^{(3)}(1;1;u) = -f_2(\alpha) u^2 \exp[-u f_1(\alpha)]. \tag{7.125}$$

Using the expressions obtained in Eqs. (7.118)–(7.125) into Eqs. (7.116) and (7.117) and performing the respective operations yields the following results:

$$\partial^3 R_c(\varphi,\psi)/\partial f_1 \partial f_1 \partial f_1 = -u_d^4 f_2(\alpha)\exp[-u_d f_1(\alpha)]; \tag{7.126}$$

$$\partial^3 R_c(\varphi,\psi)/\partial f_1 \partial f_1 \partial f_2 = u_d^3 \exp[-u_d f_1(\alpha)]. \tag{7.127}$$

### 7.5.2 Application of the 3rd-FASAM-L Methodology to Compute the 3rd-Order Sensitivities Stemming from $\partial^2 R_c(\varphi,\psi)/\partial f_1 \partial f_2 = \partial^2 R_c(\varphi,\psi)/\partial f_2 \partial f_1$

The 3rd-order sensitivities stemming from $R^{(2)}\left[2;1;\mathbf{u}^{(3)};\mathbf{f}(\alpha)\right] \triangleq \partial^2 R_c(\varphi,\psi)/\partial f_2 \partial f_1$ are obtained from the G-differential of Eq. (7.72), which will be denoted as $\left\{\delta R^{(2)}\left[2;1;\mathbf{u}^{(3)};\mathbf{v}^{(3)};\mathbf{f}(\alpha)\right]\right\}_{\alpha^0} \triangleq \left\{\delta\left[\partial^2 R_c(\varphi,\psi)/\partial f_2 \partial f_1\right]\right\}_{\alpha^0}$, and which is by definition determined as follows:

$$\left\{\delta R^{(2)}\left[2;1;\mathbf{u}^{(3)};\mathbf{v}^{(3)};\mathbf{f}(\alpha)\right]\right\}_{\alpha^0} \triangleq \left\{\delta\left[\partial^2 R_c(\varphi,\psi)/\partial f_2\,\partial f_1\right]\right\}_{\alpha^0}$$

$$\triangleq \left\{\frac{d}{d\varepsilon}\int_0^{u_{th}}\left[a_1^{(2)}(1;u)+\varepsilon\delta a_1^{(2)}(1;u)\right]\delta(u)\,du\right\}_{\alpha^0,\varepsilon=0}$$

$$= \int_0^{u_{th}} \delta a_1^{(2)}(1;u)\delta(u)\,du. \tag{7.128}$$

The function $\delta a_1^{(2)}(1;u)$ is one of the components of the 3rd-level variational function $\mathbf{v}^{(3)}(2^3;1;1;u)$, which is the solution of the 3rd-LVSS represented by Eqs. (7.101) and (7.102). To avoid the need for solving the 3rd-LVSS, the appearance of this function will be eliminated from Eq. (7.128) by deriving an alternative expression for the G-differential $\left\{\delta R^{(2)}\left[2;1;\mathbf{u}^{(3)};\mathbf{v}^{(3)};\mathbf{f}(\alpha)\right]\right\}_{\alpha^0}$ in terms of a 3rd-level adjoint sensitivity function, denoted as $\mathbf{a}^{(3)}(2^3;2;1;u) \triangleq \left[a_1^{(3)}(2;1;u),\ldots,a_8^{(3)}(2;1;u)\right]^\dagger \in \mathcal{H}_3$. The argument "2,1" of the function $\mathbf{a}^{(3)}(2^3;2;1;u)$ indicates that this (3rd-level adjoint) function corresponds to the mixed 2nd-order sensitivity of the response with respect to the "2nd and 1st" components, $(f_2, f_1)$, of the feature function $\mathbf{f}(\alpha)$.

The 3rd-level adjoint sensitivity function $\mathbf{a}^{(3)}(2^3;2;1;u)$ will be the solution of a 3rd-LASS to be constructed in the Hilbert space $\mathcal{H}_3$ by using the inner product defined in Eq. (7.104) and following the same sequence of steps as in Section 7.5.1. Thus, forming the inner product of $\mathbf{a}^{(3)}(2^3;2;1;u)$ with Eq. (7.102) yields the following relation, where the specification $\{\ \}_{\alpha^0}$ has been omitted to simplify the notation:

$$\left\langle \mathbf{a}^{(3)}(2^3;2;1;u), \mathbf{V}^{(3)}\left[2^3\times 2^3;u;\mathbf{f}\right]\mathbf{v}^{(3)}(2^3;1;1;u)\right\rangle_3 \\ = \left\langle \mathbf{a}^{(3)}(2^3;2;1;u), \mathbf{q}_V^{(3)}\left[2^3;\mathbf{u}^{(3)}(2^3;u);\mathbf{f};\delta\mathbf{f}\right]\right\rangle_3. \tag{7.129}$$

The left side of Eq. (7.129) is integrated by parts to obtain the following relation:

$$\left\langle \mathbf{a}^{(3)}(2^3;2;1;u), \mathbf{V}^{(3)}\left[2^3\times 2^3;u;\mathbf{f}\right]\mathbf{v}^{(3)}(2^3;1;1;u)\right\rangle_3 \\ = \left\langle \mathbf{v}^{(3)}(2^3;1;1;u), \mathbf{A}^{(3)}\left[2^3\times 2^3;u;\mathbf{f}\right]\mathbf{a}^{(3)}(2^3;2;1;u)\right\rangle_3, \tag{7.130}$$

where the bilinear concomitant vanished by having imposed the following boundary conditions on the components $a_i^{(3)}(2;1;u)$, $i=1,\ldots,8$, of the 3rd-level adjoint sensitivity function $\mathbf{a}^{(3)}(2^3;2;1;u)$:

$$a_1^{(3)}(2;1;u_{th})=0;\quad a_2^{(3)}(2;1;0)=0;\quad a_3^{(3)}(2;1;0)=0;\quad a_4^{(3)}(2;1;u_{th})=0;$$
$$a_5^{(3)}(2;1;0)=0;\quad a_6^{(3)}(2;1;u_{th})=0;\quad a_7^{(3)}(2;1;u_{th})=0;\quad a_8^{(3)}(2;1;0)=0. \tag{7.131}$$

The right side of Eq. (7.130) is now required to represent the G-differential $\left\{\delta R^{(2)}\left[2;1;\mathbf{u}^{(3)}\left(2^3;1;1;u\right);\mathbf{v}^{(3)}\left(2^3;1;1;u\right);\mathbf{f}(\boldsymbol{\alpha})\right]\right\}_{\alpha^0}$ by imposing the following relation:

$$\mathbf{A}^{(3)}\left[2^3\times 2^3;u;\mathbf{f}\right]\mathbf{a}^{(3)}\left(2^3;2;1;u\right)=\mathbf{s}_A^{(3)}\left(2^3;2;1;\mathbf{f}\right)\triangleq\left[0,0,0,0,0,\delta(u),0,0\right]^{\dagger}. \tag{7.132}$$

The relations provided in Eqs. (7.131) and (7.132) constitute the 3rd-LASS for the 3rd-level adjoint sensitivity function $\mathbf{a}^{(3)}\left(2^3;1;1;u\right)$, Using in Eq. (7.128) the relations in Eqs. (7.102), (7.103), (7.131), and (7.132) yields the following alternative expression for $\left\{\delta R^{(2)}\left[2;1;\mathbf{u}^{(3)}\left(2^3;1;1;u\right);\mathbf{a}^{(3)}\left(2^3;2;1;u\right);\mathbf{f}(\boldsymbol{\alpha})\right]\right\}_{\alpha^0}$, in which the function $\mathbf{v}^{(3)}\left(2^3;1;1;u\right)$ has been replaced by the function $\mathbf{a}^{(3)}\left(2^3;2;1;u\right)$:

$$\begin{aligned}
&\left\{\delta R^{(2)}\left[2;1;\mathbf{u}^{(3)}\left(2^3;1;1;u\right);\mathbf{a}^{(3)}\left(2^3;2;1;u\right);\mathbf{f}(\boldsymbol{\alpha})\right]\right\}_{\alpha^0}\\
&=\left\{\int_0^{u_{th}} a_1^{(3)}(2;1;u)\left[(\delta f_2)\delta(u)-(\delta f_1)\varphi(u)\right]du\right\}_{\alpha^0}\\
&-\left\{\int_0^{u_{th}} a_2^{(3)}(2;1;u)(\delta f_1)\psi(u)\,du\right\}_{\alpha^0}-\left\{\int_0^{u_{th}} a_3^{(3)}(2;1;u)(\delta f_1)a_1^{(1)}(u)\,du\right\}_{\alpha^0}\\
&-\left\{\int_0^{u_{th}} a_4^{(3)}(2;1;u)(\delta f_1)a_2^{(1)}(u)\,du\right\}_{\alpha^0}-\left\{\int_0^{u_{th}} a_5^{(3)}(2;1;u)(\delta f_1)a_1^{(2)}(1;u)\,du\right\}_{\alpha^0}\\
&-\left\{\int_0^{u_{th}} a_6^{(3)}(2;1;u)(\delta f_1)a_2^{(2)}(1;u)\,du\right\}_{\alpha^0}-\left\{\int_0^{u_{th}} a_7^{(3)}(2;1;u)(\delta f_1)a_3^{(2)}(1;u)\,du\right\}_{\alpha^0}\\
&-\left\{\int_0^{u_{th}} a_8^{(3)}(2;1;u)(\delta f_1)a_4^{(2)}(1;u)\,du\right\}_{\alpha^0}.
\end{aligned} \tag{7.133}$$

The 3rd-order sensitivities stemming from the relation obtained in Eq. (7.133) are the expressions that multiply the respective variations $\delta f_1$, $\delta f_2$, and are as follows:

$$\begin{aligned}\partial^3 R_c(\varphi,\psi)/\partial f_1 \partial f_2 \partial f_1 = &-\int_0^{u_{th}} a_1^{(3)}(2;1;u)\varphi(u)du - \int_0^{u_{th}} a_2^{(3)}(2;1;u)\psi(u)du \\ &-\int_0^{u_{th}} a_3^{(3)}(2;1;u)a_1^{(1)}(u)du - \int_0^{u_{th}} a_4^{(3)}(2;1;u)a_2^{(1)}(u)du \\ &-\int_0^{u_{th}} a_5^{(3)}(2;1;u)a_1^{(2)}(1;u)du - \int_0^{u_{th}} a_6^{(3)}(2;1;u)a_2^{(2)}(1;u)du \\ &-\int_0^{u_{th}} a_7^{(3)}(2;1;u)a_3^{(2)}(1;u)du - \int_0^{u_{th}} a_8^{(3)}(2;1;u)a_4^{(2)}(1;u)du;\end{aligned} \tag{7.134}$$

$$\partial^3 R_c(\varphi,\psi)/\partial f_2 \partial f_2 \partial f_1 = \int_0^{u_{th}} a_1^{(3)}(2;1;u)\delta(u)du. \tag{7.135}$$

The expressions obtained in Eqs. (7.134) and (7.135) are to be evaluated at the nominal values of parameters and state functions, but the notation $\{\ \}_{\alpha^0}$ has been omitted for simplicity.

In component form, the 3rd-LASS for the 3rd-level adjoint sensitivity function $\mathbf{a}^{(3)}(2^3;2;1;u)$ has the following expression, where the dots are used to denote zero-elements, for better visibility of the structure:

$$\begin{pmatrix} M & \bullet & \bullet & -1 & \bullet & \bullet & 1 & \bullet \\ \bullet & L & -1 & \bullet & \bullet & \bullet & \bullet & 1 \\ \bullet & \bullet & L & \bullet & 1 & \bullet & \bullet & \bullet \\ \bullet & \bullet & \bullet & M & \bullet & 1 & \bullet & \bullet \\ \bullet & \bullet & \bullet & \bullet & L & \bullet & \bullet & \bullet \\ \bullet & \bullet & \bullet & \bullet & \bullet & M & \bullet & \bullet \\ \bullet & \bullet & \bullet & \bullet & \bullet & -1 & M & \bullet \\ \bullet & \bullet & \bullet & \bullet & -1 & \bullet & \bullet & L \end{pmatrix} \begin{pmatrix} a_1^{(3)}(2;1;u) \\ a_2^{(3)}(2;1;u) \\ a_3^{(3)}(2;1;u) \\ a_4^{(3)}(2;1;u) \\ a_5^{(3)}(2;1;u) \\ a_6^{(3)}(2;1;u) \\ a_7^{(3)}(2;1;u) \\ a_8^{(3)}(2;1;u) \end{pmatrix} = \begin{pmatrix} 0 \\ 0 \\ 0 \\ 0 \\ \delta(u) \\ 0 \\ 0 \\ 0 \end{pmatrix}. \tag{7.136}$$

Solving Eq. (7.136) yields the following expressions for the components of the 3rd-level adjoint sensitivity function $\mathbf{a}^{(3)}(2^3;2;1;u)$:

$$a_1^{(3)}(2;1;u) = a_4^{(3)}(2;1;u) = a_6^{(3)}(2;1;u) = a_7^{(3)}(2;1;u) = 0;$$
$$a_5^{(3)}(2;1;u) = H(u)\exp[-uf_1(\boldsymbol{\alpha})];\quad a_2^{(3)}(2;1;u) = -u^2\exp[-uf_1(\boldsymbol{\alpha})]; \tag{7.137}$$
$$a_3^{(3)}(2;1;u) = -u\exp[-uf_1(\boldsymbol{\alpha})] = -a_8^{(3)}(2;1;u).$$

Using the expressions obtained in Eq. (7.137) into Eqs. (7.134) and (7.135) and performing the respective operations yields the following results:

$$\begin{aligned}\partial^3 R_c(\varphi,\psi)/\partial f_1\,\partial f_2\,\partial f_1 &= -\int_0^{u_{th}} a_2^{(3)}(2;1;u)\psi(u)\,du - \int_0^{u_{th}} a_3^{(3)}(2;1;u)a_1^{(1)}(u)\,du \\ &\quad - \int_0^{u_{th}} a_5^{(3)}(2;1;u)a_1^{(2)}(1;u)\,du - \int_0^{u_{th}} a_8^{(3)}(2;1;u)a_4^{(2)}(1;u)\,du \\ &= u_d^3\exp[-u_d f_1(\boldsymbol{\alpha})],\end{aligned} \tag{7.138}$$

$$\partial^3 R_c(\varphi,\psi)/\partial f_2\,\partial f_2\,\partial f_1 = 0. \tag{7.139}$$

## 7.6 CHAPTER SUMMARY

This chapter has presented an illustrative application of the "nth-Order Feature Adjoint Sensitivity Analysis Methodology for Response-Coupled Forward/Adjoint Linear Systems" (abbreviated as "nth-FASAM-L"), which has been specifically developed to be the most efficient methodology for computing exact expressions of high-order sensitivities of responses that depend simultaneously on the system's forward and adjoint state functions. The unique characteristics of the nth-FASAM-L have been illustrated in this chapter by using a paradigm model of a "contributon-flux density response" which occurs in the energy distribution of neutrons stemming from a fission source in a homogeneous mixture of materials. This analytically solvable illustrative paradigm model has been used to demonstrate the following general conclusions regarding the characteristics and applicability of the nth-FASAM-L:

i. Comparing the mathematical framework of the nth-FASAM-L methodology to the framework of the nth-CASAM-L methodology indicates that the components $f_i(\boldsymbol{\alpha}), i = 1,\dots,TF$, of the "feature function" $\mathbf{f}(\boldsymbol{\alpha}) \triangleq [f_1(\boldsymbol{\alpha}),\dots,f_{TF}(\boldsymbol{\alpha})]^\dagger$ play within the nth-FASAM-L the same role as played by the components $\alpha_j$, $j = 1,\dots,TP$, of the "vector of primary model parameters" $\boldsymbol{\alpha} \triangleq (\alpha_1,\dots,\alpha_{TP})^\dagger$ within the framework of the nth-CASAM-L. It is paramount to underscore, at the outset, that the total number of model parameters is always larger (usually by wide margin) than the total number of components of the feature function $\mathbf{f}(\boldsymbol{\alpha})$, i.e., $TP \gg TF$. The illustrative paradigm model of "neutron slowing down in a homogeneous mixture of materials" presented in this chapter comprised a feature function with two components (i.e., $TF = 2$), denoted as $f_1(\alpha)$ and $f_2(\alpha)$, which were, in turn, functions of $TP \triangleq 3M + 10$ imprecisely known model parameters (where $M$ denotes the number of materials and/or isotopes in the mixture, which is of the order of 20–50 in a nuclear reactor, depending on its service in operation).

ii. For computing the exact expressions of the *1st-order* sensitivities of a model response to the model's uncertain parameters, boundaries, and internal interfaces, both the 1st-FASAM-L and the 1st-CASAM-L methodologies require a *single* large-scale "adjoint" computation. This "large-scale" computation using either the 1st-FASAM-L or the 1st-CASAM-L methodologies involves solving the same operator equations and boundary conditions within the respective 1st-LASS; only the sources for the respective 1st-LASS differ from each other. The 1st-FASAM-L enjoys a slight computational advantage since it requires only *TF*-quadratures (one quadrature per component of the feature function), while the 1st-CASAM-L requires *TP*-quadratures (one quadrature per model parameter) for computing the 1st-order sensitivities. For the illustrative "contributon response of the neutron slowing-down" paradigm model, the computation of the 1st-order response sensitivities with respect to the model parameters required two quadratures using the 1st-FASAM-L methodology, while the 1st-CASAM-L methodology required *TP*-quadratures. Within the 1st-FASAM-L methodology, the sensitivities with respect to the primary model parameters are obtained by using the first-order sensitivities $\partial R_c / \partial f_1$ and $\partial R_c / \partial f_2$ (with respect to the components of the feature function) in conjunction with the chain rule of differentiation of the exactly known expressions of the components $f_1(\alpha)$ and $f_2(\alpha)$ in terms of the primary model parameters.

iii. Both the 2nd-FASAM-L and the 2nd-CASAM-L methodologies conceptually determine the *2nd-order* sensitivities by using the fundamental concept that "the 2nd-order sensitivities are the 1st-order sensitivities of the 1st-order sensitivities." For computing the exact expressions of the 2nd-order response sensitivities with respect to the primary model's parameters, the fundamental difference between the 2nd-FASAM-L and the 2nd-CASAM-L methodologies is as follows: the 2nd-FASAM-L requires as many large-scale "adjoint" computations as there are "feature functions of parameters" $f_i(\alpha), i = 1, \ldots, TF$ (where *TF* denotes the total number of feature functions) for solving the left side of the 2nd-LASS with *TF* distinct sources on its right side. In contradistinction, the 2nd-CASAM-L methodology requires *TP* (where *TP* denotes the total number of model parameters or non-zero 1st-order sensitivities) large-scale computations for solving the same left side of the 2nd-LASS but with *TP* distinct sources. Remarkably, the types of "large-scale" computation is the same in both the 2nd-FASAM-L and the 2nd-CASAM-L methodologies since they both solve the same operator equations and boundary conditions within the respective 2nd-LASSystems; only the sources for these adjoint systems differ from each other. Since $TF \ll TP$, the 2nd-FASAM-L methodology is considerably more efficient than the 2nd-CASAM-L methodology for computing the exact expressions of the 2nd-order sensitivities of a model response to the model's uncertain parameters, boundaries, and internal interfaces. For the illustrative contributon response paradigm model, the computation of the 2nd-order response sensitivities with respect to the model parameters using the 2nd-FASAM-L methodology requires just two large-scale computations, for solving the two 2nd-LASS that correspond to the 1st-order sensitivities, $\partial R_c/\partial f_1$ and $\partial R_c/\partial f_2$ of the contributon response with respect to the respective components, $f_1(\boldsymbol{\alpha})$ and $f_2(\boldsymbol{\alpha})$, of the model's "feature function" $f(\boldsymbol{\alpha})$. In contradistinction, computing the 2nd-order sensitivities to the model parameters using the 2nd-CASAM-L methodology requires *TP* large-scale computations, one for solving each of the 2nd-LASS that corresponds to each one of the distinct 1st-order sensitivities $\partial R_c/\partial \alpha_i$, $i = 1,\ldots,TP$, of the response with respect to the *TP* model parameters. Remarkably, only the unmixed 2nd-order sensitivity $\partial^2 R_c(\varphi,\psi)/\partial f_1 \partial f_1$ and the mixed 2nd-order sensitivity $\partial^2 R_c(\varphi,\psi)/\partial f_1 \partial f_2 = \partial^2 R_c(\varphi,\psi)/\partial f_2 \partial f_1$ are non-zero. The other unmixed 2nd-order sensitivity is identically zero, i.e., $\partial^2 R_c(\varphi,\psi)/\partial f_2 \partial f_2 \equiv 0$. In contradistinction, computing the 2nd-order sensitivities to the model parameters using the 2nd-CASAM-L methodology requires *TP* large-scale computations, one for solving each of the 2nd-LASS that corresponds to one of the distinct *TP* model parameters. None of the 2nd-order sensitivities with respect to the primary model parameters vanishes.

iv. For computing the exact expressions of the *3rd-order* response sensitivities with respect to the primary model's parameters, the 3rd-FASAM-L requires at most $TF(TF+1)/2$ large-scale "adjoint" computations for solving the 3rd-LASS with $TF(TF+1)/2$ distinct sources, while the 3rd-CASAM-L methodology requires at most $TP(TP+1)/2$ large-scale computations for solving the 3rd-LASS with $TP(TP+1)/2$ distinct sources. For the illustrative "contributon response of the neutron slowing-down" paradigm model, the computation of the 3rd-order response sensitivities with respect to the model parameters using the 3rd-FASAM-L methodology requires just *two* large-scale computations, for solving the two 3rd-LASS that correspond to the respective non-zero 2nd-order sensitivities $\partial^2 R_c(\varphi,\psi)/\partial f_1 \partial f_1$ and $\partial^2 R_c(\varphi,\psi)/\partial f_1 \partial f_2 = \partial^2 R_c(\varphi,\psi)/\partial f_2 \partial f_1$. Only the unmixed 3rd-order sensitivity $\partial^3 R_c(\varphi,\psi)/\partial f_1 \partial f_1 \partial f_1$ and the mixed 3rd-order sensitivity $\partial^3 R_c(\varphi,\psi)/\partial f_1 \partial f_1 \partial f_2$ are non-zero; all other 3rd-order sensitivities vanish identically. In contradistinction, the 3rd-CASAM-L methodology requires all $TP(TP+1)/2$ large-scale computations for solving the 3rd-LASS since all of the 2nd-order sensitivities with respect to the primary model parameters are non-zero. Furthermore, all of the 3rd-order response sensitivities with respect to the primary model parameters are non-zero.

v. The same computational-count of "large-scale computations" carries over when computing the 4th-and higher-order sensitivities, i.e., the formula for calculating the "number of large-scale adjoint computations" is formally the same for both the *n*th-FASAM-N and the *n*th-CASAM-N methodologies, but the "variable" in the formula for determining the number of adjoint computations for the *n*th-FASAM-N methodology is *TF* (i.e., total number of feature functions), while the counterpart for the formula for determining the number of adjoint computations for the *n*th-CASAM-N methodology is *TP* (i.e., total number of model parameters). Since $TF \ll TP$, it follows that the higher the order of computed sensitivities, the more efficient the *n*th-FASAM-N methodology becomes by comparison to the *n*th-CASAM-N methodology.

vi. The probability of encountering vanishing sensitivities is much higher when using the *n*th-FASAM-L methodology rather than the *n*th-CASAM-L methodology. For the illustrative "contributon response of the neutron slowing-down" paradigm model, it is evident that only a few of the response sensitivities of 4th-order (and higher-order) with respect to the components of the feature function $\mathbf{f}(\alpha)$ will *not* vanish, and the non-vanishing sensitivities will all involve the component $f_1(\alpha)$ of the feature function, since this component appears in an exponential, whereas the other component appears just as a multiplicative factor. In contradistinction, none of the higher-order response sensitivities with respect to the primary model parameters will vanish using the 2nd-CASAM-L methodology.

vii. When a model has no "feature" functions of parameters, but only comprises primary parameters, the *n*th-FASAM-L methodology becomes identical to the *n*th-CASAM-L methodology.

viii. Both the *n*th-FASAM-L and the *n*th-CASAM-L methodologies are formulated in linearly increasing higher-dimensional Hilbert spaces – as opposed to exponentially increasing parameter-dimensional spaces – thus overcoming the curse of dimensionality in sensitivity analysis of linear systems. Both the *n*th-FASAM-L and the *n*th-CASAM-L methodologies are incomparably more efficient and more accurate than any other method (statistical, finite differences, etc.) for computing exact expressions of response sensitivities (of any order) with respect to the model's uncertain parameters, boundaries, and internal interfaces.

# References


Alcouffe, R. E., Baker, R. S., Dahl, J. A., Turner, S.A., Ward, R., 2008. *PARTISN: A Time-Dependent, Parallel Neutral Particle Transport Code System*. LA-UR-08-07258, Los Alamos National Lab, Los Alamos, NM (Revised Nov. 2008).

Bellman, R. E. 1957. *Dynamic Programming*. Rand Corporation, Princeton University Press, Princeton, NJ. Republished: Bellman, R. E. 2003. *Dynamic Programming*. Courier Dover Publications, Mineola, NY.

Bonin, B. 2010. The Scientific Basis of Nuclear Waste Management, In *Handbook of Nuclear Engineering*, Cacuci, D. G., Editor. Springer Science + Business Media, New York, Vol. 5, pp. 3253–3419.

Cacuci, D. G. 1981a. Sensitivity Theory for Nonlinear Systems: I. Nonlinear Functional Analysis Approach. *J. Math. Phys. 22*, 2794–2802.

Cacuci, D. G. 1981b. Sensitivity Theory for Nonlinear Systems: II. Extensions to Additional Classes of Responses. *J. Math. Phys. 22*, 2803–2812.

Cacuci, D. G. 2015. Second-Order Adjoint Sensitivity Analysis Methodology for Computing Exactly and Efficiently First- and Second-Order Sensitivities in Large-Scale Linear Systems: I. Computational Methodology. *J. Comp. Phys. 284*, 687–699.

Cacuci, D. G. 2016. Second-Order Adjoint Sensitivity Analysis Methodology (2nd-ASAM) for Large-Scale Nonlinear Systems: I. Theory. *Nucl. Sci. Eng. 184*, 16–30.

Cacuci, D. G. 2018. *The Second-Order Adjoint Sensitivity Analysis Methodology*. CRC Press, Taylor & Francis Group, Boca Raton, FL.

Cacuci, D. G. 2019a. *BERRU Predictive Modeling: Best Estimate Results with Reduced Uncertainties*. Springer, Berlin.

Cacuci, D. G. 2019b. Towards Overcoming the Curse of Dimensionality: The Third-Order Adjoint Method for Sensitivity Analysis of Response-Coupled Linear Forward/Adjoint Systems, with Applications to Uncertainty Quantification and Predictive Modeling. *Energies 12*, 4216. https://doi.org/10.3390/en12214216

Cacuci, D. G. 2021a. Fourth-Order Comprehensive Adjoint Sensitivity Analysis of Response-Coupled Linear Forward/Adjoint Systems. I. Theoretical Framework. *Energies 14*, 3335. https://doi.org/10.3390/en14113335

Cacuci, D. G. 2021b. The nth-Order Comprehensive Adjoint Sensitivity Analysis Methodology for Response-Coupled Forward/Adjoint Linear Systems (nth-CASAM-L): I. Mathematical Framework. *Energies 14*, 8314. https://doi.org/10.3390/en14248314

Cacuci, D. G. 2022a. The nth-Order Comprehensive Adjoint Sensitivity Analysis Methodology for Nonlinear Systems (nth-CASAM-N): Mathematical Framework. *J. Nuclear Eng. 3*, 163–190. https://doi.org/10.3390/jne3030010

Cacuci, D. G. 2022b. *The nth-Order Comprehensive Adjoint Sensitivity Analysis Methodology (nth-CASAM): Overcoming the Curse of Dimensionality in Sensitivity and Uncertainty Analysis,* Volume I*: Linear Systems*. Springer Nature Switzerland, Cham, p. 362. https://doi.org/10.1007/978-3-030-96364-4

Cacuci, D. G. 2023a. *The nth-Order Comprehensive Adjoint Sensitivity Analysis Methodology (nth-CASAM): Overcoming the Curse of Dimensionality in Sensitivity and Uncertainty Analysis,* Volume III*: Nonlinear Systems*. Springer Nature Switzerland, Cham, p. 369. https://doi.org/10.1007/978-3-031-22757-8

Cacuci, D. G. 2023b. Second-Order MaxEnt Predictive Modelling Methodology. I: Deterministically Incorporated Computational Model (2nd-BERRU-PMD). *Am. J. Comp. Math. 13*, 236–266. https://doi.org/10.4236/ajcm.2023.132013

Cacuci, D. G. 2023c. Second-Order MaxEnt Predictive Modelling Methodology. II: Probabilistically Incorporated Computational Model (2nd-BERRU-PMP). *Am. J. Comp. Math. 13*, 267–294. https://doi.org/10.4236/ajcm.2023.132014

Cacuci, D. G. 2023d. Fourth-Order Predictive Modelling: I. General-Purpose Closed-Form Fourth-Order Moments-Constrained MaxEnt Distribution. *Am. J. Comp. Math. 13*, 413–438. https://doi.org/10.4236/ajcm.2023.134024

Cacuci, D. G. 2023e. Fourth-Order Predictive Modelling: II. 4th-BERRU-PM Methodology for Combining Measurements with Computations to Obtain Best-Estimate Results with Reduced Uncertainties. *Am. J. Comp. Math. 13*, 439–475. https://doi.org/10.4236/ajcm.2023.134025

Cacuci, D. G. 2023f. Computation of High-Order Sensitivities of Model Responses to Model Parameters. II: Introducing the Second-Order Adjoint Sensitivity Analysis Methodology for Computing Response Sensitivities to Functions/Features of Parameters. *Energies 16*, 6356. https://doi.org/10.3390/en16176356

Cacuci, D. G. 2024a. Introducing the nth-Order Features Adjoint Sensitivity Analysis Methodology for Nonlinear Systems (nth-FASAM-N): I. Mathematical Framework. *Am. J. Comp. Math. 14*, 11–42. https://doi.org/10.4236/ajcm.2024.141002

Cacuci, D. G. 2024b. Introducing the nth-Order Features Adjoint Sensitivity Analysis Methodology for Nonlinear Systems (nth-FASAM-N): II. Illustrative Example. *Am. J. Comp. Math. 14*, 43–95. https://doi.org/10.4236/ajcm.2024.141003

Cacuci, D. G. 2024c. The Second-Order Features Adjoint Sensitivity Analysis Methodology for Response-Coupled Forward/Adjoint Linear Systems (2nd-FASAM-L): Mathematical Framework and Illustrative Application to an Energy System. *Energies 17*, 2263. https://doi.org/10.3390/en17102263

Cacuci, D. G. 2024d. The nth-Order Features Adjoint Sensitivity Analysis Methodology for Response-Coupled Forward/Adjoint Linear Systems (nth-FASAM-L): I. Mathematical Framework. *Front. Energy Res. 12*, 1417594. https://doi.org/10.3389/fenrg.2024.1417594

Cacuci, D. G. 2024e. The nth-Order Features Adjoint Sensitivity Analysis Methodology for Response-Coupled Forward/Adjoint Linear Systems (nth-FASAM-L): II. Illustrative Application to a Paradigm Energy System. *Front. Energy Res. 12*, 1421519. https://doi.org/10.3389/fenrg.2024.1421519

Cacuci, D. G. 2025. *Advances in High-Order Predictive Modeling.* CRC Press, Boca Raton, FL, p 288.

Cacuci, D. G., Fang, R. 2023. *The nth-Order Comprehensive Adjoint Sensitivity Analysis Methodology (nth-CASAM): Overcoming the Curse of Dimensionality in Sensitivity and Uncertainty Analysis, Volume II: Application to a Large-Scale System.* Springer Nature Switzerland, Cham, p. 463. https://doi.org/10.1007/978-3-031-19635-5

Cacuci, D. G., Navon, M. I., Ionescu-Bujor, M. 2014. *Computational Methods for Data Evaluation and Assimilation.* Chapman & Hall/CRC, Boca Raton, FL.

Chadwick M. B., Herman, M., Obložinský, P., Dunn, M. E., Danon, Y., Kahler, A. C., Smith, D. L., Pritychenko, B., Arbanas, G., Brewer, R., et al. 2011. ENDF/B-VII.1: Nuclear Data for Science and Technology: Cross Sections, Covariances, Fission Product Yields and Decay Data. *Nucl. Data Sheets 112*, 2887–2996. https://doi.org/10.1016/j.nds.2011.11.002

Conlin, J. L., Parsons, D. K., Gardiner, S. J., Gray, M., Lee, M. B., White, M. C. 2013. *MENDF71X: Multigroup Neutron Cross-Section Data Tables Based upon ENDF/B-VII.1X.* Report LA-UR-15-29571; Los Alamos National Laboratory, Los Alamos, NM.

Hetrick, D. L. 1993. *Dynamics of Nuclear Reactors.* American Nuclear Society, Inc., La Grange Park, IL, pp. 164–174.

Lamarsh, J. R. 1966. *Introduction to Nuclear Reactor Theory.* Adison-Wesley Publishing Co., Reading MA, pp. 491–492.

Lewins, J. 1965. *Importance: The Adjoint Function*. Pergamon Press Ltd., Oxford, UK.

Lewis, J. M., Lakshmivarahan, S., Dhall, S. K. 2006. *Dynamic Data Assimilation: A Least Square Approach*. Cambridge University Press, Cambridge, UK.

Luo, Z., Wang, X., Liu, D. 2020. Prediction on the Static Response of Structures with Large-Scale Uncertain-But-Bounded Parameters Based on the Adjoint Sensitivity Analysis. *Struct. Multidiscip. Optim. 61*, 123–139. https://doi.org/10.1007/s00158-019-02349-w

Práger, T., Kelemen, F. D. 2014. Adjoint Methods and Their Application in Earth Sciences, In *Advanced Numerical Methods for Complex Environmental Models: Needs and Availability*, Faragó, I., Havasi, Á., Zlatev, Z., Editors. Bentham Science Publishers, Oak Park, IL, pp. 203–275.

Saltelli, A., Aleksankina, K., Becker, W., et al. 2019. Why So Many Published Sensitivity Analyses Are False: A Systematic Review of Sensitivity Analysis Practices. *Environ. Model. Softw.* https://doi.org/10.1016/j.envsoft.2019.01.012; https://www.researchgate.net/publication/321417848

Saltelli, A., Ratto, M., Andres, T., et al. 2008. *Global Sensitivity Analysis: The Primer*. John Wiley, Chichester, England.

Shannon, C. E. 1948. A Mathematical Theory of Communication. *Bell Syst. Tech. J. 27*, 379–423.

Shultis, J. K., Faw, R. E. 2000. *Radiation Shielding*. American Nuclear Society, La Grange Park, IL.

Shun, Z., McCullagh, P. 1995. Laplace Approximation of High Dimensional Integrals. *J. R. Stat. Soc. B Methodol. 57*(4), 749–760.

Stacey, W. M. 1974. *Variational Methods in Nuclear Reactor Physics*. Academic Press, New York.

Stacey, W. M. 2001. *Nuclear Reactor Physics*. John Wiley & Sons, New York.

Tang, Y., Reid, N. 2023. Laplace and Saddle Point Approximations in High Dimensions. arXiv:2107.10885v3 [math.ST]; https://doi.org/10.48550/arXiv.2107.10885. 8 Nov. 2023.

Tukey, J. W. 1957. *The Propagation of Errors, Fluctuations and Tolerances*. Technical Reports No. 10-12, Princeton University, Princeton, NJ.

Valentine, T. E. 2006. Polyethylene-Reflected Plutonium Metal Sphere Subcritical Noise Measurements, SUB-PU-METMIXED-001. In *International Handbook of Evaluated Criticality Safety Benchmark Experiments.* NEA/NSC/DOC(95)03/I-IX, Organization for Economic Co-Operation and Development (OECD), Nuclear Energy Agency (NEA), Paris, France.

Williams, M. L., Engle, W. W. 1977. The Concept of Spatial Channel Theory Applied to Reactor Shielding Analysis. *Nucl. Sci. Eng. 62*, 92.

Wilson, W. B., Perry, R. T., Shores, E. F., Charlton, W. S., Parish, T. A., Estes, G. P., Brown, T. H., Arthur, E. D., Bozoian, M., England, T. R., Madland, D. G., Stewart, J. E. 2002. SOURCES4C: A Code for Calculating ($\alpha$,n), Spontaneous Fission, and Delayed Neutron Sources and Spectra. In *12th Biennial Topical Meeting of American Nuclear Society/Radiation Protection and Shielding Division*, Santa Fe, NM, April 14–18, 2002. LA-UR-02-1839.

# Index

For Product Safety Concerns and Information please contact our EU
representative GPSR@taylorandfrancis.com
Taylor & Francis Verlag GmbH, Kaufingerstraße 24, 80331 München, Germany

www.ingramcontent.com/pod-product-compliance
Lightning Source LLC
Chambersburg PA
CBHW060824170526
45158CB00001B/75
* 9 7 8 1 0 3 2 7 6 3 5 9 0 *